machodh - Fuck your mother.

theedle ~~bhutney~~ - Bich
Buttnee

gushter - Slut

AUTHORS

Robert Alexander   Brendan Kelly

Contributing Writers

Paul Atkinson
Superintendent of Instruction
Waterloo Board of Education
Waterloo, Ontario

Maurice Barry
Vista School District
Newfoundland

Fred Crouse
Annapolis Valley Regional
   School Board
Berwick, Nova Scotia

Garry Davis
Saskatoon Public Board
   of Education
Saskatoon, Saskatchewan

George Gadanidis
Durham Board of Education
Whitby, Ontario

Liliane Gauthier
Saskatoon Board of Education
Saskatoon, Saskatchewan

Florence Glanfield
Mathematics Education
   Consultant
Edmonton, Alberta

Jim Nakamoto
Sir Winston Churchill
   Secondary School
Vancouver, British Columbia

Linda Rajotte
Georges P. Vanier
   Secondary School
Courtenay, British Columbia

Elizabeth Wood
National Sport School
Calgary Board of Education
Calgary, Alberta

Rick Wunderlich
Shuswap Junior
   Secondary School
Salmon Arm, British Columbia

Addison
Wesley

Toronto

DEVELOPMENTAL EDITORS
Claire Burnett
Lesley Haynes
Sarah Mawson

EDITORS
Santo D'Agostino
Mei Lin Cheung
Julia Cochrane
David Gargaro
Tony Rodrigues
Rajshree Shankar
Anita Smale

RESEARCHERS
Lynne Gulliver
Louise MacKenzie

DESIGN/PRODUCTION
Pronk&Associates

ART DIRECTION
Pronk&Associates

ELECTRONIC ASSEMBLY & TECHNICAL ART
Pronk&Associates

Acknowledgments appear on page 652.

Canadian Cataloguing in Publication Data

Alexander, Bob, 1941 –
      Addison-Wesley mathematics 10

Western Canadian ed.
Includes index.
Previous ed. written by Brendan Kelly, Bob Alexander, Paul Atkinson.
ISBN 0–201–34619–2

1. Mathematics.   I.Kelly, B. (Brendan), 1943 – .   II. Davis, Garry, 1963 – .   III. Title.

QA39.2.K446 1998   510   C97-932419-X

ISBN 0–201–34619–2

This book contains recycled product and is acid free.
Printed and bound in Canada

2 3 4 5 6–BP–03 02 01 00

REVIEWERS/CONSULTANTS

CONTENTS Mathematics 10

CONTENTS

Welcome to Addison-Wesley Mathematics 10

Western Canadian Edition

This book is about mathematical thinking, mathematics in the real world, and using technology to enhance mathematical understanding. We hope it helps you see that mathematics can be useful, interesting, and enjoyable.

These introductory pages illustrate how your student book works.

Mathematical Modelling

Each chapter begins with a provocative problem. Once you have explored relevant concepts, you use mathematical modelling to illustrate and solve the problem.

Consider This Situation is the first step in a four-stage modelling process. This page introduces the chapter problem. You think about the problem and discuss it in general terms.

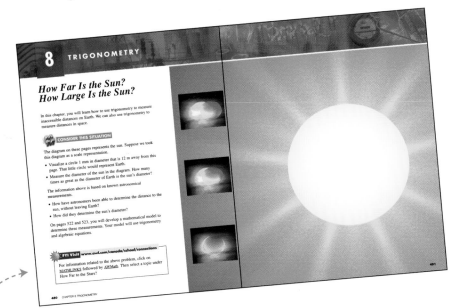

FYI Visit refers you to our web site, where you can connect to other sites with information related to the chapter problem.

You revisit the chapter problem in a Mathematical Modelling section later in the chapter.

Develop a Model suggests a graph or table, a formula or pattern, or some approximation that can represent the situation. Related exercises help you construct and investigate the model.

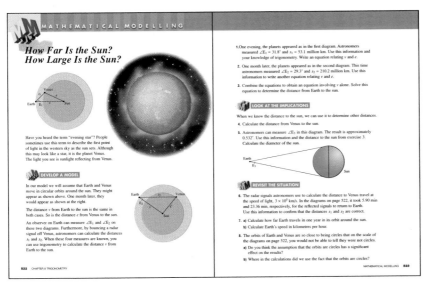

Look at the Implications encourages you to relate your findings to the original problem. What is the significance of your results?

Revisit the Situation invites you to criticize the model you developed. Does it provide a reasonable representation of the situation? Can you refine the model to obtain a closer approximation to the situation? Can you develop another problem that fits a similar model?

Short **Mathematical Modelling** boxes echo the fourth stage of the modelling process. Each box leads you back to an earlier problem to reflect on the validity of the solution, and to consider alternative ways to model the problem.

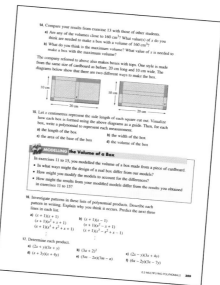

Concept Development

Here's how a typical lesson in your student book works.

By completing **Investigate**, you discover the thinking behind new concepts.

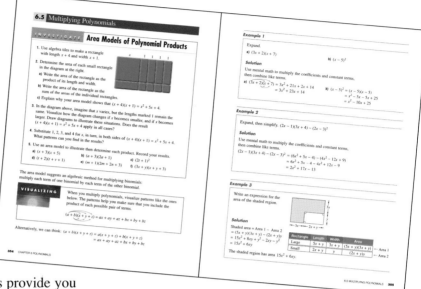

Examples with full solutions provide you with a model of new methods.

Visualizing helps you learn by seeing. It provides another way to understand and remember concepts.

Discussing the Ideas helps you clarify your understanding by talking about the preceding explanations and examples.

Communicating the Ideas helps you confirm that you understand important concepts. If you can explain it or write about it, you probably understand it!

Exercises, including end-of-chapter **Review** and **Cumulative Review** exercises, help you reinforce your understanding.

There are three levels of **Exercises** in the sections in this book.

A exercises involve the simplest skills of the lesson.

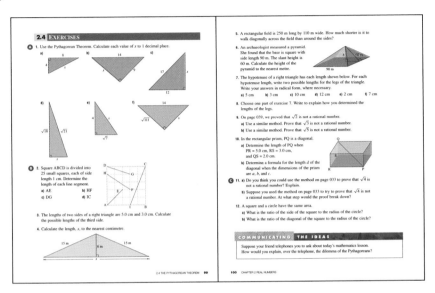

B exercises usually require several steps, and they may involve applications or problem solving.

C exercises are more thought-provoking. They may call on previous knowledge or foreshadow upcoming work.

Technology is incorporated into exercises in several ways. Special logos tell you when an exercise requires the use of technology.

This logo tells you that you need a graphing calculator to complete this exercise.

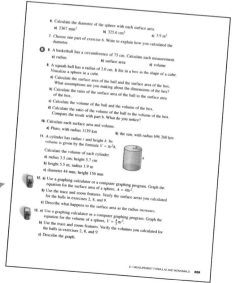

Technology

The graphing calculator and the computer are tools for learning and doing mathematics in ways that weren't possible a few years ago.

For some computer activities, you use computer applications such as spreadsheets.

This computer logo tells you that the exercise involves a spreadsheet.

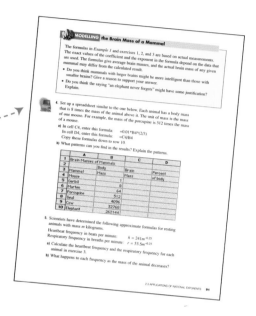

Other computer activities require the use of a computer database. The CD logo tells you that the data are available from *Addison-Wesley Mathematics 10 Template and Data Kit*. These databases provide authentic data for you to analyze. The *Template and Data Kit* also provides every spreadsheet in your student book.

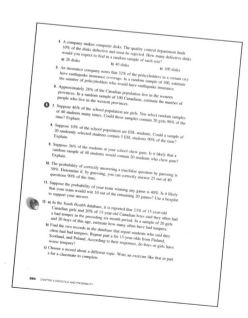

You need ClarisWorks™, Microsoft Works®, or Microsoft® Office 97 to use *Addison-Wesley Mathematics 10 Template and Data Kit*.

Computer applications are also featured in **Linking Ideas: Mathematics & Technology**. Completing these activities will show you an efficient and effective use of technology.

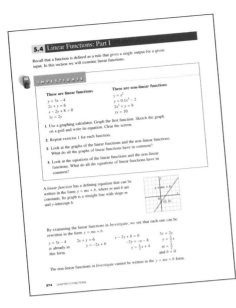

The graphing calculator is a powerful mathematical tool. It does not replace the need to develop good graphing skills, but it can enhance your mathematical understanding.

In selected Investigates and exercises, the calculator helps you explore patterns graphically, numerically, and symbolically. You develop skills to connect these different ways of looking at a situation.

Exploring with a Graphing Calculator presents a series of exercises that involve the graphing calculator and relate to a specific mathematical concept. For example, exercises that require you to examine sets of graphs may reveal a pattern that points you to a new general result for linear graphs.

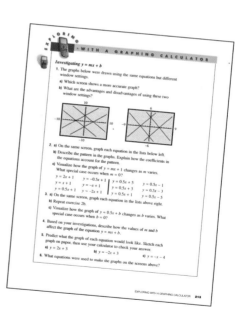

Here are some other sections you will encounter as you work through your student book.

Linking Ideas

These sections show how mathematics topics relate to other school subjects or to the world outside school.

This linking feature shows how you can use algebra to analyze bird eggs.

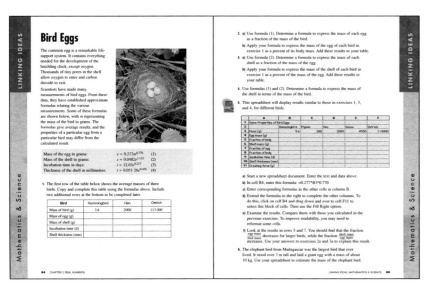

Other links

- Mathematics and Construction
- Mathematics and Geography
- Mathematics and History

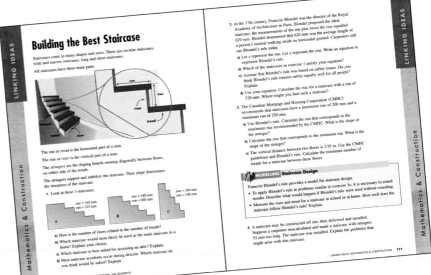

Problem Solving

Special **Problem Solving** sections introduce classic problems and puzzles, and their solutions. The accompanying exercises offer problems that may involve extensions to the introductory problem.

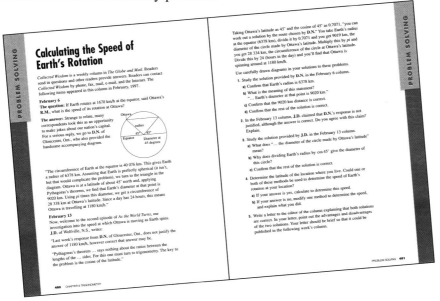

Mathematics Files

These short sections present key mathematical concepts that form part of the foundation for upcoming lessons. Selected **Mathematics Files** may present an alternative method to one already taught, or a specialized method for solving a problem.

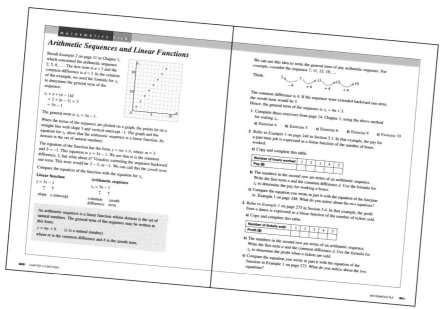

How Many Links Are Needed?

According to the movie *Six Degrees of Separation*, each of us is separated from each other by no more than six other people. For example, you probably do not know the tennis player, Monica Seles, who won the Canadian Open in 1995, 1996, and 1997. But you know A who knows B who knows C who knows D who knows E who knows F who knows Monica Seles.

 CONSIDER THIS SITUATION

Suppose you try to verify that there are no more than six people linking you with Monica Seles.

- What does it mean to say that you know someone?

- How many people do you know?

- Will six links really connect you with Monica Seles?

On pages 30 and 31, you will develop a mathematical model to explore this question. You will use a number sequence to estimate the number of links between you and any other person on Earth.

 FYI Visit **www.awl.com/canada/school/connections**

For information related to the above problem, click on <u>MATHLINKS</u> followed by <u>AWMath</u>. Then select a topic under How Many Links Are Needed?

1.1 Arithmetic Growth

INVESTIGATE

The Chinese calendar identifies years with animals. There are 12 animals, and they are always used in the order shown below. This order also shows a year associated with each name.

rat	ox	tiger	rabbit	dragon	snake	horse	ram	monkey	rooster	dog	pig
1996	1997	1998	1999	2000	2001	2002	2003	2004	2005	2006	2007

1. The year 2000 is the year of the dragon. After 2000, which three years will be the year of the dragon?

2. On July 1, 1997, Hong Kong was returned to China. 1997 was the year of the ox. After 1997, which three years will be the year of the ox?

3. After 1998, what will be three years for each animal?

   a) the horse

   b) the rooster

   c) the rat

4. Suppose you know one year that is identified with one of these animals. How can you determine other years that are identified with the same animal?

Here are some years in the Chinese system that are the year of the dog:

$$1994, 2006, 2018, 2030, \ldots$$

This sequence of numbers is an example of an *arithmetic sequence*. The numbers in a sequence are called *terms*. In an arithmetic sequence, each term after the first is calculated by adding the same number. In the sequence above, this number is 12. You can determine this number by subtracting any term from the next term. For example, in the sequence above, $2018 - 2006 = 12$.

In an arithmetic sequence, the number obtained by subtracting any term from the next term is a constant. This constant is the *common difference.*

These are arithmetic sequences.

4, 9, 14, 19, …	common difference 5
17, $17\frac{1}{2}$, 18, $18\frac{1}{2}$, …	common difference $\frac{1}{2}$
10, 9.5, 9, 8.5, …	common difference -0.5
12, 2, -8, -18, …	common difference -10

Example 1

For the arithmetic sequence 2, 9, 16, …, determine the 6th term.

Solution

Use mental math to calculate the common difference.

The common difference is $16 - 9 = 7$.

Continue to add 7 to determine all the terms up to the 6th term:

2, 9, 16, 23, 30, 37

The 6th term of the sequence is 37.

Example 2

A car salesperson receives a base salary of $275 per week, plus $250 for every car sold.

a) What is the weekly salary if 6 cars are sold? 7 cars are sold? 8 cars are sold?

b) Draw a graph to show the weekly salary for up to 10 cars sold.

Solution

a) The salary for 6 cars sold is $275 + 6 \times \$250 = \$1775$.

   The salary for 7 cars sold is $250 more, or $2025.

   The salary for 8 cars sold is $250 more, or $2275.

b) Make a table of values.

Weekly Salary

Cars sold	0	1	2	3	4	5	6	7	8	9	10
Salary ($)	275	525	775	1025	1275	1525	1775	2025	2275	2525	2775

Use the data to draw a graph.

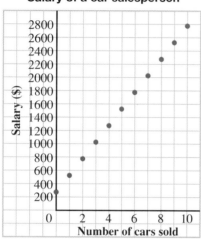

Salary of a car salesperson

Example 3

Insert two numbers between 17 and 59, so the four numbers form an arithmetic sequence.

Solution

Let d represent the common difference. Visualize the numbers in the sequence:

$$17 \underset{+d}{\frown} \blacksquare \underset{+d}{\frown} \blacksquare \underset{+d}{\frown} 59$$

To go from the 1st term to the 4th term, add d three times. Write an equation to represent this.

$$17 + 3d = 59$$
$$3d = 42$$
$$d = 14$$

The arithmetic sequence is 17, 31, 45, 59.

DISCUSSING THE IDEAS

1. In an arithmetic sequence, you always add the same number to any term to get the next term. Why is the number you add called a "difference"?

2. Look at the graph in *Example 2*.

   a) The plotted points appear to lie on a line. Why is this line not part of the graph?

   b) How would the graph change if the base salary were higher? Lower? Explain your answers.

   c) How would the graph change if the amount per car were higher? Lower? Explain your answers.

1.1 EXERCISES

A 1. State the common difference, then list the next three terms of each arithmetic sequence.

   a) 12, 15, 18, …   b) 25, 21, 17, …

   c) 27, 37, 47, …   d) 31, 21, 11, …

   e) −8, −5, −2, …   f) 1.8, 2.3, 2.8, …

2. Here is a pattern of natural numbers in three rows. Assume that the pattern continues.

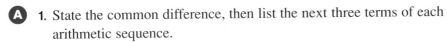

   Row 1   1   4   7   10   13   …
   Row 2   2   5   8   11   14   …
   Row 3   3   6   9   12   15   …

   a) What are the next five numbers in row 2?

   b) In which row will the number 100 appear?

   c) Describe a procedure you could use to determine in which list any given natural number appears. Test your procedure with some examples.

3. A sum of $83 was deposited in a bank on January 1. A sum of $20 is deposited in the bank on the 12th day of each month. Suppose this pattern continues. How much will be in the bank on September 1?

4. The disappearance of the dinosaurs about 65 million years ago is one of the great mysteries of science. Scientists have recently found that mass extinctions of Earth's creatures are separated by periods of roughly 26 million years.

   a) About when did other mass extinctions occur?

   b) Suppose the theory is correct. Estimate when the next mass extinction might occur.

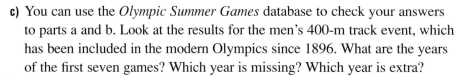

B **5. a)** Determine the indicated term of each arithmetic sequence.

   i) 6, 11, 16, … 7th   **ii)** 18, 15, 12, … 8th

   iii) 23, 34, 45, … 10th   **iv)** 45, 37, 29, … 9th

   v) 5, 10, 15, … 12th   **vi)** −10, − 4, 2, … 10th

 b) Choose one sequence from part a. Write to explain how you determined
    the term.

6. The output of a northern gold mine has remained constant at 22 000 ounces
   per year. At the end of last year, the total output of the mine was 90 000
   ounces of gold.

 a) What will the total output be at the end of this year? At the end of next
    year?

 b) Draw a graph to show the total output of the mine for the next five years.

 c) How would the graph change if the annual output is more than
    22 000 ounces per year?

7. In 1993, Canada's national debt was approximately $510 billion, and was
   growing at about $30 billion per year. At this rate, what was the national
   debt in 1994? In 1995? In 1996?

8. The Olympic Summer Games are held every four years. The first of the
   modern era were held in 1896.

 a) Beginning with 1896, list what should have been the first six Summer
    Olympic years.

 b) During one of those years, the Summer Olympics were cancelled. What
    year do you think that was? Explain your answer.

 c) You can use the *Olympic Summer Games* database to check your answers
    to parts a and b. Look at the results for the men's 400-m track event, which
    has been included in the modern Olympics since 1896. What are the years
    of the first seven games? Which year is missing? Which year is extra?

9. a) Copy then complete each arithmetic sequence.

   i) ■, 7, 12, ■, ■   **ii)** ■, ■, 29, 25, ■

   iii) 5, ■, 21, ■, ■   **iv)** 50, ■, ■, 35, ■

   v) ■, −8, ■, ■, ■, 32   **vi)** 43, ■, ■, ■, 77

 b) Choose one sequence from part a. Write to explain how you determined
    the missing terms.

10. a) Insert two numbers between 8 and 30, so the four numbers form an
    arithmetic sequence.

 b) Insert three numbers between 10 and 55, so the five numbers form an
    arithmetic sequence.

11. Determine the first five terms of each arithmetic sequence.

   a) The 2nd term is 14 and the 5th term is 23.

   b) The 3rd term is 35 and the 7th term is 55.

   c) The 5th term is 4 and the 8th term is −2.

C 12. Here is a game for two people, A and B. A begins by mentioning a single-digit number. B may add any single-digit number to the number mentioned by A. A may add any single-digit number to the result. Players alternate in this manner, always adding any single-digit number to the previous result. The winner is the first person to reach 50. Play this game with a friend, and determine a winning strategy.

13. The table shows the day of the week on which January 1 fell from 1984 to 1998.

   a) January 1 usually occurs one day later in the week than it did the year before, but sometimes it occurs two days later. Explain why this happens.

   b) Copy the table. Continue to enter the years until a pattern develops. Describe the pattern.

   c) Predict the day of the week on which January 1 will fall in each year.

   i) 2010   **ii)** 2020   **iii)** 2040   **iv)** 2100

   d) Write a set of instructions for someone to determine the day of the week on which January 1 falls, for any given year.

Sun	Mon	Tues	Wed	Thurs	Fri	Sat
1984		1985	1986	1987	1988	
1989	1990	1991	1992		1993	1994
1995	1996		1997	1998		

COMMUNICATING THE IDEAS

Suppose you know the first term and the common difference of an arithmetic sequence. How can you tell from these numbers whether the terms of the sequence become greater or smaller as you extend the sequence? Write an explanation in your journal. Support your explanation with examples.

1.2 The General Term of an Arithmetic Sequence

Aminah has $70 in her bank account. From her part-time job, she plans to save $50 each month and deposit this amount in her account. If she follows this plan, the amount of money in her account will increase as follows as the months go by:

$$70, 120, 170, 220, 270, \ldots$$

These numbers form an arithmetic sequence with a common difference of 50. We write: the 1st term, $t_1 = 70$; the 2nd term, $t_2 = 120$; the 3rd term, $t_3 = 170$; the 4th term, $t_4 = 220$; the 5th term, $t_5 = 270$, and so on. The sequence shows that in the 5th month, Aminah will have $270 in her account. This amount can be calculated by adding four differences to the first term:

$$\begin{aligned}
t_5 &= 70 + 50 + 50 + 50 + 50 \\
&= 70 + 4 \times 50 \\
&= 270
\end{aligned}$$

To calculate the amount Aminah will have in the 10th month, add nine differences to the first term:

$$\begin{aligned}
t_{10} &= 70 + 9 \times 50 \\
&= 520
\end{aligned}$$

Aminah will have $520 in her account in the 10th month.

To calculate the amount she will have in the nth month, add $(n - 1)$ differences to the first term:

$$\begin{aligned}
t_n &= 70 + (n - 1) \times 50 \\
&= 70 + 50n - 50 \\
&= 50n + 20
\end{aligned}$$

In the nth month, the amount in dollars Aminah will have is $50n + 20$.

We say that t_n is the *general term* of the sequence because we can determine any particular term by substitution. For example, to determine t_{24}, substitute 24 for n to obtain $t_{24} = 50(24) + 20$, or 1220. This represents the amount Aminah will have in her account after two years, assuming that she follows her plan.

In the general arithmetic sequence, the first term is represented by a and the common difference by d. The first few terms are:

$$\begin{aligned}
t_1 &= a \\
t_2 &= a + d \\
t_3 &= a + 2d \\
t_4 &= a + 3d \\
&\ \vdots \\
t_n &= a + (n - 1)d
\end{aligned}$$

The general term of an arithmetic sequence is given by

$$t_n = a + (n - 1)d$$

where a is the first term, n the term number, and d the common difference.

If you know the first term and the common difference of an arithmetic sequence, you can determine any other term.

Example 1

In the arithmetic sequence 2, 5, 8, …, determine each term.

a) t_n **b)** t_{20} **c)** t_{100}

Solution

Use mental math to calculate the common difference.

The first term is 2 and the common difference is 3.

a) $t_n = a + (n - 1)d$

Substitute 2 for a and 3 for d.

$$\begin{aligned} t_n &= 2 + (n - 1) \times 3 \\ &= 2 + 3n - 3 \\ &= 3n - 1 \end{aligned}$$

b) Use the formula in part a.

$$t_n = 3n - 1$$

Substitute 20 for n.

$$\begin{aligned} t_{20} &= 3(20) - 1 \\ &= 60 - 1 \\ &= 59 \end{aligned}$$

c) Substitute 100 for n in the formula in part a.

$$\begin{aligned} t_{100} &= 3(100) - 1 \\ &= 299 \end{aligned}$$

Example 2

An arithmetic sequence is 3, 10, 17, 24, ... One term in this sequence is 129. Which term is it?

Solution

The first term is $a = 3$ and the common difference is $d = 7$.

Use the formula for the nth term: $t_n = a + (n - 1)d$

Substitute 3 for a, 7 for d, and 129 for t_n.

$129 = 3 + (n - 1) \times 7$
$129 = 3 + 7n - 7$
$133 = 7n$
$\quad n = 19$

129 is the 19th term of the sequence.

Example 3

In an arithmetic sequence, the 4th term is 73 and the 10th term is 121.

a) List the sequence to show the first four terms.

b) Write the general term of the sequence.

c) How many terms of the sequence are less than 200?

Solution

a) Visualize the numbers listed in the sequence.

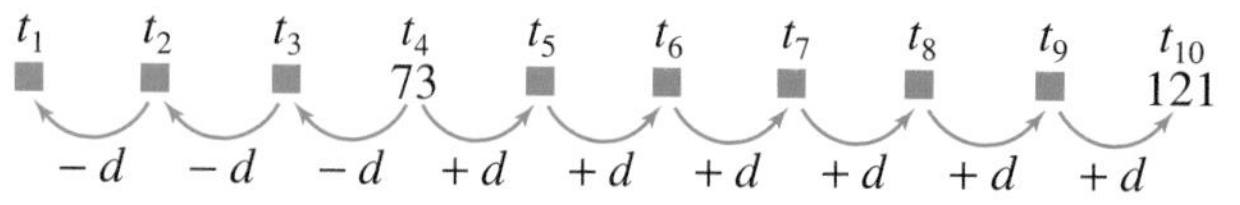

To go from the 4th term to the 10th term, we add 6 differences.

Since $121 - 73 = 48$, then six differences, or $6d = 48$

That is, $d = 8$

To go from the 1st term to the 4th term, we add 3 differences. Hence, to go back from the 4th term to the 1st term, we subtract 3 differences.

$t_1 = 73 - 3(8)$
$\quad = 73 - 24$
$\quad = 49$

The sequence is 49, 57, 65, 73, ...

b) Use the formula $t_n = a + (n - 1)d$. Substitute for a and d.

$$t_n = 49 + (n - 1)8$$
$$= 49 + 8n - 8$$
$$= 8n + 41$$

The general term of the sequence is $8n + 41$.

c) Write an inequality involving the general term.

$$8n + 41 < 200$$
$$8n < 159$$
$$n < \frac{159}{8}$$
$$n < 19.875$$

Hence, 19 terms of the sequence are less than 200.

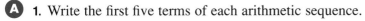

DISCUSSING THE IDEAS

1. In the formula for t_n, explain why d is multiplied by $(n - 1)$ instead of by n.

2. How could you check the answer to *Example 2*?

3. a) How could you check the answer to *Example 3c*?

   b) Why can we not round 19.875 to 20 and say that 20 terms of the sequence are less than 200?

4. Suppose the first term and the common difference of an arithmetic sequence are equal. How are all the terms related to the first term?

1.2 EXERCISES

A **1.** Write the first five terms of each arithmetic sequence.

   a) $a = 2$, $d = 3$ **b)** $a = 7$, $d = 4$ **c)** $a = -1$, $d = -3$

   d) $a = 12$, $d = -4$ **e)** $a = -8$, $d = 5$ **f)** $a = 5$, $d = -8$

2. For the arithmetic sequence 3, 5, 7, 9, ..., determine each term.

   a) t_n **b)** t_8 **c)** t_{25}

3. For the arithmetic sequence 11, 8, 5, 2, ..., determine each term.

   a) t_n **b)** t_6 **c)** t_{20}

B **4. a)** Write the general term for each arithmetic sequence.

   i) 5, 8, 11, 14, ...   **ii)** 17, 14, 11, 8, ...

   iii) 5, 7, 9, 11, ...   **iv)** 10, 8, 6, 4, ...

   b) Choose one sequence from part a. Write to explain how you determined the general term.

5. Each sequence is an arithmetic sequence. Write a formula for t_n, then use it to determine the indicated term.

   a) 1, 5, 9, 13, ... t_{17} **b)** 3, 6, 9, 12, ... t_{21}

   c) -4, 1, 6, 11, ... t_{13} **d)** 41, 35, 29, 23, ... t_{18}

   e) -2, -5, -8, -11, ... t_{10} **f)** 9, 1, -7, -15, ... t_{46}

6. A pile of bricks is arranged in rows. The number of bricks in each row forms the arithmetic sequence 65, 59, 53, ...

   a) One row contains 17 bricks. Which row is this?

   b) How many rows of bricks are there? Explain your assumptions and answer.

7. An arithmetic sequence is 8, 14, 20, 26, ... Which term is 92?

8. An Enstrom 280FX Shark helicopter is flying at an altitude of 1850 ft when it begins to climb. Its rate of climb is 1450 ft/min.

   a) List the helicopter's altitude after each of the next 5 minutes.

   b) Write a formula to describe the general term of the sequence in part a.

   c) The maximum altitude of this helicopter is 12 000 ft. After how many minutes will the helicopter reach maximum altitude?

   d) Choose a different model of helicopter from the *Helicopters* database. Write a similar exercise about the model you choose. Exchange exercises with a classmate. Check each other's work.

9. a) For each arithmetic sequence, a later term in the sequence is given. Which term is it?

   i) 5, 8, 11, ... 41   **ii)** 2, 7, 12, ... 122   **iii)** 14, 25, 36, ... 234

   iv) 75, 68, 61, ... 12   **v)** -12, -6, 0, ... 72   **vi)** 28, 20, 12, ... -164

   b) Choose one sequence from part a. Write to explain how you identified the term.

10. In an arithmetic sequence, the 3rd term is 25 and the 9th term is 43.

   a) List the sequence. Show the first four terms.

   b) Write the general term of the sequence.

   c) How many terms of the sequence are less than 100?

11. The Olympic Winter Games are held every four years. The first Winter Olympics were held in 1924.

   a) Beginning with 1924, list what should have been the first six Winter Olympic years.

   b) In two of those years, the Winter Olympics were cancelled. What years do you think they were? Explain your answer.

   c) The Winter Olympics followed the pattern of being held every four years until 1992. Beginning in 1994, another pattern of holding the Winter Olympics every four years was begun. How many Winter Olympics followed the first pattern?

   d) Why do you think the first pattern was broken in 1994?

C 12. a) Copy this pattern. Extend it horizontally and vertically to complete more of the empty spaces.

2	4	6	8	10	...
7	9	11	13	...	
12	14	16	...		
17	19	⋮			
22	⋮				
⋮					

   b) Find as many arithmetic sequences in the pattern as you can.

   c) Choose two of your sequences. Describe how they are related.

   d) Write the general term for each sequence you chose. Explain how the general terms show the relationship you described in part c.

13. Write an expression for the general term of each sequence. Is the sequence an arithmetic sequence?

   a) $1 \times 1, 3 \times 4, 5 \times 7, 7 \times 10, \ldots$

   b) $2 \times 3, 4 \times 6, 6 \times 9, 8 \times 12, \ldots$

   c) $\frac{1}{3}, \frac{2}{5}, \frac{3}{7}, \frac{4}{9}, \ldots$

   d) $\frac{1 \times 3}{2 \times 4}, \frac{3 \times 5}{4 \times 6}, \frac{5 \times 7}{6 \times 8}, \frac{7 \times 9}{8 \times 10}, \ldots$

COMMUNICATING THE IDEAS

Explain what is meant by "the general term of a sequence", and why it is useful. Use some examples to illustrate your explanation. Write your explanation in your journal.

Arithmetic Sequences in Astronomy

Throughout recorded history, comets have been associated with significant events such as famines, plagues, and floods. In 1705, the astronomer Edmund Halley noticed similarities in the records of major comets that had appeared in 1531, 1607, and 1682. He also noticed that those dates were apart by about the same number of years.

Halley concluded that the three appearances represented return visits of the same comet, which was in an elongated orbit around the sun. He predicted it would return in 1758. The comet returned in 1759, and records have since been found for every appearance of this comet since 239 B.C. If it were not for the gravitational influence of the planets, Halley's comet would reappear every 77 years.

1. Halley's comet appeared twice during the twentieth century. During which two years do you think it appeared?

2. Halley's comet will appear only once during the twenty-first century. Predict the year when this will occur.

3. The Bayeux tapestry is a long narrow embroidery that illustrates the Norman Conquest of England in 1066. It is said to have been completed by Queen Matilda of France. A portion of this tapestry shows a crowd of people pointing to a comet. Do you think the comet could be Halley's comet? Explain.

4. How many times has Halley's comet appeared since its first recorded appearance?

1.3 The Sum of an Arithmetic Series

Each of two summer jobs is for 3 months, or 12 weeks.
Job A pays $500 per month.
Job B pays $100 per week with a $5 raise each week,
if the person performs well.

Assume the person performs well.
Which is the better-paying job?

Job A

The total salary for 3 months, in dollars,
is $3 \times 500 = 1500$.

Job B

The payments, in dollars, are the first 12 terms
of an arithmetic sequence:
$$100, 105, 110, \ldots, t_{12}$$
$$t_{12} = 100 + 11(5)$$
$$= 155$$
The total salary, in dollars, is $100 + 105 + 110 + \cdots + 145 + 150 + 155$.

This expression is an example of an *arithmetic series* because it
indicates that the terms of an arithmetic sequence are to be added.
Instead of adding the twelve numbers, the sum can be found as follows.

Let S represent the sum of the series: $\qquad S = 100 + 105 + 110 + \cdots + 145 + 150 + 155$

Write the sum in reverse order: $\qquad \underline{S = 155 + 150 + 145 + \cdots + 110 + 105 + 100}$

Add the left sides and the right sides: $\quad 2S = 255 + 255 + 255 + \cdots + 225 + 225 + 255$

$$2S = 12 \times 255 \qquad \text{(since there are 12 terms)}$$
$$2S = 3060$$
$$S = 1530 \qquad \boxed{\text{Divide both sides by 2.}}$$

Job B pays a total of $1530. It is the better-paying job.

This method can be used to calculate the sum of any number of terms of an arithmetic series.

We shall calculate the sum of the first 18 terms of the arithmetic series $2 + 9 + 16 + 23 + \cdots$

Observe that $a = 2$, $d = 7$, and $n = 18$.

We determine the 18th term using the formula $t_n = a + (n - 1)d$.
$$t_{18} = 2 + 17 \times 7$$
$$= 121$$

We use the 18th term to write some preceding terms:

$t_{17} = 121 - 7$ $t_{16} = 114 - 7$
      $= 114$ $= 107$

Let S represent the sum: $S = \quad 2 + \quad 9 + \quad 16 + \cdots + 107 + 114 + 121$
Write in reverse order: $S = 121 + 114 + 107 + \cdots + \quad 16 + \quad 9 + \quad 2$
Add the equations: $2S = 123 + 123 + 123 + \cdots + 123 + 123 + 123$
                            $2S = 18 \times 123 \qquad$ (since there are 18 terms)
                            $2S = 2214$
                             $S = 1107$ **Divide by 2.**

The sum of the first 18 terms of the series is 1107.

The method explained above can be used to determine a formula for the sum of the first n terms of the general arithmetic series.

Let S_n represent the sum of the first n terms of the general arithmetic series. We use the last term t_n and the common difference d to write some preceding terms:

$t_{n-1} = t_n - d$ $t_{n-2} = t_n - 2d$

Then: $S_n = \qquad a + (a + d) + (a + 2d) + \cdots + (t_n - 2d) + (t_n - d) + t_n$
Reversing: $S_n = \qquad t_n + (t_n - d) + (t_n - 2d) + \cdots + (a + 2d) + (a + d) + a$
Adding: $2S_n = (a + t_n) + (a + t_n) + \quad (a + t_n) + \cdots + \quad (a + t_n) + (a + t_n) + (a + t_n)$
              $2S_n = (a + t_n) \times n \qquad$ (since there are n terms)
              $S_n = \dfrac{(a + t_n) \times n}{2}$

VISUALIZING

The sum of the first n terms of the general arithmetic series is
$$S_n = \frac{(a + t_n)}{2} \times n$$
$$= \text{(mean of first and last terms)} \times \text{(number of terms)}$$

Example 1

Determine the sum of the first 50 terms of the arithmetic series $3 + 4.5 + 6 + 7.5 + \cdots$.

Solution

Observe that $a = 3$, $d = 1.5$, and $n = 50$.
Determine the 50th term using the formula $t_n = a + (n - 1)d$.

$t_{50} = 3 + 49(1.5)$
$\quad = 76.5$ ◄───── This is the last term of the series.

Use the formula $S_n = \frac{(a + t_n)}{2} \times n$.

$$S_{50} = \frac{(3 + 76.5)}{2} \times 50$$
$$= 39.75 \times 50$$
$$= 1987.5$$

The sum of the first 50 terms of the series is 1987.5.

Example 2

Determine the sum of the arithmetic series $6 + 10 + 14 + \cdots + 50$.

Solution

To use the formula for S_n, we must know the number of terms.
Observe that $a = 6$, $d = 4$, and $t_n = 50$.

Use the formula $t_n = a + (n - 1)d$ to calculate n, the number of terms.

$$50 = 6 + (n - 1)4$$
$$= 6 + 4n - 4$$
$$4n = 48$$
$$n = 12$$ ◄───── There are 12 terms in the series.

Use the formula $S_n = \frac{(a + t_n)}{2} \times n$.

$$S_{12} = \frac{(6 + 50)}{2} \times 12$$
$$= 28 \times 12$$
$$= 336$$

The sum of the series is 336.

1. When you learned to solve equations in an earlier grade, you may have used the model of a balance. In the solution of the example on page 17, the left sides and the right sides of two equations were added. How can the step of adding the sides of two equations be justified using the model of a balance?

2. **a)** In the solution of *Example 1*, what does the number 39.75 represent?

   b) Why does multiplying this number by 50 give the sum of the series?

3. On page 18, you visualized the sum of the first n terms of an arithmetic series as the mean of the first and last terms multiplied by the number of terms.

   a) What other way could you visualize the sum?

   b) Solve *Examples 1* and *2* using this other way of visualizing the sum.

1.3 EXERCISES

A 1. Determine the sum of each arithmetic series.

   a) $3 + 12 + 21 + 30 + 39 + 48$

   b) $19 + 31 + 43 + 55 + 67 + 79 + 91$

   c) $6 + 13 + 20 + 27 + 34 + 41 + 48 + 55$

   d) $25 + 31 + 37 + 43 + 49 + 55 + 61 + 67 + 73$

B 2. For the arithmetic series $6 + 8 + 10 + 12 + \cdots$

   a) Determine the 20th term.   **b)** Determine the sum of the first 20 terms.

3. **a)** Determine the sum of the first 10 terms of each arithmetic series.

   i) $3 + 7 + 11 + \cdots$ **ii)** $5 + 11.5 + 18 + \cdots$

   iii) $2 + 8 + 14 + \cdots$ **iv)** $45 + 39 + 33 + \cdots$

   v) $6 + 16.8 + 27.6 + \cdots$ **vi)** $21 + 15.1 + 9.2 + \cdots$

   b) Choose one series from part a. Write to explain how you determined the sum.

4. Tasty Treats finds that its profit from the sale of ice cream increases by $5 per week during the 15-week summer season. Suppose the profit for the first week is $30. Determine the profit for the season.

5. For three summer months (12 weeks), Job A pays $325 per month with a monthly raise of $100. Job B pays $50 per week with a weekly raise of $10. Which is the better-paying job?

6. For three summer months (12 weeks), Job A pays $400 per month with a monthly raise of $20. Job B pays $100 per week with a weekly raise of $5. Do the jobs pay the same total amount over the summer, or does one pay more than the other? Explain your answer.

7. In a supermarket, apple juice cans are stacked in a display arranged in layers. The numbers of cans in the layers form an arithmetic sequence. There are 48 cans in the bottom layer, and 20 cans in the top layer. There are 8 layers. How many cans are in the display?

8. Raji's annual salary is in a range from $25 325 in the 1st year to $34 445 in the 7th year.

   a) The salary range is an arithmetic sequence with seven terms. Determine the raise Raji can expect each year.

   b) What is her salary in the fourth year?

   c) In which year does her salary exceed $30 000 for the first time?

   d) What is the total amount Raji will earn in the seven years?

   e) Choose one of parts a to d. Write to explain how you calculated the answer.

9. A pile of bricks is arranged in rows. The numbers of bricks in the rows form an arithmetic sequence. There are 35 bricks in the 4th row and 20 bricks in the 9th row.

   a) How many bricks are in the first row?

   b) How many rows of bricks are there?

   c) How many bricks are in the pile?

   Explain any assumptions you made.

10. Determine the sum of each arithmetic series.

    a) $1 + 2 + 3 + \cdots + 9$ b) $1 + 2 + 3 + \cdots + 99$

    c) $1 + 2 + 3 + \cdots + 999$ d) $1 + 2 + 3 + \cdots + 9999$

11. Determine a formula for the sum of the first n natural numbers, $1 + 2 + 3 + \cdots + n$.

12. a) Determine the sum of each arithmetic series.

    i) $2 + 7 + 12 + \cdots + 62$ ii) $4 + 11 + 18 + \cdots + 88$

    iii) $3 + 5.5 + 8 + \cdots + 133$ iv) $20 + 14 + 8 + \cdots + (-40)$

    b) Choose one series from part a. Write to explain how you determined the sum.

13. This sentence is called a "snowball sentence."

I do not know where family doctors acquired illegibly perplexing handwriting; nevertheless, extraordinary pharmaceutical intellectuality, counterbalancing indecipherability, transcendentalizes intercommunications' incomprehensibleness.

a) Why is the name "snowball sentence" appropriate?

b) How many letters are in this snowball sentence?

14. In the popular TV quiz show "Jeopardy!", a contestant gives each response as a question to a clue hidden behind a panel that shows an amount of money. When the contestant's response is correct, the contestant wins the money. Answer the questions below in as many different ways as you can.

a) In "Jeopardy!" (below left), what is the total amount of money shown?

JEOPARDY!

WORLD ORIGINS	OCEANS	SCIENCE	MOVIES	MODERN POETRY	THIS & THAT
$100	$100	$100	$100	$100	$100
$200	$200	$200	$200	$200	$200
$300	$300	$300	$300	$300	$300
$400	$400	$400	$400	$400	$400
$500	$500	$500	$500	$500	$500

DOUBLE JEOPARDY!

FOOD	TV QUIZ SHOWS	SPORTS	MATH	DRAMA	ODDS & ENDS
$200	$200	$200	$200	$200	$200
$400	$400	$400	$400	$400	$400
$600	$600	$600	$600	$600	$600
$800	$800	$800	$800	$800	$800
$1000	$1000	$1000	$1000	$1000	$1000

b) In "Double Jeopardy!" (above right), what is the total amount shown?

C **15.** The sum of the first 5 terms of an arithmetic series is 85. The sum of the first 6 terms is 123. Determine the first four terms of the series.

16. The arithmetic series $3 + 7 + 11 + 15 + \cdots$ is given.

a) Determine t_{20} and t_n.   **b)** Determine S_{20} and S_n.

c) How many terms are less than 500?

d) How many terms have a sum less than 500?

COMMUNICATING THE IDEAS

Explain the difference between a "sequence" and a "series." Make up some examples to use in your explanation. Write your explanation in your journal.

1.4 Geometric Growth

INVESTIGATE

You will need a sheet of blank paper and a ruler.

1. Fold the paper in half, giving 2 layers of paper.
   Fold it in half again, giving 4 layers.
   Continue to fold in half as
   long as you can.

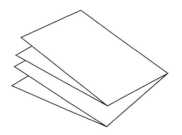

2. After each step, the folded paper is rectangular.
   Calculate the area of the rectangle after each step.

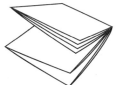

   a) Record your results in a table like this.

Number of folds	Number of layers	Area of rectangle
0	1	
1		
2		
3		

   b) Describe the patterns you see in the table.

3. Suppose you extend the patterns in the table to 10 folds.

   a) How many layers of paper would there be?

   b) What would the area of the rectangle be?

   c) About how thick would the resulting wad of paper be?

   d) What assumptions are you making in your answers to parts b and c?

In *Investigate*, you should have found the numbers of layers follow this pattern:

   1, 2, 4, 8, 16, …

The areas of the rectangles follow a pattern similar to the one below. The first number depends on the size of the piece of paper you started with, and whether you used centimetres or millimetres as the unit of length. For this example, the units are square centimetres.

   600, 300, 150, 75, 37.5, …

These two sequences of numbers are examples of *geometric sequences*. In a geometric sequence, each term after the first term is found by multiplying by the same number. To determine the number by which you multiply, divide any term by the term before it. In the first sequence above, this number is $16 \div 8 = 2$. In the second sequence, the number is $75 \div 150 = \frac{1}{2}$.

In a geometric sequence, the ratio formed by dividing any term by the preceding term is a constant. This constant is the *common ratio*.

These are geometric sequences.

1, 3, 9, 27, 81, …	common ratio 3
2, 10, 50, 250, …	common ratio 5
12, 6, 3, 1.5, 0.75, …	common ratio 0.5
3, −12, 48, −192, …	common ratio −4

Example 1

For the geometric sequence 2, 6, 18, …, determine the 6th term.

Solution

Use mental math to calculate the common ratio.

The common ratio is 3. Continue to multiply by 3 to determine all the terms up to the 6th term.

2, 6, 18, 54, 162, 486

The 6th term of the sequence is 486.

Example 2

A store is conducting a Dutch auction. It will take 10% off the cost of an item each day.

a) Suppose an item originally costs $250. Determine its cost for each of the next five days.

b) Show the results on a graph.

Solution

a) The cost each day is 90% of the previous day's cost.

Cost on first day:	$250.00 \times 0.9 = \$225.00$
Cost on second day:	$225.00 \times 0.9 = \$202.50$
Cost on third day:	$202.50 \times 0.9 = \$182.25$
Cost on fourth day:	$182.25 \times 0.9 = \$164.03$
Cost on fifth day:	$164.03 \times 0.9 = \$147.63$

b) Use the data from part a to draw a graph.

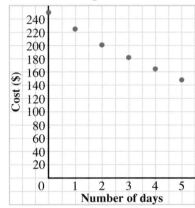

How the cost changes in a Dutch auction

Example 3

The number of insects in a colony doubles every month. There are currently about 1000 insects in the colony.

a) How many insects will there be after 6 months?

b) Show the results on a graph.

c) Write an expression to represent the number of insects after n months.

Solution

a) Number of insects after

month 1:	$1000 \times 2 = 2000$
month 2:	$1000 \times 2 \times 2 = 1000 \times 2^2 = 4000$
month 3:	$1000 \times 2 \times 2 \times 2 = 1000 \times 2^3 = 8000$
month 4:	$1000 \times 2^4 = 16\,000$
month 5:	$1000 \times 2^5 = 32\,000$
month 6:	$1000 \times 2^6 = 64\,000$

After 6 months, there will be about 64 000 insects.

b) Use the data from part a to draw a graph.

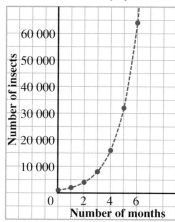

Growth of insect population

c) Look at the patterns in part a. After each month, the number of insects is 1000 times a power of 2, in which the exponent is the month number. After n months, the number of insects is 1000×2^n.

Example 4

Insert two numbers between 5 and 320, so the four numbers form a geometric sequence.

Solution

Let r represent the common ratio.
Visualize the numbers in this sequence.
To go from the 1st term to the 4th term, multiply by r three times.
Write an equation to represent this.

t_1 t_2 t_3 t_4
5   ■   ■   320
$\times r$ $\times r$ $\times r$

$$5 \times r^3 = 320$$
$$r^3 = \frac{320}{5}$$
$$= 64$$
$$r = 4$$

We know that 4 x 4 x 4=64.

The geometric sequence is 5, 20, 80, 320.

1. Suppose we solved *Example 2* by calculating the discount each day. Do you think the discounts would form a geometric sequence? Explain your reasoning. Then find out if you were correct.

2. Look at the graph in *Example 2*.
   a) The plotted points appear to lie on a curve. Why is this curve not part of the graph?
   b) How would the graph change if the original cost of the item had been more than $250? Less than $250? Explain your answers.
   c) How would the graph change if the discount each day had been more than 10%? Less than 10%? Explain.

3. In *Example 4*, suppose only one number is inserted between 5 and 320 so the three numbers form a geometric sequence.
   a) What two possibilities are there for this number?
   b) Why are there two values of r in this case, but only one in *Example 4*?

4. Suppose the first term and the common ratio of a geometric sequence are equal. How are all the terms related to the first term?

1.4 EXERCISES

A 1. State the common ratio, then list the next three terms of each geometric sequence.
   a) 2, 20, 200, …
   b) 48, 24, 12, …
   c) 5, −10, 20, …
   d) 3, 9, 27, …

2. A colony contains 100 insects. Suppose the population doubles every month. How many insects are there after each time?
   a) 2 months
   b) 4 months
   c) 6 months

B 3. a) Determine the indicated term of each geometric sequence.
   i) 3, 6, 12, … t_6
   ii) 3, −6, 12, … t_6
   iii) 200, 20, 2, … t_5
   iv) 5, 20, 80, … t_7

   b) Choose one sequence from part a. Write to explain how you determined the term.

4. Each year, the value of a car depreciates to 70% of its value the previous year. A car was bought new for $20 000.
   a) Determine its approximate value after 5 years.

b) Draw a graph to show how its value drops during the five years.

c) Write an expression to represent its value after *n* years.

5. A ball is dropped from a height of 2 m. After each bounce, it rises to 75% of its previous height.

  a) What height does the ball reach after each of the first five bounces?

  b) After how many bounces does the ball reach a height of only 20 cm?

  c) Show the results on a graph.

6. In exercise 5, explain how the graph would change in each case.

  a) The ball was dropped from a height greater than 2 m.

  b) The ball rises to a lower percent of its previous height on each bounce.

7. Strep throat is an infection caused by bacteria called streptococci. After you have been infected, it is possible for these bacteria to double in number every 20 min.

  a) Suppose a single bacterium began reproducing at noon. About how many bacteria would be present at each time?

  i) 1 p.m.   **ii)** 2 p.m.   **iii)** 3 p.m.

  b) Show the results on a graph.

 MODELLING the Growth of Bacteria

In exercise 7, the way you calculate the number of bacteria every 20 min and the graph of the results are *models* of the growth of the bacteria.

• Give some reasons why the actual number of bacteria present every 20 min might be different from the numbers predicted by the model.

• Do you think the growth could continue indefinitely? Explain.

8. Copy and complete each geometric sequence.

  a) ▨, 8, 16, ▨, ▨   **b)** ▨, ▨, 12, 4, ▨

  c) 1, ▨, ▨, −27, ▨   **d)** 1, ▨, 16, ▨, ▨

  e) ▨, 160, ▨, ▨, ▨, 10   **f)** 10, ▨, ▨, ▨, 6250

9. **a)** Insert two numbers between 2 and 54, so the four numbers form a geometric sequence.

  b) Insert three numbers between 4 and 2500, so the five numbers form a geometric sequence.

  c) Choose one of part a or b. Write to explain how you determined the numbers.

10. The advertisement below appeared in a local newspaper.

Item	Retail Price	Week 1 Price	Week 2 Price	Week 3 Price	Week 4 Price
15' Chrysler Tri Hull	$2000	$1800	$1620	$1458	$1312
14' Starcraft, 20 Yamaha, Yachter Trailer	$7000	$6300	$5670	$5103	$4593
16' Wilker 1979, 100 Javalin, Boat Trailer	$9000				
19' Steury, 105 h.p. Chrysler	$5699				

BOAT WORLD BIG TEN SALE

10% will be taken off each boat's price for the next four weeks. 60 day warranty on all items. First come, first served.

a) In the first two rows, check that the prices for each week are correct.

b) Calculate the prices in the other two rows.

c) Let x dollars represent the retail price of a boat. Write an expression to represent the price of the boat in each of the first four weeks.

C **11.** In exercise 10, let x dollars represent the retail price of a boat. Write an expression to represent the total cost of the boat after each week, including taxes. Use the current GST tax rate and the current sales tax rate in your province.

12. Suppose a cottage is bought for $40 000. It appreciates in value by about 7% each year. What is its approximate value after each time?

a) 5 years   **b)** 10 years   **c)** n years

13. The general geometric sequence is represented by a, ar, ar^2, ar^3, ...

a) Write an expression for the general term t_n.

b) Compare your expression with the expression for the general term of an arithmetic sequence. In what ways are they similar? In what ways are they different?

COMMUNICATING THE IDEAS

Suppose you know the first term and the common ratio of a geometric sequence. How can you tell from these numbers whether the terms of the sequence become greater or smaller as you extend the sequence? What other possibilities are there for what happens to the terms as you extend the sequence? How can you tell when these will happen? Write an explanation in your journal. Support your explanation with examples.

How Many Links Are Needed?

On page 2, it was suggested that there are only six people linking you with Monica Seles. Could this really be true?

DEVELOP A MODEL

Suppose you know 100 people. Any one of these people could be person A in the photographs above. That is, there are 100 possibilities for person A.

Think about these 100 people. How many people do *they* know? Suppose each of them also knows 100 people. Some of these people will also know you. Suppose 20 of these people do not know you. Hence, there are $100 \times 20 = 2000$ possibilities for B. You are linked to each of these people by A.

Suppose each of these 2000 people knows 20 people who do not know you. Hence, there are $2000 \times 20 = 40\,000$ possibilities for C. You are linked to each of these people by A and B.

1. Use the same reasoning. How many possibilities are there for each person?

   a) D   b) E   c) F   d) G

2. Record your results in a table like this.

Person	Number of links	Number of people
A	0	100
B	1	2000
C	2	
D		
E		
F		
G		

 LOOK AT THE IMPLICATIONS

3. The world's population in 1996 was approximately 6 billion. Was your answer to exercise 1d greater than 6 billion? If it was, then Monica Seles must be one of these people.

4. Suppose you want to find out who the people are who link you to Monica Seles.

   a) Who might some of these people be?

   b) Do you think it would be possible to find out who the six people are who link you to Monica Seles?

 REVISIT THE SITUATION

The way you solved this problem is called a *mathematical model*. The model was based on these assumptions.

- You know 100 people.
- Each person knows 20 people who do not know any of the previous people.
- You start with 100 and multiply by 20 to calculate the number of people linked to you after each step: 100, 2000, 40 000, …

5. **a)** Do you think these assumptions are reasonable? Explain.

   b) If your answer to part a was no, what changes would you make to the model? How would your changes affect the answer to the problem?

6. How many links would be needed to link you with any other person in each place?

   a) your school   **b)** your province   **c)** Canada

1.5 Revisiting the Exponent Laws

A geometric sequence involves repeated multiplication by the same number, the common ratio. When the first term and the common ratio of a geometric sequence are equal, the terms of the sequence are powers. For example, the terms of the geometric sequence with first term 2 and common ratio 2 are powers of 2.

Numeral form: 2   4   8   16   32   ...

Power form: 2^1 2^2 2^3 2^4 2^5 ...

Recall that the definition of a power depends on whether the exponent is a positive integer, zero, or a negative integer.

Positive Integral Exponent

$a^n = a \cdot a \cdot a \cdot \cdots \cdot a$

$\qquad n$ factors

Zero Exponent

a^0 is defined to be equal to 1.

$$a^0 = 1, \quad (a \neq 0)$$

Negative Integral Exponent

a^{-n} is defined to be the reciprocal of a^n.

$$a^{-n} = \frac{1}{a^n}, \quad (a \neq 0)$$

These definitions permit us to extend the sequence of powers of 2 above to the left.

Multiplying each term by 2 increases the exponent by 1.

$\rightarrow$

... $\frac{1}{16}$ $\frac{1}{8}$ $\frac{1}{4}$ $\frac{1}{2}$ 1   2   4   8   16   ...

... 2^{-4} 2^{-3} 2^{-2} 2^{-1} 2^0 2^1 2^2 2^3 2^4 ...

$\leftarrow$

Dividing each term by 2 decreases the exponent by 1.

We use these definitions to evaluate a power with any integral exponent.

Example 1

Simplify each power.

a) 4^3 **b)** 3^{-2} **c)** $(-2)^{-3}$ **d)** $\left(\frac{1}{2}\right)^0$

Solution

Use mental math.

a) $4^3 = 4 \times 4 \times 4$
$ = 64$

b) $3^{-2} = \frac{1}{3^2}$
$\phantom{3^{-2}} = \frac{1}{9}$

c) $(-2)^{-3} = \frac{1}{(-2)^3}$
$\phantom{(-2)^{-3}} = -\frac{1}{8}$

d) $\left(\frac{1}{2}\right)^0 = 1$

The definitions of integral exponents lead to some basic laws for working with exponents. These examples will help you recall the laws they illustrate.

Expression	**Exponent Law**
$x^3 \cdot x^2 = x \cdot x \cdot x \cdot x \cdot x$ $= x^{3+2}$, or x^5	$x^m \cdot x^n = x^{m+n}$
$x^5 \div x^3 = \frac{x \cdot x \cdot x \cdot x \cdot x}{x \cdot x \cdot x}$ $= x^{5-3}$, or x^2	$x^m \div x^n = x^{m-n} \quad (x \neq 0)$
$(x^3)^2 = (x \cdot x \cdot x)(x \cdot x \cdot x)$ $= x^{3 \times 2}$, or x^6	$(x^m)^n = x^{mn}$
$(xy)^3 = xy \cdot xy \cdot xy$ $= x \cdot x \cdot x \cdot y \cdot y \cdot y$ $= x^3 y^3$	$(xy)^n = x^n y^n$
$\left(\frac{x}{y}\right)^2 = \frac{x}{y} \cdot \frac{x}{y}$ $= \frac{x^2}{y^2}$	$\left(\frac{x}{y}\right)^n = \frac{x^n}{y^n} \quad (y \neq 0)$

We use the exponent laws to simplify products and quotients involving powers.

Example 2

Simplify.

a) $(x^3y^2)(x^2y^4)$ **b)** $\dfrac{a^5b^3}{a^2b^2}$ **c)** $\left(\dfrac{x^2}{z^3}\right)^2$

Solution

Use mental math.

a) $(x^3y^2)(x^2y^4)$
$= x^3 \cdot y^2 \cdot x^2 \cdot y^4$
$= x^5y^6$

b) $\dfrac{a^5b^3}{a^2b^2} = \dfrac{a^5}{a^2} \cdot \dfrac{b^3}{b^2}$
$= a^3b$

c) $\left(\dfrac{x^2}{z^3}\right)^2 = \dfrac{x^2}{z^3} \cdot \dfrac{x^2}{z^3}$
$= \dfrac{x^4}{z^6}$

Example 3

Simplify.

a) $x^{-3} \cdot x^5$ **b)** $m^2 \div m^{-3}$ **c)** $(n^{-2})^{-3}$

Solution

Use mental math.

a) $x^{-3} \cdot x^5 = x^{-3 + 5}$
$= x^2$

b) $m^2 \div m^{-3} = m^{2 - (-3)}$
$= m^5$

c) $(n^{-2})^{-3} = n^{(-2) \times (-3)}$
$= n^6$

Example 4

The number of insects in a colony doubles every month. There are 1000 insects in the colony now. About how many were there in the colony three months ago?

Solution

Let x represent the number of insects 3 months ago. Then, after 3 successive doublings the colony grows to 1000 insects.

$x \times 2^3 = 1000$
$x = \dfrac{1000}{2^3}$
$= 125$

There were about 125 insects in the colony 3 months ago.

1. Suppose the first term and the common ratio of a geometric sequence are not equal. Is it still possible for the terms of the sequence to be powers? Explain.

2. Is it possible for all the terms of an arithmetic sequence to be powers? Explain.

3. When you *multiply* using the exponent law $x^m \cdot x^n = x^{m+n}$, you *add* the exponents. What are other examples in mathematics where you carry out an operation by performing a different operation?

4. Look at *Example 3* on page 25. How could you use the result of part c of this example to solve *Example 4* above?

1.5 EXERCISES

A 1. Simplify.

a) 2^4

b) 5^{-2}

c) 3^{-1}

d) $\left(\frac{1}{4}\right)^{-1}$

e) $\left(\frac{2}{3}\right)^{-1}$

f) $\left(\frac{3}{4}\right)^{-2}$

g) 0.5^{-1}

h) 1.5^0

2. Simplify.

a) 10^0

b) $(-3)^{-2}$

c) $\left(-\frac{1}{2}\right)^3$

d) $\left(-\frac{2}{3}\right)^{-1}$

e) $\left(-\frac{3}{5}\right)^{-2}$

f) $(-1)^{-4}$

g) 0.1^{-4}

h) $\frac{1}{2^{-3}}$

3. Choose one part of exercise 1 or 2. Write to explain how you simplified the expression.

4. Simplify.

a) $x^3 \cdot x^4$

b) $a^2 \cdot a^5$

c) $b^3 \cdot b^5 \cdot b$

d) $m^2 \cdot m^3 \cdot m^4$

5. Simplify.

a) $\frac{x^4}{x^2}$

b) $\frac{y^7}{y^3}$

c) $\frac{n^6}{n^5}$

d) $\frac{a^8}{a^5}$

6. Simplify.

a) $(x^3)^3$

b) $(y^2)^3$

c) $(a^2b^2)^3$

d) $(xy^3)^2$

7. Simplify.

a) $\frac{x^2}{x^5}$

b) $\frac{c^3}{c^4}$

c) $\frac{y^2}{y^7}$

d) $\frac{a^2}{a^6}$

8. Choose one part of exercises 4 to 7. Write to explain how you simplified the expression.

9. A colony of 10 000 insects doubles in number every month. How many insects were there at each time?

    a) 2 months ago   **b)** 5 months ago

B **10.** Simplify.

    a) $x^{-3} \cdot x^4$ **b)** $d^{-4} \cdot d^{-1}$ **c)** $a^6 \cdot a^{-2}$ **d)** $y^4 \cdot y^{-4}$

    e) $x^{-5} \cdot x^{-1} \cdot x^{-3}$ **f)** $b^4 \cdot b^{-3} \cdot b^2$ **g)** $k^8 \cdot k^{-2} \cdot k^{-6}$ **h)** $p^{-1} \cdot p^7 \cdot p^{-6}$

11. Simplify.

    a) $\dfrac{x^{-5}}{x^2}$ **b)** $\dfrac{r^3}{r^{-2}}$ **c)** $\dfrac{s^5}{s^{-5}}$ **d)** $\dfrac{t^{-4}}{t^4}$

    e) $\dfrac{c^{-1}}{c^{-2}}$ **f)** $\dfrac{x^{-2}}{x^{-4}}$ **g)** $\dfrac{b^{-8}}{b^{-3}}$ **h)** $\dfrac{t^4}{t^{-7}}$

12. Simplify.

    a) $(x^{-2})^3$ **b)** $(y^{-1})^{-2}$ **c)** $(m^{-3})^2$ **d)** $(c^3)^{-3}$

    e) $(a^4)^{-1}$ **f)** $(x^{-1}y^2)^{-1}$ **g)** $(x^2y^{-3})^2$ **h)** $(a^{-2}b^2)^{-2}$

13. Simplify.

    a) $(a^{-2}b^4)(a^2b^{-5})$ **b)** $\dfrac{(x^{-2})^3}{(x^3)^{-2}}$ **c)** $\dfrac{x^2y^{-2}}{y^{-1}}$

    d) $(x^{-1}y^2)^{-3}(x^2)^{-1}$ **e)** $\dfrac{(c^{-3}d)^{-1}}{(c^2d)^{-2}}$ **f)** $(m^2n^{-2})(m^{-1}n^2)^{-1}$

14. a) Evaluate each expression for $x = -1$ and $y = 2$.

    i) $(x^3y^2)(x^2y^3)$ **ii)** $\dfrac{x^{-4}y^5}{xy^3}$ **iii)** $(x^3y^2)^3$

    iv) $(x^{-1}y^{-2})(x^{-2}y^{-3})$ **v)** $\dfrac{x^{-3}y^{-2}}{x^2y^{-6}}$ **vi)** $(x^{-4}y^{-3})^{-2}$

    b) Choose one expression from part a. Write to explain how you evaluated it.

15. Make a table of powers of 5 from 5^{-8} to 5^{10}. Use your table to simplify each expression without doing the arithmetic.

    a) 125×625 **b)** 625^2 **c)** $0.0016 \times 78\ 125$

    d) $125 \div 0.000\ 32$ **e)** $78\ 125^{-1}$ **f)** $390\ 625 \div 15\ 625$

C **16.** Simplify.

    a) $\left(\dfrac{c^4}{d}\right)^{-2} \times \dfrac{c^8}{d}$ **b)** $\left(\dfrac{-5x^3}{2y}\right)^{-2} \times \left(\dfrac{x}{y^{-4}}\right)^2$ **c)** $\left(\dfrac{a^{-3}b}{c^2}\right)^{-4} \times \left(\dfrac{c^5}{a^4b^{-3}}\right)^{-1}$

COMMUNICATING THE IDEAS

Suppose your friend telephones you to ask about today's mathematics lesson. How would you explain, over the telephone, how geometric sequences and powers are related?

1.6 Working with Data in Tables

INVESTIGATE

Chai Kim works at a music store. She is paid every two weeks. Her most recent earnings statement is shown below.

STATEMENT OF EARNINGS					
	Regular Hours	Regular Pay	Overtime Hours	Overtime Pay	Weekly Total
August 18–24	40	288.00	0	0.00	288.00
August 25–31	35	252.00	0	0.00	252.00
				Gross Pay	540.00

Use the information in the earnings statement to complete each exercise.

1. a) What is Chai Kim's hourly rate of pay? Explain how you know.
   b) Suppose you know how many regular hours Chai Kim works in a week. How could you calculate her regular pay?
   c) Let x represent the number of regular hours. Write an algebraic expression for the regular pay for the week.
2. Suppose Chai Kim's rate of pay had been \$7.50/h. Make a table like the one above, then calculate her gross pay for the two weeks.
3. Suppose Chai Kim had worked 4 h overtime at time and a half during the first week. What do you think this means? Calculate her gross pay for the two weeks.

When you make purchases, you pay sales tax. The federal government collects a goods and services tax (GST). In all provinces except Alberta, the provincial government collects a provincial sales tax (PST). Unless stated otherwise, the taxes are calculated on the prices before taxes. The table shows the sales tax rates in 1997.

Province or Territory	GST %	PST%
Alberta	7%	no tax
British Columbia	7%	7%
Manitoba	7%	7%
Northwest Territories	7%	no tax
Ontario	7%	8%
Prince Edward Island	7%	10%
Quebec	7%	7.5%
Saskatchewan	7%	7%
Yukon	7%	no tax

New Brunswick, Nova Scotia, and Newfoundland and Labrador have a harmonized sales tax of 15%, which is usually included in the selling price and not added separately at the cash register. In Prince Edward Island and Quebec, PST is calculated on the price of an item after the GST has been added.

Example 1

Suppose you live in British Columbia, Manitoba, or Saskatchewan where the PST is 7%.

a) Copy and complete the table.

Item	Price ($)	GST ($)	PST ($)	Total ($)
Musical tape	4.00			
CD	16.75			

b) Let x dollars represent the selling price. Write expressions for the GST, the PST, and the total cost.

Solution

a) Express 7% as a decimal: 0.07.

Multiply each price by 0.07 to determine the GST and the PST.

Item	Price ($)	GST ($)	PST ($)	Total ($)
Musical tape	4.00	0.28	0.28	4.56
CD	16.75	1.17	1.17	19.09

b) When the selling price is x dollars:

the GST is $0.07x$

the PST is $0.07x$

so, the total cost is $x + 0.07x + 0.07x = 1.14x$

When there is a discount, the discount is normally applied before the sales tax.

Example 2

Suppose you live in British Columbia, Manitoba, or Saskatchewan. You buy two CDs, with regular prices of $15.99 and $12.99. They are on sale for 20% off.

a) Create a table to show the calculation of discount, sales taxes, and total cost.

b) Let x dollars represent the regular price. Determine an expression for the total cost.

Solution

a) Multiply each regular price by 0.20 to calculate the discount.

Subtract the discount from the regular price to calculate the net price.

Multiply each net price by 0.07 to calculate the taxes.

Regular price ($)	Discount ($)	Net price ($)	GST ($)	PST ($)	Total ($)
15.99	3.20	12.79	0.90	0.90	14.59
12.99	2.60	10.39	0.73	0.73	11.85

b) When the selling price is x dollars:

the discount is $0.20x$

the net price is $x - 0.20x = 0.80x$

the GST is $0.07 \times 0.80x = 0.056x$

the PST is $0.07 \times 0.80x = 0.056x$

the total cost is $0.80x + 0.056x + 0.056x = 0.912x$

Example 3

Repeat *Example 2a*. Calculate the sales taxes before the discount.

Solution

Multiply each regular price by 0.07 to calculate the taxes.

Add the regular price and the taxes to calculate the subtotal.

Multiply the subtotal by 0.20 to calculate the discount.

Regular price ($)	GST ($)	PST ($)	Subtotal ($)	Discount ($)	Total ($)
15.99	1.12	1.12	18.23	3.65	14.58
12.99	0.91	0.91	14.81	2.96	11.85

1. In *Example 1*, how does the result of part b suggest a different way to calculate the total cost of the two items in part a? Do the two methods always give the same result?

2. In *Example 1*, what expression would represent the total cost in part b for sales in each place?

   a) Alberta, Northwest Territories, or Yukon

   b) Ontario

   c) Prince Edward Island

   d) Quebec

3. In *Example 2*, what expression would represent the total cost in part b for sales in Alberta, Northwest Territories, or Yukon?

4. a) In *Examples 2* and *3*, explain why the amount the customer pays is the same, regardless of whether the discount is applied before or after the taxes are added.

   b) Is the amount the government collects in the two situations the same?

   c) Is the amount the store receives in the two situations the same?

1.6 EXERCISES

A 1. For the table below

Price ($)	GST ($)	PST ($)	Total ($)
80.00	5.60	8.56	94.16
124.75	8.73	13.35	146.83

   a) What is the GST rate? How is the GST calculated?

   b) What is the PST rate? How is the PST calculated?

   c) What is the total GST on the two items?

   d) What is the total PST on the two items?

   e) Choose part c or d. Write to explain how you calculated the tax.

B 2. Refer to *Investigate* on page 37. During the week of September 1-7, Chai Kim worked 15 regular hours and 2 overtime hours at time and a half. During the week of September 8-14, she worked 8 regular hours and no overtime. Make a table similar to the one on her earnings statement for these two weeks. Assume that her rate of pay is $7.20/h.

3. Here are the final standings for the Western Conference of the National Hockey League on April 14, 1997.

	GP	W	L	T	Pts
Colorado	82	49	24	9	107
Dallas	82	48	26	8	104
Detroit	82	38	26	18	94
Anaheim	82	36	33	13	85
Phoenix	82	38	37	7	83
St. Louis	82	36	35	11	83
Edmonton	82	36	37	9	81
Chicago	82	34	35	13	81
Vancouver	82	35	40	7	77
Calgary	82	32	41	9	73
Toronto	82	30	44	8	68
Los Angeles	82	28	43	11	67
San Jose	82	27	47	8	62

a) How are the points for each team calculated?

b) What would happen to the standings if wins were worth 3 points and ties 1 point?

c) What would happen to the standings if wins were worth 2 points and ties 0 points?

4.

Price ($)	GST ($)	PST ($)	Total ($)
50.00	3.50	3.50	57.00
243.29	17.03	17.03	277.35

a) What is the PST rate?

b) Modify the table to provide for a PST of 8.5%.

c) Suppose the total cost of the first item after both taxes is $59.75. What is the PST rate?

5. A CD is priced at $12.99. The total cost, including PST and 7% GST, is $15.00. What is the PST rate?

6. Suppose you have $20.00 to spend. What is the greatest selling price an item could have so that you could pay for it including taxes in your province or territory? Write to explain how you calculated the selling price.

7. This table shows information about new car sales in Canada and the United States from 1991 to 1996.

	Canada car sales		United States car sales	
	Number of car dealers	Sales per dealer	Number of car dealers	Sales per dealer
1991	3967	319	24 272	511
1992	3930	307	23 361	553
1993	3872	301	22 845	610
1994	3856	320	22 451	673
1995	3819	296	22 426	657
1996	3714	316	22 288	677

a) For each year, calculate the total number of cars sold in each country. Record your results in a table.

b) The population of the United States is approximately 10 times that of Canada. Some people say this means that data for the two countries can be compared approximately by multiplying Canada's data by 10. Is this claim supported by:

   i) the number of car dealers in the table above?

   ii) the number of cars sold each year that you calculated in part a?

   Give possible explanations for your answers.

8. A table similar to the one below may appear in the newspaper.

Foreign Exchange

	Canadian dollar	U.S. dollar	British Pound	German mark	Japanese yen	French franc	Italian lira
Canadian dollar	—	1.3672	2.2117	0.8054	0.011280	0.2386	0.000807
U.S. dollar	0.7314	—	1.6177	0.5891	0.008250	0.1745	0.000590
British pound	0.4521	0.6182	—	0.3642	0.005100	0.1079	0.000365
German mark	1.2416	1.6975	2.7461	—	0.014005	0.2963	0.001002
Japanese yen	88.65	121.21	196.07	71.40	—	21.15	0.071543
French franc	4.1911	5.7301	9.2695	3.3755	0.047276	—	0.003382
Italian lira	1239.16	1694.18	2740.64	998.02	13.977695	295.66	—

To convert money from one currency to another, follow these steps:

- Find the column for the currency you have.
- Go down that column to the row of the currency you require.
- Multiply the number you find by the amount you have.

a) Use the table to convert $100 Canadian to each currency.

   i) U.S. dollars   ii) German marks   iii) Japanese yen

b) Convert $450 Canadian to each currency.

   i) U.S. dollars   ii) British pounds   iii) Italian lira

c) Convert 10 000 Japanese yen to each currency.

   i) Canadian dollars   ii) French francs   iii) German marks

9. Use the table in exercise 8.

   a) Assume you require $100 U.S. How much do you need in each currency?

      i) Canadian dollars   ii) French francs   iii) Japanese yen

   b) Assume you require $450 U.S. How much do you need in each currency?

      i) Canadian dollars   ii) German marks   iii) Italian lira

   c) Assume you require £500. How much do you need in each currency?

      i) Canadian dollars   ii) U.S. dollars   iii) Japanese yen

   d) Write the steps to convert one currency to another, but you have to start with the currency you require, and the amount of that currency you need.

C 10. Refer to *Examples 2* and *3*. Let x dollars represent the price of an item. Use an algebraic expression to show each situation.

   a) The customer pays the same regardless of when the discount is calculated.

   b) The store receives more if the discount is calculated before the taxes.

COMMUNICATING THE IDEAS

Suppose someone has a certain amount of money to spend. In your journal, write a set of instructions for her to determine the greatest selling price an item could have so that she could pay for it including sales taxes. Use the sales tax rate for your province.

How Many Kernels of Wheat?

An ancient story tells of a servant who worked for an empress. As a reward for her work, the empress asked the servant what she would like to have. The servant asked for a quantity of wheat. The amount of wheat was to be determined using a chessboard as follows.

Place one kernel of wheat on the first square of the chessboard, two on the second square, four on the third square, eight on the fourth square, and so on. Each square was to contain twice the number of kernels that were put on the preceding square. The empress was surprised and delighted that the servant had asked for so little. She ordered another servant to fetch a bag of wheat so that the kernels could be counted out.

Some facts about wheat
One bushel holds about 1 million kernels of wheat.
It takes about 32 500 000 kernels of wheat to make one tonne.
Canada's annual production: 30 million tonnes
World's production in 1996:   581 million tonnes

1. Make a table similar to this to record the number of the square and the number of kernels on each square.

Square number	Kernels on square
1	1
2	2
3	4
4	8
⋮	⋮

2. **a)** Complete the first 10 rows of the table. This will indicate the number of kernels needed up to the 10th square.

   b) What patterns can you find in the table?

   c) Suppose you know the square number. How can you determine the number of kernels needed for that square?

   d) How many kernels are needed for the 64th square? Use your calculator to express the number in scientific notation.

3. **a)** On which square of the chessboard would one bushel of wheat be needed?

   b) How much wheat would be needed for the next three squares after this?

4. **a)** On which square of the chessboard would Canada's annual wheat production be needed?

   b) How many years' production of Canada's wheat would be needed for the next three squares after this?

5. **a)** On which square of the chessboard would the world's annual production of wheat be needed?

   b) How many years' production would be needed for the last square of the chess board?

6. All your calculations have concerned the number of kernels needed for a particular square.

   a) Modify your table to include a third column with the heading "Total kernels." Complete this column for several rows.

   b) What patterns can you find in this column?

   c) How is the number of kernels needed to fill a particular square related to the number of kernels on all the previous squares? What implications does this have for your answers to exercises 3b, 4b, and 5b?

7. The empress' servant would have to count the kernels put on each square. Suppose the servant could count 10 kernels each second.

   a) How long would it take the servant to count the kernels for the 64th square?

   b) Which is the first square that would take more than one year to fill?

8. Approximately how many kernels would have to be counted each second to be able to finish the job in one year?

Chai Kim saved $1000 last year, and invested it in an investment that pays 6% annual interest. How much will her investment be worth after 5 years? Follow these steps to answer this question.

1. Make a table similar to the one below. Include the information shown, with rows down to Year 5.

Year	Opening balance ($)	Interest rate (%)	Interest earned ($)	Closing balance ($)
1	1000.00	6	60.00	1060.00
2	1060.00	6	63.60	1123.60
3	1123.60	6		

  a) Verify that the numbers in the table are correct.

  b) Calculate the interest and closing balance in Year 3, then enter them in the table.

  c) Complete the next two rows of the table.

  d) How much will Chai Kim's investment be worth at the end of Year 5?

  e) How much interest did Chai Kim earn in the 5 years?

2. Suppose Chai Kim had invested $2000 instead of $1000. How much would her investment be worth after 5 years?

3. Suppose the interest rate had been 5% instead of 6%. How much would Chai Kim's $1000 investment be worth after 5 years at that rate?

The table on the next page shows how Chai Kim's investment of $1000 would grow during 5 years at an interest rate of 7%.

Year	Opening balance ($)	Interest rate (%)	Interest earned ($)	Closing balance ($)
1	1000.00	7	70.00	1070.00
2	1070.00	7	74.90	1144.90
3	1144.90	7	80.14	1225.04
4	1225.04	7	85.75	1310.79
5	1310.79	7	91.76	1402.55

Observe how the numbers in each row are calculated.

Opening balance × Interest rate = Interest earned

Opening balance + Interest earned = Closing balance

Opening balance = Closing balance from the previous row

The calculations in each row after the first depend on the result from the previous row. We say that the rows are *related recursively*.

Many tables involving financial calculations contain rows that are related recursively.

Example 1

Use the data in the table above.

a) What is the total interest earned during the 5 years?

b) Let A dollars represent the initial investment. Write expressions for the interest earned and the closing balance in Year 1.

c) Suppose the interest rate increased to 9% in Year 4, and returned to 7% in Year 5. What would the investment be worth at the end of Year 5?

Solution

a) Add the numbers in the Interest earned column:

$70.00 + 74.90 + 80.14 + 85.75 + 91.76 = 402.55$

The total interest earned during the 5 years is $402.55.

b) For an initial investment of A dollars:

the interest earned in Year 1 is $A \times 0.07$, or $0.07A$

the closing balance in Year 1 is $A + 0.07A$, or $1.07A$

c) Recalculate the last two rows of the table.

Year	Opening balance ($)	Interest rate (%)	Interest earned ($)	Closing balance ($)
4	1225.04	9	110.25	1335.29
5	1335.29	7	93.47	1428.76

At the end of Year 5, the investment would be worth $1428.76.

INVESTIGATE

Joe saves $1000 every year. At the end of the year, he always invests this amount in an investment that pays 6% annual interest. How much will his investments be worth after 5 years? Follow these steps to answer this question.

1. Make a table similar to the one below. Include the information shown. Joe invests $1000 initially, and at the end of each year. Hence, he makes six investments of $1000, but the last one does not earn any interest in Year 5.

Year	Opening balance ($)	Interest rate (%)	Interest earned ($)	Annual investment ($)	Closing balance ($)
1	1000.00	6	60.00	1000.00	2060.00
2	2060.00	6	123.60	1000.00	3183.60
3	3183.60	6		1000.00	
4		6		1000.00	
5		6		1000.00	

   a) Verify that the numbers in the table are correct.
   b) Complete the table.
   c) How much will Joe's investments be worth at the end of Year 5?
   d) How much interest has Joe earned after 5 years?
   e) How are the rows related recursively?

2. Suppose Joe had invested $2000 every year instead of $1000. How much would his investments be worth after 5 years?

3. Suppose the interest rate had been 5% instead of 6%. How much would Joe's six $1000 investments be worth after 5 years?

This table shows how Joe's annual investments of $1000 would grow during 5 years at an interest rate of 7%.

Year	Opening balance ($)	Interest rate (%)	Interest earned ($)	Annual investment ($)	Closing balance ($)
1	1000.00	7	70.00	1000.00	2070.00
2	2070.00	7	144.90	1000.00	3214.90
3	3214.90	7	225.04	1000.00	4439.94
4	4439.94	7	310.80	1000.00	5750.74
5	5750.74	7	402.55	1000.00	7153.29

Observe how the numbers in each row are calculated.

Opening balance × Interest rate = Interest earned

Opening balance + Interest earned + Annual investment = Closing balance

Opening balance = Closing balance from previous row

___

Example 2

Use the data in the table above.

a) What is the total interest earned during the 5 years?

b) Suppose Joe invested an additional $1500 in Year 3. How much would his investments be worth at the end of Year 5?

Solution

a) Six deposits of $1000 were made, totalling $6000. Since the closing balance after 5 years was $7153.29, the total interest earned was $7153.29 − $6000 = $1153.29.

b) Enter $2500 as the investment in Year 3. Recalculate Year 3's closing balance, and the last two rows of the table.

Year	Opening balance ($)	Interest rate (%)	Interest earned ($)	Annual investment ($)	Closing balance ($)
3	3214.90	7	225.04	2500.00	5939.94
4	5939.94	7	415.80	1000.00	7355.74
5	7355.74	7	514.90	1000.00	8870.64

At the end of Year 5, the investments will be worth $8870.64.

1. In *Example 1*, why are the numbers in the Interest earned column increasing?

2. In *Example 1*, what other way is there to determine the total interest earned?

3. In *Example 1c*, only the interest rate in Year 4 changed. Why did we need to recalculate the last two rows of the table instead of just the fourth row?

4. Look at the table on page 47. The numbers in each column form a sequence.
    a) Two columns contain numbers that are terms of an arithmetic sequence. Which columns are they?
    b) Does any column contain numbers that are terms of a geometric sequence? Explain.

5. What other way is there to solve *Example 2b*?

6. In *Example 2*
    a) Explain why the number of $1000 investments is 1 more than the number of years.
    b) How would you change the table so that the number of $1000 investments is the same as the number of years?

7. Look at the table at the top of page 49.
    a) Which columns contain numbers that are terms of an arithmetic sequence?
    b) Does any column contain numbers that are terms of a geometric sequence? Explain.

1.7 EXERCISES

A 1. Partial data for three different investments are shown below. Determine the numbers that would appear in the spaces in each table. Assume that the interest rate does not change.

a)

Year	Opening balance ($)	Interest rate (%)	Interest earned ($)	Closing balance ($)
1	3000.00	6	180.00	3180.00
2				

b)

Year	Opening balance ($)	Interest rate (%)	Interest earned ($)	Closing balance ($)
4	3108.07	9	279.73	3387.80
5				

c)

Year	Opening balance ($)	Interest rate (%)	Interest earned ($)	Closing balance ($)
2	4568.75	7.5	342.66	4911.41
3				

2. Partial data for three investments are shown below. Determine the numbers that would appear in the spaces in each table. Assume that the interest rate and the annual investment do not change.

a)

Year	Opening balance ($)	Interest rate (%)	Interest earned ($)	Annual investment ($)	Closing balance ($)
1	2000.00	5.5	110.00	2000.00	4110.00
2					

b)

Year	Opening balance ($)	Interest rate (%)	Interest earned ($)	Annual investment ($)	Closing balance ($)
4	4942.56	7	345.98	1500.00	6788.54
5					

c)

Year	Opening balance ($)	Interest rate (%)	Interest earned ($)	Annual investment ($)	Closing balance ($)
2	3416.00	6.75	230.58	3200.00	6846.58
3					

B 3. Use the data in this table.

Year	Opening balance ($)	Interest rate (%)	Interest earned ($)	Closing balance ($)
1	2500.00	6	150.00	2650.00
2	2650.00	6	159.00	2809.00
3	2809.00	6	168.54	2977.54
4	2977.54	6	178.65	3156.19

a) Calculate the total interest earned during the 4 years.

b) Let A dollars represent the initial investment. Write expressions for the interest earned and the closing balance in Year 1.

c) Suppose the interest rate increased to 9% in Year 4. What would the closing balance be at the end of that year?

d) Suppose the interest rate increased to 9% in both Year 3 and Year 4. What would the closing balance be at the end of Year 4?

e) Choose one of parts a to d. Write to explain how you completed it.

4. Suppose you invest $1000.00 at 8% for 3 years. Use a table to calculate the value of your investment at the end of Year 3.

5. Use the data in this table.

Year	Opening balance ($)	Interest rate (%)	Interest earned ($)	Annual investment	Closing balance ($)
1	500.00	8.75	43.75	500.00	1043.75
2	1043.75	8.75	91.33	500.00	1635.08
3	1635.08	8.75	143.07	500.00	2278.15
4	2278.15	8.75	199.34	500.00	2977.49

a) Calculate the total interest earned during the 4 years.

b) Suppose an additional $2000.00 is invested in Year 3. How much will the investments be worth at the end of Year 4?

6. Suppose you invest $500.00 now, and another $500.00 at the end of every year for 4 years. Assume that your investments earn 7.25% interest annually. Use a table to calculate the value of your investments after Year 4. Write to explain how you completed the table for Year 3.

MODELLING the Growth of an Investment

In the examples and exercises in this section we have assumed that the interest rates remain constant during the years the money is invested. However, interest rates change frequently, and most investments are affected by the changes.

- How would the results of the calculations be affected if interest rates rise?
- When interest rates are very low, would you be willing to invest money for a long period of time at a constant interest rate? Explain.

C **7.** Suppose an amount A dollars is invested at an annual interest rate of $i\%$.

a) Use a table to determine expressions for the values of the investment after 1, 2, 3, and 4 years.

b) Write an expression for the value of the investment after n years.

COMMUNICATING THE IDEAS

Suppose someone plans to invest the same amount of money each year for a number of years. Suppose you know the interest rate she hopes to earn. In your journal, write a set of instructions for her to estimate the value of the investments after a number of years.

How Long Will It Take for an Investment to Double in Value?

1. This spreadsheet will calculate the future value of an investment when a certain amount is invested now at a fixed interest rate.

	A	B	C	D	E
1	Calculating the Value of Investments				
2					
3	Year	Opening	Interest	Interest	Closing
4		balance	rate	earned	balance
5	1			=B5*C5	=B5+D5
6	=A5+1	=E5	=C5	=B6*C6	=B6+D6

a) Start a new spreadsheet document. Enter the text and formulas above. Format columns B, D, and E to show numbers to 2 decimal places. Format Column C to show percents to 2 decimal places.

b) Explain the purpose of each formula in rows 5 and 6.

c) Extend the spreadsheet down several rows, by selecting a block of cells starting with row 6 and choosing the Fill Down option.

d) To check that you have entered everything correctly, enter 1000 in cell B5 and 7% in cell C5. The results in rows 5 to 9 should be the same as those in the table on page 47.

e) In column E, look for a closing balance that is as close as possible to $2000.00. Approximately how many years will it take for the investment to double in value?

2. a) Enter other interest rates in cell C5, then repeat exercise 1e. Use rates from 4% to 14%. Summarize the results in a table.

Interest rate (%)	Approximate doubling time (years)

b) As the interest rate increases, what happens to the approximate doubling time?

c) Suppose the interest rate is doubled or tripled. What happens to the approximate doubling time?

d) Multiply each interest rate by the approximate doubling time. What do you notice about the results?

e) State a rule to estimate the doubling time for any interest rate.

Mathematics & Technology

The owner of the music store where Chai Kim works needs to borrow
$25 000.00 from the bank to expand her business. One loan she is
considering is for 5 years at 9%, with an annual payment of $6427.31.
How much interest would the owner pay on this loan? Follow these steps
to answer this question.

1. Make a table similar to the one below. Include the information shown,
   with rows down to Year 5.

Year	Opening balance ($)	Interest rate (%)	Interest charged ($)	Annual payment ($)	Closing balance ($)
1	25 000.00	9	2250.00	6427.31	20 822.69
2	20 822.69	9	1874.04	6427.31	16 269.42
3	16 269.42	9		6427.31	

a) Verify that the numbers in the table are correct.

b) Calculate the interest charged and the closing balance in Year 3,
   then enter them in the table.

c) Complete the next two rows of the table.

d) Calculate the total interest for the loan. Try to find two different
   ways to do this.

2. Suppose the owner needs to borrow $50 000.00 instead of $25 000.00.

a) What would the annual payment be?

b) How much interest would the owner pay on the loan during the
   5 years?

The owner of the music store where Chai Kim works is also considering a
10-year loan. The table on the next page shows how a loan of $25 000.00 at 9%
is repaid with annual payments over a 10-year period. The annual payment is
$3895.50.

Year	Opening balance ($)	Interest rate (%)	Interest charged ($)	Annual payment ($)	Closing balance ($)
1	25 000.00	9	2250.00	3895.50	23 354.50
2	23 354.50	9	2101.91	3895.50	21 560.91
3	21 560.91	9	1940.48	3895.50	19 605.89
4	19 605.89	9	1764.53	3895.50	17 474.92
5	17 474.92	9	1572.74	3895.50	15 152.16
6	15 152.16	9	1363.69	3895.50	12 620.35
7	12 620.35	9	1135.83	3895.50	9 860.68
8	9 860.68	9	887.46	3895.50	6 852.64
9	6 852.64	9	616.74	3895.50	3 573.88
10	3 573.88	9	321.65	3895.50	0.03

Observe how the numbers in each row are calculated.

Opening balance × Interest rate = Interest charged

Opening balance + Interest charged − Annual payment = Closing balance

Opening balance = Closing balance from the previous row

We will use the data from this table in *Examples 1, 2,* and *3.*

Example 1

a) Calculate the total interest for this loan.

b) How much of the annual payment in Year 1 goes toward repaying the loan?

c) How much of the annual payment in Year 5 goes toward repaying the loan?

d) Suppose you know the interest charged in any year. How can you determine the amount of that annual payment that goes toward repaying the loan?

e) Suppose the interest rate went up to 12.5% in Year 10. How much would still be owing at the end of that year? Assume that the annual payment remains the same.

Solution

a) There are 10 payments of $3895.50, totalling $38 955.00.
Since the amount borrowed is $25 000.00, the total interest is
$38 955.00 − $25 000.00 = $13 955.00.

b) Year 1's payment is $3895.50, but $2250.00 interest is charged. Hence, the amount that goes toward repaying the loan is $3895.50 − $2250.00 = $1645.50.

c) In Year 5, the amount that goes toward repaying the loan is
$3895.50 − $1572.74 = $2322.76.

d) Let *I* dollars represent the interest charged in any year. The amount of a payment that goes toward repaying the loan is $3895.50 − *I*.

e) At the start of Year 10, $3573.88 is owing.
The interest on this amount at 12.5% is 0.125 × $3573.88 = $446.74.
The amount owing at the end of the year is
$3573.88 + $446.74 − $3895.50 = $125.12.

Example 2

The loan agreement allows the borrower to make an extra payment in any year. Suppose business is good in the third year. The owner makes an extra payment of $10 000.00 at the end of Year 3.

a) Show how this affects the loan.

b) Calculate the total interest paid.

c) How much interest is saved by making this extra payment?

Solution

a) Insert another column in the table to allow for extra payments. Copy the numbers from the previous table down to the row where the extra payment is made.

Year	Opening balance ($)	Interest rate (%)	Interest charged ($)	Annual payment ($)	Extra payment ($)	Closing balance ($)
1	25 000.00	9	2250.00	3895.50		23 354.50
2	23 354.50	9	2101.91	3895.50		21 560.91
3	21 560.91	9	1940.48	3895.50	10 000.00	

Recalculate the numbers from here until the closing balance is reduced to 0.

Year	Opening balance ($)	Interest rate (%)	Interest charged ($)	Annual payment ($)	Extra payment ($)	Closing balance ($)
1	25 000.00	9	2250.00	3895.50		23 354.50
2	23 354.50	9	2101.91	3895.50		21 560.91
3	21 560.91	9	1940.48	3895.50	10 000.00	9 605.89
4	9 605.89	9	864.53	3895.50		6 574.92
5	6 574.92	9	591.74	3895.50		3 271.16
6	3 271.16	9	294.40	3565.56		0

At the beginning of Year 6, the amount owing is $3271.16, and the interest on this amount is $294.40. The payment required for this year is the sum of these amounts, $3565.56. The loan is paid off 4 years earlier than before, with a smaller payment in the sixth year.

b) The total amount paid is
$5 \times \$3895.50 + \$3565.56 + \$10\,000.00 = \$33\,043.06$.
The total interest paid is $\$33\,043.06 - \$25\,000.00 = \$8043.06$.

c) Using the answer from *Example 1a,* the interest saved is
$\$13\,955.00 - \$8043.06 = \$5911.94$.

Example 3

What extra payment at the end of Year 5 would pay off the loan at the end of Year 8?

Solution

Paying off the loan at the end of Year 8 means to pay it off 2 years earlier. Look at row 5 in the table on page 55. The closing balance is $15 152.16. To pay off the loan 2 years earlier, visualize skipping 2 rows in the table. The closing balance after Year 7 is $9860.68. This is the closing balance that is needed after Year 5. Hence, the extra payment required at the end of Year 5 is $15 152.16 − $9860.68 = $5291.48.

1. In *Example 1*, why are the numbers in the Interest charged column decreasing?

2. In *Example 1a*, what other way is there to determine the total interest paid?

3. In *Example 1b* and *c*, what other way is there to determine the amount of the annual payment that goes toward repaying the loan?

4. In *Example 1*, why is the Closing balance at the end of Year 10 not 0? What simple adjustment would make it 0?

5. In the table on page 55, which columns contain numbers that are terms of a geometric sequence? Explain.

6. In *Example 2*, suppose the $10 000.00 extra payment had been made at the end of Year 2 instead of Year 3. Would the total interest paid be greater than, the same as, or less than the amount in *Example 2b*? Explain.

7. At the end of Year 5, would the extra payment needed to pay off the loan at the end of Year 7 be greater than, the same as, or less than the answer in *Example 3*? Explain.

1.8 EXERCISES

A 1. Partial data for three different loans are shown. What number belongs in each space?

a)

Year	Opening balance ($)	Interest rate (%)	Interest charged ($)	Annual payment ($)	Closing balance ($)
1	5000.00	6	300.00	679.34	4620.66
2		6	277.24	679.34	

b)

Year	Opening balance ($)	Interest rate (%)	Interest charged ($)	Annual payment ($)	Closing balance ($)
6	10 904.43	7.5	817.83	2695.19	9027.07
7	9 027.07	7.5		2695.19	

c)

Year	Opening balance ($)	Interest rate (%)	Interest charged ($)	Annual payment ($)	Closing balance ($)
4	27 448.24	9			24 464.88

2. Partial data for three loans are shown. Calculate the number that belongs in each space.

a)

Year	Opening balance ($)	Interest rate (%)	Interest charged ($)	Annual payment ($)	Closing balance ($)
1	43 000.00	8.25	3 547.50	6 480.71	40 066.79
2					

b)

Year	Opening balance ($)	Interest rate (%)	Interest charged ($)	Annual payment ($)	Closing balance ($)
7	34 242.00	11	3766.62	11 037.09	26 971.53
8					

c)

Year	Opening balance ($)	Interest rate (%)	Interest charged ($)	Annual payment ($)	Closing balance ($)
5	4970.59	5.5	273.38	995.01	4248.96
6					

3. For each loan in exercise 2, how much of the annual payment goes toward repaying the loan?

a) in the first year in the table   b) in the year you calculated

This table provides data on the repayment of a $100 000 farm loan. The farmer has negotiated for one annual payment to be made each year after the harvest. Use this table to answer exercises 4 to 9.

Year	Opening balance ($)	Interest rate (%)	Interest charged ($)	Annual payment ($)	Extra payment ($)	Closing balance ($)
1	100 000.00	7.5	7500.00	17 072.70		90 427.30
2	90 427.30	7.5	6782.05	17 072.70		80 136.65
3	80 136.65	7.5	6010.25	17 072.70		69 074.20
4	69 074.20	7.5	5180.57	17 072.70		57 182.07
5	57 182.07	7.5	4288.66	17 072.70		44 398.03
6	44 398.03	7.5	3329.85	17 072.70		30 655.18
7	30 655.18	7.5	2299.14	17 072.70		15 881.62
8	15 881.62	7.5	1191.12	17 072.70		0.04

4. a) What is the period of the loan?

   b) What is the annual payment?

   c) What is the annual interest rate?

5. a) What is the total amount the farmer pays for the loan?

   b) How much interest is paid?

B **6.** How much of the annual payment at the end of each of these years went toward the opening balance?

a) Year 1 **b)** Year 2 **c)** Year 5

7. How much would be owing at the end of Year 8 in each case?

a) if the interest rate increases to 9.25% in Year 8

b) if the interest rate increases to 9.25% in both Year 7 and Year 8

8. The farmer makes an extra payment of $20 000.00 at the end of Year 4.

a) Show how this affects the loan.

b) Calculate the total interest paid.

c) How much interest is saved by making this extra payment?

9. What extra payment at the end of Year 3 would pay off the loan at the end of each of these years?

a) Year 7 **b)** Year 6 **c)** Year 5

10. On a 5-year loan of $6000 at 12% annually, the annual payment is $1664.46.

a) Calculate the total amount paid during the 5 years.

b) How much interest was paid during the 5 years?

11. Calculate the total interest paid for each loan.

a) a 6-year loan of $10 000 at 7.5% annually, with an annual payment of $2130.45

b) a 25-year loan of $120 000 at 8% annually, with an annual payment of $11 241.45

12. This table provides data on the repayment of a loan. The loan will be completely repaid after 3 more years.

Year	Opening balance ($)	Interest rate (%)	Interest charged ($)	Annual payment ($)	Extra payment ($)	Closing balance ($)
1	50 000.00	8.5	4250.00	10 980.35		43 269.65
2	43 269.65	8.5	3677.92	10 980.35		35 967.22
3	35 967.22	8.5	3057.21	10 980.35		28 044.08
4						
5						
6						

a) How much of the first year's payment went toward the opening balance?

b) Copy and complete the last three rows of the table.

c) Suppose the interest rate rises to 10% in Year 6. How much will be owing at the end of Year 6?

d) What extra payment at the end of Year 1 would pay off the loan at the end of Year 3?

13. On a 4-year loan of $25 000 at 8% annually, the annual payment is $7548.02. Use a table to determine each amount.

a) the balance owing after each year

b) the amount that is still owing after Year 4 if the interest rate goes up to 12% in Year 3

c) the extra payment at the end of Year 2 that would pay off the loan at the end of Year 3

14. In 1997, the sales of a particular CD doubled every month. The CD was released in June with sales of 15 000 that month.

a) Make a table to illustrate the monthly sales figures from June to December.

b) The demand for the CD peaked in December that year. Starting in January 1998, and every month thereafter, sales dropped to one-quarter of what they were the previous month. How many CDs were sold in May 1998?

c) How many CDs were sold during the 12-month period from June 1997 to May 1998?

C 15. Amira deposited $50 in her bank account, then withdrew half the balance. The next day she deposited $50, then withdrew half the balance. On the third day she deposited $50 again, then withdrew half the balance. The balance was then $500.

a) What was the balance at the beginning?

b) How much money did Amira withdraw in all?

COMMUNICATING THE IDEAS

Suppose a farmer has borrowed a sum of money at a fixed interest rate, then pays it back with regular annual payments. How could you determine the total amount of interest paid on the loan? How could you determine the amount of any year's payment that goes toward reducing the loan? In your journal, use an example to illustrate your explanations.

Calculating Loan Payments

Open the file LOANPMTS from the *Template and Data Kit*. This spreadsheet will appear. It contains the same information as the example on page 55, involving the music store owner's loan.

	A	B	C	D	E	F
1	CALCULATING LOAN PAYMENTS					
2						
3	Principal		25,000.00			
4	Annual interest rate (%)		9.00			
5	Number of years		10			
6						
7	Payment	Opening	Interest	Interest	Annual	Closing
8	number	balance	rate (%)	charged	payment	balance
9	1	25,000.00	9.00	2,250.00	3,895.50	23,354.50
10	2	23,354.50	9.00	2,101.91	3,895.50	21,560.91
11						

You can use the program to calculate the annual payments for any loan. Enter the principal (the amount of the loan), the interest rate, and the number of years in cells C3, C4, and C5, respectively.

1. Predict whether the annual payment on each loan will be greater than or less than the annual payment above. Use the computer to confirm your prediction.

   a) Borrow $25 000.00 at 7% annually and repay over 10 years.

   b) Borrow $25 000.00 at 9% annually and repay over 8 years.

2. Calculate the annual payment for each loan.

   a) Borrow $5000.00 at 6.5% annually and repay over 5 years.

   b) Borrow $29 000.00 at 11.25% annually and repay over 9 years.

Copy the formulas in row 18 down to row 33. A quick way to do this is to select a block of cells starting with row 18 and use the Fill Down option.

3. Use the spreadsheet to calculate the annual payment for each loan.

   a) Borrow $25 000.00 at 9% annually and repay over 25 years.

   b) Borrow $100 000.00 at 8.25% annually and repay over 20 years.

Enter these data in the indicated cells:

G7: Total   G8: Interest   G9: =D9   G10: =G9+D10

Copy the formula in cell G10 down column G.

4. Calculate the total interest paid for each loan.

   a) Borrow $25 000.00 at 9% annually and repay over 25 years.

   b) Borrow $100 000.00 at 8.25% annually and repay over 20 years.

Insert a new column between columns E and F. Label it "Extra Payment."
Click on cell G9 and change the formula to: =B9+D9–E9–F9
Copy this formula down column G.

5. a) Verify the solution of *Example 2* on page 57.

   b) Verify the solution of *Example 3* on page 57.

6. Enter 25000 in cell C3, 9 in cell C4, and 10 in cell C5 to change the
spreadsheet back to the example on page 55. That loan is for 10 years,
but it is unusual for interest rates to be the same for this long. Suppose
after 5 years the interest rate increases to 12%. At that time, the loan
has been reduced to $15 152.16, and there are 5 years remaining to
repay the loan.

   a) These options are available to the music store owner. Use your
      spreadsheet to answer each question.

      Option 1: Repay the balance over the remaining years.
               What is the annual payment?

      Option 2: Repay the balance over a longer period such as 7 years.
               What is the annual payment?

      Option 3: Continue to pay the annual payment each year.
               How does this affect the loan?

   b) What are the advantages and disadvantages of each option? Which
      do you think is the best?

7. Enter 100000 in cell C3, 7.5 in cell C4, and 8 in cell C5 to change the
spreadsheet to the exercise about the farm loan on page 59. Suppose
after 4 years the interest rate decreases to 6%. At that time, the loan
has been reduced to $57 182.07, and there are 4 years remaining to
repay the loan.

   a) The options in exercise 6a are available to the farmer. Use your
      spreadsheet to answer each question in exercise 6a for the farmer.

   b) What are the advantages and disadvantages of each option? Which
      do you think is the best?

Many loans provide for monthly instead of annual payments. To modify
your spreadsheet to accommodate monthly payments, follow these steps.

Click on cell E7, then change the data to Monthly.

Click on cell C9 and change the formula to: =C$4/12

Click on cell E9 and change the formula to:
=ROUND(PMT(C9/100,C$5*12,–C$3),2)

If you are using Microsoft Works for Windows 3.1, change the formula in cell E9 to: =ROUND(PMT(C$3,C9/100,C$5*12),2)

Copy the formulas down many rows. For 4-year loans, which are common when financing the purchase of a car, copy the formulas down to row 56. For 25-year loans, which are common when financing the purchase of a home, copy the formulas down to row 308.

8. When she purchased her new car, Karol borrowed $9200.00 at 4.9% annually over 48 months.

   a) Use your modified spreadsheet to determine her monthly payment.

   b) How much interest does she pay during the period of the loan?

   c) How much does Karol still owe after each time?

       i) 12 months   ii) 24 months   iii) 36 months

   d) Explain why the amount Karol still owes after 24 months is greater than one-half the amount she borrowed.

9. When he purchased his new car, Kareem borrowed $8350.00 and agreed to repay the loan over 36 months. His monthly payments were $258.78. Use your spreadsheet to determine the interest rate the finance company charged him.

10. To finance his new home purchase, Ray took a 25-year mortgage of $100 000 at 7.7% annually.

   a) Use your spreadsheet to determine his monthly payment.

   b) How much interest would Ray pay during the period of the loan?

   c) How much would Ray owe after 5 years?

   d) How much interest would Ray pay during the first 5 years?

11. In exercise 10, the finance company allows Ray to make an extra payment at the end of any year. In each situation below, estimate how much Ray can save by making the extra payments.

   a) Pay an extra $3000 at the end of the first year.

   b) Pay an extra $3000 at the end of the fifth year.

   c) Pay an extra $3000 at the end of each year up to the fifth year.

12. In exercise 10, Ray must refinance his mortgage at the end of the fifth year. What would his monthly payments be if the interest rate increased to 12.25%? Assume that the new mortgage runs for 20 years.

1. a) Insert two numbers between 5 and 44, so the four numbers form an arithmetic sequence.

b) Insert three numbers between 99 and 167, so the five numbers form an arithmetic sequence.

2. a) Determine the indicated term of each geometric sequence.

   i) 1, 2, 4, ... t_6 **ii)** 1, 2, 4, ... t_{12} **iii)** 1, 2, 4, ... t_{13} **iv)** 1, 2, 4, ... t_{14}

b) Choose one sequence from part a. Write to explain how you determined the term.

3. Look at the results of exercise 2.

a) Compare the results of parts a i) and ii). Write to explain why t_{12} is not twice t_6.

b) How are t_{12}, t_{13}, and t_{14} related? Is the same relationship true for all geometric sequences?

4. Simplify.

a) $x^5 \cdot x^{-2}$ **b)** $x^{-1} \cdot x^{-2}$ **c)** $x^6 \cdot x^{-4}$ **d)** $x^3 \cdot x^7$

e) $x^2 \cdot x^{-1} \cdot x^{-4}$ **f)** $x^{-2} \cdot x^4 \cdot x^{-2}$ **g)** $x^7 \cdot x^{-3} \cdot x^{-4}$ **h)** $x^{-7} \cdot x^5 \cdot x^{-2}$

5. Simplify.

a) $\dfrac{x^4}{x^5}$ **b)** $\dfrac{x^{-2}}{x^2}$ **c)** $\dfrac{x^3}{x^{-3}}$ **d)** $\dfrac{x^{-1}}{x^{-1}}$

6. Use the data in this table.

Year	Opening balance ($)	Interest rate (%)	Interest earned ($)	Annual investment ($)	Closing balance ($)
1	1000.00	10	100.00	500.00	1600.00
2	1600.00	10	160.00	500.00	2260.00
3	2260.00	10	226.00	500.00	2986.00
4	2986.00	10	298.60	500.00	3784.60
5	3784.60	10	378.46	500.00	4663.06

a) Calculate the total interest earned.

b) Recalculate the table for each interest rate.   **i)** 8%   **ii)** 12%

c) Suppose the original table were changed in each of two ways.

   i) The interest rate in Year 1 is 15% instead of 10%.

   ii) The interest rate in Year 5 is 15% instead of 10%.

   In which case do you think the closing balance will be greater? Verify your answer by calculating a new table for each case.

2 REAL NUMBERS

Population Growth and Deforestation

 CONSIDER THIS SITUATION

From 1960 to 1990:

- The world's population rose from 3.2 billion to 5.3 billion.

- Rain forests were reduced in area from 2.10 billion hectares to 1.75 billion hectares.

As the population grows, greater and greater demands are made on Earth's capacity to produce food and other products.

- Some people believe that deforestation is one sign of the dangers of high world population. What are other ways in which people affect their environment?

- Do you think it is possible to measure the greatest number of people who can survive on Earth? Discuss your ideas.

On pages 84 and 85, you will develop a mathematical model to explore patterns in population growth and deforestation. Your model will use number sequences and rational exponents.

 FYI Visit **www.awl.com/canada/school/connections**

For information related to the above problem, click on MATHLINKS followed by AWMath. Then select a topic under World Population or Environmental Issues and Deforestation.

2.1 Radicals

Gold leaf is so thin that one dollar's worth would cover a square with an approximate area of 3600 cm^2.

Since $60^2 = 3600$, each side of the square is 60 cm long.

We say that 60 is a square root of 3600, and we write $\sqrt{3600} = 60$.

The gold produced in Canada in one day would almost fill a cube with an approximate volume of 8000 cm^3.

Since $20^3 = 8000$, each edge of the cube is 20 cm long.

We say that 20 is a cube root of 8000, and we write $\sqrt[3]{8000} = 20$.

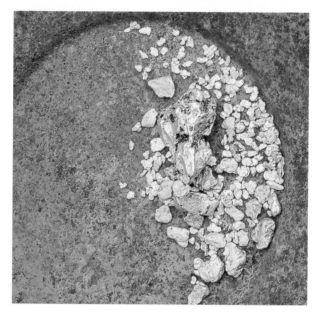

VISUALIZING

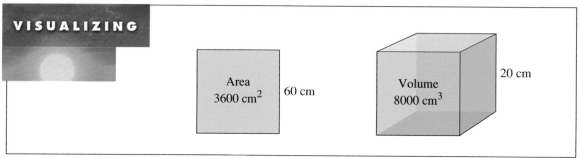

Area 3600 cm^2 60 cm

Volume 8000 cm^3 20 cm

Expressions such as $\sqrt{3600}$ and $\sqrt[3]{8000}$ are *radicals*. We use radicals when we work with square roots and cube roots of numbers.

Square Roots

A number r is a *square root* of a number x if $r^2 = x$.

A positive number always has two square roots, one positive, the other negative. Since the squares of both positive and negative numbers are positive, it is impossible to obtain a negative number when a number is squared. Hence, the square root of a negative number is not defined.

The *radical sign*, $\sqrt{}$, always denotes the positive square root.

$\sqrt{x}$ means the positive square root of x, where $x \geq 0$.

Example 1

Determine each square root, without using a calculator.

a) $\sqrt{1600}$ **b)** $\sqrt{2.25}$ **c)** $\sqrt{0.09}$

Solution

a) $\sqrt{1600} = 40$,
  since $40^2 = 1600$

b) $\sqrt{2.25} = 1.5$,
  since $1.5^2 = 2.25$

c) $\sqrt{0.09} = 0.3$,
  since $0.3^2 = 0.09$

In *Example 1*, the square roots were exact. Many numbers do not have exact square roots, but you can use a calculator to determine approximations of them. Since the calculator displays only a fixed number of digits, the number displayed may only be an approximation of the square root. For example, if you use your calculator to determine $\sqrt{7}$, you will obtain 2.645 751 311 (assuming your calculator has 10-digit accuracy). You can obtain different approximations of $\sqrt{7}$ by rounding or by truncating.

Example 2

Sharon used her calculator to determine $\sqrt{7}$. She obtained 2.645 751 311. Write approximations to 1, 2, and 3 decimal places that are obtained by rounding and by truncating. Use your calculator to check your approximations.

Solution

Use a table to record the results.

Number of decimal places	Rounding	Check	Truncating	Check
1	$\sqrt{7} \doteq 2.6$	$2.6 \times 2.6 = 6.76$	$\sqrt{7} \doteq 2.6$	$2.6 \times 2.6 = 6.76$
2	$\sqrt{7} \doteq 2.65$	$2.65 \times 2.65 = 7.0225$	$\sqrt{7} \doteq 2.64$	$2.64 \times 2.64 = 6.9696$
3	$\sqrt{7} \doteq 2.646$	$2.646 \times 2.646 = 7.001\ 316$	$\sqrt{7} \doteq 2.645$	$2.645 \times 2.645 = 6.996\ 025$

In *Example 2*, observe that if you include more decimal places when you estimate the square root, you get closer to 7 when you square the estimate. Furthermore, rounding and truncating provide either the same estimate, or one that differs only by 1 in the final digit.

Example 3

A square has an area of 30 cm². Determine the perimeter of the square in exact form, and in approximate form to 2 decimal places.

Solution

Since $\sqrt{30} \times \sqrt{30} = 30$, the length of each side is $\sqrt{30}$ cm.
Hence, the perimeter is $4 \times \sqrt{30}$ cm, which we write as $4\sqrt{30}$ cm.
In approximate form:

$4\sqrt{30} \doteq 4 \times 5.4772$
$\qquad \doteq 21.9089$

To 2 decimal places, the perimeter of the square is 21.91 cm.

Area
30 cm²

Cube Roots

A number r is a *cube root* of a number x if $r^3 = x$.

The cube root of a positive number is positive and the cube root of a negative number is negative.

$\sqrt[3]{x}$ means the cube root of x.

Example 4

Determine each cube root.

a) $\sqrt[3]{125}$ b) $\sqrt[3]{-64}$ c) $\sqrt[3]{18}$

Solution

a) Use mental math.
$\sqrt[3]{125} = 5$,
since $5^3 = 125$

b) Use mental math.
$\sqrt[3]{-64} = -4$,
since $(-4)^3 = -64$

c) Use the cube-root function on your calculator. Consult your manual if necessary.
$\sqrt[3]{18} \doteq 2.621$ rounded to 3 decimal places

Higher Roots

Similarly, the *fourth roots* of 16 are 2 and -2, since $2^4 = 16$, and $(-2)^4 = 16$.

We write $\sqrt[4]{16} = 2$ to indicate the positive fourth root of 16.

And the *fifth root* of -32 is -2, since $(-2)^5 = -32$.

We write $\sqrt[5]{-32} = -2$.

> An expression of the form $\sqrt[n]{x}$, where n is a natural number, is a *radical*.
> If n is even, the expression represents only the positive root.

Roots of numbers are useful for working with geometric sequences.

Example 5

Insert two numbers between 10 and 60, so that the four numbers form a geometric sequence. List the first four terms in exact form, then in approximate form to 2 decimal places.

Solution

The first term is $a = 10$.
Let r represent the common ratio.
The fourth term is represented by $t_4 = ar^3$.
Since the fourth term is 60,

$$10r^3 = 60$$
$$r^3 = 6$$

Take the cube root of each side.

$$r = \sqrt[3]{6}$$

In exact form:
The second term is:
$10 \times \sqrt[3]{6} = 10\sqrt[3]{6}$
The third term is:
$10 \times \sqrt[3]{6} \times \sqrt[3]{6} = 10(\sqrt[3]{6})^2$
The first four terms of the sequence are:
$10, 10\sqrt[3]{6}, 10(\sqrt[3]{6})^2, 60$.

In approximate form:
Use the cube-root function on your calculator.
$r \doteq 1.8171$
The common ratio is approximately 1.8171.
The second term is $10 \times 1.8171 \doteq 18.171$
The third term is $18.171 \times 1.8171 \doteq 33.0185$
To 2 decimal places, the first four terms are:
10, 18.17, 33.02, and 60.

1. Why do you think the radical sign is defined to denote the positive square root and not the negative square root (or both square roots)?

2. Is the square root of a number always less than the number? Use the radicals in *Example 1* to support your answer.

3. What is the difference between rounding and truncating? Use the data in *Example 2* to explain your answer.

4. In the first Check column in the solution of *Example 2*, why is one number less than 7 and the others greater than 7? Will this ever happen in the second Check column?

5. In *Example 5*, two numbers were inserted between 10 and 60 so that the numbers were in geometric sequence. How would the solution change if:

   a) one number is inserted?   b) three numbers are inserted?

   c) four numbers are inserted?

2.1 EXERCISES

A 1. Determine the square roots of each number.

   a) 49   b) 81   c) 121   d) 400   e) 529   f) 625

2. Simplify without using a calculator.

   a) $\sqrt{64}$ b) $\sqrt{100}$ c) $\sqrt{144}$ d) $\sqrt{900}$ e) $\sqrt{1600}$

   f) $\sqrt{0.25}$ g) $\sqrt{0.04}$ h) $\sqrt{0.01}$ i) $\sqrt{0.0016}$ j) $\sqrt{0.000\,025}$

3. Use a calculator to determine each square root, to 3 decimal places.

   a) $\sqrt{2}$ b) $\sqrt{3}$ c) $\sqrt{52.3}$ d) $\sqrt{128.5}$ e) $\sqrt{471}$

4. Simplify without using a calculator.

   a) $\sqrt[3]{8}$ b) $\sqrt[3]{-27}$ c) $\sqrt[4]{81}$

   d) $\sqrt[5]{32}$ e) $\sqrt[5]{243}$ f) $\sqrt[3]{0.001}$

B 5. Use the area of each square. Determine the length of a side and the perimeter. Give the answer to 1 decimal place, where necessary.

   a)

   Area
   9.61 m²

   b)

   Area
   6.4 m²

   c)

   Area
   8.5 m²

6. Use the volume of each cube. Determine the length of an edge and the area of a face. Give the answer to 1 decimal place, where necessary.

a)
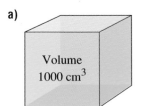
Volume
1000 cm^3

b)
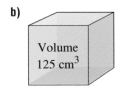
Volume
125 cm^3

c)

Volume
250 cm^3

7. Use your calculator. Determine each root to 2 decimal places. Use your calculator to check your approximations.

a) $\sqrt{6}$ **b)** $\sqrt{11}$ **c)** $\sqrt[3]{23}$ **d)** $\sqrt{124}$ **e)** $\sqrt[3]{139}$ **f)** $\sqrt[3]{254}$

8. Write each number in square root form.

a) 5   **b)** 3   **c)** 2   **d)** 4   **e)** 7   **f)** 1

9. a) Use your calculator to determine $\sqrt{60}$. Write approximations to 1, 2, and 3 decimal places that are obtained by rounding and by truncating. Use your calculator to check your approximations.

b) Suppose you have two estimates of the square root of a number, one obtained by rounding to a certain number of decimal places and the other by truncating to the same number of decimal places. How can you tell which was obtained by rounding and which by truncating? Use your results from part a to support your answer.

c) Suppose you have only one estimate of the square root of a number. Can you tell if it was obtained by rounding or by truncating? Explain.

10. Simplify the radicals in each list without using a calculator. What patterns can you find in the results? Predict the next line in each list.

a) $\sqrt{9}$
$\sqrt{900}$
$\sqrt{90\,000}$

b) $\sqrt[3]{8}$
$\sqrt[3]{8000}$
$\sqrt[3]{8\,000\,000}$

c) $\sqrt[4]{16}$
$\sqrt[4]{160\,000}$
$\sqrt[4]{1\,600\,000\,000}$

11. Simplify without using a calculator.

a) $\sqrt[3]{64}$ **b)** $\sqrt[3]{125}$ **c)** $\sqrt[4]{16}$ **d)** $\sqrt[5]{-1}$ **e)** $\sqrt[3]{216}$

f) $\sqrt[3]{-1000}$ **g)** $\sqrt[4]{256}$ **h)** $\sqrt[4]{10\,000}$ **i)** $\sqrt[3]{7^3}$ **j)** $\sqrt[5]{10^5}$

12. Choose one part of exercise 11. Write to explain how you simplified the radical.

13. A 1-L can of paint covers an area of 12 m^2.

a) Suppose one can of paint is used to cover a square area with one coat. How long is its side?

b) Suppose one can is used to cover a square area with two coats. Assume each coat uses the same amount. How long is the side of the square?

14. In one day a gold mine might produce approximately 160 cm³ of gold. Suppose five days' gold production is cast into a cube. About how long is its edge?

15. Write each number in cube root form.

a) 2   b) −3   c) 1   d) −5   e) 4   f) −1

16. Choose one part of exercise 15. Write to explain how you wrote it as a cube root.

17. The Ladner Creek Gold Mine in British Columbia has a projected reserve of at least 263 800 ounces of pure gold. The company plans to produce the gold at the rate of 53 000 ounces per year. One ounce of gold has a volume of approximately 1.5 cm³. Suppose the gold is cast into a cube. What are the volume and edge length of the cube that represents each mass of gold?

a) one year's production

b) one day's production

c) the projected reserve

18. a) The ancient Egyptians made gold sheets that were only 0.0005 cm thick. Suppose 1 cm³ of gold is made into a square sheet this thick. How long is its side?

b) Modern methods produce far thinner sheets. Suppose 1 cm³ of gold is made into a square sheet that is 100 times as thin. How long is its side?

19. Simplify.

a) $\sqrt{16+9}$

b) $\sqrt{4+9+36}$

c) $\sqrt[3]{27+64+125}$

d) $\sqrt{64}+\sqrt[3]{64}$

e) $\sqrt[3]{\sqrt{64}}$

f) $\sqrt{6-\sqrt{4}}$

g) $\sqrt{25}+\sqrt[3]{27}-\sqrt[4]{16}$

h) $5\sqrt{9}-4\sqrt[3]{-8}$

i) $\sqrt[3]{0.001}-\sqrt{0.01}$

20. Choose one part of exercise 19. Write to explain how you simplified the expression.

21. Insert one number between 5 and 10 so that the three numbers form a geometric sequence.

22. a) Insert two numbers between 3 and 12 so the four numbers form a geometric sequence. List the first four terms in exact form, and in approximate form to 2 decimal places.

b) List the next three terms in exact form and in approximate form.

23. Insert three numbers between 4 and 20 so the five numbers form a geometric sequence. List the first four terms in exact form, and in approximate form to 3 decimal places.

24. This excerpt is from *The Sciences* magazine, 1995.

In March the sea ice begins growing at twenty-two square miles a minute, the "greatest seasonal event on earth"

In the Antarctic summer— December through February — the continent is rimmed with a mere 1,500,000 square miles of sea ice. But in March a phenomenon takes place that is called "the greatest seasonal event on earth." The air temperature drops far below freezing, sometimes as low as −40°F, and the sea ice, which freezes at 29°F, begins growing at the incredible average rate of 22 square miles a minute. By the end of the Antarctic winter, in September, the ice pack has expanded to cover more than 7,334,000 square miles— nearly twice the area of the United States—in a layer usually no more than three feet thick.

This article is reprinted by permission of *The Sciences* and is from the July/August 1995 issue. Individual subscriptions are $28 per year. Write to: The Sciences, 2 East 63rd Street, New York, NY 10021.

a) Calculate to confirm that sea ice that grows from 1 500 000 square miles to 7 334 000 square miles in six months is growing at about 22 square miles a minute.

b) Visualize a square of ice with an area of 22 square miles. Calculate the side length of this square in miles and in kilometres.
(Use 1 mile ≐ 1.6 km.)

c) The sea ice is approximately 1 m thick. Calculate the volume of ice formed each minute, in cubic metres. Visualize a cube with this volume. Calculate its edge length.

MODELLING the Growth of Antarctic Sea Ice

In exercise 24, you modelled the growth of sea ice in Antarctica in two different ways.
- What assumptions about the growth of sea ice are used in the models?
- Describe the model in part b.
- Describe the model in part c.
- In what ways does the growth of sea ice in Antarctica differ from the models?

COMMUNICATING THE IDEAS

Explain what is meant by a *radical*. Use examples to illustrate your explanation. In your journal, include examples to illustrate how the value of $\sqrt[n]{x}$ depends on the sign of x and on whether n is even or odd.

INVESTIGATE

A power with a natural number exponent is defined using repeated multiplication; for example, 3^4 means $3 \times 3 \times 3 \times 3$.

A power with a rational exponent, such as $3^{\frac{1}{2}}$ or $3^{0.5}$, has no meaning according to this definition. It has been defined in another way. You can use your calculator to discover the definition.

The keystrokes are for the TI-34 calculator. If you use a different calculator, consult your manual for the keystrokes.

1. a) To determine $3^{\frac{1}{2}}$, press 3 $\boxed{y^x}$ .5 $\boxed{=}$. Record the result.

   b) Can you tell how $3^{\frac{1}{2}}$ is defined? If so, use your calculator to confirm this.

2. a) To determine $3^{\frac{1}{3}}$, press 3 $\boxed{y^x}$ $\boxed{(}$ 1 $\boxed{\div}$ 3 $\boxed{)}$ $\boxed{=}$. Record the result.

   b) Can you tell how $3^{\frac{1}{3}}$ is defined? If so, use your calculator to confirm this.

3. Copy and complete the tables. Use the results. How do you think $x^{\frac{1}{2}}$ and $x^{\frac{1}{3}}$ are defined?

4. Use the results of exercise 3. How would you define $x^{\frac{1}{n}}$?

x	$x^{\frac{1}{2}}$
1	
2	
3	
4	
9	
16	
25	

x	$x^{\frac{1}{3}}$
1	
2	
3	
8	
27	
64	
125	

5. Predict how $x^{-\frac{1}{2}}$ is defined. Copy the first table. Use your calculator to complete it. Did the results agree with your prediction?

6. Copy the second table. Use your calculator to complete it. Use the results. How do you think $x^{\frac{2}{3}}$ is defined?

x	$x^{-\frac{1}{2}}$
1	
2	
3	
4	
9	
16	
25	

x	$x^{\frac{2}{3}}$
1	
2	
3	
8	
27	
64	
125	

7. Use the results of exercises 5 and 6. How would you define $x^{-\frac{1}{n}}$ and $x^{\frac{m}{n}}$?

To give meaning to powers such as $3^{\frac{1}{2}}$ and $3^{-\frac{1}{2}}$, we *extend* the exponent law $x^m \times x^n = x^{m+n}$ so that it applies when m and n are rational numbers.

By extending the law:

$$3^{\frac{1}{2}} \times 3^{\frac{1}{2}} = 3^{\frac{1}{2}+\frac{1}{2}}$$
$$= 3^1$$
$$= 3$$

But: $\sqrt{3} \times \sqrt{3} = 3$

Therefore, $3^{\frac{1}{2}} = \sqrt{3}$

By extending the law:

$$3^{-\frac{1}{2}} \times 3^{\frac{1}{2}} = 3^{-\frac{1}{2}+\frac{1}{2}}$$
$$= 3^0$$
$$= 1$$

Therefore, $3^{-\frac{1}{2}}$ and $3^{\frac{1}{2}}$ are reciprocals.

$$3^{-\frac{1}{2}} = \frac{1}{3^{\frac{1}{2}}}$$
$$= \frac{1}{\sqrt{3}}$$

These examples and the results of *Investigate* suggest that $x^{\frac{1}{n}}$ should be defined as the nth root of x, and $x^{-\frac{1}{n}}$ as its reciprocal.

$x^{\frac{1}{n}} = \sqrt[n]{x}$	n is a natural number, $x \geq 0$ if n is even.
$x^{-\frac{1}{n}} = \dfrac{1}{\sqrt[n]{x}}$	n is a natural number, $x \neq 0$, $x > 0$ if n is even.

Example 1

Determine each exact value without using a calculator.

a) $27^{\frac{1}{3}}$

b) $\left(\dfrac{9}{16}\right)^{-\frac{1}{2}}$

Solution

a) $27^{\frac{1}{3}} = \sqrt[3]{27}$
$= 3$

b) $\left(\dfrac{9}{16}\right)^{-\frac{1}{2}} = \dfrac{1}{\left(\dfrac{9}{16}\right)^{\frac{1}{2}}}$

$$= \dfrac{1}{\sqrt{\dfrac{9}{16}}}$$

$$= \dfrac{1}{\dfrac{3}{4}}$$

$$= \dfrac{4}{3}$$

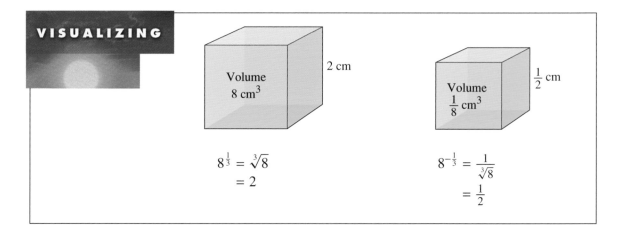

$$8^{\frac{1}{3}} = \sqrt[3]{8}$$
$$= 2$$

$$8^{-\frac{1}{3}} = \frac{1}{\sqrt[3]{8}}$$
$$= \frac{1}{2}$$

To give meaning to a power such as $3^{\frac{2}{3}}$ we extend the exponent law $(x^m)^n = x^{mn}$ so that it applies when m and n are rational numbers.

By extending the law:

$$3^{\frac{2}{3}} = (3^{\frac{1}{3}})^2 \qquad \text{or} \qquad 3^{\frac{2}{3}} = (3^2)^{\frac{1}{3}}$$
$$= (\sqrt[3]{3})^2 \qquad\qquad\qquad = \sqrt[3]{3^2}$$

These examples and the results of *Investigate* suggest the following definitions for $x^{\frac{m}{n}}$.

$$x^{\frac{m}{n}} = (\sqrt[n]{x})^m$$
$$= \sqrt[n]{x^m} \qquad n \text{ is a natural number, } x \geq 0 \text{ if } n \text{ is even.}$$

To give meaning to a power such as $3^{-\frac{2}{3}}$, we use the law $x^m \cdot x^n = x^{m+n}$, which we extended to rational exponents previously.

$$3^{-\frac{2}{3}} \times 3^{\frac{2}{3}} = 3^{-\frac{2}{3} + \frac{2}{3}}$$
$$= 3^0$$
$$= 1$$

Therefore, $3^{-\frac{2}{3}}$ and $3^{\frac{2}{3}}$ are reciprocals.

$$3^{-\frac{2}{3}} = \frac{1}{3^{\frac{2}{3}}}$$
$$= \frac{1}{(\sqrt[3]{3})^2} \qquad \text{or} \qquad \frac{1}{\sqrt[3]{3^2}}$$

This example suggests the following definitions for $x^{-\frac{m}{n}}$.

$$x^{-\frac{m}{n}} = \frac{1}{(\sqrt[n]{x})^m}$$

$$= \frac{1}{\sqrt[n]{x^m}} \qquad n \text{ is a natural number, } x \neq 0, \, x > 0 \text{ if } n \text{ is even.}$$

Example 2

Determine each exact value without using a calculator.

a) $27^{\frac{2}{3}}$

b) $\left(\frac{9}{16}\right)^{-\frac{3}{2}}$

Solution

a) $27^{\frac{2}{3}} = (\sqrt[3]{27})^2$

$= 3^2$

$= 9$

b) $\left(\frac{9}{16}\right)^{-\frac{3}{2}} = \dfrac{1}{\left(\frac{9}{16}\right)^{\frac{3}{2}}}$

$$= \dfrac{1}{\left(\sqrt{\frac{9}{16}}\right)^3}$$

$$= \dfrac{1}{\left(\frac{3}{4}\right)^3}$$

$$= \dfrac{1}{\frac{27}{64}}$$

$$= \frac{64}{27}$$

VISUALIZING

means square of
the cube root

means
reciprocal

$$8^{\frac{2}{3}} = (\sqrt[3]{8})^2$$

$= 2^2$

$= 4$

$$8^{-\frac{2}{3}} = \frac{1}{8^{\frac{2}{3}}}$$

$$= \frac{1}{(\sqrt[3]{8})^2}$$

$$= \frac{1}{2^2}$$

$$= \frac{1}{4}$$

You can use the laws of exponents to simplify expressions involving radicals and rational exponents.

Example 3

Simplify. Write each expression as a power and as a radical.

a) $\sqrt[6]{x^3}$

b) $\sqrt{\sqrt[3]{x^5}}$

c) $(\sqrt[3]{x^4})(\sqrt{x^3})$

Solution

a) $\sqrt[6]{x^3} = (x^3)^{\frac{1}{6}}$ Using the law: $\sqrt[n]{x^m} = (x^m)^{\frac{1}{n}}$

$= x^{3 \times \frac{1}{6}}$ Using the law: $(x^m)^n = x^{mn}$

$= x^{\frac{1}{2}}$, or $\sqrt{x}$

b) $\sqrt{\sqrt[3]{x^5}} = (\sqrt[3]{x^5})^{\frac{1}{2}}$ Using the law: $\sqrt[n]{x^m} = (x^m)^{\frac{1}{n}}$

$= ((x^5)^{\frac{1}{3}})^{\frac{1}{2}}$ Using the law: $\sqrt[n]{x^m} = (x^m)^{\frac{1}{n}}$ again

$= x^{5 \times \frac{1}{3} \times \frac{1}{2}}$

$= x^{\frac{5}{6}}$, or $\sqrt[6]{x^5}$

c) $(\sqrt[3]{x^4})(\sqrt{x^3}) = (x^4)^{\frac{1}{3}} \times (x^3)^{\frac{1}{2}}$ Using the law: $\sqrt[n]{x^m} = (x^m)^{\frac{1}{n}}$ twice

$= x^{\frac{4}{3}} \times x^{\frac{3}{2}}$ Using the law: $(x^m)^n = x^{mn}$

$= x^{\frac{8}{6} + \frac{9}{6}}$ Using the law: $x^m \times x^n = x^{m+n}$

$= x^{\frac{17}{6}}$, or $(\sqrt[6]{x})^{17}$

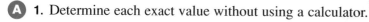

DISCUSSING   THE   IDEAS

1. Think about the definition of a rational exponent. Why is there a restriction on x if n is even? Use some examples to illustrate your explanation.

2. Determine another way to solve *Example 1b* and *Example 2b* that involves fewer steps.

2.2 EXERCISES

A **1.** Determine each exact value without using a calculator.

a) 8^0 **b)** $8^{\frac{1}{3}}$ **c)** $8^{\frac{2}{3}}$ **d)** $8^{\frac{3}{3}}$ **e)** $8^{\frac{4}{3}}$

f) $8^{-\frac{1}{3}}$ **g)** $8^{-\frac{2}{3}}$ **h)** $8^{-\frac{3}{3}}$ **i)** $8^{-\frac{4}{3}}$ **j)** $8^{-\frac{5}{3}}$

2. Determine each exact value without using a calculator.

a) $16^{\frac{1}{2}}$ **b)** $36^{\frac{1}{2}}$ **c)** $100^{\frac{1}{2}}$ **d)** $32^{\frac{1}{5}}$ **e)** $64^{\frac{1}{3}}$

f) $27^{\frac{1}{3}}$ **g)** $(-64)^{\frac{1}{3}}$ **h)** $81^{\frac{1}{4}}$ **i)** $(-27)^{\frac{1}{3}}$ **j)** $(-1000)^{\frac{1}{3}}$

3. Determine each exact value without using a calculator.

a) $4^{-\frac{1}{2}}$ **b)** $9^{-\frac{1}{2}}$ **c)** $27^{-\frac{1}{3}}$ **d)** $64^{-\frac{1}{3}}$ **e)** $(-64)^{-\frac{1}{3}}$

4. Write each expression using radicals.

   a) $4^{\frac{1}{5}}$ **b)** $4^{\frac{2}{5}}$ **c)** $4^{\frac{3}{5}}$ **d)** $4^{\frac{4}{5}}$ **e)** $4^{\frac{5}{5}}$

   f) $4^{-\frac{1}{5}}$ **g)** $4^{-\frac{2}{5}}$ **h)** $4^{-\frac{3}{5}}$ **i)** $4^{-\frac{4}{5}}$ **j)** $4^{-\frac{5}{5}}$

5. Choose one part of exercise 4. Write to explain how you wrote the power as a radical.

6. Determine each exact value without using a calculator.

   a) $9^{\frac{3}{2}}$ **b)** $27^{\frac{2}{3}}$ **c)** $4^{\frac{3}{2}}$ **d)** $25^{\frac{3}{2}}$ **e)** $32^{\frac{2}{5}}$

   f) $(-27)^{\frac{2}{3}}$ **g)** $36^{\frac{3}{2}}$ **h)** $(-64)^{\frac{2}{3}}$ **i)** $100^{\frac{3}{2}}$ **j)** $(-8000)^{\frac{2}{3}}$

7. Determine each exact value without using a calculator.

   a) $27^{-\frac{2}{3}}$ **b)** $32^{-\frac{3}{5}}$ **c)** $9^{-\frac{3}{2}}$ **d)** $16^{-\frac{3}{4}}$ **e)** $100^{-\frac{3}{2}}$

8. Write each number as a power with an exponent of $\frac{1}{2}$.

   a) 3   **b)** 2   **c)** 4   **d)** 1   **e)** 10   **f)** 8

9. Choose one part of exercise 8. Write to explain how you wrote the number as a power.

B 10. Determine each exact value without using a calculator.

   a) $27^{-\frac{4}{3}}$ **b)** $16^{-1.5}$ **c)** $81^{0.75}$ **d)** $32^{-0.4}$ **e)** $49^{\frac{3}{2}}$

   f) $\left(\frac{9}{16}\right)^{\frac{1}{2}}$ **g)** $\left(\frac{25}{49}\right)^{\frac{3}{2}}$ **h)** $\left(-\frac{1}{32}\right)^{0.8}$ **i)** $\left(\frac{8}{27}\right)^{-\frac{2}{3}}$ **j)** $\left(\frac{81}{16}\right)^{-\frac{3}{4}}$

11. Choose one part of exercise 10. Write to explain how you determined the exact value.

12. Use a calculator. Determine each value to 3 decimal places.

   a) $10^{\frac{1}{4}}$ **b)** $30^{0.7}$ **c)** $7^{\frac{2}{3}}$ **d)** $15^{1.4}$ **e)** $\sqrt[8]{2.17}$

13. **a)** A cube has a volume of 8000 cm³. Determine the length of each edge and the area of each face.

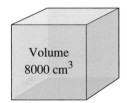
Volume
8000 cm³

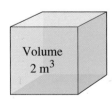
Volume
2 m³

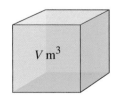
V m³

   b) Repeat part a for a cube with a volume of 2 m³. Express your answers in radical form.

   c) Repeat part a for a cube with a volume of V cubic metres.

14. a) The area of each face of a cube is 25 cm². Determine the length of each edge and the volume of the cube.

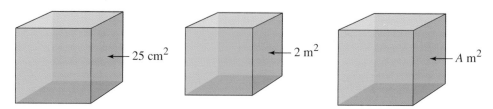

b) Repeat part a for a cube with each face with area 2 m². Express your answers in radical form.

c) Repeat part a for a cube with each face with area A square metres.

15. Write each number as a power with an exponent of $\frac{1}{3}$.

a) 3   **b)** −1   **c)** −2   **d)** −4   **e)** 1   **f)** −3

16. Simplify. Write each expression as a power ~~and as a radical.~~

a) $x \cdot x^{\frac{1}{2}}$ **b)** $m^{\frac{1}{3}} \cdot m$ **c)** $y^{\frac{3}{2}} \cdot y^{\frac{1}{2}}$ **d)** $b^{\frac{5}{2}} \cdot b^{\frac{3}{2}}$

e) $x \div x^{\frac{1}{2}}$ **f)** $m \div m^{\frac{1}{3}}$ **g)** $d^{\frac{3}{2}} \div d^{\frac{1}{2}}$ **h)** $p^{\frac{3}{5}} \div p^{\frac{1}{5}}$

17. People feel colder when it is windy. We call this effect *wind chill*. You can calculate the wind chill equivalent temperature w, using the formula below, where t represents the temperature and s the wind speed. The formula is valid for temperatures less than 5°C and wind speeds between 8 km/h and 80 km/h.

$$w = 33 - \frac{(12.1 + 6.12s^{\frac{1}{2}} - 0.32s)(33 - t)}{27.8}$$

St. John's, Sydney, and Charlottetown are among the windiest cities in Canada. They have average annual wind speeds of 24 km/h, 20 km/h, and 19 km/h, respectively. You can use this information and the *Weather* database to investigate the effect of wind chill.

a) In many Canadian cities, February is the coldest month of the year. Find the 21 records with monthly temperature data for St. John's. Sort these records by the February field. Record the February temperature from the middle record. Use this median as a typical February temperature. Repeat the procedure to find typical February temperatures for Sydney and Charlottetown.

b) List the cities from coldest to warmest, using your answer to part a.

c) Use the formula and wind speeds above. Calculate the wind chill equivalent temperature for each city. Use this additional information. How does your answer to part b change?

d) Choose another month with typical temperatures less than 5°C. Repeat parts a to c.

18. Simplify. Write each expression as a radical and as a power.

    a) $\sqrt[4]{x^2}$
            b) $\sqrt{x^4}$
            c) $\sqrt{\sqrt{x}}$

    d) $\sqrt[3]{\sqrt{x}}$
           e) $\sqrt[3]{\sqrt{3x^5}}$
         f) $\sqrt[3]{\sqrt{x^5}}$

19. Simplify. Write each expression as a radical and as a power.

    a) $(\sqrt{x^3})(\sqrt{x^2})$
      b) $(\sqrt[4]{x^3})(\sqrt[4]{x^2})$
      c) $(\sqrt{x^3})(\sqrt[5]{x^2})$

    d) $(\sqrt[5]{x^3})(\sqrt[3]{x^2})$
      e) $(\sqrt{x^3})(\sqrt[6]{x^2})$
      f) $(\sqrt[6]{x^3})(\sqrt[4]{x^2})$

20. Choose one part of exercise 19. Write to explain how you simplified the expression.

21. This display shows some radicals related to $\sqrt{2}$ in decimal, radical, and power forms.

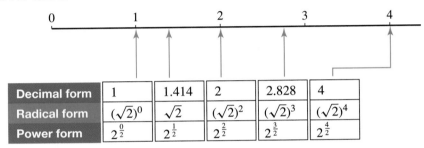

    a) Find as many patterns as you can in the display. Describe and explain each pattern.

    b) Which numbers in the display are approximations? Which numbers are exact?

    c) Which of the expressions in radical form can be written in other ways?

    d) Which of the expressions in power form can be written in other ways?

22. Make a display similar to the one in exercise 21, but which extends to the left to include numbers between 0 and 1. Your display should show the decimal, radical, and power forms for 1, $\dfrac{1}{\sqrt{2}}$, $\dfrac{1}{(\sqrt{2})^2}$, $\dfrac{1}{(\sqrt{2})^3}$, and $\dfrac{1}{(\sqrt{2})^4}$.

C 23. Use your calculator to confirm that $10^{0.3} \doteq 1.995$. Without using your calculator, determine each value.

    a) $10^{1.3}$
        b) $10^{2.3}$
        c) $10^{3.3}$
        d) $10^{4.3}$

    e) $10^{-0.7}$
       f) $10^{-1.7}$
       g) $10^{-2.7}$
      h) $10^{-3.7}$

COMMUNICATING THE IDEAS

Explain a *rational exponent*. In your journal, include some examples to illustrate your explanation.

Population Growth and Deforestation

Page 66 presented information on the world's population and the area of the rain forests for 1960 and 1990. What will these data be in 2010?

 **DEVELOP A MODEL**

Visualize how the population changed every 10 years from 1960 to 1990.

Year	1960	1970	1980	1990	2000	2010
Population (billions)	3.2			5.3		

Consider that the populations are terms of either an arithmetic sequence or a geometric sequence. Which is more reasonable? Why?

1. **a)** Copy the population table. Use the type of sequence you selected. Complete the table.

   b) What is your prediction for the world's population in 2010?

   c) Draw a graph to show the growth in population from 1960 to 2010.

2. Consider the data for the areas of rain forests around the world.

Year	1960	1970	1980	1990	2000	2010
Rain forest area (billion hectares)	2.10			1.75		

   a) Do you think the areas of rain forests represent an arithmetic sequence or a geometric sequence? Explain.

   b) Copy and complete the table, using the type of sequence you selected.

   c) Draw a graph to show the decline of the rain forests from 1960 to 2010.

LOOK AT THE IMPLICATIONS

3. Some scientists believe that Earth will not be able to support the world's population if it exceeds 8 billion. Does your prediction for 2010 exceed this number? If so, what measures might be taken to alter the pace of growth? How does this affect your model in exercise 1?

4. Compare the two graphs you drew in exercises 1 and 2. Describe how you think population growth may be influencing the destruction of the rain forests.

5. Some deforestation relates to the demand for exotic woods, such as teak. In 1990 the annual production of all wood products, worldwide, was 0.10 m^3 per capita. This is projected to increase to 0.15 m^3 per capita by 2010.

   a) Use the 1990 world population and the 1990 production figure above. Determine the total volume of wood products produced in 1990.

   b) Use your population prediction for 2010. Determine the anticipated total volume of wood products for 2010.

   c) Consider that Earth's rain forests are diminishing. How might this affect the demand for wood products?

REVISIT THE SITUATION

6. Use your tables and graphs. Estimate the annual rate of population growth and the annual rate of destruction of the rain forests.

7. A 1990 World Resources report was based on satellite observations. The report suggested that the rate of rain forest destruction had increased to between 16 and 20 million hectares annually. Use this information. Predict the area of the rain forests remaining in 2010. How does this prediction compare with your previous prediction?

8. Ecologists point out the dangers of clearing rain forests. These include erosion of soil and the extinction of plant and animal species. How might these changes affect the forests' ecological system? Research other ecological systems that might be adversely affected by the clearing of rain forests.

2.3 Applications of Rational Exponents

Some formulas contain powers with rational exponents in decimal form. We can explain the meaning of the powers by using the definition of rational exponents.

Example 1

In studies of mammals, scientists have discovered an approximate formula relating the mass of the brain to the mass of the body. The formula is $b = 0.01m^{0.7}$, where b is the mass of the brain in kilograms and m is the mass of the body in kilograms.

a) Calculate the mass of the brain of a 360-kg moose and a 120-g golden hamster.

b) Express each mass in part a as a fraction of the mass of the body, in decimal form and as a percent.

c) Determine a formula that expresses the mass of a mammal's brain as a fraction of the mass of its body.

Solution

a) For the moose:

Substitute 360 for m in the formula.

$b = 0.01 \times 360^{0.7}$

Use your calculator.

$b \doteq 0.01 \times 61.576$

$\doteq 0.615\ 76$

The mass of the moose's brain is about 0.62 kg, or 620 g.

For the golden hamster:

The mass is 120 g = 0.12 kg.

Substitute 0.12 for m in the formula.

$b = 0.01 \times 0.12^{0.7}$

$\doteq 0.01 \times 0.226\ 68$

$\doteq 0.002\ 266\ 8$

The mass of the hamster's brain is about 0.0023 kg, or 2.3 g.

b) Fraction of the body for the moose:

$\frac{0.62}{360} \doteq 0.001\ 722$

$\doteq 0.001\ 722 \times 100\%$

$\doteq 0.17\%$

The mass of the brain is about 0.17% of the mass of the body.

Fraction of the body for the hamster:

$\frac{0.0023}{0.12} \doteq 0.019\ 17$

$\doteq 0.019\ 17 \times 100\%$

$\doteq 1.9\%$

The mass of the brain is about 1.9% of the mass of the body.

c) The mass of the brain is $b = 0.01m^{0.7}$. To express b as a fraction of the mass of the body, divide each side by m.

$\frac{b}{m} = \frac{0.01m^{0.7}}{m}$ — To divide the powers, subtract their exponents.

$= 0.01m^{-0.3}$

Rational exponents are very useful for modelling growth using geometric sequences.

Example 2

Strep throat is caused by bacteria called streptococci. After you have been infected, it is possible for these bacteria to double in number every 20 min.

a) Suppose 100 bacteria are present now. Calculate to complete the table. Show the number of bacteria every 5 min for the next 40 min.

Growth of Bacteria

Time (min)	0	5	10	15	20	25	30	35	40
Population	100				200				

b) Show the results on a graph.

Solution

a) Since the population doubles every 20 min, there are 200 bacteria after 20 min. Assume that the populations are terms of a geometric sequence with first term a and common ratio r.

The first term is $a = 100$.
The fifth term is $ar^4 = 200$.

Substitute 100 for a: $\quad 100r^4 = 200$ —— Divide each side by 100.

$$r^4 = 2$$
$$r = \sqrt[4]{2}$$ —— Take the fourth root of each side.
$$= 2^{\frac{1}{4}}$$

Use your calculator.

Population after 5 min: $\quad 100 \times 2^{\frac{1}{4}} \doteq 119$

Population after 10 min: $\quad 100 \times (2^{\frac{1}{4}})^2 = 100 \times 2^{\frac{2}{4}} \doteq 141$

Population after 15 min: $\quad 100 \times (2^{\frac{1}{4}})^3 = 100 \times 2^{\frac{3}{4}} \doteq 168$

Population after 20 min: $\quad 100 \times (2^{\frac{1}{4}})^4 = 100 \times 2^{\frac{4}{4}} = 100 \times 2^1 = 200$

Population after 25 min: $\quad 100 \times (2^{\frac{1}{4}})^5 = 100 \times 2^{\frac{5}{4}} \doteq 238$

Population after 30 min: $\quad 100 \times (2^{\frac{1}{4}})^6 = 100 \times 2^{\frac{6}{4}} \doteq 283$

Population after 35 min: $\quad 100 \times (2^{\frac{1}{4}})^7 = 100 \times 2^{\frac{7}{4}} \doteq 336$

Population after 40 min: $\quad 100 \times (2^{\frac{1}{4}})^8 = 100 \times 2^{\frac{8}{4}} = 100 \times 2^2 = 400$

Here is the completed table.

Growth of Bacteria

Time (min)	0	5	10	15	20	25	30	35	40
Population	100	119	141	168	200	238	283	336	400

b) Use the data from part a to plot the points. Draw a smooth curve through the plotted points.

Growth of Bacteria

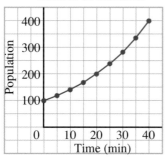

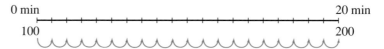

VISUALIZING

In *Example 2*, the population doubles from 100 to 200 in 20 min. Visualize multiplying the number 100 by 2 to obtain 200.

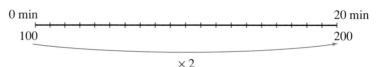

To determine the population after any number of minutes, visualize replacing the number 2 with 20 equal numbers whose product is 2. Each number is $\sqrt[20]{2}$. Hence, the population increases by a factor of $\sqrt[20]{2} = 2^{\frac{1}{20}}$ each minute.

You can use this result to estimate the population after any number of minutes. For example, to estimate the population after 12 min, multiply 100 by $(2^{\frac{1}{20}})^{12} = 2^{\frac{12}{20}}$. The result is $100 \times 1.515\ 72 \doteq 152$.

Example 3

The golden hamster provides an amazing story of species survival. This rodent was unknown until 1930, when a small colony of 13 golden hamsters was discovered. Within ten years, the population of the golden hamster had risen from 13 to several million! A conservative estimate of this number is 3 000 000. Calculate to estimate the population for each year from 1931 to 1934.

Solution

In 10 years, the population increased by a factor of $\frac{3\ 000\ 000}{13} \doteq 230\ 769$.

We assume that the populations are terms of a geometric sequence.

Each year, the population increased by a factor of $\sqrt[10]{230\ 769} = (230\ 769)^{\frac{1}{10}}$.

1931 population: $13 \times (230\ 769)^{\frac{1}{10}} \doteq 45$

1932 population: $13 \times (230\ 769)^{\frac{2}{10}} \doteq 154$

1933 population: $13 \times (230\ 769)^{\frac{3}{10}} \doteq 528$

1934 population: $13 \times (230\ 769)^{\frac{4}{10}} \doteq 1816$

Some formulas contain powers with rational exponents in decimal form. We can explain the meaning of the powers by using the definition of rational exponents.

Example 4

The height, h metres, of a Douglas fir tree can be estimated from the formula $h = 34.1 \times d^{0.67}$, where d is the diameter, in metres, at the base. Use this formula. Estimate the height of a Douglas fir with a base diameter of 4.35 m.

Solution

Substitute 4.35 for d in the formula.

$h = 34.1 \times (4.35)^{0.67}$

Use your calculator.

$h \doteq 34.1 \times 2.677\ 86$

$ \doteq 91.315$

The Douglas fir is approximately 91 m high.

1. In *Example 1*, what does the expression $m^{0.7}$ mean? Use a radical to explain your answer.

2. In *Example 2*, how do the populations compare at these times? Explain.

   a) after 5 min, and 20 min later

   b) after 10 min, and 20 min later

3. In *Example 2*, how could you determine the points on the graph corresponding to times of 2.5 min, 7.5 min, 12.5 min, … ?

2.3 EXERCISES

A 1. The formula in *Example 1* does not apply to humans. For humans, the formula relating brain mass to body mass is $b = 0.085m^{0.66}$.

   a) Calculate the mass of the brain of a person with a mass of 45 kg.

   b) Estimate the mass of your brain.

2. Use the formula in *Example 1*.

   a) Calculate the mass of the brain of a 900-kg steer and a 30-g mouse.

   b) Express each mass in part a as a percent of the body mass.

   c) Solve part b a different way.

B 3. Some scientists believe that the exponent in the formula in *Example 1* should be $\frac{2}{3}$. Use the formula $b = 0.01m^{\frac{2}{3}}$ for brain mass.

   a) Calculate the mass of the brain of each mammal in this list.

Elephant	6400 kg
Cat	6.4 kg
Shrew	0.0064 kg

   b) How are the masses of the three animals in part a related? How are the masses of their brains related?

   c) Express the mass of each brain in part a as a percent of the mass of the body.

   d) How are the percents in part c related? What can you conclude from the result?

   e) Use the formula $b = 0.01m^{\frac{2}{3}}$. Determine a formula that expresses the mass of a mammal's brain as a percent of the mass of its body.

 **MODELLING** the Brain Mass of a Mammal

The formulas in *Example 1* and exercises 1, 2, and 3 are based on actual measurements. The exact values of the coefficient and the exponent in the formula depend on the data that are used. The formulas give average brain masses, and the actual brain mass of any given mammal may differ from the calculated result.

- Do you think mammals with larger brains might be more intelligent than those with smaller brains? Give a reason to support your answer.

- Do you think the saying "an elephant never forgets" might have some justification? Explain.

4. Set up a spreadsheet similar to the one below. Each animal has a body mass that is 8 times the mass of the animal above it. The unit of mass is the mass of one mouse. For example, the mass of the porcupine is 512 times the mass of a mouse.

   a) In cell C4, enter this formula: =0.01*B4^(2/3)
      In cell D4, enter this formula: =C4/B4
      Copy these formulas down to row 10.

   b) What patterns can you find in the results? Explain the patterns.

	A	B	C	D
1	Brain Masses of Mammals			
2	Mammal	Body	Brain	Percent
3		mass	mass	of body
4	Mouse	1		
5	Gerbil	8		
6	Marten	64		
7	Porcupine	512		
8	Seal	4096		
9	Cow	32768		
10	Elephant	262144		

5. Scientists have determined the following approximate formulas for resting animals with mass m kilograms.

   Heartbeat frequency in beats per minute: $h = 241m^{-0.25}$
   Respiratory frequency in breaths per minute: $r = 53.5m^{-0.25}$

   a) Calculate the heartbeat frequency and the respiratory frequency for each animal in exercise 3.

   b) What happens to each frequency as the mass of the animal decreases?

c) On average, about how many heartbeats does an animal have during each breath?

d) Does the result in part c depend on the mass of the animal? Explain.

6. An approximate formula for the surface area of the body in square metres is $A = 0.096m^{0.7}$, where m is the person's mass in kilograms.

a) Calculate the surface area of a newborn baby with mass 3.8 kg.

b) Calculate the surface area of a person with mass 140 kg.

c) Calculate your surface area.

7. Another formula for skin area also takes into account the person's height. The formula is $A = 0.025h^{0.42}m^{0.5}$, where A is in square metres, h is the height in centimetres, and m is the mass in kilograms.

a) Calculate the surface area of a newborn baby whose length and mass are 50 cm and 3.8 kg, respectively.

b) Calculate the surface area of a person whose height and mass are 170 cm and 140 kg, respectively.

c) Use this formula to calculate your surface area.

8. About 400 years ago, the astronomer Johann Kepler developed a formula to determine the time it takes each planet to travel once around the sun (called the *period*). The formula, known as *Kepler's Third Law*, is $T \doteq 0.2R^{\frac{3}{2}}$, where T is the period in Earth days and R is the mean distance from the planet to the sun in millions of kilometres.

a) Use Kepler's Third Law. Calculate the period of each planet.

b) Use your results to draw a graph to show how the period depends on the distance from the sun.

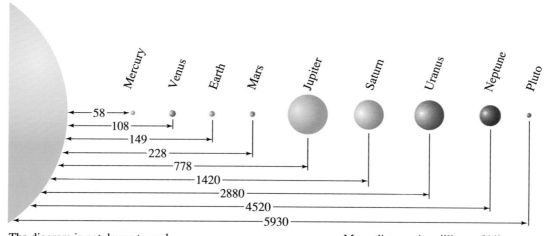

The diagram is not drawn to scale.

Mean distance in millions of kilometres

9. a) Refer to *Example 3*. Complete the calculations for all years from 1930 to 1940. Show your results in a table.

b) Use your data to draw a graph to show how the numbers of golden hamsters grew from 1930 to 1940.

10. A herd of caribou, known as the Kaminuriak, ranges in the Northwest Territories west of Hudson Bay. After decades of decline, a census in 1980 indicated a population of 39 000. During the next five years the population increased dramatically, growing to 320 000 animals in 1985. Copy the table. Calculate to complete the table, showing the results to the nearest thousand.

Size of Kaminuriak Caribou Herd

Year	1980	1981	1982	1983	1984	1985	1986
Population (thousands)	39					320	

C **11.** On average, animals tend to live for approximately 1.5 billion heartbeats.

a) Use the first formula in exercise 5. Determine a formula for the average lifetime in years of an animal with mass m kilograms.

b) Estimate the average lifetime of each animal in exercise 3.

12. In *Example 3* and exercise 9, you used a geometric sequence to model the growth of the golden hamster population. Use this model to predict the golden hamster population for this year. Do you think the result is reasonable? Explain.

13. When a satellite is h kilometres above Earth, the time, T minutes, for one revolution around Earth is given by the formula $T = 1.66 \times 10^{-4}(6370 + h)^{1.5}$. This time is called the *period* of the satellite.

a) Calculate the period of a satellite at each altitude. Express the answers in both minutes and hours.

i) 200 km   **ii)** 600 km   **iii)** 35 848.52 km

b) What special feature does a satellite have at an altitude of 35 848.52 km? How might such a satellite be used?

COMMUNICATING THE IDEAS

Suppose you know the population of a group of people or animals in two different non-consecutive years. In your journal, write an explanation of how you could estimate the population in a year between the two given years. Your explanation should be understandable by someone who has not done the examples or exercises in this section. Make up an example to use in your explanation.

Bird Eggs

The common egg is a remarkable life-support system. It contains everything needed for the development of the hatchling chick, except oxygen. Thousands of tiny pores in the shell allow oxygen to enter and carbon dioxide to exit.

Scientists have made many measurements of bird eggs. From these data, they have established approximate formulas relating the various measurements. Some of these formulas are shown below, with m representing the mass of the bird in grams. The formulas give average results, and the properties of a particular egg from a particular bird may differ from the calculated result.

Mass of the egg in grams:	$e = 0.277m^{0.770}$	(1)
Mass of the shell in grams:	$s = 0.0482e^{1.132}$	(2)
Incubation time in days:	$i = 12.03e^{0.217}$	(3)
Thickness of the shell in millimetres:	$t = 0.051\ 26e^{0.456}$	(4)

1. The first row of the table below shows the average masses of three birds. Copy and complete this table using the formulas above. Include two additional rows at the bottom to be completed later.

Bird	Hummingbird	Hen	Ostrich
Mass of bird (g)	3.6	2000	113 000
Mass of egg (g)			
Mass of shell (g)			
Incubation time (d)			
Shell thickness (mm)			

2. a) Use formula (1). Determine an expression for the mass of each egg as a fraction of the mass of the bird.

b) Apply your expression to write the mass of the egg of each bird in exercise 1 as a percent of its body mass. Add these results to your table.

3. a) Use formula (2). Determine an expression for the mass of each shell as a fraction of the mass of the egg.

b) Apply your expression to write the mass of the shell of each bird in exercise 1 as a percent of the mass of the egg. Add these results to your table.

4. Use formulas (1) and (2). Determine an expression for the mass of the shell in terms of the mass of the bird.

5. This spreadsheet will display results similar to those in exercises 1, 3, and 4, for different birds.

	A	B	C	D	E	F
1	Some Properties of Bird Eggs					
2		Hummingbird	Pigeon	Hen	Goose	Ostrich
3	Mass (g)	3.6	280	2000	4500	113000
4	Egg mass (g)					
5	Fraction of bird					
6	Shell mass (g)					
7	Fraction of egg					
8	Incubation time (d)					
9	Shell thickness (mm)					

a) Start a new spreadsheet document. Enter the text and data above.

b) In cell B4, enter this formula: =0.277*B3^0.770

c) Enter corresponding formulas in the other cells in column B.

d) Extend the formulas to the right to complete the other columns. To do this, click on cell B4 and drag down and over to cell F9 to select this block of cells. Then use the Fill Right option.

e) Examine the results. Compare them with those you calculated in the previous exercises. To improve readability, you may need to reformat some cells.

f) Look at the results in rows 5 and 7. You should find that the fraction $\frac{\text{egg mass}}{\text{bird mass}}$ decreases for larger birds, while the fraction $\frac{\text{shell mass}}{\text{egg mass}}$ increases. Use your answers to exercises 2a and 3a to explain this result.

6. The elephant bird from Madagascar was the largest bird that ever lived. It stood over 3 m tall and laid a giant egg with a mass of about 10 kg. Use your spreadsheet to estimate the mass of the elephant bird.

Mathematics & Science

Clay tablets reveal that some time between 1900 BC and 1600 BC, the ancient Babylonians knew how the sides of a right triangle are related. This relationship, known as the *Pythagorean Theorem*, is named after the Greek mathematician, Pythagoras, who discovered it independently more than a thousand years later. Pythagoras achieved more than the Babylonians, because he also explained why this relationship is true for all right triangles.

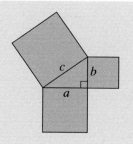

Pythagorean Theorem

In any right triangle, the area of the square on the hypotenuse is equal to the sum of the areas of the squares on the other two sides.

$$c^2 = a^2 + b^2$$

Recall the work you did in grades 8 and 9. The Pythagorean Theorem is important because you can use it to calculate the length of the third side of a right triangle when you know the lengths of the other two sides. For this reason, the Pythagorean Theorem is used in many applications of mathematics.

Example

Calculate each value of x to 1 decimal place.

a)

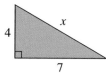

b)

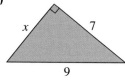

Solution

a) $x^2 = 4^2 + 7^2$
$= 16 + 49$
$= 65$

Since $(\sqrt{65})^2 = 65$ and $(-\sqrt{65})^2 = 65$, x could be $\sqrt{65}$ or $-\sqrt{65}$.

However, since x represents a length, it must be positive.

Hence, $x = \sqrt{65}$
$\doteq 8.1$

b) $9^2 = x^2 + 7^2$
$81 = x^2 + 49$
$81 - 49 = x^2$
$32 = x^2$

Since $x > 0$,
$x = \sqrt{32}$
$\doteq 5.7$

There is another reason why the Pythagorean Theorem is important. About 2500 years ago, the ancient Greeks used it to make a discovery that confused them because they did not realize that they had found a new kind of number. The followers of Pythagoras, known as the Pythagoreans, discovered that $\sqrt{2}$ is not a rational number. They were able to demonstrate that $\sqrt{2}$ cannot be expressed in the form $\frac{m}{n}$, where m and n are natural numbers. Here is how they did it:

Step 1

Assume that $\sqrt{2}$ is a rational number.

Then there are natural numbers m and n such that $\sqrt{2} = \frac{m}{n}$, where m and n are in lowest terms.

Step 2

Square each side to obtain: $\quad (\sqrt{2})^2 = \left(\frac{m}{n}\right)^2$

$$2 = \frac{m^2}{n^2} \quad \longleftarrow \boxed{\text{Multiply each side by } n^2.}$$

$$2 \times n^2 = \frac{m^2}{n^2} \times n^2$$

$$2n^2 = m^2$$

Step 3

Since the left side of this equation is even, the right side is even. Hence, m must be an even number. Represent this even number by $2p$. Substitute $2p$ for m:

$$2n^2 = (2p)^2$$
$$2n^2 = 4p^2$$
$$n^2 = 2p^2$$

Step 4

Since the right side of this equation is even, the left side is even. Hence, n must be an even number. That is, both m and n are even. This means that the fraction $\frac{m}{n}$ is not in lowest terms, although we assumed in Step 1 that it is in lowest terms. This contradicts the assumption in Step 1 that $\sqrt{2}$ can be written as a fraction in lowest terms.

Step 5

The assumption in Step 1 that $\sqrt{2}$ is a rational number is incorrect. Hence, $\sqrt{2}$ is not a rational number.

The method of proof used above is known as "indirect proof" or "proof by contradiction." By assuming that $\sqrt{2}$ is a rational number in lowest terms, the Pythagoreans concluded that it is not in lowest terms. That meant that $\sqrt{2}$ could not be a rational number.

The discovery that $\sqrt{2}$ is not a rational number baffled the Pythagoreans because they thought all numbers were rational. They also knew that all lengths can be represented by numbers.

The Dilemma of the Pythagoreans

1. They thought that all numbers were rational, and the rational numbers filled up all the points on the number line, with no points left over.

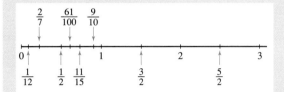

2. They knew that the length of the diagonal of a square with sides 1 unit long was $\sqrt{2}$, but this was not a rational number.

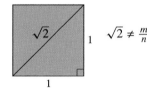

3. That meant there was a point at $\sqrt{2}$ on the number line that could not be represented by a rational number.

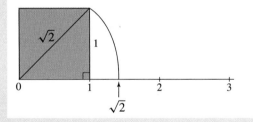

4. They were puzzled by the apparent contradiction: the number $\sqrt{2}$ existed because it could be plotted on the number line. But how could $\sqrt{2}$ exist when it was not a rational number?

DISCUSSING THE IDEAS

1. For the proof that $\sqrt{2}$ is not a rational number:

   a) In Step 1, why do we assume that it is a rational number?

   b) In Step 1, why can we assume that $\frac{m}{n}$ is in lowest terms?

   c) In Step 3, how do we know the left side of the equation is even?

2. Why does a square with sides 1 unit long have a diagonal with length $\sqrt{2}$?

2.4 EXERCISES

A 1. Use the Pythagorean Theorem. Calculate each value of x to 1 decimal place.

a)
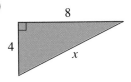
8

4

x

b)
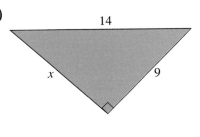
14

x

9

c)
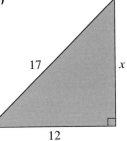
17

x

12

d)
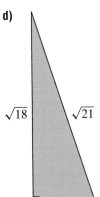
$\sqrt{18}$

$\sqrt{21}$

x

e)
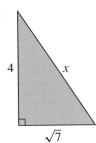
4

x

$\sqrt{7}$

f)
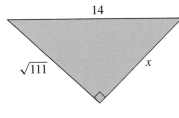
14

$\sqrt{111}$

x

B 2. Square ABCD is divided into 25 small squares, each of side length 1 cm. Determine the length of each line segment.

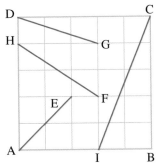

a) AE   b) HF

c) DG   d) IC

3. The lengths of two sides of a right triangle are 5.0 cm and 3.0 cm. Calculate the possible lengths of the third side.

4. Calculate the length, x, to the nearest centimetre.

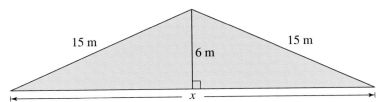

15 m   15 m

6 m

x

5. A rectangular field is 250 m long by 110 m wide. How much shorter is it to walk diagonally across the field than around the sides?

6. An archaeologist measured a pyramid. She found that the base is square with side length 90 m. The slant height is 60 m. Calculate the height of the pyramid to the nearest metre.

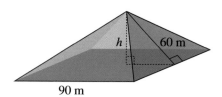

7. The hypotenuse of a right triangle has each length shown below. For each hypotenuse length, write two possible lengths for the legs of the triangle. Write your answers in radical form, where necessary.

a) 5 cm   **b)** 3 cm   **c)** 10 cm   **d)** 12 cm   **e)** 2 cm   **f)** 7 cm

8. Choose one part of exercise 7. Write to explain how you determined the lengths of the legs.

9. On page 97, we proved that $\sqrt{2}$ is not a rational number.

a) Use a similar method. Prove that $\sqrt{3}$ is not a rational number.

b) Use a similar method. Prove that $\sqrt{5}$ is not a rational number.

10. In the rectangular prism, PQ is a diagonal.

a) Determine the length of PQ when PR = 5.0 cm, RS = 3.0 cm, and QS = 2.0 cm.

b) Determine a formula for the length d of the diagonal when the dimensions of the prism are a, b, and c.

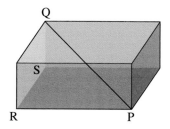

C **11. a)** Do you think you could use the method on page 97 to prove that $\sqrt{4}$ is not a rational number? Explain.

b) Suppose you used the method on page 97 to try to prove that $\sqrt{4}$ is not a rational number. At what step would the proof break down?

12. A square and a circle have the same area.

a) What is the ratio of the side of the square to the radius of the circle?

b) What is the ratio of the diagonal of the square to the radius of the circle?

COMMUNICATING THE IDEAS

Suppose your friend telephones you to ask about today's mathematics lesson. How would you explain, over the telephone, the dilemma of the Pythagoreans?

2.5 Irrational Numbers

When you used your calculator to determine the square root of a number, you may have noticed that the decimals probably neither repeated nor terminated. Another number like this is π, which occurs in the formulas for the circumference and area of a circle.

$$C = \pi d \qquad\qquad A = \pi r^2$$

Earlier in your study of mathematics, you probably used $\pi \doteq 3.14$ or $\pi \doteq 3.1416$ in your calculations with these formulas. You can get a few more digits using the $\boxed{\pi}$ key on your calculator: $\pi \doteq 3.141\ 592\ 654$.

Today, a home computer with the proper software can calculate π to one thousand decimal places in a fraction of a second, producing the following result.

3.14159265358979323846264338327950288419716939937510582097494459230781640628620899862803482534211706798214808651328230664709384460955058223172535940812848111745028410270193852110555964462294895493038196442881097566593344612847564823378678316527120190914564856692346034861045432664821339360726024914127372458700660631558817488152092096282925409171536436789259036001133053054882046652138414695194151160943305727036575959195309218611738193261179310511854807446237996274956735188575272489122793818301194912983367336244065664308602139494639522473719070217986094370277053921717629317675238467481846766940513200056812714526356082778577134275778960917363717872146844090122495343014654958537105079227968925892354201995611212902196086403441815981362977477130996051870721134999999837297804995105973173281609631859502445945534690830264252230825334468503526193118817101000313783875288658753320838142061717766914730359825349042875546873115956286388235378759375195778185778053217122680661300192787661119590921642019893

Computer companies frequently use the computation of π to test their powerful new computers. If the result is correct to millions of decimal places, the computer has done billions of operations without making an error. In 1995, π was calculated to more than 4 billion decimal places. It would take about one thousand books like this one to print π to this many decimal places using the same size type as in the above result.

Mathematicians know that no matter how many decimal places are calculated for π, the numbers will never terminate or repeat. This means that π cannot be expressed in the form $\frac{m}{n}$, where m and n are integers. For this reason, π is an irrational number.

Any number that cannot be expressed in the form $\frac{m}{n}$, where m and n are integers, and $n \neq 0$, is an *irrational number*. The decimal representation of an irrational number neither terminates nor repeats.

Example 1

Does each number appear to be rational or irrational?

a) $x = 0.123456789101112\ldots$

b) $y = 2.41331133113311331\ldots$

c) $\sqrt{2} = 1.4142135623730950488016887242096980785696718753769\ldots$

Solution

a) $x = 0.123456789101112\ldots$

Although there is a pattern in the decimal representation of x, there is no repetition of a sequence of digits. Hence, x appears to be irrational.

b) $y = 2.41331133113311331\ldots$

y appears to be rational since the sequence 1331 repeats.

c) $\sqrt{2} = 1.4142135623730950488016887242096980785696718753769\ldots$

There is no repeating sequence of digits. Therefore, $\sqrt{2}$ appears to be irrational.

In *Example 1a*, we cannot be certain that the number is irrational, since a sequence of digits that repeats could occur farther along in the decimal expansion.

We know that the Pythagoreans demonstrated $\sqrt{2}$ could not be expressed as a rational number. They proved that $\sqrt{2}$ is an irrational number, even though they did not know that irrational numbers existed. By using methods similar to theirs, we can prove many radicals are irrational numbers.

Suppose x is a positive number that is not the square of a rational number. Then the number $\sqrt{x}$ is irrational.

Example 2

Which of the following numbers are irrational?

a) $\sqrt{3}$ **b)** $\sqrt{16}$ **c)** $\sqrt{20}$ **d)** $\sqrt{1.44}$ **e)** $\sqrt{\frac{4}{9}}$ **f)** $\sqrt{\frac{4}{5}}$

Solution

a) $\sqrt{3}$ is irrational since 3 is not a perfect square.

b) $\sqrt{16}$ is rational since 16 is a perfect square; $\sqrt{16} = 4$

c) $\sqrt{20}$ is irrational since 20 is not a perfect square.

d) $\sqrt{1.44}$ is rational since 1.44 is a perfect square; $\sqrt{1.44} = 1.2$

e) $\sqrt{\frac{4}{9}}$ is rational since $\frac{4}{9}$ is a perfect square; $\sqrt{\frac{4}{9}} = \frac{2}{3}$

f) $\sqrt{\frac{4}{5}}$ is irrational since $\frac{4}{5}$ is not a perfect square.

Any number that can be plotted on a number line can be expressed as a decimal. These numbers are grouped into two sets:

The Real Numbers

Rational numbers	Irrational numbers
These numbers have decimal representations that terminate or repeat.	These numbers have decimal representations that neither terminate nor repeat.

Rational numbers: $\frac{1}{2}$ 0.65 $-\frac{7}{3}$ $5.\overline{21}$

> **Integers, I**
> -8 $\quad$ -51
>> **Whole numbers, W**
>> 0
>>> **Natural numbers, N**
>>> 5 $\quad$ 144

Irrational numbers: $\sqrt{2}$ $\quad$ $\sqrt{51} - 1$ $\quad$ π $\quad$ $-\sqrt[3]{10}$ $\quad$ 1.717 717 77...

There are no other possibilities. All the numbers represented above are *real numbers*. The set of real numbers consists of the rational numbers and the irrational numbers. It is the set of all the numbers that can be represented as decimals and plotted on a number line.

1. In *Example 1a*, is it possible for x to be rational? In *Example 1b*, is it possible for y to be irrational? Explain your answers.

2. In the solution of *Example 1b*, we stated that the sequence of digits 1331 repeats. Is this the only sequence of digits that repeats, or are there others?

3. Use the fact that the decimal approximation of π neither terminates nor repeats. Explain how this means that π cannot be expressed in the form $\frac{m}{n}$, where m and n are integers.

2.5 EXERCISES

 1. Does each number appear to be rational or irrational?

   a) 2.1474747474…   b) −6.132133134…   c) 72.04129647…

   d) 0.161661666…   e) −2.236067977…   f) −4.317495…

2. Using a calculator, $\sqrt{434} \doteq 20.832\ 666\ 66$.

   a) Is the number 20.832 666 66 rational or irrational? Explain.

   b) Is the number $\sqrt{434}$ rational or irrational? Explain.

3. Examine the decimal expansion of π on page 101.

   a) The longest sequence of repeating digits is at the right end of the fifth last line. Identify the repeating digits in this sequence.

   b) Suppose you had calculated the decimal expansion up to and including those repeating digits, but no farther. Would this be enough to prove that π is a rational number? Explain.

4. Look at the diagram on page 103.

   a) Explain why the natural numbers rectangle lies inside the integers rectangle.

   b) Explain why the integers rectangle lies inside the rational numbers rectangle.

5. Which of these numbers are irrational?

   a) $\sqrt{21}$ b) $\sqrt{16}$ c) $\sqrt{2\frac{1}{4}}$ d) $\sqrt{200}$ e) $\sqrt{2.5}$

6. Choose one part of exercise 5. Write to explain how you knew whether the number was irrational.

B 7. Which of these numbers are irrational?

   a) $\sqrt{3}$ b) $\sqrt{24}$ c) $\sqrt{25}$ d) $2 + \sqrt{36}$

   e) $2\sqrt{36}$ f) $\sqrt{36} + \sqrt{64}$ g) $\sqrt{36 + 64}$ h) $\sqrt{2 + \sqrt{4}}$

8. Which numbers below belong to each set?

   a) natural numbers   b) integers

   c) rational numbers   d) irrational numbers

   $\dfrac{3}{5}$ $0.2\overline{17}$ -6 $41\ 275$ $3\sqrt{2}$

   6π $-2\dfrac{1}{4}$ $\sqrt[3]{8}$ $\sqrt{121}$ $6.121\ 121\ \dots$

9. a) Are all integers rational numbers?

   b) Are all natural numbers integers?

   c) Are all whole numbers natural numbers?

   d) Are some rational numbers integers?

   e) Are some rational numbers irrational?

10. For each question in exercise 9, write to explain your answer giving examples where they help with your explanation.

11. Write a number that is:

   a) an integer but not a natural number

   b) a whole number but not a natural number

   c) a rational number and an integer

   d) a rational number but not an integer

12. Draw a large diagram similar to that on page 103. Write each number below in its proper place on the diagram.

   23 0.35 $\dfrac{4}{3}$ 10^6 $-32.141\ 414$

   $\sqrt{7}$ $\sqrt{\dfrac{1}{4}}$ -2π $\dfrac{\pi}{2}$ $\sqrt{9} + \sqrt{16}$

C 13. Determine if it is possible for the square root of an irrational number to be a rational number. Explain your thinking.

COMMUNICATING THE IDEAS

Write a letter to the Pythagoreans to help them unravel their dilemma. Explain the error in their thinking.

Significant Digits

All measurements are approximate. For example, if you measure the length of
this rectangle, you will find that it is approximately
3.8 cm. This means that it is closer to 3.8 cm than to
either 3.7 cm or 3.9 cm. We say that the length is
3.8 cm *to the nearest tenth of a centimetre.* This
means that the length is between 3.75 cm and
3.85 cm. We say that the digits 3 and 8 in the
measurement 3.8 cm are *significant digits.*

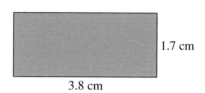

1.7 cm

3.8 cm

Similarly, the width of the rectangle is 1.7 cm to the nearest tenth of a
centimetre. This means that the width is between 1.65 cm and 1.75 cm.

Suppose we multiply the length by the width to calculate the area.

$$A = 3.8 \text{ cm} \times 1.7 \text{ cm}$$
$$= 6.46 \text{ cm}^2 \qquad\qquad (1)$$

Observe that the calculated area contains three digits, although the
measurements themselves contain only two digits. Since the length and width
are measurements, they are not exact numbers. Hence, we cannot say for
certain that the area is 6.46 cm^2.

The length could be as short as 3.75 cm.
The width could be as short as 1.65 cm.
When we use these measurements, the area
could be as small as:

$$A = 3.75 \text{ cm} \times 1.65 \text{ cm}$$
$$= 6.1875 \text{ cm}^2 \qquad (2)$$

The length could be as long as 3.85 cm.
The width could be as long as 1.75 cm.
When we use these measurements, the area
could be as great as:

$$A = 3.85 \text{ cm} \times 1.75 \text{ cm}$$
$$= 6.7375 \text{ cm}^2 \qquad (3)$$

Hence, all we know for certain is that the area is between 6.1875 cm^2 and
6.7375 cm^2. To accommodate this discrepancy, we use the following convention:

> When we calculate with measurements, the final answer should be written with the
> same number of significant digits used in the least accurate measurement.

After we multiply the measurements as in (1) above, we should round the area
to express it with the same number of significant digits used in the
measurements. Since the measurements have two significant digits, we write the
area as 6.5 cm^2.

Accuracy

The convention does not always give accurate results. For example, the rectangle could have dimensions closer to 3.75 cm and 1.65 cm than to 3.85 cm and 1.75 cm. To two significant digits, the area could be 6.4 cm^2, 6.3 cm^2, or 6.2 cm^2.

Suppose the measurements do not have the same number of significant digits. The area should be written with the same number of digits as the least significant measurement.

Leading zeros

Suppose the dimensions of the rectangle on page 106 are expressed in metres.

$$A = 0.038 \text{ m} \times 0.017 \text{ m}$$
$$= 0.000\ 646 \text{ m}^2$$

The area is approximately 0.000 65 m^2.

Observe that the leading zeros are not counted as significant digits.

Measurements

Significant digits are used *only* when the numbers involved are measurements. For example, we can visualize a rectangle that is *exactly* 3.8 cm long and *exactly* 1.7 cm wide. The area of this rectangle is *exactly* 6.46 cm^2, and it should be written this way without using the convention.

Irrational numbers

When irrational numbers such as π and $\sqrt{2}$ are used in calculations, decimal approximations are required. Since these numbers are not measurements, the approximations can have more significant digits than the measurements in the calculations.

1. The measured dimensions of a rectangular carton are 57 cm by 43 cm by 25 cm, to the nearest centimetre.

   a) Use these measurements. Calculate the volume of the carton. Write the volume using the number of significant digits determined by the convention.

   b) Use the least possible value of each measurement. Calculate the least possible volume.

   c) Use the greatest possible value of each measurement. Calculate the greatest possible volume.

   d) Express each volume in parts a, b, and c to:

      i) 2 significant digits   ii) 1 significant digit

   e) To how many significant digits should the volume be expressed, to be representative of all possible values?

2.6 Relating the Sides of Special Triangles

Irrational numbers such as π, $\sqrt{2}$, and $\sqrt{3}$ frequently occur in problems involving circles and triangles.

Example 1

---

a) Calculate the area of an equilateral triangle with sides 2 units long.

b) Determine a formula for the area of an equilateral triangle with sides $2s$ units long.

Solution

a) Draw an equilateral ΔABC with sides 2 units long. Construct the perpendicular from A to BC, meeting BC at D. By symmetry, D has to be the midpoint of BC.

Apply the Pythagorean Theorem to ΔABD.

$$AB^2 = BD^2 + AD^2$$
$$4 = 1 + AD^2$$
$$AD^2 = 3$$
$$AD = \sqrt{3}$$

The area of ΔABC is

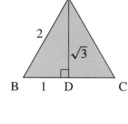

$$\frac{1}{2} \times BC \times AD = \frac{1}{2} \times 2 \times \sqrt{3}$$
$$= \sqrt{3}$$

The area of the equilateral triangle is $\sqrt{3}$ square units.

b) Draw an equilateral triangle similar to the one above, but with sides $2s$ units long. Since all equilateral triangles are similar, all lengths in this triangle are s times as long as those in the preceding triangle. The area of this triangle is

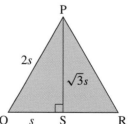

$$\frac{1}{2} \times 2s \times \sqrt{3}s = \sqrt{3}s^2$$

The area of this triangle is $\sqrt{3}s^2$ square units.

In *Example 1a*, the area of the triangle is exactly $\sqrt{3}$ square units. Since $\sqrt{3}$ is irrational and cannot be written as a terminating decimal, the area can be written in decimal form but only as an approximation. Since $\sqrt{3} \doteq 1.732\,05$, the area of the triangle is approximately 1.732 square units to 3 decimal places.

Consider the diagram in *Example 1b*. Recall that each angle in an equilateral triangle is 60°. By symmetry, PS bisects ∠P. Hence, the angles in ΔPQS are 30°, 60°, and 90°. The diagram shows that the sides of a triangle with these angles are related in a certain way.

We can use these results to write a property for all triangles similar to ΔABD.

The 30-60-90 Property

In a right triangle with angles 30°, 60°, and 90°, the sides are related as follows.

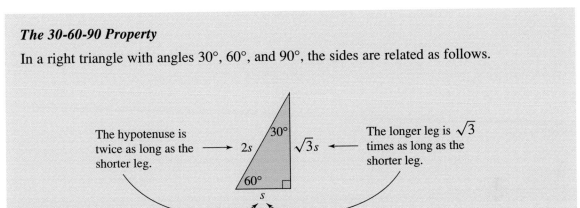

The hypotenuse is twice as long as the shorter leg.

The longer leg is $\sqrt{3}$ times as long as the shorter leg.

Example 2

An equilateral triangle is inscribed in a circle with radius 10.0 cm. Calculate the area of the triangle. Express the area in exact form, and in decimal form to the nearest tenth of a square centimetre.

Solution

Draw a diagram, where O is the centre of the circle, and OD ⊥ BC.

Since the diagram is symmetrical, O lies on AD and the radii from O bisect the angles of the triangle.

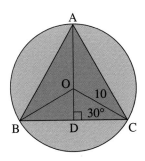

Hence, ∠OCD = 30°, and ΔOCD is a 30-60-90 triangle with hypotenuse 10 cm.

By the 30-60-90 property, OC is twice as long as OD. That is, OD is half as long as OC.

Hence, OD = 5 cm

By the 30-60-90 property, DC is $\sqrt{3}$ times as long as OD.

Hence, DC = $5\sqrt{3}$ cm

BC = 2DC

    = $10\sqrt{3}$ cm

AD = AO + OD

    = 10 cm + 5 cm

    = 15 cm

The area of △ABC is $\frac{1}{2} \times$ BC $\times$ AD = $\frac{1}{2} \times 10\sqrt{3} \times 15$

$$= 75\sqrt{3}$$
$$\doteq 75 \times 1.732\ 05$$
$$\doteq 129.904$$

The area of the equilateral triangle is $75\sqrt{3}$ cm², or approximately 129.9 cm².

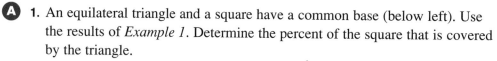

DISCUSSING THE IDEAS

1. In *Example 1*, suppose the triangle had sides 1 unit long in part a and *s* units long in part b. What would the results have been?

2. How do the lengths of the sides of a 45-45-90 triangle compare? Describe a 45-45-90 property that is similar to the 30-60-90 property.

2.6 EXERCISES

A 1. An equilateral triangle and a square have a common base (below left). Use the results of *Example 1*. Determine the percent of the square that is covered by the triangle.

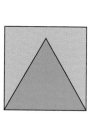

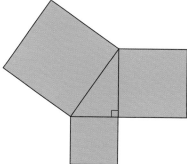

2. Squares are drawn on the sides of a 30-60-90 triangle (above right). How do the areas of the squares compare?

B **3.** Determine each value of x and y to 2 decimal places, where necessary.

a)

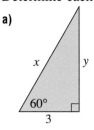

60°
3

b)

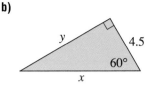

y
4.5
60°
x

c)

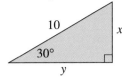

10
x
30°
y

d)

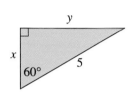

y
x
60°
5

e)

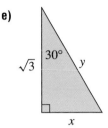

30°
$\sqrt{3}$ y
x

f)

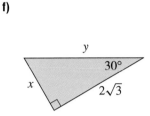

y
30°
x
$2\sqrt{3}$

4. Choose one part of exercise 3. Write to explain how you determined x and y.

5. How could you check your answers in exercise 3?

6. Determine each value of x to 2 decimal places, where necessary.

a)

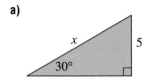

x
5
30°

b)

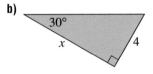

30°
x
4

c)

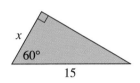

x
60°
15

d)

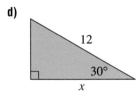

12
30°
x

e)

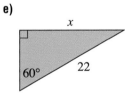

x
60°   22

f)

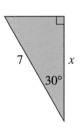

7   x
30°

7. Each measurement is the length of the shorter leg of a 30-60-90 triangle. For each shorter leg, write the lengths of the longer leg and the hypotenuse. Write the lengths in radical form, if necessary.

a) 1 cm   b) 5 cm   c) 3 cm   d) 9 cm   e) 2 cm   f) 10 cm

8. Suppose each measurement in exercise 7 is the length of the hypotenuse of a 30-60-90 triangle. For each hypotenuse, write the lengths of the legs of the triangle. Write the lengths in radical form, if necessary.

9. Each measurement is the length of the longer leg of a 30-60-90 triangle. For each longer leg, write the lengths of the shorter leg and the hypotenuse.

a) $\sqrt{3}$ cm   b) $2\sqrt{3}$ cm   c) $4\sqrt{3}$ cm   d) $\dfrac{\sqrt{3}}{4}$ cm   e) $\dfrac{\sqrt{3}}{6}$ cm

10. Choose one part of exercise 9. Write to explain how you determined the lengths of the other two sides of the triangle.

11. In the diagram (below left), the smaller circle has a radius of 3 cm.

   a) Determine the radius of the larger circle.

   b) Determine the side length of the equilateral triangle.

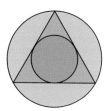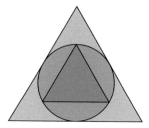

12. In the diagram (above right), the radius of the circle is 5 cm. Determine the lengths of the sides of both equilateral triangles.

13. An equilateral triangle is inscribed in a circle with radius 8.0 cm. Calculate the area of the triangle. Express the area in exact form, and in decimal form to the nearest tenth of a square centimetre.

14. Write to explain how you calculated the area in exercise 13.

15. Three cylindrical logs with radius 10 cm are piled as shown (below left). Determine the height of the top of the pile above the ground. Express your answer in exact form, and in decimal form to 3 decimal places.

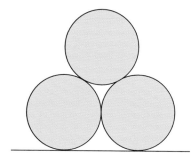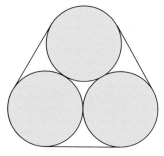

16. The logs in exercise 15 are strapped together as shown (above right). Determine the length of strapping required if 5 cm is needed for overlapping. Express your answer to the nearest centimetre.

17. An equilateral triangle is inscribed in a circle. The area of the circle is 64π. Determine the area of the triangle. Express your answer in exact form, and in decimal form to 3 decimal places.

18. An equilateral triangle has area 24 cm². Determine its side length and height. Express your answers in decimal form to 3 decimal places.

19. The small square has side length 6 cm. Calculate the side length and the area of the large square. Express your answers in exact form, and in decimal form to 3 decimal places, where necessary.

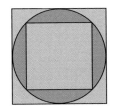

C **20.** The honeybee is the only insect that produces food for humans. Its honey is stored in a honeycomb with hexagonal cells. Each cell has sides approximately 2.9 mm long. A standard honeycomb measures 43.8 cm by 20.6 cm. Approximately how many cells does it contain? Remember a comb has two sides.

MODELLING the Storage of Honey in a Comb

When bees fill a comb with honey, they rarely do a perfect job. They will fill between 90% and 95% of the cells in a comb with honey. A cell is about 1.1 cm high.

- Use the information from exercise 20 to calculate the volume of honey in a comb.
- The model in exercise 20 ignores the thickness of the wall of the cell, which is about 0.1 mm. Suppose you consider this thickness in your calculations for the number of cells and the volume of honey in a comb. What are better estimates for the number of cells and the volume of honey?

COMMUNICATING THE IDEAS

Suppose you know the length of one leg of a 45-45-90 triangle. In your journal, write a set of instructions for someone to calculate the lengths of the other two sides of this triangle. Use a diagram to illustrate your instructions.

Suppose you know the length of the shorter leg of a 30-60-90 triangle. In your journal, write a set of instructions for someone to calculate the lengths of the other two sides of this triangle. Use a diagram to illustrate your instructions.

Squaring the Circle

1

As far back as 1800 BC, the ancient Egyptians were interested in the following problem. First they drew a circle. Then they wanted to draw a square that had the same area as the circle. The Egyptians thought they had solved the problem by drawing the side of the square as $\frac{8}{9}$ of the diameter of the circle.

3

The ancient Greeks tried to "square the circle." First, they drew a circle. Using only a straightedge and compasses, they tried to construct a square that had the same area as the circle. No matter how hard they tried, they could not find a way to do this.

2

In the 5th century BC, the ancient Greeks established rules for constructing geometric figures. Since they considered the line and the circle to be the basic figures, only compasses and an unmarked straightedge could be used.

4

For many centuries, mathematicians and non-mathematicians tried in vain to "square the circle" using a straightedge and compasses.

Finally, it was proved in 1882 that the construction is *impossible*. It cannot be done using only compasses and a straightedge. A problem that originated almost 4000 years ago was finally solved, but in an unexpected way.

Although it is impossible to "square the circle" using a straightedge and compasses, you can use your calculator to determine the approximate length of the side of a square that has the same area as any given circle.

Solve each problem.

1. This circle has a radius of 2.5 cm.

   a) Calculate the area of the circle.

   b) Visualize a square that has the same area as the circle. Calculate the side length of the square.

2. A circle has radius r centimetres. Determine an expression in terms of r for each measurement.

   a) the area of the square that has the same area as the circle

   b) the side length of the square that has the same area as the circle

3. The coloured figure in the diagram is made up of arcs of circles with radii r centimetres. Determine an expression in terms of r for each area.

   a) the area of the enclosing square

   b) the area of the coloured figure

4. This tessellation is based on a grid of circles whose centres are 1 cm apart.

   a) Visualize the circles that were used to create the tessellation.

   b) Determine the area of each figure in the tessellation.

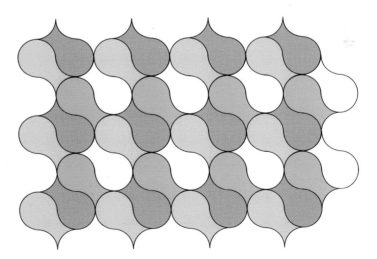

5. The ancient Egyptians constructed a square whose sides were $\frac{8}{9}$ as long as the diameter of a given circle. Calculate to determine if the area of the square they constructed was smaller than, equal to, or greater than the area of the circle.

2.7 Multiplying Radicals

Earlier in your study of mathematics you encountered fractions, decimals, integers, and rational numbers. You learned how to add, subtract, multiply, and divide each type of number. In the remainder of this chapter we will consider how to carry out these operations with radicals. We begin with multiplication because this is the simplest operation for radicals.

INVESTIGATE

1. Does $\sqrt{4} \times \sqrt{9} = \sqrt{4 \times 9}$? Explain.

2. a) Use your calculator. Determine $\sqrt{3}$, $\sqrt{5}$, and $\sqrt{3 \times 5}$. Write down the results.

   b) In your calculator, enter the decimal approximation of $\sqrt{3}$. Multiply it by the decimal approximation of $\sqrt{5}$. Does $\sqrt{3} \times \sqrt{5} = \sqrt{3 \times 5}$?

3. Make up some other examples similar to those in exercises 1 and 2.

4. Make a prediction about the expressions $\sqrt{a} \times \sqrt{b}$ and $\sqrt{a \times b}$. Use other examples to test your prediction.

To explain why $\sqrt{3} \times \sqrt{5} = \sqrt{3 \times 5}$, square each expression.

Left side:
$$\begin{aligned} (\sqrt{3} \times \sqrt{5})^2 &= \sqrt{3} \times \sqrt{5} \times \sqrt{3} \times \sqrt{5} \\ &= \sqrt{3} \times \sqrt{3} \times \sqrt{5} \times \sqrt{5} \\ &= 3 \times 5 \\ &= 15 \end{aligned}$$

Right side:
$$\begin{aligned} (\sqrt{3 \times 5})^2 &= \sqrt{3 \times 5} \times \sqrt{3 \times 5} \\ &= 3 \times 5 \\ &= 15 \end{aligned}$$

Since the results are the same, we can conclude that $\sqrt{3} \times \sqrt{5} = \sqrt{3 \times 5}$. The following property can be justified in the same way.

Multiplication Property

$$\sqrt{a} \times \sqrt{b} = \sqrt{ab} \quad (a \geq 0, b \geq 0)$$

When we multiply radicals, we frequently encounter expressions such as $3\sqrt{2}$, which means $3 \times \sqrt{2}$, just as $3n$ means $3 \times n$.

Example 1

Consider the expression $3\sqrt{2} \times 4\sqrt{5}$.

a) Determine a decimal approximation to 3 decimal places.

b) Determine an equivalent form of the expression.

c) Use the expression in part b. Determine a decimal approximation.

Solution

a) $3\sqrt{2} \times 4\sqrt{5} \doteq 3 \times 1.4142 \times 4 \times 2.2361$
$$\doteq 37.948$$

b) Do not use a calculator.

$3 \times \sqrt{2} \times 4 \times \sqrt{5} = 3 \times 4 \times \sqrt{2} \times \sqrt{5}$ **Use the Multiplication Property.**
$$= 12 \times \sqrt{10}$$
$$= 12\sqrt{10}$$

c) $12\sqrt{10} \doteq 12 \times 3.162\ 28$
$$\doteq 37.947$$

In *Example 1*, fewer keystrokes are required to calculate a decimal approximation of the given expression for part c than for part a. From this example, you can see that you can reduce the likelihood of making errors by simplifying an expression before calculating.

Example 2

Express as a product of radicals without using a calculator.

a) $\sqrt{21}$ b) $\sqrt{30}$ c) $\sqrt{20}$

Solution

a) $\sqrt{21} = \sqrt{7 \times 3}$ b) $\sqrt{30} = \sqrt{6 \times 5}$ c) $\sqrt{20} = \sqrt{4 \times 5}$
        $= \sqrt{7} \times \sqrt{3}$ $= \sqrt{6} \times \sqrt{5}$ $= \sqrt{4} \times \sqrt{5}$

In *Example 2c*, since $\sqrt{4} = 2$, we can write $\sqrt{20} = \sqrt{4} \times \sqrt{5} = 2\sqrt{5}$.

An expression of the form $\sqrt{x}$ is called an *entire radical*.

An expression of the form $a\sqrt{x}$ is called a *mixed radical*.

$$\sqrt{20} = 2\sqrt{5}$$

entire radical   mixed radical

To determine if a radical can be expressed as a mixed radical, look for a factor that is a perfect square.

Example 3

Express as a mixed radical, if possible, without using a calculator.

a) $\sqrt{45}$ b) $\sqrt{70}$ c) $\sqrt{72}$

Solution

a) 45 has 9 as a perfect-square factor.

$$\sqrt{45} = \sqrt{9} \times \sqrt{5}$$
$$= 3\sqrt{5}$$

b) Since 70 does not have a perfect-square factor, $\sqrt{70}$ cannot be expressed as a mixed radical.

c) 72 has 36 as a perfect-square factor. $\sqrt{72} = \sqrt{36} \times \sqrt{2}$
$$= 6\sqrt{2}$$

In *Example 3c*, since 72 has more than one perfect-square factor, $\sqrt{72}$ can be written in these ways:

$$\sqrt{72} = \sqrt{4 \times 18} \qquad \sqrt{72} = \sqrt{9 \times 8} \qquad \sqrt{72} = \sqrt{36 \times 2}$$
$$= \sqrt{4} \times \sqrt{18} \qquad\quad = \sqrt{9} \times \sqrt{8} \qquad\quad = \sqrt{36} \times \sqrt{2}$$
$$= 2\sqrt{18} \qquad\qquad\quad = 3\sqrt{8} \qquad\qquad\quad = 6\sqrt{2}$$

All three results are different ways of writing the same radical, $\sqrt{72}$.

We say that $6\sqrt{2}$ is in *simplest form* because the radical contains no perfect-square factor.

The radicals in *Example 3* involve square roots. You can express a radical involving a cube root as a mixed radical in a similar way. Instead of looking for a factor that is a perfect square, look for a factor that is a perfect cube.

Example 4

Express $\sqrt[3]{40}$ as a mixed radical without using a calculator.

Solution

Observe that $40 = 8 \times 5$, and 8 is a perfect cube since $2^3 = 8$.

$$\sqrt[3]{40} = \sqrt[3]{8 \times 5}$$
$$= \sqrt[3]{8} \times \sqrt[3]{5}$$
$$= 2\sqrt[3]{5}$$

DISCUSSING THE IDEAS

1. Describe in words the property $\sqrt{a} \times \sqrt{b} = \sqrt{ab}$.

2. Following *Example 2*, we wrote $\sqrt{20}$ as a mixed radical, $2\sqrt{5}$. Could we have written the other radicals in *Example 2* as mixed radicals? Explain your answer.

3. Which of these radicals do *you* think is the simplest: $\sqrt{72}$, $2\sqrt{18}$, $3\sqrt{8}$, or $6\sqrt{2}$? Justify your choice.

4. a) Can every mixed radical be expressed as an entire radical?

   b) Can every entire radical be expressed as a mixed radical?
      Illustrate your answers with examples.

5. How does expressing radicals as mixed radicals compare with expressing fractions in lowest terms?

2.7 EXERCISES

A 1. Multiply without using a calculator.

   a) $\sqrt{7} \times \sqrt{8}$
   b) $\sqrt{11} \times \sqrt{14}$
   c) $\sqrt{8} \times (-\sqrt{18})$

   d) $2\sqrt{5} \times \sqrt{2}$
   e) $-\sqrt{2} \times 3\sqrt{8}$
   f) $(-7\sqrt{3})(-5\sqrt{8})$

2. Express as a product of two radicals without using a calculator.

   a) $\sqrt{24}$
   b) $\sqrt{18}$
   c) $\sqrt{45}$
   d) $\sqrt{28}$
   e) $\sqrt{72}$
   f) $\sqrt{60}$

   g) $\sqrt{39}$
   h) $\sqrt{65}$
   i) $\sqrt{96}$
   j) $\sqrt{120}$
   k) $\sqrt{126}$
   l) $\sqrt{105}$

3. Express as a mixed radical in simplest form without using a calculator.

   a) $\sqrt{32}$
   b) $\sqrt{50}$
   c) $\sqrt{27}$
   d) $\sqrt{96}$
   e) $\sqrt{8}$
   f) $\sqrt{75}$

   g) $\sqrt{63}$
   h) $\sqrt{54}$
   i) $\sqrt{76}$
   j) $\sqrt{80}$
   k) $\sqrt{88}$
   l) $\sqrt{200}$

4. Write a set of instructions for another student to express an entire radical as a mixed radical.

5. Express as an entire radical without using a calculator.

a) $3\sqrt{2}$ b) $2\sqrt{3}$ c) $2\sqrt{5}$ d) $5\sqrt{2}$ e) $3\sqrt{5}$ f) $5\sqrt{3}$

g) $4\sqrt{2}$ h) $2\sqrt{4}$ i) $3\sqrt{4}$ j) $4\sqrt{3}$ k) $4\sqrt{5}$ l) $5\sqrt{4}$

6. Choose one part of exercise 5. Write to explain how you wrote the mixed radical as an entire radical.

7. Express as a mixed radical in simplest form without using a calculator.

a) $\sqrt[3]{16}$ b) $\sqrt[3]{54}$ c) $\sqrt[3]{24}$ d) $\sqrt[3]{375}$ e) $\sqrt[3]{108}$ f) $\sqrt[3]{250}$

g) $\sqrt[3]{192}$ h) $\sqrt[3]{128}$ i) $\sqrt[3]{32}$ j) $\sqrt[3]{81}$ k) $\sqrt[3]{40}$ l) $\sqrt[3]{135}$

B 8. a) In the diagram, each dot is a vertex of a square. Use the diagram to show that $\sqrt{8} = 2\sqrt{2}$.

b) Draw a similar diagram to show that $\sqrt{18} = 3\sqrt{2}$.

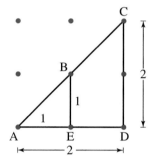

9. Use your calculator. Determine the decimal approximation to 4 decimal places for each radical in each list. What patterns can you find in the results? For each pattern, explain why you think it occurs. Then predict the next two lines in each list.

a) $\sqrt{6}$
$\sqrt{60}$
$\sqrt{600}$
$\sqrt{6000}$
$\vdots$

b) $\sqrt{2.5}$
$\sqrt{25}$
$\sqrt{250}$
$\sqrt{2500}$
$\vdots$

c) $\sqrt{10}$
$\sqrt{20}$
$\sqrt{40}$
$\sqrt{80}$
$\vdots$

10. For each expression below

i) Determine a decimal approximation to 3 decimal places.

ii) Determine an equivalent form of the expression.

iii) Use the expression in part ii to determine the decimal approximation.

a) $2\sqrt{3} \times \sqrt{2}$ b) $5\sqrt{7} \times 2\sqrt{5}$ c) $4\sqrt{12} \times 3\sqrt{6}$

11. Multiply without using a calculator.

a) $2\sqrt{6} \times 3\sqrt{2}$ b) $3\sqrt{5} \times 7\sqrt{10}$ c) $-8\sqrt{6} \times 6\sqrt{8}$

d) $5\sqrt{10} \times 4\sqrt{6}$ e) $(-7\sqrt{12})(-2\sqrt{6})$ f) $11\sqrt{3} \times 5\sqrt{6}$

12. Multiply without using a calculator.

   a) $12\sqrt{3} \times (-3\sqrt{18})$ b) $(-3\sqrt{5})(-5\sqrt{3})$ c) $-7\sqrt{\frac{6}{35}} \times 2\sqrt{\frac{5}{9}}$

   d) $-3\sqrt{\frac{6}{15}} \times \sqrt{\frac{10}{9}}$ e) $-5\sqrt{0.3} \times 2\sqrt{0.7}$ f) $4\sqrt{9} \times 11\sqrt{0.4}$

13. Choose one part of exercise 12. Write to explain how you found the product.

14. Determine the simplest form of each expression without using a calculator.

   a) $\sqrt{24} \times \sqrt{18}$ b) $2\sqrt{24} \times 5\sqrt{6}$

   c) $3\sqrt{20} \times 2\sqrt{5}$ d) $2\sqrt{6} \times 7\sqrt{8} \times 5\sqrt{2}$

   e) $3\sqrt{7} \times 2\sqrt{6} \times 5\sqrt{2}$ f) $4\sqrt{8} \times 3\sqrt{6} \times 7\sqrt{3}$

15. Write each radical as the product of two mixed radicals without using a calculator.

   a) $6\sqrt{15}$ b) $8\sqrt{42}$ c) $30\sqrt{5}$ d) $12\sqrt{10}$ e) $24\sqrt{3}$ f) $10\sqrt{21}$

16. Write each number as the product of two mixed radicals without using a calculator.

   a) 18   b) 24   c) 12   d) 60   e) 30   f) 21

17. Compare your answers to exercises 15 and 16 with those of a classmate. Were your answers different? If your answer was yes, did one of you calculate incorrectly? Explain.

18. Show that each sequence is a geometric sequence. Then determine the next two terms.

   a) $\sqrt{3}, \sqrt{12}, \sqrt{48}, \ldots$ b) $\sqrt{6}, \sqrt{12}, \sqrt{24}, \ldots$

   c) $\sqrt{2}, \sqrt{6}, \sqrt{18}, \ldots$ d) $\sqrt{5}, \sqrt{50}, \sqrt{500}, \ldots$

19. Arrange in order from least to greatest.

   a) $7\sqrt{2}, 3\sqrt{7}, 2\sqrt{15}, 4\sqrt{6}$

   b) $\sqrt{43}, 4\sqrt{3}, 5\sqrt{2}, 2\sqrt{10}, 2\sqrt{13}$

   c) $6\sqrt{2}, 3\sqrt{7}, 2\sqrt{17}, 4\sqrt{5}, 2\sqrt{21}$

20. Choose one part of exercise 19. Write to explain how you ordered the numbers.

21. Arrange in order from least to greatest. Use a method different from the one you used in exercise 19.

   a) $5, 4\sqrt{2}, 2\sqrt{6}, 3\sqrt{3}$

   b) $3\sqrt{5}, \sqrt{31}, 2\sqrt{7}, 4\sqrt{2}$

   c) $4\sqrt{5}, 5\sqrt{3}, 2\sqrt{19}, 6\sqrt{2}, 3\sqrt{10}$

22. Suppose some radicals are given. Write a set of instructions for another student to arrange them in order from greatest to least.

23. The largest square in this diagram has side length 6 cm. Calculate the side length and the area of each of the two smaller squares. Try to do this in more than one way. Express your answers in exact form, and in decimal form to 3 decimal places.

24. The diagram in exercise 23 is part of this pattern. Visualize how the pattern can be extended both outward and inward with more squares.

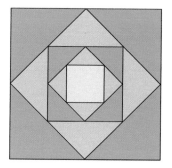

a) Copy and complete a table for the squares in the pattern. Begin with the smallest square. Express the results in exact form.

Side length (cm)	Area (cm²)
6	

b) Find as many patterns in the table as you can.

c) Extend your table by two more rows.

C 25. Given that $\sqrt{3} \doteq 1.7321$, determine a decimal approximation for each radical, without using a calculator.

a) $\sqrt{300}$, $\sqrt{30\ 000}$, $\sqrt{3\ 000\ 000}$

b) $\sqrt{0.03}$, $\sqrt{0.0003}$

c) $\sqrt{12}$, $\sqrt{27}$, $\sqrt{48}$, $\sqrt{75}$

d) $\sqrt{\frac{3}{4}}$, $\sqrt{\frac{1}{3}}$

26. In the diagram, what is the area of square WXYZ?

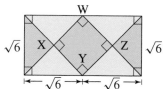

COMMUNICATING THE IDEAS

Suppose you are given a radical of the form $\sqrt{a}$, where a is a positive integer. How can you tell whether it can be written in the form $b\sqrt{c}$, where b and c are positive integers? Write an explanation in your journal. Support your explanation with an example that can be written in the form $b\sqrt{c}$ and with an example that cannot be written in that form.

2.8 Dividing Radicals

1. Simplify $\dfrac{\sqrt{36}}{\sqrt{4}}$, $\sqrt{\dfrac{36}{4}}$, and $\sqrt{9}$. Does $\dfrac{\sqrt{36}}{\sqrt{4}} = \sqrt{9}$?

2. Use your calculator to simplify $\dfrac{\sqrt{21}}{\sqrt{3}}$, $\sqrt{\dfrac{21}{3}}$, and $\sqrt{7}$. Does $\dfrac{\sqrt{21}}{\sqrt{3}} = \sqrt{7}$?

3. Make up some other examples similar to those in exercises 1 and 2.

4. Use the above examples as a guide. Make a prediction about the expressions $\dfrac{\sqrt{a}}{\sqrt{b}}$ and $\sqrt{\dfrac{a}{b}}$. Test your prediction with other examples.

We can explain why $\dfrac{\sqrt{21}}{\sqrt{3}} = \sqrt{7}$ because division is the inverse of multiplication.

We know that $\dfrac{56}{8} = 7$ because $7 \times 8 = 56$.

Similarly, $\dfrac{\sqrt{21}}{\sqrt{3}} = \sqrt{7}$ because $\sqrt{7} \times \sqrt{3} = \sqrt{21}$.

This suggests the following property, which can be justified in the same way.

Division Property

$$\frac{\sqrt{a}}{\sqrt{b}} = \sqrt{\frac{a}{b}} \quad (a \geq 0,\ b > 0)$$

Example 1

Determine an equivalent form for each expression without using a calculator.

a) $\dfrac{2\sqrt{15}}{\sqrt{3}}$

b) $\dfrac{\sqrt{36}}{\sqrt{2}}$

c) $\dfrac{9\sqrt{24}}{2\sqrt{18}}$

Solution

a)
$$\frac{2\sqrt{15}}{\sqrt{3}}$$
$$= 2\sqrt{\frac{15}{3}}$$
$$= 2\sqrt{5}$$

b)
$$\frac{\sqrt{36}}{\sqrt{2}} = \sqrt{\frac{36}{2}}$$
$$= \sqrt{18}$$
$$= 3\sqrt{2}$$

or $\frac{\sqrt{36}}{\sqrt{2}} = \frac{6}{\sqrt{2}}$

c)
$$\frac{9\sqrt{24}}{2\sqrt{18}}$$
$$= \frac{9 \times \sqrt{4} \times \sqrt{6}}{2 \times \sqrt{9} \times \sqrt{2}}$$
$$= \frac{9 \times 2 \times \sqrt{6}}{2 \times 3 \times \sqrt{2}}$$
$$= \frac{3\sqrt{6}}{\sqrt{2}}$$
$$= 3\sqrt{\frac{6}{2}}$$
$$= 3\sqrt{3}$$

The answers to *Example 1b*, found by two different methods, must be equal. You can show they are by multiplying the numerator and the denominator of $\frac{6}{\sqrt{2}}$ by $\sqrt{2}$.

$$\frac{6}{\sqrt{2}} = \frac{6}{\sqrt{2}} \times \frac{\sqrt{2}}{\sqrt{2}}$$
$$= \frac{6 \times \sqrt{2}}{\sqrt{2} \times \sqrt{2}}$$
$$= \frac{6\sqrt{2}}{2}$$
$$= 3\sqrt{2}$$

This procedure is called *rationalizing the denominator*. That is, we write the denominator as a rational number, to replace the irrational number.

Example 2

Consider the expression $\frac{12}{\sqrt{3}}$.

a) Determine a decimal approximation to 3 decimal places.

b) Determine an equivalent form for the expression that does not have a radical in the denominator.

c) Use the expression in part b to determine a decimal approximation.

Solution

a) $\frac{12}{\sqrt{3}} \doteq \frac{12}{1.7321}$
$\doteq 6.928$

b) $\frac{12}{\sqrt{3}} = \frac{12}{\sqrt{3}} \times \frac{\sqrt{3}}{\sqrt{3}}$
$= \frac{12\sqrt{3}}{3}$
$= 4\sqrt{3}$

c) $4\sqrt{3} \doteq 4 \times 1.7321$
$\doteq 6.928$

Consider how the expression $\frac{12}{\sqrt{3}}$ was evaluated in *Example 2*. Before calculators were developed, a table of square roots and paper-pencil computation were used to evaluate such an expression. This meant that it was necessary to evaluate $\frac{12}{1.7321}$ by hand. To avoid the tedious long division, the expression was converted to the equivalent form $4\sqrt{3}$ by rationalizing the denominator. It was much easier to evaluate 4×1.7321. With a calculator, it makes no difference which form is used, as *Example 1* illustrates.

Rationalizing denominators has declined in importance as a computational tool. However, it is still important because it helps us understand that an expression involving radicals can have different forms. That is, $\frac{12}{\sqrt{3}}$ and $4\sqrt{3}$ represent the same number.

Example 3

Determine an equivalent form for each expression by rationalizing the denominator.

a) $\dfrac{5}{\sqrt{5}}$

b) $\dfrac{3}{\sqrt{18}}$

Solution

a) $\dfrac{5}{\sqrt{5}} = \dfrac{5}{\sqrt{5}} \times \dfrac{\sqrt{5}}{\sqrt{5}}$

$= \dfrac{5\sqrt{5}}{5}$

$= \sqrt{5}$

b) $\dfrac{3}{\sqrt{18}} = \dfrac{3}{\sqrt{18}} \times \dfrac{\sqrt{2}}{\sqrt{2}}$

$= \dfrac{3\sqrt{2}}{\sqrt{36}}$

$= \dfrac{3\sqrt{2}}{6}$

$= \dfrac{\sqrt{2}}{2}$

Example 4

A circle is inscribed in an equilateral triangle with sides 12.0 cm long. Calculate the circumference and the area of the circle. Express the results in exact form, and in decimal form to the nearest tenth.

Solution

Draw a diagram, where O is the centre of the circle, and OD ⊥ BC.

By symmetry, observe that ΔOBD is a 30-60-90 triangle.

According to the 30-60-90 property, BD is $\sqrt{3}$ times as long as OD.

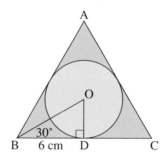

$$BD = \sqrt{3} \times OD$$
$$6 = \sqrt{3} \times OD$$
$$OD = \frac{6}{\sqrt{3}}$$
$$= \frac{6}{\sqrt{3}} \times \frac{\sqrt{3}}{\sqrt{3}}$$
$$= \frac{6\sqrt{3}}{3}$$
$$= 2\sqrt{3}$$

Hence, the radius of the circle is $2\sqrt{3}$ cm.

The circumference is $C = 2\pi r$

$$= 2\pi(2\sqrt{3})$$
$$= 4\pi\sqrt{3}$$
$$\doteq 4\pi \times 1.7321$$
$$\doteq 21.8$$

The area is $A = \pi r^2$

$$= \pi(2\sqrt{3})^2$$
$$= \pi \times 2^2 \times (\sqrt{3})^2$$
$$= \pi \times 4 \times 3$$
$$= 12\pi$$
$$\doteq 37.7$$

The circumference is $4\pi\sqrt{3}$ cm, or approximately 21.8 cm.

The area is 12π cm², or approximately 37.7 cm².

DISCUSSING THE IDEAS

1. Describe in words the property $\frac{\sqrt{a}}{\sqrt{b}} = \sqrt{\frac{a}{b}}$.

2. In *Example 1b*, we showed that the two answers were equal by multiplying $\frac{6}{\sqrt{2}}$ by $\frac{\sqrt{2}}{\sqrt{2}}$.

   What do you think would happen if you multiplied the other answer, $3\sqrt{2}$, by $\frac{\sqrt{2}}{\sqrt{2}}$? Try it to see if your prediction is correct.

3. Explain the result in *Example 3a* using the meaning of $\sqrt{5}$.

4. In *Example 3b*, how did we know to multiply by $\frac{\sqrt{2}}{\sqrt{2}}$? By what other expression could we have multiplied $\frac{3}{\sqrt{18}}$? Try it to see if you get the same result.

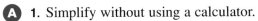

A **1.** Simplify without using a calculator.

a) $\dfrac{\sqrt{24}}{\sqrt{3}}$ b) $\dfrac{\sqrt{56}}{\sqrt{8}}$ c) $\dfrac{\sqrt{72}}{\sqrt{6}}$ d) $\dfrac{3\sqrt{35}}{\sqrt{7}}$ e) $\dfrac{6\sqrt{18}}{2\sqrt{6}}$

f) $\dfrac{5\sqrt{30}}{2\sqrt{15}}$ g) $\dfrac{4\sqrt{20}}{2\sqrt{5}}$ h) $\dfrac{3\sqrt{12}}{6\sqrt{3}}$ i) $\dfrac{\sqrt{24}}{7\sqrt{12}}$ j) $\dfrac{\sqrt{8}}{6\sqrt{18}}$

2. Determine an equivalent form for each expression without using a calculator.

a) $\dfrac{3\sqrt{20}}{4\sqrt{5}}$ b) $\dfrac{\sqrt{12}}{2\sqrt{6}}$ c) $\dfrac{\sqrt{18}}{3\sqrt{3}}$ d) $\dfrac{\sqrt{21}}{5\sqrt{3}}$ e) $\dfrac{\sqrt{8}}{2\sqrt{24}}$

f) $\dfrac{3\sqrt{30}}{\sqrt{5}}$ g) $\dfrac{\sqrt{7}}{4\sqrt{35}}$ h) $\dfrac{6\sqrt{8}}{7\sqrt{2}}$ i) $\dfrac{4\sqrt{27}}{9\sqrt{32}}$ j) $\dfrac{5\sqrt{18}}{3\sqrt{50}}$

B **3.** For each expression below

  i) Determine a decimal approximation to 3 decimal places.

  ii) Determine an equivalent form that does not have a radical in its denominator.

  iii) Use the expression in part ii to determine a decimal approximation.

a) $\dfrac{\sqrt{40}}{\sqrt{8}}$ b) $\dfrac{10}{\sqrt{2}}$ c) $\dfrac{3}{\sqrt{6}}$ d) $\dfrac{2\sqrt{5}}{\sqrt{10}}$ e) $\dfrac{2\sqrt{12}}{3\sqrt{8}}$

4. Show that each sequence is a geometric sequence, then determine the next two terms.

a) $\dfrac{1}{\sqrt{5}}, \dfrac{1}{\sqrt{20}}, \dfrac{1}{\sqrt{80}}, \dots$ b) $\dfrac{2}{\sqrt{2}}, \dfrac{2}{\sqrt{18}}, \dfrac{2}{\sqrt{162}}, \dots$

5. Choose one part of exercise 4. Write to explain how you completed it.

6. For each expression, determine an equivalent form that does not have a radical in its denominator.

a) $\dfrac{1}{\sqrt{6}}$ b) $\dfrac{3}{\sqrt{5}}$ c) $\dfrac{12}{\sqrt{18}}$ d) $\dfrac{10}{\sqrt{5}}$

e) $\dfrac{6}{\sqrt{12}}$ f) $\dfrac{3}{2\sqrt{7}}$ g) $\dfrac{3\sqrt{20}}{4\sqrt{12}}$ h) $\dfrac{8\sqrt{18}}{3\sqrt{75}}$

7. Each mixed radical below is the quotient after the division of two mixed radicals. In each case, write what those two mixed radicals could be.

a) $3\sqrt{5}$ b) $4\sqrt{2}$ c) $6\sqrt{7}$ d) $2\sqrt{10}$ e) $5\sqrt{6}$ f) $2\sqrt{2}$

Compare your answers with those of a classmate. Were your answers different? If they were, did one of you calculate incorrectly? Explain.

8. Choose one part of exercise 7. Write to explain how you determined the two radicals.

9. Determine x in exact form, and to 2 decimal places.

a)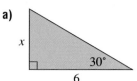

b)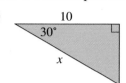

c)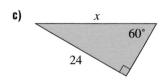

10. In the diagram (below left), the length of each side of the triangle is 4.0 cm. Determine the radii of the two circles.

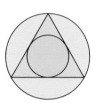
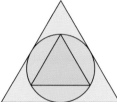

11. In the diagram (above right), the length of each side of the larger triangle is 12.0 cm.

a) Determine the radius of the circle.

b) Determine the length of each side of the smaller triangle.

12. An equilateral triangle has an area of 30 cm². Determine the length of its base and its height to 2 decimal places.

13. Determine an equivalent form for each expression.

a) $\dfrac{3\sqrt{48}}{2\sqrt{27}}$

b) $\dfrac{6\sqrt{50}}{5\sqrt{18}}$

c) $\dfrac{4\sqrt{54}}{3\sqrt{12}}$

d) $\dfrac{3\sqrt{20}}{2\sqrt{10}}$

e) $\dfrac{5\sqrt{24}}{2\sqrt{18}}$

f) $\dfrac{7\sqrt{32}}{5\sqrt{63}}$

g) $\dfrac{10\sqrt{27}}{3\sqrt{20}}$

h) $\dfrac{16\sqrt{24}}{4\sqrt{96}}$

i) $\dfrac{4\sqrt{45}}{3\sqrt{54}}$

j) $\dfrac{3\sqrt{60}}{2\sqrt{27}}$

C **14.** Is the quotient of two irrational numbers always an irrational number? Give examples to support your answer.

15. Determine if it is possible to find positive integers a and b to satisfy each condition. Explain your answer.

a) Both $\sqrt{a} \times \sqrt{b}$ and $\dfrac{\sqrt{a}}{\sqrt{b}}$ are rational.

b) Both $\sqrt{a} \times \sqrt{b}$ and $\dfrac{\sqrt{a}}{\sqrt{b}}$ are irrational.

c) Only one of $\sqrt{a} \times \sqrt{b}$ and $\dfrac{\sqrt{a}}{\sqrt{b}}$ is rational.

COMMUNICATING THE IDEAS

Suppose your friend telephones you to ask about tonight's homework. How would you explain, over the telephone, what "rationalizing the denominator" means?

INVESTIGATE

1. Does $\sqrt{9} + \sqrt{16} = \sqrt{25}$? Explain.

2. a) Use your calculator. Determine $\sqrt{3}$, $\sqrt{5}$, and $\sqrt{8}$.
   b) Does $\sqrt{3} + \sqrt{5} = \sqrt{8}$?

3. Based on the examples in exercises 1 and 2, what can you conclude about $\sqrt{a} + \sqrt{b}$ and $\sqrt{a+b}$?

4. The dots in the diagrams below are one unit apart horizontally and vertically.
   a) In the first diagram, AB + BC = AC. Write a radical for each of these three lengths. Use the results to write an addition statement involving radicals.
   b) Do the same for the second diagram, using DE + EF = DF.
   c) Do the same for the third diagram, using GH + HI = GI.

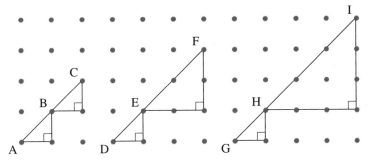

5. Suppose the pattern in exercise 4 is extended to show another diagram. Write the corresponding addition statement.

6. Summarize your results from exercises 4 and 5 in tables. In each case, write a single radical for the expression. Extend the pattern for two more rows.

Expression	Radical
$\sqrt{2} + \sqrt{2}$	
$\sqrt{2} + 2\sqrt{2}$	
$\sqrt{2} + 3\sqrt{2}$	
$\sqrt{2} + 4\sqrt{2}$	
⋮	

Expression	Radical
$\sqrt{2} + \sqrt{8}$	
$\sqrt{2} + \sqrt{18}$	
$\sqrt{2} + \sqrt{32}$	
⋮	

7. Use your calculator to confirm some of the above results.

Just as x and $2x$ are like terms, radicals such as $\sqrt{2}$ and $2\sqrt{2}$ are *like radicals*.

In *Investigate*, you should have discovered that $\sqrt{2} + 2\sqrt{2} = 3\sqrt{2}$.

Sometimes, radicals that appear to be unlike radicals can be expressed in equivalent forms that are like radicals.

For example, consider $\sqrt{2} + \sqrt{8}$. We can write $\sqrt{8} = 2\sqrt{2}$.

This means that, although $\sqrt{2}$ and $\sqrt{8}$ are not like radicals, you can add them by expressing $\sqrt{8}$ in a different form.

$$\begin{aligned} \sqrt{2} + \sqrt{8} &= \sqrt{2} + 2\sqrt{2} \\ &= 3\sqrt{2} \end{aligned}$$

You can add like radicals just as you add like terms.

Radicals such as $\sqrt{3}$ and $\sqrt{5}$ are not like radicals, and they cannot be expressed as like radicals. The only way you can add them is by using decimal approximations.

Example 1

Write each expression in terms of a single radical, if possible.

a) $2\sqrt{3} + 3\sqrt{3}$ b) $6\sqrt{2} - 4\sqrt{2} + \sqrt{2}$ c) $4\sqrt{6} + 2\sqrt{10}$

Solution

a) $\quad 2\sqrt{3} + 3\sqrt{3}$
$\quad = (2 + 3)\sqrt{3}$
$\quad = 5\sqrt{3}$

b) $\quad 6\sqrt{2} - 4\sqrt{2} + \sqrt{2}$
$\quad = (6 - 4 + 1)\sqrt{2}$
$\quad = 3\sqrt{2}$

c) $4\sqrt{6}$ and $2\sqrt{10}$ cannot be added because they are not like radicals and they cannot be expressed as like radicals.

Example 2

Consider the expression $4\sqrt{18} - \sqrt{8}$.

a) Determine a decimal approximation to 3 decimal places.

b) Determine an equivalent form for the expression.

c) Use the expression in part b to determine a decimal approximation.

Solution

a) $\begin{aligned} 4\sqrt{18} - \sqrt{8} &\doteq 4 \times 4.2426 - 2.8284 \\ &\doteq 14.142 \end{aligned}$

b) $4\sqrt{18} - \sqrt{8} = 4 \times \sqrt{9 \times 2} - \sqrt{4 \times 2}$
$$= 4 \times \sqrt{9} \times \sqrt{2} - \sqrt{4} \times \sqrt{2}$$
$$= 4 \times 3\sqrt{2} - 2\sqrt{2}$$
$$= 12\sqrt{2} - 2\sqrt{2}$$
$$= 10\sqrt{2}$$

c) $10\sqrt{2} \doteq 10 \times 1.4142$
$$\doteq 14.142$$

Example 3

Determine an equivalent form for each expression.

a) $\sqrt{18} - \sqrt{2}$ **b)** $2\sqrt{98} + \sqrt{10} - 5\sqrt{8} - 3\sqrt{40}$

Solution

a) $\sqrt{18} - \sqrt{2} = \sqrt{9} \times \sqrt{2} - \sqrt{2}$
$$= 3\sqrt{2} - \sqrt{2}$$
$$= 2\sqrt{2}$$

b) $2\sqrt{98} + \sqrt{10} - 5\sqrt{8} - 3\sqrt{40} = 2 \times \sqrt{49} \times \sqrt{2} + \sqrt{10} - 5 \times \sqrt{4} \times \sqrt{2} - 3 \times \sqrt{4} \times \sqrt{10}$
$$= 2 \times 7\sqrt{2} + \sqrt{10} - 5 \times 2\sqrt{2} - 3 \times 2\sqrt{10}$$
$$= 14\sqrt{2} + \sqrt{10} - 10\sqrt{2} - 6\sqrt{10}$$
$$= 14\sqrt{2} - 10\sqrt{2} + \sqrt{10} - 6\sqrt{10}$$
$$= 4\sqrt{2} - 5\sqrt{10}$$

Radicals involving cube roots can also be added and subtracted using similar methods.

Example 4

Determine the exact value of $\sqrt[3]{16} + 5\sqrt[3]{54}$.

Solution

$\sqrt[3]{16} + 5\sqrt[3]{54} = \sqrt[3]{8 \times 2} + 5\sqrt[3]{27 \times 2}$
$$= \sqrt[3]{8} \times \sqrt[3]{2} + 5 \times \sqrt[3]{27} \times \sqrt[3]{2}$$
$$= 2 \times \sqrt[3]{2} + 5 \times 3 \times \sqrt[3]{2}$$
$$= 2\sqrt[3]{2} + 15\sqrt[3]{2}$$
$$= 17\sqrt[3]{2}$$

1. How can you tell that the expression $4\sqrt{18} - \sqrt{8}$ in *Example 2* can be simplified to an equivalent form, but the expression $4\sqrt{6} + 2\sqrt{10}$ in *Example 1c* cannot be simplified to an equivalent form?

2. In *Example 3b*, why can we not combine $4\sqrt{2}$ and $5\sqrt{10}$?

2.9 EXERCISES

A 1. Simplify.

a) $5\sqrt{7} - 3\sqrt{7}$ b) $11\sqrt{6} + 5\sqrt{6}$ c) $2\sqrt{13} - 8\sqrt{13}$

d) $6\sqrt{19} - 31\sqrt{19}$ e) $4\sqrt{3} + 29\sqrt{3}$ f) $7\sqrt{15} - 2\sqrt{15}$

2. Simplify.

a) $4\sqrt{5} - 11\sqrt{5} + 3\sqrt{5}$ b) $2\sqrt{10} + 7\sqrt{10} - 6\sqrt{10}$

c) $5\sqrt{2} - 16\sqrt{2} + 29\sqrt{2}$ d) $2\sqrt{6} - 6\sqrt{2} + 11\sqrt{6}$

e) $4\sqrt{10} - 10\sqrt{10} + 3\sqrt{5}$ f) $3\sqrt{5} - 9\sqrt{2} + 5\sqrt{5} - 2\sqrt{2}$

3. Determine an equivalent form for each expression.

a) $\sqrt{40} + \sqrt{90}$ b) $\sqrt{32} + \sqrt{8}$ c) $\sqrt{12} - \sqrt{75}$ d) $\sqrt{20} - \sqrt{45}$

e) $\sqrt{50} - \sqrt{18}$ f) $\sqrt{24} - \sqrt{96}$ g) $3\sqrt{20} - 2\sqrt{80}$ h) $\sqrt{54} + \sqrt{150}$

4. Choose one part of exercise 3. Write to explain how you found an equivalent form.

5. Determine the length of the hypotenuse in this triangle. Explain why this example shows that $\sqrt{2} + \sqrt{8} \neq \sqrt{10}$.

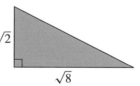

B 6. For each expression below

i) Determine a decimal approximation to 3 decimal places.

ii) Determine an equivalent form for the expression.

iii) Use the expression in part ii to determine a decimal approximation.

a) $3\sqrt{27} + 4\sqrt{3}$ b) $5\sqrt{32} - 2\sqrt{8}$ c) $10\sqrt{20} - 7\sqrt{45}$

7. Choose one expression in exercise 6. Write to explain how you completed the three parts of the exercise.

8. Determine an equivalent form for each expression.

a) $\sqrt{54} + \sqrt{150} - \sqrt{6}$　　b) $\sqrt{28} - \sqrt{63} + \sqrt{112}$

c) $\sqrt{80} + \sqrt{45} - \sqrt{125}$　　d) $\sqrt{12} + \sqrt{27} + \sqrt{48}$

e) $\sqrt{75} - \sqrt{3} + \sqrt{147}$　　f) $\sqrt{98} - \sqrt{72} - \sqrt{50}$

9. Each mixed radical below is the sum of two unlike mixed radicals. For each sum, write what the two mixed radicals could be.

a) $10\sqrt{2}$　　b) $5\sqrt{3}$　　c) $8\sqrt{7}$　　d) $6\sqrt{5}$　　e) $10\sqrt{11}$　　f) $9\sqrt{6}$

10. Suppose each mixed radical in exercise 9 is the difference of two unlike mixed radicals. For each difference, write what the two radicals could be.

11. Choose one part of exercise 9 or 10. Write to explain how you determined the two radicals.

12. Determine the exact value of each expression.

a) $\sqrt[3]{16} + \sqrt[3]{54}$　　　　b) $\sqrt[3]{128} + \sqrt[3]{250}$　　　　c) $2\sqrt[3]{24} - \sqrt[3]{3}$

d) $4\sqrt[3]{375} - 6\sqrt[3]{81}$　　e) $5\sqrt[3]{40} + 2\sqrt[3]{135}$　　f) $7\sqrt[3]{500} - 3\sqrt[3]{32}$

13. A rectangle has a width of 2 cm and a diagonal of 6 cm. Determine each measurement.

a) the length　　　　b) the area　　　　c) the perimeter

14. Determine the area and perimeter of a rectangle with diagonal 8 cm and one side 6 cm.

15. Simplify.

a) $2\sqrt{3} + 4\sqrt{12}$　　　　　b) $5\sqrt{48} - 7\sqrt{3}$

c) $3\sqrt{8} + 6\sqrt{18}$　　　　　d) $4\sqrt{50} - 7\sqrt{32}$

e) $2\sqrt{24} + 3\sqrt{54}$　　　　f) $6\sqrt{20} - 2\sqrt{45}$

g) $3\sqrt{8} + 5\sqrt{18} - 6\sqrt{2}$　　h) $5\sqrt{28} - 3\sqrt{63} + 2\sqrt{112}$

i) $8\sqrt{24} - 2\sqrt{54} - \sqrt{28}$　　j) $2\sqrt{63} - 8\sqrt{8} - 2\sqrt{32}$

16. Simplify.

a) $5\sqrt{12} - 2\sqrt{48} - 7\sqrt{75}$　　　　b) $3\sqrt{7} + 2\sqrt{11} - \sqrt{11} + 4\sqrt{7}$

c) $\sqrt{48} - \sqrt{20} - \sqrt{27} - \sqrt{45}$　　　d) $4\sqrt{18} - 2\sqrt{63} + \sqrt{175} + 5\sqrt{98}$

e) $2\sqrt{12} + 3\sqrt{50} - 2\sqrt{75} - 6\sqrt{32}$　　f) $7\sqrt{24} + 3\sqrt{28} + 9\sqrt{54} + 6\sqrt{175}$

g) $3\sqrt{27} - 2\sqrt{50} - 5\sqrt{75} + 2\sqrt{32}$　　h) $2\sqrt{112} - 3\sqrt{18} - 2\sqrt{175} - \sqrt{98}$

17. Choose one part of exercise 16. Write to explain how you simplified the expression.

18. Bay City is 80 km due west of Keyport. The towns are linked by a straight stretch of railroad track. The road from Bay City to Keyport passes through Grenville, which is 10 km east and 10 km north of Bay City. How much farther is it by road than by train from Bay City to Keyport?

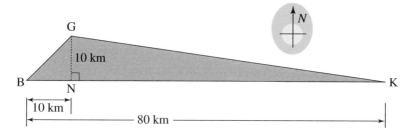

19. A straight stretch of railroad track connects Goshen to Humber, 64 km due west. The highway between the two towns passes through Ironton, 8 km east and 8 km north of Humber. How much farther is it to drive from Humber to Goshen than to take the train?

C **20.** For all real numbers x, does $\sqrt{x^2} = x$? Explain your answer.

21. Each right triangle (below left) has a hypotenuse 4 cm and the shortest side 2 cm. Determine the perimeter of the figure.

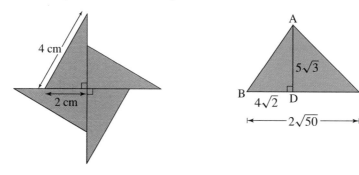

22. a) In the diagram (above right), determine if $\angle BAC = 90°$.

b) Suppose P is a point on AD such that $\angle BPC = 90°$. Determine the length of PD.

23. Find out if there are any values of a and b such that $\sqrt{a} + \sqrt{b}$ and $\sqrt{a + b}$ are equal.

2.10 Combined Operations with Radicals

To simplify expressions involving radicals, you can use either decimal approximations or exact forms.

Example 1

Determine an equivalent form for the expression $\sqrt{2}(\sqrt{10} - \sqrt{2})$.

Solution

Use the Distributive Law.

$$\sqrt{2}(\sqrt{10} - \sqrt{2}) = \sqrt{2} \times \sqrt{10} - \sqrt{2} \times \sqrt{2}$$
$$= \sqrt{20} - 2$$

In *Example 1*, observe that $\sqrt{20} = \sqrt{4 \times 5}$, or $2\sqrt{5}$.

Hence, another equivalent form is $2\sqrt{5} - 2$.

Since 2 is a common factor in both terms of this expression, another equivalent form is $2(\sqrt{5} - 1)$.

That is, all these expressions are equivalent:

$$\sqrt{2}(\sqrt{10} - \sqrt{2}) \qquad \sqrt{20} - 2 \qquad 2\sqrt{5} - 2 \qquad 2(\sqrt{5} - 1)$$

Example 2

Simplify the expression $(3\sqrt{5} - \sqrt{2})(\sqrt{5} + 4\sqrt{2})$.

Solution

The pattern at the right shows how the products are obtained.

$$(3\sqrt{5} - \sqrt{2})(\sqrt{5} + 4\sqrt{2})$$
$$= 3\sqrt{5} \times \sqrt{5} + 3\sqrt{5} \times 4\sqrt{2} - \sqrt{2} \times \sqrt{5} - \sqrt{2} \times 4\sqrt{2}$$
$$= 3 \times \sqrt{5} \times \sqrt{5} + 3 \times 4 \times \sqrt{5} \times \sqrt{2} - \sqrt{10} - 4 \times \sqrt{2} \times \sqrt{2}$$
$$= 3 \times 5 + 12 \times \sqrt{10} - \sqrt{10} - 4 \times 2$$
$$= 15 + 12\sqrt{10} - \sqrt{10} - 8$$
$$= 7 + 11\sqrt{10}$$

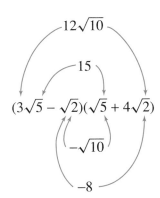

Example 3

Determine an equivalent form for each expression.

a) $(4 + \sqrt{5})(4 - \sqrt{5})$ **b)** $(\sqrt{6} - 2\sqrt{3})(\sqrt{6} + 2\sqrt{3})$

Solution

a) $(4 + \sqrt{5})(4 - \sqrt{5}) = 16 - 4\sqrt{5} + 4\sqrt{5} - 5$

$$= 11$$

b) $(\sqrt{6} - 2\sqrt{3})(\sqrt{6} + 2\sqrt{3}) = 6 + 2\sqrt{18} - 2\sqrt{18} - 4 \times 3$

$$= 6 - 12$$

$$= -6$$

In *Example 3*, we found that the products $(4 + \sqrt{5})(4 - \sqrt{5})$ and $(\sqrt{6} - 2\sqrt{3})(\sqrt{6} + 2\sqrt{3})$ are rational numbers. Observe in each case that the expressions that are multiplied differ only in the sign of one term. The fact that products like these are rational numbers can be used to simplify radical expressions with two terms in the denominator.

Example 4

Consider the expression $\dfrac{2}{\sqrt{5} - \sqrt{3}}$.

a) Determine a decimal approximation to 3 decimal places.

b) Determine an equivalent form for the expression by rationalizing the denominator.

c) Use the expression in part b to determine a decimal approximation.

Solution

a) $\dfrac{2}{\sqrt{5} - \sqrt{3}} \doteq \dfrac{2}{2.2361 - 1.7321}$

$$\doteq \dfrac{2}{0.5040}$$

$$\doteq 3.968$$

b) The denominator is $\sqrt{5} - \sqrt{3}$. Since the product $(\sqrt{5} - \sqrt{3})(\sqrt{5} + \sqrt{3})$
is a rational number, multiply the numerator and denominator by $\sqrt{5} + \sqrt{3}$.

$$\frac{2}{\sqrt{5} - \sqrt{3}} = \frac{2}{\sqrt{5} - \sqrt{3}} \times \frac{\sqrt{5} + \sqrt{3}}{\sqrt{5} + \sqrt{3}}$$

$$= \frac{2(\sqrt{5} + \sqrt{3})}{(\sqrt{5} - \sqrt{3})(\sqrt{5} + \sqrt{3})}$$

$$= \frac{2(\sqrt{5} + \sqrt{3})}{5 + \sqrt{15} - \sqrt{15} - 3}$$

$$= \frac{2(\sqrt{5} + \sqrt{3})}{2}$$

$$= \sqrt{5} + \sqrt{3}$$

c) $\sqrt{5} + \sqrt{3} \doteq 2.2361 + 1.7321$
$\doteq 3.968$

DISCUSSING THE IDEAS

1. How could you show that the four expressions after *Example 1* are equivalent? Which
form do you think is the simplest? Why?

2. In grade 9, you learned how to expand a product such as $(3x - y)(x + 4y)$.

a) How does this product compare with the product in *Example 2*?

b) Expand the product and compare the result with the solution to *Example 2*. In what
ways are the results similar? In what ways are the results different?

3. In grade 9, you also learned how to expand a product such as $(x + y)(x - y)$.

a) How does this product compare with the products in *Example 3*?

b) Expand the product and compare the result with the solution to *Example 3*. In what
ways are the results similar? In what ways are the results different?

4. a) In *Example 4*, why was the expression $\dfrac{\sqrt{5} + \sqrt{3}}{\sqrt{5} + \sqrt{3}}$ introduced in part b?

b) How could you check the results of *Example 4 a* and *b*?

5. In elementary school, you learned that two division statements result from one
multiplication statement. For example, since $3 \times 4 = 12$, you know that $\dfrac{12}{3} = 4$ and
$\dfrac{12}{4} = 3$. Think about this as you look back at the solution to *Example 4*. Discuss how
you could determine $\dfrac{2}{\sqrt{5} + \sqrt{3}}$ in both decimal form and in exact form using only the
calculations completed in the solution of *Example 4*.

2.10 EXERCISES

A 1. Determine an equivalent form for each expression.
   a) $\sqrt{2}(\sqrt{5} - \sqrt{7})$
   b) $\sqrt{3}(\sqrt{11} + \sqrt{2})$
   c) $\sqrt{6}(\sqrt{13} - \sqrt{5})$
   d) $2\sqrt{3}(3\sqrt{5} + 3\sqrt{7})$
   e) $\sqrt{5}(\sqrt{6} - \sqrt{10})$
   f) $4\sqrt{2}(3\sqrt{11} + 5\sqrt{13})$

2. Expand and simplify.
   a) $\sqrt{13}(\sqrt{3} + \sqrt{13})$
   b) $3\sqrt{3}(\sqrt{3} - 2\sqrt{6})$
   c) $3\sqrt{2}(2\sqrt{2} - 5\sqrt{8})$
   d) $2\sqrt{5}(3\sqrt{2} + 4\sqrt{3})$
   e) $6\sqrt{6}(3\sqrt{2} - 4\sqrt{3})$
   f) $2\sqrt{6}(3\sqrt{6} - 5\sqrt{8})$

3. Rationalize the denominator.
   a) $\dfrac{2}{\sqrt{5}}$
   b) $\dfrac{7}{\sqrt{11}}$
   c) $\dfrac{4}{\sqrt{3}}$
   d) $\dfrac{5\sqrt{2}}{2\sqrt{7}}$
   e) $\dfrac{6\sqrt{10}}{\sqrt{3}}$
   f) $\dfrac{12\sqrt{7}}{7\sqrt{5}}$
   g) $\dfrac{18\sqrt{5}}{3\sqrt{2}}$
   h) $\dfrac{20\sqrt{7}}{4\sqrt{3}}$

B 4. Determine a decimal approximation for each expression, to 3 decimal places.
   a) $\sqrt{3}(\sqrt{7} + \sqrt{5})$
   b) $\sqrt{5}(\sqrt{11} - \sqrt{3})$
   c) $(\sqrt{3} - \sqrt{5})(\sqrt{5} + \sqrt{7})$
   d) $(2\sqrt{3} - 3\sqrt{5})(3\sqrt{3} - 2\sqrt{5})$
   e) $(3\sqrt{6} + 2\sqrt{10})(2\sqrt{2} + 3\sqrt{5})$
   f) $5\sqrt{8}(3\sqrt{2} + 2\sqrt{3} - \sqrt{5})$

5. Simplify each expression.
   a) $(\sqrt{3} + 2)(\sqrt{3} + 1)$
   b) $(\sqrt{7} - 6)(\sqrt{7} + 1)$
   c) $(\sqrt{8} - 5)(\sqrt{2} - 3)$
   d) $(2 + \sqrt{12})(4 - \sqrt{3})$
   e) $(\sqrt{5} + \sqrt{2})(\sqrt{5} - \sqrt{2})$
   f) $(\sqrt{6} - \sqrt{3})(\sqrt{12} - \sqrt{6})$
   g) $(3 - \sqrt{2})^2$
   h) $(1 + \sqrt{3})^2$
   i) $(5 - 2\sqrt{2})^2$

6. Choose one part of exercise 5. Write to explain how you simplified the expression.

7. Determine each product. Investigate patterns in each list of products. Describe each pattern. Predict the next three products in each pattern.
   a) $(\sqrt{3} + 1)(\sqrt{3} + 2)$
      $(\sqrt{3} + 2)(\sqrt{3} + 3)$
      $(\sqrt{3} + 3)(\sqrt{3} + 4)$
      $\vdots$
   b) $(1 + \sqrt{2})(1 + \sqrt{2})$
      $(1 + \sqrt{3})(1 + \sqrt{3})$
      $(1 + \sqrt{4})(1 + \sqrt{4})$
      $\vdots$

8. Determine an equivalent form for each expression.
   a) $(5 + 3\sqrt{2})(4 - \sqrt{2})$
   b) $(6 - 4\sqrt{2})(2 - 5\sqrt{2})$
   c) $(2\sqrt{3} + 1)(\sqrt{3} - 4)$
   d) $(\sqrt{7} - 3)(4\sqrt{7} + 1)$
   e) $(4 + 2\sqrt{3})(4 - 2\sqrt{3})$
   f) $(2\sqrt{5} + \sqrt{2})(2\sqrt{5} - \sqrt{2})$
   g) $(3\sqrt{7} - \sqrt{2})^2$
   h) $(\sqrt{6} + 2\sqrt{3})^2$
   i) $(2\sqrt{2} - 3\sqrt{10})^2$

9. Simplify.

a) $(\sqrt{3} + \sqrt{2})^2$

b) $(\sqrt{3} + \sqrt{2})(\sqrt{3} - \sqrt{2})$

c) $(\sqrt{8} + \sqrt{2})^2$

d) $(\sqrt{8} - \sqrt{2})(\sqrt{8} + \sqrt{2})$

e) $(\sqrt{7} - 5)^2$

f) $(\sqrt{7} + 5)(\sqrt{7} - 5)$

g) $(3\sqrt{5} + 2\sqrt{3})^2$

h) $(3\sqrt{5} - 2\sqrt{3})(3\sqrt{5} + 2\sqrt{3})$

i) $(5\sqrt{3} + \sqrt{12})^2$

j) $(5\sqrt{3} + \sqrt{12})(5\sqrt{3} - \sqrt{12})$

10. Choose two parts of exercise 9: one from each column. Write to explain how you simplified each expression.

11. Determine a decimal approximation for each expression, to 3 decimal places.

a) $\dfrac{3}{\sqrt{7} - \sqrt{5}}$

b) $\dfrac{5}{\sqrt{10} + \sqrt{3}}$

c) $\dfrac{-6}{\sqrt{3} - \sqrt{5}}$

12. Rationalize each denominator. Investigate patterns in each list of quotients. Describe each pattern. Predict the next three quotients in each pattern.

a) $\dfrac{1}{\sqrt{5} - \sqrt{3}}$

$\dfrac{1}{\sqrt{6} - \sqrt{3}}$

$\dfrac{1}{\sqrt{7} - \sqrt{3}}$

$\vdots$

b) $\dfrac{\sqrt{2}}{\sqrt{2} + \sqrt{1}}$

$\dfrac{\sqrt{3}}{\sqrt{3} + \sqrt{2}}$

$\dfrac{\sqrt{4}}{\sqrt{4} + \sqrt{3}}$

$\vdots$

13. Rationalize the denominator of each expression.

a) $\dfrac{2\sqrt{3} + 4}{\sqrt{3}}$

b) $\dfrac{5\sqrt{7} - 3}{\sqrt{7}}$

c) $\dfrac{4\sqrt{5} - 2}{\sqrt{5}}$

d) $\dfrac{6\sqrt{2} - \sqrt{3}}{\sqrt{3}}$

e) $\dfrac{8\sqrt{6} + \sqrt{5}}{\sqrt{5}}$

f) $\dfrac{3\sqrt{10} - \sqrt{2}}{\sqrt{2}}$

g) $\dfrac{5\sqrt{8} + 2\sqrt{3}}{\sqrt{6}}$

h) $\dfrac{3\sqrt{12} - 4\sqrt{3}}{2\sqrt{2}}$

14. Rationalize the denominator of each expression.

a) $\dfrac{3}{\sqrt{5} - \sqrt{2}}$

b) $\dfrac{5}{\sqrt{7} + \sqrt{3}}$

c) $\dfrac{11}{7 - \sqrt{5}}$

d) $\dfrac{\sqrt{5}}{\sqrt{6} + 1}$

e) $\dfrac{\sqrt{6}}{\sqrt{12} - \sqrt{5}}$

f) $\dfrac{\sqrt{7}}{\sqrt{15} - \sqrt{10}}$

g) $\dfrac{3}{\sqrt{5} + \sqrt{2}}$

h) $\dfrac{6}{5 + \sqrt{5}}$

15. Rationalize the denominator of each expression.

a) $\dfrac{\sqrt{2} + 1}{\sqrt{2} - 1}$

b) $\dfrac{3 + \sqrt{5}}{3 - \sqrt{5}}$

c) $\dfrac{\sqrt{5} - \sqrt{3}}{\sqrt{5} + \sqrt{3}}$

d) $\dfrac{3\sqrt{2} + \sqrt{3}}{2\sqrt{3} + \sqrt{2}}$

e) $\dfrac{5\sqrt{3} + \sqrt{2}}{2\sqrt{3} - \sqrt{2}}$

f) $\dfrac{5\sqrt{3} - 3\sqrt{5}}{\sqrt{5} - \sqrt{3}}$

16. Choose one part of exercise 15. Write to explain how you simplified the expression.

17. Write each expression as a single fraction.

a) $\dfrac{1}{\sqrt{5}} + \dfrac{1}{\sqrt{3}}$ b) $\dfrac{1}{\sqrt{2}} - \dfrac{1}{\sqrt{6}}$ c) $\dfrac{1}{\sqrt{3}} + \dfrac{1}{\sqrt{6}}$ d) $\dfrac{2}{\sqrt{7}} - \dfrac{3}{\sqrt{5}}$

e) $\dfrac{3}{\sqrt{12}} + \dfrac{2}{\sqrt{18}}$ f) $\dfrac{5}{\sqrt{8}} - \dfrac{2}{\sqrt{6}}$ g) $\dfrac{7}{\sqrt{20}} - \dfrac{4}{\sqrt{12}}$ h) $\dfrac{3\sqrt{2}}{\sqrt{12}} - \dfrac{5\sqrt{3}}{\sqrt{8}}$

18. Choose one part of exercise 17. Write to explain how you simplified the expression.

19. In rectangle ABCD, points E and F are located on AB and BC such that ΔEBF and ΔFCD are congruent 30-60-90 triangles, and EB = FC = 5 cm. Determine each exact value.

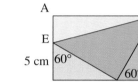

a) the lengths AD and AE

b) the areas of ΔAED and ΔDEF

c) the area of rectangle ABCD

20. A square with sides 6 cm long is inscribed in an equilateral triangle (below left).

a) Determine the length of the sides of the triangle.

b) Determine the area of the triangle.

Express your answers in exact form, and to 3 decimal places.

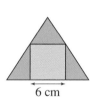   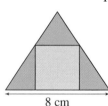

C **21.** A square is inscribed in an equilateral triangle with sides 8 cm long (above right).

a) Determine the length of the sides of the square.

b) Determine the area of the square.

Express your answers in exact form, and to 3 decimal places.

COMMUNICATING THE IDEAS

In some ways radicals are similar to fractions, and operations with radicals are similar to operations with fractions. In your journal, write to describe some of these similarities. Use examples to illustrate your explanations.

1. Simplify without using a calculator.

a) $\sqrt{5^2}$ b) $\sqrt[3]{3.7^3}$ c) $\sqrt[8]{9^4}$ d) $(\sqrt{3})^6$ e) $(\sqrt[3]{2})^{12}$ f) $\sqrt{2 + \sqrt{2^2}}$

2. Simplify without using a calculator.

a) $\left(\frac{16}{9}\right)^{-\frac{1}{2}}$ b) $32^{\frac{3}{5}}$ c) $100\,000^{-\frac{2}{5}}$ d) $\left(\frac{16}{0.0001}\right)^{-\frac{1}{4}}$ e) $\left(\frac{256}{81}\right)^{\frac{3}{4}}$

3. Choose one part of exercise 2. Write to explain how you simplified the expression.

4. Przewalski's Wild Horse is the only known species of wild horse left in the world. In 1956, there were only 36 of them. They were taken into protective captivity and there was an international effort toward registered breeding. By 1976, there were 250 wild horses. Copy and complete the table. Assume the populations are terms of a geometric sequence.

Year	1956	1961	1966	1971	1976	1981	1986	1991	1996
Population	36				250				

5. Write to explain how you completed exercise 4. Include any assumptions you made.

6. Calculate each value of x and y to 2 decimal places.

a)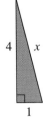

b)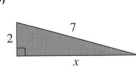

c)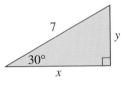

d)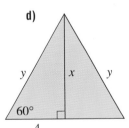

7. Use this diagram to prove the Pythagorean Theorem. (Hint: First show that the angles in the smaller quadrilateral are all 90°. Then write two expressions for the area of the large square, equate the expressions, and simplify.)

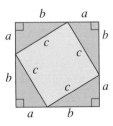

8. a) Is $\sqrt{5}$ rational or irrational? Explain.

b) If you said $\sqrt{5}$ is rational, express it as a ratio of two integers. If you said $\sqrt{5}$ is irrational, prove it.

9. Simplify by writing as a mixed radical.

a) $\sqrt{3} \times \sqrt{6}$ b) $\sqrt{243}$ c) $\sqrt{500}$ d) $\sqrt{343}$

10. Multiply.

a) $2\sqrt{5} \times 3\sqrt{10}$ b) $(-3\sqrt{6})(2\sqrt{24})$ c) $5\sqrt{48} \times (-\sqrt{375})$

d) $\frac{\sqrt{3}}{4} \times \frac{18}{\sqrt[3]{27}}$ e) $\frac{5}{\sqrt{8}} \times \frac{\sqrt{98}}{3}$ f) $\frac{\sqrt{2}}{3} \times \frac{\sqrt{3}}{4} \times \frac{\sqrt{4}}{5} \times \frac{\sqrt{5}}{6}$

11. For each geometric sequence, calculate its common ratio then determine the next two terms. If the sequence is not geometric, say so and predict the next two terms.

a) $\frac{\sqrt{2}}{3}, \frac{2}{3\sqrt{3}}, \frac{2\sqrt{2}}{9}, \ldots$ b) $\frac{1}{\sqrt{3}}, \frac{1}{\sqrt{5}}, \frac{\sqrt{3}}{5}, \ldots$

c) $\left(\frac{1}{2}\right)^{\frac{1}{7}}, \left(\frac{1}{2}\right)^{\frac{2}{7}}, \left(\frac{1}{2}\right)^{\frac{3}{7}}, \ldots$ d) $\frac{\sqrt{2}}{3}, \frac{\sqrt{3}}{4}, \frac{\sqrt{4}}{5}, \ldots$

12. Choose one part of exercise 11. Write to explain how you determined or predicted the next two terms.

13. Determine a decimal approximation for each expression, to 3 decimal places.

a) $\sqrt{3}(\sqrt{2} - 1)$ b) $\sqrt{2}(1 + \sqrt{2})$

c) $\sqrt{5}(\sqrt{3} - \sqrt{6})$ d) $(2\sqrt{2} - 1)(\sqrt{3} - \sqrt{2})$

e) $(3\sqrt{7} - 2\sqrt{6})(2\sqrt{5} - 3\sqrt{2})$ f) $(1 + \sqrt{2} + \sqrt{3})(1 - \sqrt{2} + \sqrt{3})$

14. Simplify.

a) $3\sqrt{5} + 2\sqrt{5}$ b) $5\sqrt{3} - 2\sqrt{3} + 4\sqrt{3}$ c) $5\sqrt{7} - \sqrt{63}$

d) $\sqrt{30} + \sqrt{270}$ e) $\sqrt{45} - \sqrt{80} + \sqrt{125}$ f) $3\sqrt{2} + 7\sqrt{32} - 4\sqrt{8}$

15. Determine a decimal approximation for each expression, to 3 decimal places.

a) $\frac{1}{\sqrt{2} + \sqrt{3}}$ b) $\frac{3}{\sqrt{6} - \sqrt{2}}$ c) $\frac{4\sqrt{2} + 1}{\sqrt{5} + \sqrt{3}}$

16. Rationalize the denominator of each expression.

a) $\frac{3\sqrt{2} - 1}{\sqrt{3}}$ b) $\frac{5\sqrt{3} - \sqrt{5}}{\sqrt{2}}$ c) $\frac{6\sqrt{2} - \sqrt{3}}{\sqrt{3} - 1}$ d) $\frac{3\sqrt{2} - 5\sqrt{3}}{\sqrt{5} - \sqrt{3}}$

17. Choose one part of exercise 16. Write to explain how you rationalized the denominator.

18. Rationalize the denominator of each expression.

a) $\frac{\sqrt{3} - 2}{\sqrt{3} + 2}$ b) $\frac{2\sqrt{3} - 3\sqrt{2}}{5\sqrt{2} - 2\sqrt{3}}$ c) $\frac{\sqrt{6} - \sqrt{3}}{\sqrt{6} + \sqrt{3}}$ d) $\frac{2\sqrt{3} - \sqrt{2}}{\sqrt{3} - \sqrt{2}}$

19. Write each expression as a single fraction.

a) $\frac{1}{\sqrt{2}} + \frac{1}{\sqrt{3}}$ b) $\frac{3}{\sqrt{7}} - \frac{2}{\sqrt{6}}$ c) $\frac{\sqrt{5}}{3\sqrt{2}} + \frac{3}{\sqrt{3}}$ d) $\frac{\sqrt{3}}{3\sqrt{2}} - \frac{\sqrt{2}}{4\sqrt{3}}$

1. Copy and complete each arithmetic sequence.

   a) 3, ■, ■, ■, 3   b) 3, ■, ■, ■, 23   c) 3, ■, ■, ■, 43

   d) 3, ■, ■, ■, 103   e) 3, ■, ■, ■, 303   f) 3, ■, ■, ■, 3303

2. Write a formula for the general term of each sequence.

   a) 1, 3, 5, …   b) 1, 5, 9, …   c) 1, 9, 17, …

   d) 4, 7, 10, …   e) 6, 10, 14, …   f) 11, 9, 7, …

3. Determine the sum of the first 10 terms of each arithmetic series.

   a) $1 + 2 + 3 + \cdots$ b) $2 + 4 + 6 + \cdots$ c) $3 + 6 + 9 + \cdots$

   d) $4 + 8 + 12 + \cdots$ e) $5 + 10 + 15 + \cdots$ f) $6 + 12 + 18 + \cdots$

4. Determine the indicated term of each geometric sequence.

   a) 1, 2, 4, … t_5 b) 2, 6, 18, … t_9 c) 3, −9, 27, … t_{10} d) 16, −8, 4, … t_{11}

5. Simplify.

   a) $(x^3y^{-2})(xy^3)(x^2y^2)$ b) $(x^{-1})^2(y^{-2})^{-2}(xy^2)^2$ c) $(x^2y^3)^2(xy^{-1})^3$

   d) $\dfrac{x^4y^2}{xy^3}$ e) $\dfrac{(x^2y^{-1})(xy^2)^2}{x^2y^3}$ f) $\dfrac{(x^{-2}y^3)(x^3y^{-1})^2(x^2y^{-4})^3}{(x^4y^{-1})^2(x^3y^3)^2(xy^2)^2}$

6. One way to measure the growth of the Internet is by the number of computers offering information, called *hosts*. Copy the table below. Calculate to complete the table. Assume that the numbers of hosts are terms of a geometric sequence.

   Number of Internet Hosts

Date (January each year)	1993	1994	1995	1996	1997	1998	1999
Number of hosts (millions)	1.3				16.2		

7. In the diagram, the equilateral triangle has sides of length 2 cm.

   a) Determine the radius R_1 of the inscribed circle.

   b) Determine the radius R_2 of the circumscribed circle.

   c) Copy the diagram. Circumscribe a large equilateral triangle about the circle with radius R_2. Then circumscribe a circle with radius R_3 about that triangle. Determine R_3.

   d) Continue the construction to determine R_4.

   e) Describe any pattern in your answers to parts a to d. What kind of sequence do they illustrate? Predict the radii of the next two circles.

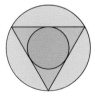

Who Is the World's Fastest Human?

At the 1996 Olympic Summer Games, Canadian sprinter Donovan Bailey won the 100-m sprint in a world-record time of 9.84 s. Five days later, American sprinter Michael Johnson won the 200-m sprint in a world-record time of 19.32 s.

 CONSIDER THIS SITUATION

At the time of the 1996 Summer Olympics, there was much controversy about which sprinter should be declared the world's fastest human. Was it Donovan Bailey or Michael Johnson?

Examine the information above. Is it sufficient for you to decide who was faster at the 1996 Olympic Games?

• If your answer is yes, who do you think was faster? Why?

• If your answer is no, what additional information do you need to make a decision? Why do you need that information?

• What is each person's average speed for the race?

You will examine this situation in Sections 3.3 and 3.4. On pages 180 and 181, you will develop a mathematical model to explore the question. Your model will involve line segments on a coordinate grid.

 FYI Visit www.awl.com/canada/school/connections

> For information related to the above problem, click on <u>MATHLINKS</u> followed by <u>AWMath</u>. Then select a topic under World's Fastest Human.

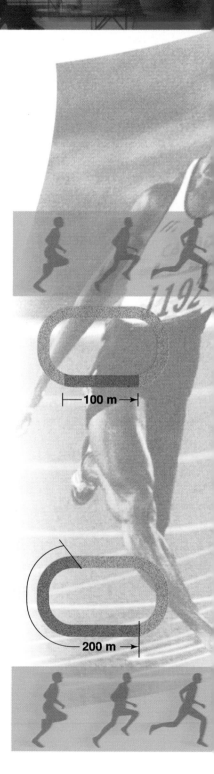

100 m
9.84

200 m
19.32

3.1 Length of a Line Segment

INVESTIGATE **Length**

1. Use the number line below. Determine the length of each line segment.

   a) BC   **b)** AB   **c)** AC

   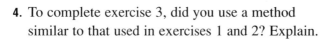

2. **a)** In exercise 1, how did you find the length of each segment?

   b) Point B is at 2 and point C is at 8. How could you find the length of BC by using the numbers 2 and 8?

   c) Use only the numbers −5 and 2 to determine the length of AB.

   d) Is the order important when you work with the numbers in part c? Explain.

3. Use the diagram. Determine the length of each line segment.

   a) AB   **b)** BC

   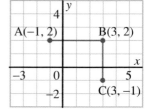

4. To complete exercise 3, did you use a method similar to that used in exercises 1 and 2? Explain.

5. Visualize line segment AC. What kind of triangle is △ABC? Use the lengths AB and BC. Determine the length of AC.

6. What theorem did you use to find the length of AC? Express AC in terms of AB and BC.

7. Use the diagram. Consider line segment AB. Visualize a right triangle that enables you to calculate the length of AB. What are the coordinates of the third vertex of the triangle? Use the sides of this right triangle. Determine the length of AB.

   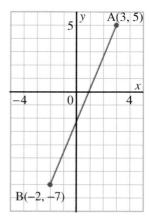

8. Find another right triangle with AB as its hypotenuse that could also be used to determine the length of AB. What are the coordinates of the third vertex of this triangle?

9. Use grid paper. Plot the points P(−3, 4) and Q(5, 1). Calculate the length of PQ. Compare your solution with that of another student. Did you both draw the same diagram? Did you both obtain the same length? Explain any differences.

We use coordinate systems to identify the locations of stars in the sky, the locations of cities on Earth's surface, and the locations of streets and buildings within cities. The Cartesian coordinate system is useful for describing the positions of points in the plane. In this system, the distance between any two points is easily calculated.

Example 1

Calculate the length of line segment PQ.

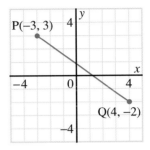

Solution

Draw a right $\triangle$PQR for which PQ is the hypotenuse.

Use the Pythagorean Theorem.

$$PQ^2 = PR^2 + RQ^2$$
$$= [(-2) - 3]^2 + [4 - (-3)]^2$$
$$= 5^2 + 7^2$$
$$= 25 + 49$$
$$PQ = \sqrt{74}$$
$$\doteq 8.6$$

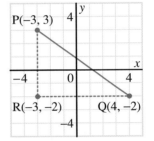

PQ is approximately 8.6 units long.

The method in *Example 1* can be used to develop a formula for the distance between any two points $P_1(x_1, y_1)$ and $P_2(x_2, y_2)$.

In the diagram, $\triangle P_1NP_2$ is a right triangle.

The coordinates of N are (x_2, y_1).

$P_1N = x_2 - x_1$ or $x_1 - x_2$ (whichever is positive)
$NP_2 = y_2 - y_1$ or $y_1 - y_2$ (whichever is positive)

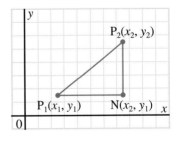

Use the Pythagorean Theorem.

$$P_1P_2{}^2 = P_1N^2 + NP_2{}^2$$
$$= (x_2 - x_1)^2 + (y_2 - y_1)^2$$
$$P_1P_2 = \sqrt{(x_2 - x_1)^2 + (y_2 - y_1)^2}$$

The distance between any two points $P_1(x_1, y_1)$ and $P_2(x_2, y_2)$ is given by this formula.
$$P_1P_2 = \sqrt{(x_2 - x_1)^2 + (y_2 - y_1)^2}$$

Example 2

The points L(1, 5), M(−3, 1), and N(6, −4) are the vertices of △LMN.

a) Draw △LMN on a grid. Calculate the lengths of its sides, to one decimal place.

b) Use the lengths of its sides to classify △LMN.

Solution

a) $LM = \sqrt{(-3-1)^2 + (1-5)^2}$

$= \sqrt{(-4)^2 + (-4)^2}$

$= \sqrt{16 + 16}$

$= \sqrt{32}$

$\doteq 5.7$

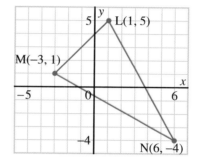

$MN = \sqrt{(6+3)^2 + (-4-1)^2}$

$= \sqrt{81 + 25}$

$= \sqrt{106}$

$\doteq 10.3$

$LN = \sqrt{(6-1)^2 + (-4-5)^2}$

$= \sqrt{25 + 81}$

$= \sqrt{106}$

$\doteq 10.3$

b) Since two sides of △LMN have the same length, it is an isosceles triangle.

DISCUSSING THE IDEAS

1. In the solution to *Example 1*, the right triangle used to calculate the length of PQ was drawn under PQ. Is there another position where a right triangle could be drawn to calculate the length of PQ? If so, would the calculations be the same as or different from those for the triangle under PQ? Explain.

2. On page 147, we wrote the expression $(x_2 - x_1)^2 + (y_2 - y_1)^2$ for $P_1P_2{}^2$. Would it have made a difference if we had used $(x_1 - x_2)^2 + (y_1 - y_2)^2$? Explain.

3. The distance formula involves the radical sign, which indicates the positive square root of a number. Why is this appropriate here?

A 1. Plot each pair of points. Calculate the distance between them.

   a) A(7, 3), B(7, −2)   b) C(−3, −5), D(−3, 2)   c) E(−4, 3), F(7, 3)

   d) G(6, −4), H(−6, −4)   e) O(0, 0), J(0, −8)   f) K(12.5, 5.0), L(−1.5, 5.0)

2. Determine the length of each line segment on the grid below.

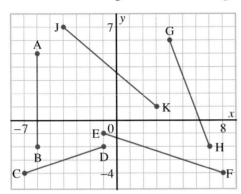

3. Plot each pair of points. Estimate the distance between them.
   Calculate the distance between each pair of points.

   a) M(5, 4), N(1, 2)   b) P(−2, 3), Q(4, −1)   c) R(−8, 9), S(−3, 4)

   d) T(−3, 0), U(8, −4)   e) O(0, 0), V(−5, 2)   f) W(3.5, 5.0), X(7.5, 8.5)

4. a) Calculate the lengths of the sides of each right triangle on the grid.

   b) Calculate the area of each triangle in part a.

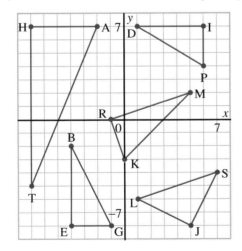

5. Calculate the length of the line segment with each pair of endpoints.

   a) A(5, 9), B(−2, 2)   b) C(1, −6), D(5, 2)

   c) E(−3, 7), F(5, 2)   d) G(−9, −3), H(−3, 4)

B 6. On a grid, draw the triangle with each set of vertices. Classify each triangle as scalene, isosceles, or equilateral.

   a) A(−4, 3), B(−2, −4), C(3, 5)

   b) P(−1, 2), Q(1, −5), R(5, −2)

   c) J(−2, 5), K(4, −1), L(6, 7)

7. Choose one part of exercise 6. Write to explain how you classified the triangle.

8. On a grid, draw the rectangle with each set of vertices. For each rectangle, determine the lengths of its sides and its perimeter.

   a) A(−6, −3), B(3, −3), C(3, 5), D(−6, 5)

   b) J(−3, 3), K(0, −6), L(3, −5), M(0, 4)

   c) P(−8, −1), Q(−4, −7), R(8, 1), S(4, 7)

9. For each rectangle in exercise 8, determine the lengths of its diagonals and its area.

10. Determine the length of each line segment AB, BC, and AC. Compare the lengths AB + BC and AC. What do you conclude about the points A, B, and C? Explain.

   a) A(9, 7), B(5, 4), C(1, 1)

   b) A(−4, 8), B(2, 0), C(5, −4)

   c) A(−5, −2), B(3, 2), C(7, 4)

11. A circle has centre M(2, 1) and radius 5. Which of the following points are on the circle?

   a) A(6, 4)   b) B(3, 6)   c) C(−3, 1)   d) D(6, −1)

12. Choose one part of exercise 11. Write to explain how you determined whether the point lies on the circle.

13. Look at the diagram. Alissa at A wants to travel to Brent at B. Each block has dimensions 60 m by 60 m. Alissa uses the roads and travels the shortest route.

   a) How many different paths can Alissa take?

   b) What is the length of each path? (Assume that the width of the road is negligible.)

   c) Suppose Alissa were able to travel in a straight line to Brent. What is the length of this path?

14. a) Plot the point E(2, 5). Calculate the length of the line segment joining E to each point.

    i) A(2, 1)   **ii)** B(3, 1)   **iii)** C(4, 1)   **iv)** D(5, 1)

  b) The line segment joining the point F(a, 1) to E has the same length as EC. Determine the value of a.

  c) The line segment joining the point G(-2, b) to E has the same length as ED. Determine two values of b.

  d) Use a diagram to explain and justify your answers to parts b and c.

  e) Use calculations to explain and justify your answers to parts b and c.

15. Look at the diagram. Chantelle at C wants to travel to her brother Daniel at D. Each block has dimensions 50 m by 50 m. Chantelle uses the roads.

  a) How many different paths of length 250 m can Chantelle take?

  b) What is the length of the straight-line path between Daniel and Chantelle?

16. a) For each given length of a line segment, determine the possible coordinates of the endpoints of a line segment with that length.

    i) 1   **ii)** 5   **iii)** 7.5   **iv)** 13   **v)** 20   **vi)** 10.5

  b) Compare your answers to part a with those of a classmate. Did both of you get the same answers? Explain any differences.

17. a) For each given length of a line segment, determine the possible coordinates of the endpoints of a line segment with that length.

    i) $\sqrt{2}$ **ii)** $\sqrt{5}$ **iii)** $\sqrt{13}$ **iv)** $\sqrt{41}$ **v)** $\sqrt{29}$ **vi)** $\sqrt{20}$

  b) Compare your answers to part a with those of a classmate. Did both of you get the same answers? Explain any differences.

18. Heather and Kim took part in a scavenger hunt. They begin the hunt at the same starting point. Heather walks 20 m north, then 50 m east to find her first treasure. Kim walks 70 m south, then 30 m west and stops.

  a) How far is each girl from the starting point?

  b) How far are the girls from each other?

  c) Suppose Heather looks directly at Kim. Will the starting point be in her direct line of sight? Explain.

19. A person in a lighthouse gives the location of a ship as an ordered pair. The first coordinate is the distance, in nautical miles, measured due east from the lighthouse to the ship. The second coordinate is the distance, in nautical miles, measured due north from the lighthouse to the ship. An ocean freighter sends a distress signal from a location F(240, 160). A second freighter at S(50, 420) and a coast guard cutter at C(520, 100) hear the distress call. Which vessel is closer to the ship in distress?

20. A city is located at A(100, 200) on a navigational map marked in nautical miles. A pilot charters her Bell 230 helicopter to a group of 5 geologists and mining executives. They plan to fly from the city to a geological site at B(40, 80) for a 3-h visit. From there, the pilot will take them to a field office at C(100, 40), where they will remain for a couple of weeks. The pilot will have to fly from C back to A without paying customers. Sketch a grid to represent this situation. Use your sketch and information from the *Helicopters* database to complete this exercise.

a) Calculate the distances from A to B, B to C, and C to A, to the nearest nautical mile.

b) Assume the pilot flies at the airspeed that gives her the maximum range at 4000 ft. Calculate the approximate flying time for each leg of the trip.

c) What is the minimum the pilot should charge for the trip to recoup her operating costs? Explain your reasoning.

d) Suppose you were the pilot. Would you refuel on this trip, if possible? Explain.

e) From the database, choose another helicopter that could carry 5 passengers. Complete parts b to d for this helicopter.

21. A surveyor establishes the corner points of a field as A(-75, 375), B(-75, -100), C(150, -100), and D(250, 150). The numbers represent the distances, in metres, from a pair of coordinate axes, with their origin at a well, W(0, 0). Find the length of each diagonal, the perimeter, and the area of the field.

22. Susan and Paul are playing pool on a table that measures 1.4 m by 2.6 m. To establish the position of a ball, consider a coordinate system on a square grid, where each square has side length 0.2 m. The origin, O(0, 0), is at the bottom left corner of the table. The cue ball is at C(3, 10) and the target ball is at T(5, 1).

a) What are the coordinates of the other three corners of the table?

b) Suppose Susan can hit the cue ball directly toward the target ball. How far must her cue ball travel before it hits the target ball?

C **23.** Look at the pool table in exercise 22. Suppose Paul places a ball in such a way that it blocks the direct path of Susan's ball. To hit the target ball, Susan decides to bounce the cue ball off the side of the table labelled P.

a) What are the coordinates of the point Susan must hit?

b) What is the total distance the ball must travel in this play?

24. Which points are equidistant from A(4, 0) and B(0, 2)?

a) P(0, −3)   b) Q(3, 7)   c) R(5, 7)   d) S(2, 1)

25. Determine the coordinates of the points on the x-axis that are 5 units from B(5, 4).

26. Determine the coordinates of the point on the y-axis that is equidistant from each pair of points.

a) P(3, 0) and Q(3, 6)   b) R(4, 0) and S(2, 6)   c) T(5, 0) and U(1, 6)

27. Choose one part of exercise 26. Write to explain how you determined the coordinates.

28. In the Bell telephone system, long-distance charges are based on the straight-line distance between the caller and the receiver. Each telephone exchange in North America is given a set of coordinates. The distances are calculated using the formula for the distance between two points.

a) Calculate the phone distance from Ottawa to each of Vancouver, Quebec City, and Washington.

b) Would it be cheaper to call Calgary or Washington from Vancouver?

City	Coordinates
Calgary	(3850, 2800)
Ottawa	(2241, 4388)
Quebec City	(1930, 3638)
Vancouver	(4500, 3000)
Washington	(1489, 5605)

COMMUNICATING THE IDEAS

In your journal, describe a method to determine the distance between any two points in the coordinate plane without having to plot the points. Use an example to explain why your method works.

Shortest Networks

In a new housing development, four houses at the vertices of a 100-m square are to be connected with cable television lines. Where should the cables be located so that the total length of cable is a minimum?

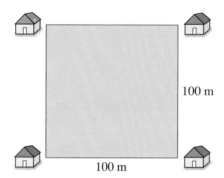

100 m

100 m

Two possibilities are shown below, but each possibility uses more cable than is necessary to connect the four houses.

Diagrams like these, which show connected points, are *networks*. A physical model can be used to find a network that has a shorter length than either network above.

This diagram shows two parallel sheets of Plexiglas joined by four rods at the vertices of a square. When the model is dipped in a soap solution, a thin soap film connects the four rods in the manner shown.

When viewed from above, the pattern formed by the film indicates the shortest network joining the four points. The least amount of cable is used when it follows a pattern like this.

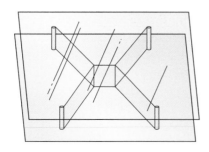

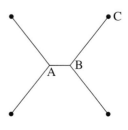

The solution to problems like this has applications in the design and construction of networks such as railways, roads, power lines, and computer chips.

You can use a computer to solve the problem on the preceding page.

1. On grid paper, draw a diagram similar to this.

2. Visualize how the diagram changes if point B starts at the centre of the square and moves to the right. As it does so, point A also moves to the left so that the diagram is symmetrical at all times.
   How do the coordinates of point B change as point B moves to the right from the centre of the square?

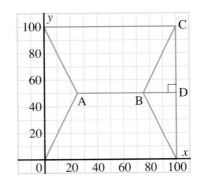

3. This spreadsheet will display the total length of the cable for any position of point B.

	A	B	C
1	Shortest Network Problem		
2	x-coordinate of B		55
3	Length of AB		=2*(C2-50)
4	Length of BC		=SQRT((100-C2)^2+50^2)
5	Length of cable		=C3+4*C4

   a) Start a new spreadsheet document. Enter the text and formulas above.

   b) Explain the purpose of each formula in cells C3, C4, and C5.

4. Enter numbers in cell C2 that represent possible x-coordinates for point B. Find an x-coordinate that makes the length of cable in cell C5 as short as possible. Determine this coordinate to 1 decimal place.

5. Draw a large diagram on grid paper similar to the one above. Plot point B as accurately as you can, using the x-coordinate corresponding to the minimum network length you calculated. Plot point A the same distance to the left of the centre of the square as point B is to the right. Use a ruler to draw the network and △BCD. Measure the angles at which the segments meeting at point B intersect. What do you notice? What are the measures of the angles in △BCD?

Mathematics & Technology

3.2 Midpoint of a Line Segment

INVESTIGATE **Midpoint**

1. Use the number line below. Determine the midpoint of line segment AB.

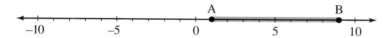

2. **a)** In exercise 1, how did you find the midpoint of the segment?

   b) How could you determine the midpoint of AB by using the numbers 9 and 1?

3. Use this diagram. Determine the coordinates of the midpoint of AB and the midpoint of BC.

4. To find the midpoints of AB and BC, did you use a method similar to that used in exercises 1 and 2? Explain.

5. Visualize line segment AC. Determine the coordinates of the midpoint of AC. How do these coordinates compare to the coordinates of the midpoints in exercise 3?

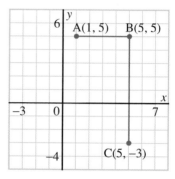

From *Investigate*, you may have discovered that the coordinates of the midpoint of a line segment are related to the coordinates of the endpoints of the segment.

Example 1

Determine the coordinates of the midpoint, M, of the line segment joining A(2, 1) and B(8, 5).

Solution

Complete the right △ABC for which AB is the hypotenuse.

Draw the perpendicular from E, the midpoint of AC. Draw the perpendicular from D, the midpoint of BC. The perpendiculars meet at M, the midpoint of AB.

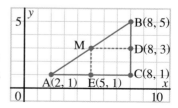

In the diagram, the coordinates of M are the means of the coordinates of A and B.

x-coordinate of M $= \dfrac{2 + 8}{2}$

$= 5$

y-coordinate of M $= \dfrac{1 + 5}{2}$

$= 3$

The coordinates of M are (5, 3).

When M is the midpoint of a line segment having endpoints $P_1(x_1, y_1)$ and $P_2(x_2, y_2)$, the coordinates of M are:

$$\left(\frac{x_1 + x_2}{2}, \ \frac{y_1 + y_2}{2} \right)$$

Example 2

a) Determine the coordinates of the midpoint, M, of the line segment joining P(−5, 7) and Q(3, −3).

b) Calculate the distance from M to P.

Solution

a) x-coordinate of M $= \dfrac{-5 + 3}{2}$

$= \dfrac{-2}{2}$

$= -1$

y-coordinate of M $= \dfrac{7 + (-3)}{2}$

$= \dfrac{4}{2}$

$= 2$

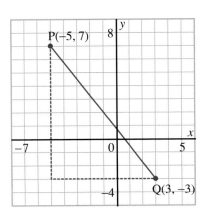

The coordinates of M are (−1, 2).

b) Use the distance formula to calculate the length of MP.

$MP = \sqrt{(-5 + 1)^2 + (7 - 2)^2}$

$= \sqrt{(-4)^2 + 5^2}$

$= \sqrt{16 + 25}$

$= \sqrt{41}$

$\doteq 6.4$

The distance MP is approximately 6.4 units.

1. In *Example 2*, can you think of another way to calculate the distance from P to the midpoint of segment PQ? Explain.

2. In *Example 2*, if we write the *x*-coordinates of P, M, and Q in order, we obtain −5, −1, 3.
   a) What type of sequence do these numbers form?
   b) Do the *y*-coordinates form the same type of sequence?
   c) Would you always get this type of sequence if you wrote this list of numbers for other examples? Explain.

3. You can determine the coordinates of the midpoint of a line segment from a diagram or by using the midpoint formula. What are the advantages and disadvantages of each method?

3.2 EXERCISES

A 1. State the coordinates of the midpoint of each line segment.

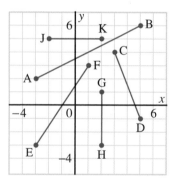

2. Determine the coordinates of the midpoint of the line segment with each pair of endpoints.
   a) A(0, 4), B(6, 4)
   b) R(−2, 5), S(−2, −1)
   c) K(1, 1), L(7, 9)
   d) F(5, 3), G(−3, 0)
   e) M(−6, 5), N(3, −2)
   f) P(0, 7), Q(−4, −2)
   g) U(−3, −6), V(7, 5)
   h) C(−9, 0), D(5, −4)
   i) T(−1, −2), R(−6, 5)

3. On a grid, draw the triangle with vertices J(0, 3), K(8, 1), and L(6, 9). Determine the coordinates of the midpoint of each side. Solve this problem using the midpoint formula; then solve this problem by graphing. Compare your solutions.

B **4.** The endpoints of a diameter of a circle are C(5, 5) and D(−1, −3).

   a) Determine the coordinates of the centre of the circle.

   b) Determine the length of the radius.

   Write to explain an alternative method to solve this problem.

5. The coordinates of the vertices of a rectangle are given. Draw each rectangle on a grid. Determine the coordinates of the midpoint of each diagonal. What do you notice? Explain.

   a) A(1, 4), B(1, −2), C(11, −2), D(11, 4)

   b) P(6, 10), Q(−2, 6), R(4, −6), S(12, −2)

6. A parallelogram has vertices E(0, −3), F(7, −1), G(10, 4), and H(3, 2). Draw the parallelogram on a grid. Determine the coordinates of the midpoint of each diagonal. What do you notice? Explain.

7. A right triangle has vertices A(−2, 8), B(6, 4), and C(4, 0).

   a) Draw the triangle on a grid. Determine the coordinates of the midpoint M of the hypotenuse.

   b) Calculate the lengths of AM, BM, and CM. What do you notice?

8. A triangle has vertices A(6, 8), B(−2, 2), and C(6, 2). The midpoints of AC and BC are M and N, respectively.

   a) Graph the triangle. Determine the coordinates of M and N.

   b) Calculate the lengths of MN and AB. What do you notice?

9. A triangle has vertices D(5, 2), E(−3, 3), and F(2, −5).

   a) Graph the triangle. Determine the coordinates of the midpoint of each side.

   b) Join the midpoints to form a triangle. Write to explain how this triangle compares with the original triangle.

10. A triangle has vertices P(−2, 4), Q(−4, −4), and R(6, 0).

   a) Graph the triangle. Draw its three medians.

   b) Calculate the lengths of the three medians of △PQR.

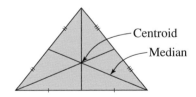

11. A line segment has endpoints A(2, 7) and B(−4, −2). Determine the coordinates of the points that divide AB into three equal parts.

12. Write to explain how you completed exercise 11.

13. a) Determine the coordinates of the midpoint of the line segment with each pair of endpoints.

    i) A(−2, −4), B(2, −5)   **ii)** C(−3, −4), D(3, −5)   **iii)** E(−4, −4), F(4, −5)

    b) For each midpoint in part a, determine the coordinates of possible endpoints of a different line segment with that midpoint.

14. Draw an example of a line segment AB that has one endpoint A on the *x*-axis and midpoint M on the *y*-axis. How are the coordinates of B related to those of A and M?

15. Each point A is an endpoint and M the midpoint of a line segment. Locate A and M on a grid. Determine the coordinates of the other endpoint of the segment. Think of two ways to solve this problem.

    a) A(−2, 4), M(2, −1)   **b)** A(8, −1), M(5, −5)

    c) A(3, 5), M(−1.5, 1.5)   **d)** A(−6, −4), M(−3, −2)

16. Choose one part of exercise 15. Write to explain one way you determined the coordinates.

17. a) The length of a line segment and the coordinates of its midpoint are given. Determine some possible coordinates of the endpoints of each segment.

    i) $2\sqrt{2}$; (3, 5) **ii)** $2\sqrt{2}$; (5, −4) **iii)** $6\sqrt{2}$; (−3, −4)   **iv)** $4\sqrt{2}$; (−5, 2)

    b) Compare your answers to part a with those of a classmate. Did you get the same answers? Explain any differences.

18. For each pair of points A and B, determine the coordinates of the three points that divide line segment AB into four equal parts. Think of another way to solve this problem.

    a) A(−4, 4), B(8, −12)   **b)** A(5, −7), B(−9, 5)

    c) A(−2, −6), B(8, 4)   **d)** A(−5, −3), B(5, 3)

19. Points P, Q, and R divide a line segment into four equal parts. Determine the coordinates of the other endpoint when those of P and the adjacent endpoint are as given below.

    a) P(5, 2), (2, 4)   **b)** P(−2, 1), (−6, −2)

    c) P(4, −1.5), (8, −5)   **d)** P(−3, −6), (−4, −8)

20. BC and AD are diameters of a circle. The coordinates are A(-3, 5), B(3, 9), and C(1, -1).

   a) Determine the coordinates of the centre of the circle.

   b) Determine the coordinates of D.

   c) Determine the length of the radius.

21. A map's numerical coordinates are in kilometres. The town of Plimville is at (4.5, 6.1) and Valley View is at (13.3, 2.7). A road is to be constructed on a direct line joining Plimville and Valley View. Each community is responsible for the cost of the construction of the road to the midpoint. The cost for construction is $23 000 per kilometre.

   a) Determine the coordinates of the midpoint of the road.

   b) Determine the cost of the construction for Valley View.

   c) What is the total cost of the road?

   Think of another way you could solve the problem.

C **22.** A line segment has endpoints A(x_1, y_1) and B(x_2, y_2). Determine an expression for the coordinates of the points that divide AB into:

   a) 3 equal parts   **b)** 4 equal parts

23. A square has vertices P(4, 3), Q(11, 6), R(8, 13), and S(1, 10).

   a) Calculate the mean of the x-coordinates and the mean of the y-coordinates of the vertices.

   b) Use the results of part a as the coordinates of a point C. By graphing, determine how point C is related to the square. Why do you think this result occurs?

24. Triangle ABC has vertices A(-6, 2), B(8, -2), and C(4, 6). Determine the coordinates of its centroid. Think of another way to solve this problem, using the ideas in exercise 23.

25. The midpoints of the sides of a triangle have coordinates G(3, 1), H(-1, 2), and J(1, -3). Determine the coordinates of the vertices of the triangle.

COMMUNICATING THE IDEAS

Write a definition of the midpoint of a line segment. Since you are defining the term *midpoint*, you must not use the word "midpoint" in your definition. Read your definition to a partner. Work together to ensure your definition is clear and concise. When it is, write the definition in your journal.

3.3 Slope of a Line Segment

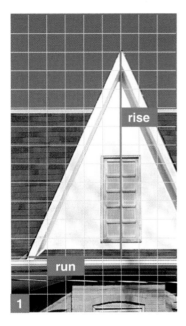

1. a) Which roof is the steepest?

   b) Which roof is least steep?

   c) List the roofs in order of steepness.

2. The first picture above shows what is meant by the rise and the run of a roof.

   a) Use the grid on each picture. Estimate the rise and the run of each roof. Copy and complete this table.

House	rise	run	$\dfrac{\text{rise}}{\text{run}}$
1			
2			
3			

   b) Use your calculator. Divide the rise by the run. Write the results in the fourth column.

   c) Compare the numbers in the fourth column with your answers to exercise 1. What do you notice?

The term *slope* is often used to describe steepness.

$$\text{slope} = \frac{\text{rise}}{\text{run}}$$

The pitch of a roof, the steepness of a ski run, or the gradient of a mountain road are all examples of slope. In each case, the slope is the ratio of the rise to the run.

___

Example 1

These diagrams show how the rise and run of a water wave are defined. Oceanographers have found that a wave tends to fall over, or *break*, when its slope becomes greater than $\frac{2}{7}$. For each wave shown, determine if it would break.

a)
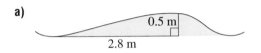
2.8 m

b)
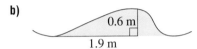
1.9 m

Solution

Determine the slope of each wave.

a) $\text{slope} = \frac{\text{rise}}{\text{run}}$
$= \frac{0.5}{2.8}$
$\doteq 0.179$

b) $\text{slope} = \frac{\text{rise}}{\text{run}}$
$= \frac{0.6}{1.9}$
$\doteq 0.316$

Since $\frac{2}{7} \doteq 0.286$, the wave in part b would break, and the wave in part a would not.

___

In a coordinate system, we can determine the slope of any line segment from its endpoints $P_1(x_1, y_1)$ and $P_2(x_2, y_2)$.

From P_1 to P_2:

- the rise is the difference in the *y*-coordinates: $y_2 - y_1$
- the run is the difference in the *x*-coordinates: $x_2 - x_1$

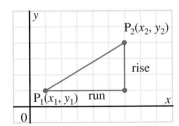

The slope of the line segment joining $P_1(x_1, y_1)$ and $P_2(x_2, y_2)$ is given by the formula:

$$\text{slope of } P_1P_2 = \frac{y_2 - y_1}{x_2 - x_1}, \ (x_2 \neq x_1)$$

Example 2

Graph each line segment. Determine its slope.

a) A(2, 1), B(5, 3)   **b)** M(−3, 4), N(−1, −2)

c) R(−2, 4), S(5, 4)   **d)** J(1, −2), K(1, 3)

Explain the result in part d.

Solution

a) Slope of AB $= \dfrac{3 - 1}{5 - 2}$

$= \dfrac{2}{3}$

b) Slope of MN $= \dfrac{-2 - 4}{(-1) - (-3)}$

$= \dfrac{-6}{2}$

$= -3$

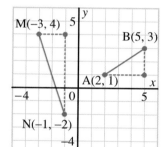

c) Slope of RS $= \dfrac{4 - 4}{5 - (-2)}$

$= \dfrac{0}{7}$

$= 0$

d) Slope of JK $= \dfrac{3 - (-2)}{1 - 1}$

$= \dfrac{5}{0}$

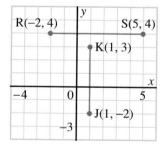

The denominator in part d is 0 and division by 0 is not defined. Therefore, the slope of JK cannot be defined as a real number.

We say that the slope of JK is *undefined*.

When we determine the slope of a line segment, it does not matter which point is taken as the first point and which is taken as the second point. In *Example 2a*, the slope of segment AB could also be found as follows:

$$\text{Slope of BA} = \frac{1-3}{2-5}$$
$$= \frac{-2}{-3}$$
$$= \frac{2}{3}$$

VISUALIZING

On a coordinate grid:

The slope of any horizontal segment is zero.

The slope of any vertical segment is not defined as a real number.

A line segment *rising* to the right has a *positive* slope.

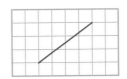

A line segment *falling* to the right has a *negative* slope.

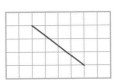

DISCUSSING   THE IDEAS

1. a) In the formula for the slope of a line segment, why is $x_2 \neq x_1$ included? Describe the positions of the two points when $x_2 = x_1$.

   b) What is the slope of a line segment if $y_2 = y_1$? Describe the positions of the two points when $y_2 = y_1$.

2. AB is any line segment on a coordinate grid. Visualize what happens if point B begins to rotate around point A, while point C is fixed. As point B rotates around point A, how does each of the following change?

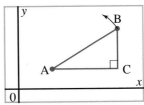

   a) the shape and position of $\triangle ABC$ b) the rise from point A to point B

   c) the run from point A to point B   d) the slope of AB

3. Draw a diagram to show segment AB from exercise 2 with point B in different positions. Label each position of point B with an estimate of the slope of AB. Discuss with a partner how the slope changes as point B is rotated through one complete turn.

A **1.** This diagram represents a roof. Determine its slope.

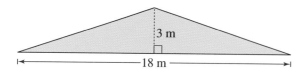

2. A section of a roller-coaster track falls 25 m in a horizontal distance of 15 m. What is the slope of this section of track?

3. Use the diagram to determine the slope of each line segment.

a)

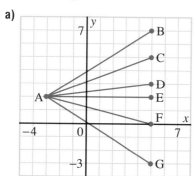

b)

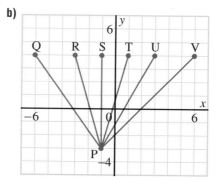

4. Write to explain the method you used to complete exercise 3. Compare your method with that of a classmate. If your classmate's method was different from yours, explain it.

5. The slope of a line segment is 3. What is a possible rise and run?

6. The slope of a line segment is 4. The run is 6. What is the rise?

B **7.** Graph each line segment. Determine its slope from the diagram.

a) A(−2, 7), B(6, −4)   b) C(3, −5), D(8, 10)

c) E(1, 6), F(5, −4)   d) G(−3, 7), H(−3, −7)

e) J(−4, −3), K(8, 5)   f) L(2, −7), M(7, −7)

8. The slope of a line segment is 5. The line segment has endpoints E(6, 3) and F(4, k). What is the value of k? Write to explain how you determined the value of k.

9. The coordinates of the vertices of triangles are given. Graph each triangle. Determine the slope of each side.

a) A(5, −1), B(0, 4), C(−2, −5)   b) R(−3, 4), S(6, 7), T(2, −3)

c) L(4, −2), M(−4, 8), N(4, 8)   d) E(−2, −1), F(−1, −6), G(5, 6)

10. Two points on a line segment are G(5, 6) and H(2, −3). Determine the coordinates of two other points on the line segment.

11. a) For each given slope, determine the coordinates of the endpoints of a line segment with that slope.

    i) 3   **ii)** −5   **iii)** $\frac{3}{5}$ **iv)** $-\frac{5}{3}$ **v)** −3   **vi)** 5

    b) Compare your answers to part a with those of a classmate. Explain any differences.

12. One safety standard for a ladder is that the distance from the wall to the base of the ladder should be one-quarter of the distance up the wall to the top of the ladder.

    a) What is the slope of a ladder that is leaning safely?

    b) A ladder is 2.6 m long. The top of the ladder touches a window 2.5 m above the ground. What is the slope of the ladder? Is the ladder safe?

    c) Choose either part a or b. Write to explain how you determined the slope.

13. This graph was constructed using data from the *Helicopters* database. It shows the heights, in feet, of 3 different helicopters as they climb to their maximum altitudes.

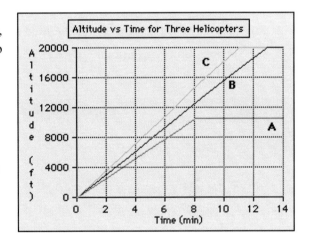

    a) Without calculating, rank the helicopters from the least rate of climb to the greatest rate of climb.

    b) For each helicopter, estimate its maximum altitude; then calculate its approximate rate of climb.

    c) Is the helicopter with the greatest rate of climb the first to reach its maximum altitude? Explain.

    d) Use the *Helicopters* database to identify each helicopter.

    e) Choose 3 other helicopters from the *Helicopters* database. Construct a graph of altitude against time similar to the graph above. Have a classmate use your graph to complete parts a to d of this exercise.

14. Use the information about Donovan Bailey and Michael Johnson from page 144. On graph paper, start a *distance-time* graph. Draw axes with Time in seconds along the horizontal axis, and Distance in metres along the vertical axis. Label the origin O.

  a) Plot point D to represent Bailey's time for the 100 m. Join OD. Calculate the slope of OD. What does this slope represent?

  b) Plot point M to represent Johnson's time for the 200 m. Join OM. Calculate the slope of OM. What does this slope represent?

  c) Is this information enough for you to decide who was faster at the 1996 Olympics? Write to explain your answer.

  Note: Save your graphs for use later.

15. Here is one way to compare Bailey's and Johnson's 1996 races. Consider Johnson's results for each 100-m segment of his race. He reached the 100-m tape at 10.12 s.

  a) Suppose Johnson's graph from exercise 14 was redrawn to account for this information. How would the graph change?

  b) How does Johnson's time for the first 100 m compare with Bailey's time?

  c) Suppose Bailey had run a 200-m sprint at the same average speed as his 100-m sprint. How would his time compare with Johnson's time?

MODELLING to Determine the World's Fastest Human

In exercises 14 and 15, you used slopes of line segments to model Donovan Bailey's and Michael Johnson's world-record sprints.

• Are the points between the endpoints of these line segments accurate representations of the positions of the runners during the two races? Explain your answer.

• Would it be fair to use either model in exercise 15 to decide who was faster? Explain your answer.

---

16. Use a coordinate grid.

  a) If possible, draw a triangle so that all three sides have positive slopes.

  b) If possible, draw a quadrilateral so that three sides have positive slopes.

  c) If possible, draw a quadrilateral so that all four sides have positive slopes.

17. The line segment joining each pair of points has the given slope. Determine each value of x. Draw each segment on a grid.

  a) A$(-1, 2)$, B$(x, 6)$ slope $\frac{1}{2}$ **b)** G$(0, -2)$, H$(x, 3)$ slope 1

  c) R$(x, 0)$, S$(-2, 4)$ slope $-\frac{2}{3}$ **d)** U$(x, -3)$, V$(4, 9)$ slope 3

18. The line segment joining each pair of points has the given slope. Determine each value of y. Draw each segment on a grid.

a) C(1, 3), D(5, y)   slope $\frac{3}{4}$

b) Q(0, y), R(4, 6)   slope 2

c) L(-2, 6), M(5, y)   slope -1

d) E(0, y), F(4, -2)   slope $-\frac{3}{2}$

19. Choose one part of exercise 17 or 18. Write to explain how you determined the value of x or y.

20. Look at this diagram.

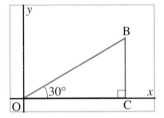

a) You are given BC = 4. Determine the length of OC, the slope of OB, and the length of OB.

b) You are given OB = 16. Determine the length of OC, the length of BC, and the slope of OB.

c) You are given OC = 12. Determine the length of BC, the slope of OB, and the length of OB.

C **21.** One endpoint of a line segment is A(4, 6). The other endpoint is on the x-axis. Determine the coordinates of the endpoint on the x-axis for each given slope of the segment.

a) 1   **b)** 2   **c)** 3   **d)** $\frac{1}{2}$ **e)** -2 **f)** $-\frac{1}{2}$

22. One endpoint of a line segment is B(-3, 4). The other endpoint is on the y-axis. Determine the coordinates of the endpoint on the y-axis for each given slope of the segment.

a) 3   **b)** -2 **c)** 1   **d)** $\frac{1}{2}$ **e)** $-\frac{1}{4}$

23. Two line segments have a common endpoint N(0, 4). They have slopes 2 and $-\frac{1}{2}$. The other endpoints are on the x-axis. Determine their coordinates.

24. A line segment has length 10. Its endpoints are on the coordinate axes. The slope of the line segment is $-\frac{3}{4}$. Determine the possible coordinates of its endpoints.

COMMUNICATING THE IDEAS

Look up the word *slope* in the dictionary. Is there more than one definition? Are any of the definitions similar in meaning to the mathematical term? Record your ideas in your journal.

Building the Best Staircase

Staircases come in many shapes and sizes. There are circular staircases, wide and narrow staircases, long and short staircases.

All staircases have three main parts.

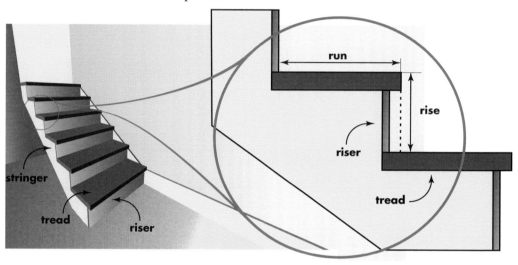

The run or *tread* is the horizontal part of a stair.

The rise or *riser* is the vertical part of a stair.

The *stringers* are the sloping boards running diagonally between floors, on either side of the treads.

The stringers support and stabilize the staircase. Their slope determines the steepness of the staircase.

1. Look at these 3 staircases.

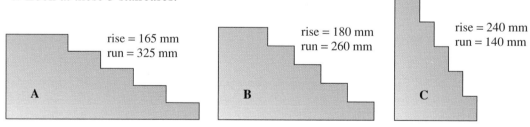

rise = 165 mm
run = 325 mm

rise = 180 mm
run = 260 mm

rise = 240 mm
run = 140 mm

A   B   C

a) How is the number of risers related to the number of treads?

b) Which staircase would most likely be used as the main staircase in a home? Explain your choice.

c) Which staircase is best suited for accessing an attic? Explain.

d) Most staircase accidents occur during descent. Which staircase do you think would be safest? Explain.

Mathematics & Construction

2. In the 17th century, Francois Blondel was the director of the Royal Academy of Architecture in Paris. Blondel proposed the ideal staircase: the measurements of the run plus twice the rise equalled 620 mm. Blondel determined that 620 mm was the average length of a person's normal walking stride on horizontal ground. Carpenters still use Blondel's rule today.

   a) Let x represent the run. Let y represent the rise. Write an equation to represent Blondel's rule.

   b) Which of the staircases in exercise 1 satisfy your equation?

   c) Assume that Blondel's rule was based on safety issues. Do you think Blondel's rule ensures safety equally well for all people? Explain.

   d) Use your equation. Calculate the rise for a staircase with a run of 520 mm. Where might you find such a staircase?

3. The Canadian Mortgage and Housing Corporation (CMHC) recommends that staircases have a maximum rise of 200 mm and a minimum run of 250 mm.

   a) Use Blondel's rule. Calculate the run that corresponds to the maximum rise recommended by the CMHC. What is the slope of the stringer?

   b) Calculate the rise that corresponds to the minimum run. What is the slope of the stringer?

   c) The vertical distance between two floors is 2.95 m. Use the CMHC guidelines and Blondel's rule. Calculate the minimum number of treads for a staircase between these floors.

MODELLING Staircase Design

Francois Blondel's rule provides a model for staircase design.

- To apply Blondel's rule in problems similar to exercise 3c, it is necessary to round results. Describe what would happen if Blondel's rule were used without rounding.

- Measure the riser and tread for a staircase at school or at home. How well does the staircase follow Blondel's rule? Explain.

4. A staircase may be constructed off site, then delivered and installed. Suppose a carpenter miscalculated and made a staircase with stringers 55 mm too long. The staircase was installed. Explain the problems that might arise with this staircase.

LINKING IDEAS

Mathematics & Construction

Length, Midpoint, and Slope of a Line Segment

1. This spreadsheet will determine the length, the coordinates of the midpoint, and the slope of any line segment with given endpoints.

	A	B	C
1	Length, Midpoint, and Slope of a Line Segment		
2	Endpoint	3	2
3	Endpoint	7	8
4	Rise	=C3-C2	
5	Run	=B3-B2	
6	Length	=SQRT(B4^2+B5^2)	
7	Midpoint	=(B2+B3)/2	=(C2+C3)/2
8	Slope	=B4/B5	

a) Start a new spreadsheet document. Enter the text and formulas shown above.

b) Explain the purpose of the formulas in cells B4 and B5.

c) Explain the purpose of the formulas in cells B6, B7, C7, and B8.

2. To test your spreadsheet, enter the coordinates of one endpoint of a line segment in cells B2 and C2. Enter the coordinates of the other endpoint in cells B3 and C3. Use examples from previous sections of this chapter, or make up your own examples. Include some horizontal and vertical line segments.

3. Document the spreadsheet so that someone else can use it without assistance.

Investigation 1

4. **a)** On a sheet of grid paper, draw a diagram like the one below.

   b) Visualize a point P moving along the vertical line through point A in the positive direction. Visualize how line segment OP changes as point P moves farther from point A. Visualize what happens to the coordinates of the midpoint of OP, to the slope of OP, and to the length of OP.

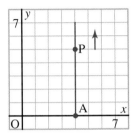

5. Copy this table in your notebook.

Coordinates of P		Midpoint of OP		Slope of OP	Length of OP
x-coord	y-coord	x-coord	y-coord		
4	0				
4	1				

   a) Use your spreadsheet from the preceding page. Complete the table for several positions of point P on the vertical line through point A.

   b) Compare the coordinates of the midpoint of OP, the slope of OP, and the length of OP with the coordinates of point P. What patterns can you find? Explain the patterns.

6. **a)** Visualize a point P on the vertical line through point A so that OP is twice as long as OA. Use your spreadsheet. Determine the coordinates of this point.

   b) Plot this point on your grid, and label it P. Use a ruler to join OP. Use a protractor to measure ∠POA. What do you notice? Explain.

Investigation 2

In this diagram, segment OA has length 5 units. Visualize this segment rotating counterclockwise around the origin O. Point A will trace out a circle with diameter 10 units. This diagram shows a few positions of this segment. Observe that the outside endpoints form a polygon with 28 sides. In this investigation, you will calculate the perimeter of this polygon and use the result to determine an approximation of π.

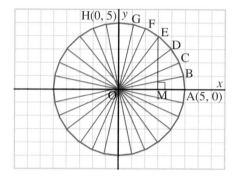

7. **a)** Calculate the circumference of the circle traced by point A as it moves around O.

   b) How does the perimeter of the polygon compare with the circumference of the circle?

8. **a)** Observe that the *y*-coordinate of E is 4 and that △OEM is a right triangle. Calculate the length of OM, which is the *x*-coordinate of E.

   b) Use a similar method to calculate the *x*-coordinates of B, C, and D.

   c) Use your spreadsheet from page 172. Determine the lengths of segments AB, BC, CD, and DE.

   d) Determine the lengths of segments EF, FG, and GH.

9. Use the results of exercise 8. Calculate the perimeter of the 28-sided polygon formed by the outside endpoints of the line segments.

10. Compare your answer in exercise 9 with the circumference of the circle you calculated in exercise 7a.

    a) Use the result of exercise 9. Determine an approximation of π. Explain your method.

    b) Is your approximation less than π or greater than π? Use the diagram to explain your answer.

    c) What is the difference between your approximation of π and the approximation of π on page 101? Express this difference as a percent of the approximation on page 101.

3.4 Slopes of Parallel Line Segments

1. Mark equal segments AB and DE on opposite edges of a square piece of paper. Visualize the paper in different positions on a coordinate grid.

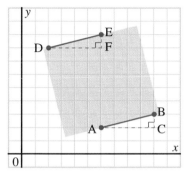 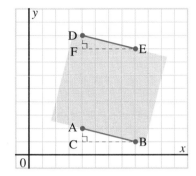

a) How are line segments AB and DE related? Explain how you know.

b) How does the rise from A to B compare with the rise from D to E?

c) How does the run from A to B compare with the run from D to E?

d) How are the slopes of AB and DE related? Explain how you know.

2. Would your answers to exercise 1 change if AB and DE had different lengths? Explain.

3. Write a general conclusion about the relationship between the slopes of parallel line segments.

In this diagram, line segments AB and CD are parallel.

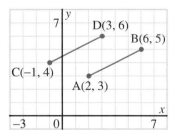

$$\text{Slope of AB} = \frac{5-3}{6-2} \qquad\qquad \text{Slope of CD} = \frac{6-4}{3-(-1)}$$

$$= \frac{2}{4} \qquad\qquad\qquad\qquad\qquad = \frac{2}{4}$$

$$= \frac{1}{2} \qquad\qquad\qquad\qquad\qquad = \frac{1}{2}$$

This example illustrates a fundamental property of slope.

If the slopes of two line segments are equal, the segments are parallel.
If two non-vertical line segments are parallel, their slopes are equal.

Example 1

Determine whether the quadrilateral with vertices A(0, −6), B(2, −1), C(−1, 5), and D(−3, 0) is a parallelogram.

Solution

Draw the quadrilateral on a grid.

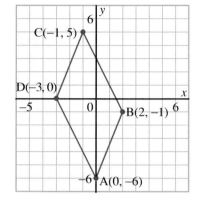

Slope of AB = $\dfrac{-1-(-6)}{2-0}$

$= \dfrac{5}{2}$

Slope of DC = $\dfrac{5-0}{-1-(-3)}$

$= \dfrac{5}{2}$

Slopes of AB and DC are equal, therefore AB ∥ DC.

Slope of AD = $\dfrac{0-(-6)}{-3-0}$

$= \dfrac{6}{-3}$

$= -2$

Slope of BC = $\dfrac{5-(-1)}{-1-2}$

$= \dfrac{6}{-3}$

$= -2$

Slopes of AD and BC are equal; therefore AD ∥ BC.

Since both pairs of opposite sides are parallel, ABCD is a parallelogram.

Example 2

The points P(2, −5), Q(−2, 1), and R(3, −1) are given. Determine the coordinates of a point S on the y-axis so that line segment RS is parallel to line segment PQ.

Solution

Graph the points P, Q, and R.
Let S(0, y) be the point on the y-axis.

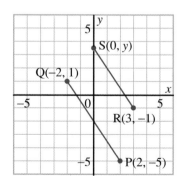

Slope of PQ = $\dfrac{1-(-5)}{-2-2}$

$= \dfrac{6}{-4}$

$= -\dfrac{3}{2}$

Slope of RS = $\dfrac{y-(-1)}{0-3}$

$= \dfrac{y+1}{-3}$

Since PQ ∥ RS, the slopes are equal:

$$\dfrac{y+1}{-3} = -\dfrac{3}{2}$$ Multiply each side by −3.

$$\dfrac{y+1}{-3} \times \dfrac{-3}{1} = \dfrac{-3}{2} \times \dfrac{-3}{1}$$

$$y + 1 = 4.5$$

$$y = 4.5 - 1$$

$$= 3.5$$

The coordinates of S are (0, 3.5).

1. Look at the last sentence on page 175. Explain why the word "non-vertical" is included.

2. In *Example 1*, do you have to check that both pairs of opposite sides are parallel? Would it be sufficient to check that only one pair is parallel? Explain.

3. In *Example 2*, what are the coordinates of the midpoint of PQ? Label this point M. How does the slope of PM compare with the slope of RS? Explain.

3.4 EXERCISES

 1. Which pairs of line segments are parallel? Explain how you know.

a)
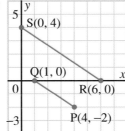

b)

c)
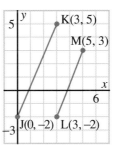

2. The coordinates of the endpoints of pairs of line segments are given. Graph each pair of line segments. Determine their slopes. Which line segments are parallel?

   a) A(−2, −1), B(1, 5) and C(2, −1), D(4, 3)

   b) E(−3, 2), F(5, 5) and O(0, 0), H(5, 2)

   c) R(−1, 4), S(7, −2) and T(3, 4), U(9, 0)

3. The coordinates of the vertices of quadrilaterals are given. Draw each quadrilateral on a grid. Determine whether it is a parallelogram.

   a) A(5, 3), B(−3, −3), C(−2, −8), D(6, −2)

   b) P(−6, 1), Q(−2, −6), R(10, 2), S(7, 9)

   c) J(−4, 5), K(−2, −1), L(6, −4), M(4, 2)

4. Choose one part of exercise 3. Write to explain how you determined whether the quadrilateral was a parallelogram.

5. a) Determine the slope of the line segment joining each point to the origin.

    i) A(3, 2)   **ii)** B(6, 4)   **iii)** C(9, 6)   **iv)** D(12, 8)

  b) What patterns do you see in part a?

  c) Use the patterns in part a. Write the coordinates of 3 other points that lie on segment OD.

6. Graph each pair of line segments. Calculate the slopes only for the pairs of line segments that are parallel.

  a) A(−4, 12), B(−2, 2) and C(4, −6), D(2, 6)

  b) E(3, 0), F(9, 9) and G(2, 5), H(11, −1)

  c) J(2, 10), K(−7, −2) and L(7, −4) and M(16, 8)

  d) N(−5, −8), P(8, −7) and Q(−4, −10) and R(11, −10)

B **7.** A triangle has vertices A(2, 5), B(−4, 1), and C(6, −3). The midpoints of AC and BC are M and N, respectively.

  a) Graph the triangle. Determine the coordinates of M and N.

  b) Determine the slopes of MN and AB. What do you notice?

  c) Let P be the midpoint of AB. Determine the slopes of PM, BC, PN, and AC. What do you notice?

8. A quadrilateral has vertices A(−4, 0), B(8, 2), C(2, 8), and D(−2, 4).

  a) Graph the quadrilateral. Locate the midpoint of each side.

  b) Draw the quadrilateral formed by the midpoints in part a. What kind of quadrilateral is it?

  c) Use slopes to confirm your prediction in part b.

9. The points A(−2, 0), B(6, 4), and C(−3, 4) are given. Determine the coordinates of a point D on the y-axis so that line segment CD is parallel to AB.

10. The points A(6, 3), B(2, 9), and C(2, 3) are given. Determine the coordinates of a point D so that CD is parallel to AB when D is on the y-axis.

11. Use the information on page 144. At the start of a race, it takes a certain time for runners to react to the sound of the starter's pistol. Suppose this time is subtracted from the runners' times for the two races.

  a) About how long do you think the reaction time might be?

  b) Use your estimate from part a to modify the graphs in Section 3.3 exercise 14. Repeat the slope calculations. Does this change your opinion about who was faster?

  c) Explore the effect of using different reaction times. What would the reaction time have to be for this model to favour Donovan Bailey?

In exercise 11, you re-examined the graphical models of Donovan Bailey's and Michael Johnson's world-record sprints.

- Do you think the method of exercise 11 introduces a fairer way to compare the runners? Do you still need more information? Explain.

12. a) The coordinates of the endpoints of a line segment are given. For each line segment, write the coordinates of the endpoints of a parallel line segment.

i) A(7, 6), B(−6, 3)   **ii)** C(−3, 7), D(1, −5)

iii) E(2, 3), F(−2, −7)   **iv)** G(−4, 2), H(6, −4)

b) Compare your answers for part a with those of a classmate. Write to explain any differences.

13. Determine the value of x so that the line segment with endpoints A(x, 3) and B(1, 7) is parallel to the line segment with endpoints C(2, −4) and D(5, −2).

14. Determine the value of y so that the line segment with endpoints A(2, y) and B(8, 6) is parallel to the line segment with endpoints C(−7, 6) and D(3, 1).

C **15.** The coordinates of three vertices of a parallelogram are given. Determine all possible coordinates of the fourth vertex.

a) A(−4, 1), T(−3, −4), G(5, 0)   **b)** W(−4, 4), B(1, −2), M(8, −5)

16. The coordinates of the midpoints of the sides of a triangle are given. Determine the coordinates of the vertices of the triangle.

a) A(5, 0), B(2, 3), C(7, 3)   **b)** P(−3, 3), Q(−3, 8), R(5, 5)

COMMUNICATING THE IDEAS

Suppose your friend telephones you to discuss tonight's homework. How would you explain, over the telephone, how to tell if two line segments are parallel? Use examples to illustrate your explanations. Record your answer in your journal.

Who Is the World's Fastest Human?

It is difficult to compare Donovan Bailey's and Michael Johnson's 1996 results, because they ran races of different lengths. All the models used so far assume that each sprinter ran at a constant speed, but this is not the case.

High-speed cameras revealed that each sprinter accelerated from 0 m/s to top speed in the first part of the race, then reduced speed gradually to the end of the race.

DEVELOP A MODEL

These tables show the times for each 10-m segment of the race, for each sprinter.

Bailey's time (s)	0.17	1.9	3.1	4.1	4.9	5.6	6.5	7.2	8.1	9.0	9.84
Distance (m)	0	10	20	30	40	50	60	70	80	90	100

Johnson's time (s)	0.16	1.9	3.3	4.5	5.5	6.3	7.0	7.7	8.4	9.2	10.12
Distance (m)	0	10	20	30	40	50	60	70	80	90	100
Time (s)		11.1	12.1	13.1	14.0	14.8	15.6	16.4	17.3	18.2	19.32
Distance (m)		110	120	130	140	150	160	170	180	190	200

The times for the second half of Johnson's race are not official, but they are good estimates based on the work of Professor Tibshirani of the Statistics Department at the University of Toronto.

Suppose you were to graph the data for each sprinter. What might each graph look like?

How could you interpret the graphs? How might that interpretation help you determine the world's fastest human?

1. Plot the data for each sprinter on a distance-time graph. Join adjacent points with line segments.

2. **a)** Identify the steepest segment on each graph. Determine its slope.

   b) What does the slope of each segment represent?

Another way to examine the sprinters' data is to graph speed against time. The average speed for each 10-m interval was calculated for each runner. Here are the results.

Bailey's time (s)	0.17	1.9	3.1	4.1	4.9	5.6	6.5	7.2	8.1	9.0	9.84
Speed (m/s)	0	5.78	8.33	10.00	12.50	14.29	11.11	14.29	11.11	11.11	11.90

Johnson's time (s)	0.16	1.9	3.3	4.5	5.5	6.3	7.0	7.7	8.4	9.2	10.12
Speed (m/s)	0	5.75	7.14	8.33	10.00	12.50	14.29	14.29	14.29	12.50	10.87
Time (s)		11.1	12.1	13.1	14.0	14.8	15.6	16.4	17.3	18.2	19.32
Speed (m/s)		10.20	10.0	10.0	11.1	12.5	12.5	12.5	11.1	11.1	8.93

3. Plot the data for each sprinter on a speed-time graph. Join adjacent points with line segments. You may do this using a spreadsheet.

 LOOK AT THE IMPLICATIONS

4. Do your results in exercise 2 enable you to decide who was faster at the 1996 Summer Olympics? Explain.

5. Use the graphs you drew in exercises 1 and 3. Compare the performances of Johnson and Bailey. What do you conclude?

6. Following the fastest-human controversy in the summer of 1996, Bailey and Johnson agreed to race in a 150-m sprint in June, 1997. Who do you think should have won this race? Explain your choice.

 REVISIT THE SITUATION

7. In their 150-m race in June, 1997, Donovan Bailey won in a time of 14.99 s. Michael Johnson pulled up lame and failed to finish.

a) How does this result compare with your prediction in exercise 6?

b) What shortcomings are there in concluding that Donovan Bailey is the fastest human?

8. Create a new problem related to the situation on these pages. Solve it by applying or modifying the model you developed above.

Solving Equations of the Form $\dfrac{a}{b} = \dfrac{c}{d}$

In the diagram, R(0, −3) is a fixed point on the y-axis. Point S is on the x-axis. Visualize how the slope of RS changes as S moves along the x-axis. Suppose we want to determine the coordinates of S when the slope of RS is $\frac{2}{5}$.

Slope of RS $= \dfrac{0 - (-3)}{x - 0}$

$\qquad\qquad = \dfrac{3}{x}$

Since the slope is $\frac{2}{5}$:

$$\frac{2}{5} = \frac{3}{x} \quad ①$$

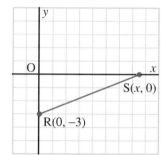

Both sides of this equation are in fraction form. That is, the equation has the form $\frac{a}{b} = \frac{c}{d}$. We solve this equation in the same way that we solve any other equation. The same operation must be performed on each side.

Multiply each side by x.

$$\frac{2}{5} \times x = \frac{3}{x} \times x$$

$$\frac{2x}{5} = 3$$

Multiply each side by 5.

$$\frac{2x}{5} \times 5 = 3 \times 5$$

$$2x = 15 \quad ②$$

Divide each side by 2.

$$x = \frac{15}{2}$$

$$\quad = 7.5$$

Look how we solved the equation. In the first step, we multiplied each side by one of the denominators. In the second step, we multiplied each side by the other denominator. By comparing equation ② with equation ①, observe how these two steps can be completed together. That is, we can solve the equation using this shortcut, as follows.

$$\frac{2}{5} = \frac{3}{x}$$
$$2 \times x = 3 \times 5$$
$$2x = 15$$
$$x = \frac{15}{2}$$
$$= 7.5$$

This shortcut can only be used with equations of the form $\frac{a}{b} = \frac{c}{d}$. It is a convenient way to multiply each side by the two denominators in the same operation.

1. Solve each equation.

a) $\frac{x}{10} = \frac{3}{5}$ b) $\frac{a}{12} = \frac{3}{4}$ c) $\frac{3}{n} = \frac{2}{5}$ d) $\frac{2}{7} = \frac{1}{x}$

e) $\frac{3}{15} = \frac{4}{m}$ f) $\frac{8}{5} = \frac{y}{4}$ g) $\frac{x}{2} = 5$ h) $\frac{6}{a} = 12$

2. Solve only those equations that can be solved using the shortcut.

a) $\frac{5}{4} = \frac{3}{n} + 2$ b) $\frac{7}{x} = \frac{3}{4}$ c) $1 + \frac{x}{2} = \frac{3}{5}$ d) $\frac{c}{5} = 4$

e) $\frac{x}{2} = \frac{1}{3} + 6$ f) $\frac{8}{12} = \frac{m}{9}$ g) $\frac{2}{3} = \frac{1}{r} - 4$ h) $\frac{m}{15} = \frac{5}{6}$

3. Why can the shortcut only be used to solve equations of the form $\frac{a}{b} = \frac{c}{d}$? Use examples to illustrate your answer.

4. Solve each equation.

a) $\frac{n}{3} = -7$ b) $\frac{3}{4} = \frac{-3}{x}$ c) $\frac{4}{-5} = \frac{2a}{9}$ d) $\frac{s}{10} = \frac{-3}{4}$

5. The slopes of two parallel line segments are given. Determine each value of k.

a) $\frac{2}{3}, \frac{4}{k}$ b) $\frac{3}{2}, \frac{k}{-4}$ c) $\frac{-1}{5}, \frac{2}{k}$ d) $\frac{4}{7}, \frac{-k}{2}$

e) $\frac{k}{4}, 2$ f) $\frac{-k}{5}, \frac{3}{2}$ g) $\frac{k}{2}, \frac{3}{2}$ h) $\frac{-k}{3}, \frac{-2}{7}$

6. Determine the coordinates of a point G on the x-axis so that when it is joined to the point F(0, 5) on the y-axis, the slope of FG has each value.

a) $\frac{2}{3}$ b) $-\frac{2}{3}$ c) $\frac{5}{2}$ d) $-\frac{5}{2}$

7. The following equation occurred in the solution of *Example 2* on page 176.
$$\frac{y + 1}{-3} = -\frac{3}{2}$$

a) Solve this equation using the shortcut. Compare your solution with the solution on page 176.

b) Explain why the equation can be solved using the shortcut.

8. Refer to exercise 10 on page 178. Solve a similar problem in which D is on the x-axis.

We discovered in the preceding section that parallel line segments have the same slope. Is there a relationship between the slopes of perpendicular line segments?

INVESTIGATE

1. Mark equal segments AB and DE on adjacent edges of a square piece of paper. Visualize the paper in different positions on a coordinate grid.

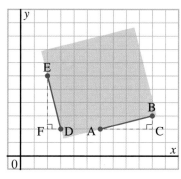 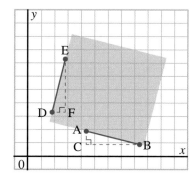

   a) How are line segments AB and DE related? Explain how you know.

   b) How does the rise from A to B compare with the run from D to E?

   c) How does the run from A to B compare with the rise from D to E?

   d) How are the slopes of AB and DE related? Explain how you know.

2. Would your answers to exercise 1 change if AB and DE had different lengths? Explain.

3. Write a general conclusion about the relationship between the slopes of perpendicular line segments.

Here is another way to illustrate the relationship between the slopes of perpendicular line segments.

Draw a right triangle and rotate it 90°.

In the diagram, right △ONA is rotated 90° counterclockwise about O(0, 0) to △ON′A′.

The coordinates of A are (5, 3).

The coordinates of A′ are (−3, 5), and OA ⊥ OA′.

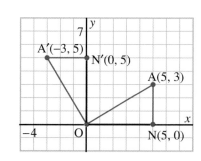

Slope of OA $= \frac{3 - 0}{5 - 0}$

$\qquad = \frac{3}{5}$

Slope of OA′ $= \frac{5 - 0}{-3 - 0}$

$\qquad = -\frac{5}{3}$

The numbers $\frac{3}{5}$ and $-\frac{5}{3}$ are *negative reciprocals*. The product of negative reciprocals is -1. This example suggests the relation between slopes of perpendicular line segments.

If the slopes of two line segments are negative reciprocals, the segments are perpendicular.

If two line segments are perpendicular (and neither one is vertical), their slopes are negative reciprocals.

Example 1

A triangle has vertices A(-2, 3), B(8, -2), and C(4, 6). Determine whether it is a right triangle.

Solution

Graph the triangle.

From the graph, $\angle C$ appears to be a right angle.

Calculate the slopes of AC and BC.

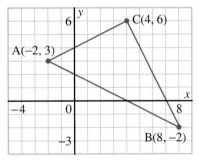

Slope of AC $= \frac{6 - 3}{4 - (-2)}$ $\qquad$ Slope of BC $= \frac{6 - (-2)}{4 - 8}$

$\qquad = \frac{3}{6}$ $\qquad\qquad\qquad\qquad = \frac{8}{-4}$

$\qquad = \frac{1}{2}$ $\qquad\qquad\qquad\qquad = -2$

Since $\left(\frac{1}{2}\right)(-2) = -1$, AC is perpendicular to BC, and $\triangle$ABC is a right triangle.

Example 2

A line segment has endpoints A(5, 4) and B(1, −2).

a) Determine the coordinates of a point C so that line segment AC is perpendicular to AB.

b) What are the coordinates of C when C is on the *x*-axis?

Solution

a) Draw segment AB on a grid.

$$\text{Slope of AB} = \frac{-2-4}{1-5}$$

$$= \frac{-6}{-4}$$

$$= \frac{3}{2}$$

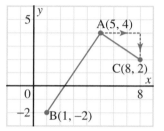

The slope of a line segment perpendicular to AB is $\frac{-2}{3}$. Begin at A. Move 3 right and 2 down. A possible point is C(8, 2).

b) Let C(x, 0) represent the point on the *x*-axis.

$$\text{Slope of AC} = \frac{4-0}{5-x}$$

$$= \frac{4}{5-x}$$

Since AC is perpendicular to AB, the slope of AC is $-\frac{2}{3}$.

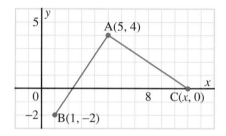

$$-\frac{2}{3} = \frac{4}{5-x}$$

$$-2(5-x) = 3 \times 4$$

$$-10 + 2x = 12$$

$$2x = 22$$

$$x = 11$$

The coordinates of C are (11, 0).

DISCUSSING THE IDEAS

1. In the second sentence in the display on page 185, explain why the words "and neither one is vertical" are included.

2. In *Example 1*, what are the advantages of drawing the graph?

3. In *Example 2*, what other possibilities are there for point C?

4. What other ways are there to solve the equation in *Example 2b*?

A **1.** Which pairs of line segments are perpendicular? Explain how you know.

a)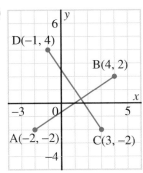

b)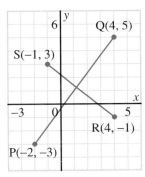

c)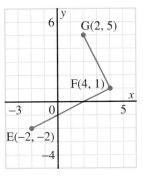

2. Determine the slope of a line segment perpendicular to a segment with each given slope.

a) $\frac{2}{3}$ b) $\frac{5}{8}$ c) $-\frac{3}{4}$ d) $-\frac{1}{2}$ e) $-\frac{1}{3}$

3. Which pairs of numbers are slopes of perpendicular line segments?

a) $\frac{3}{4}, -\frac{4}{3}$ b) $\frac{2}{3}, \frac{3}{2}$ c) $\frac{4}{5}, -\frac{4}{5}$ d) $-4, \frac{1}{4}$ e) $3, \frac{1}{3}$ f) $2, -\frac{1}{2}$

4. The coordinates of the endpoints of pairs of line segments are given. Graph each line segment. Determine its slope. Which line segments are perpendicular?

a) O(0, 0), B(6, 4) and C(5, −1), D(1, 5)

b) H(−3, 1), I(6, 4) and J(2, 0), K(0, 6)

c) L(5, −3), M(1, 4) and N(1, −1), P(6, 2)

5. Choose one part of exercise 4. Write to explain how you determined whether the line segments were perpendicular.

6. Graph each pair of line segments. Calculate the slopes only for the pairs of line segments that are perpendicular.

a) A(−4, 12), B(−2, 2) and C(7, 3), D(−11, 1)

b) E(3, 0), F(9, 9) and G(13, 0), H(4, 9)

c) J(2, 10), K(−7, −2) and L(7, 0) and M(−5, 9)

d) N(−5, −8), P(8, −7) and P(8, −7) and R(7, −5)

B **7.** The coordinates of the vertices of triangles are given. Draw each triangle on a grid. Determine whether it is a right triangle.

a) D(−2, 2), E(−6, 2), F(−6, −1)   b) A(3, 0), B(−4, 4), C(−1, −2)

c) P(−3, 1), Q(3, −3), R(7, 3)   d) K(3, 2), L(−5, −1), M(−2, −8)

8. The coordinates of the vertices of quadrilaterals are given. Draw each quadrilateral on a grid. Determine whether it is a rectangle.

   a) A(5, 4), B(−4, −2), C(−2, −5), D(7, 1)

   b) J(−3, 2), K(−2, −3), L(6, −2), M(5, 3)

   c) P(5, 1), Q(−4, 4), R(−6, −2), S(3, −5)

9. The coordinates of the endpoints of line segments are given. Determine the slope of a line segment parallel to each segment.

   a) A(0, 4), B(2, 0)   b) C(−1, 1), D(3, 3)   c) E(4, −2), F(−1, 3)

   d) G(0, 1), H(−5, −1)   e) J(2, 3), K(−1, 3)   f) K(3, 5), M(3, −2)

10. Determine the slope of a line segment perpendicular to each segment in exercise 9.

11. The slopes of perpendicular line segments are given. Calculate each value of k.

   a) $3, k$ b) $\frac{-1}{2}, k$ c) $\frac{k}{2}, 2$ d) $\frac{k}{3}, -3$ e) $\frac{k}{4}, -4$

   f) $\frac{k}{2}, \frac{1}{4}$ g) $\frac{6}{k}, \frac{-2}{3}$ h) $\frac{3}{5}, \frac{k}{6}$ i) $\frac{1}{3}, \frac{k}{2}$ j) $\frac{-1}{7}, \frac{k}{5}$

12. The coordinates of the endpoints of line segments are given. Graph each segment. Determine the coordinates of a point C so that AC is perpendicular to AB.

   a) A(3, 2), B(6, 8)   b) A(0, 5), B(5, 3)   c) A(1, 3), B(1, −2)

   d) A(−2, 4), B(4, 1)   e) A(0, 0), B(7, 2)   f) A(4, 3), B(−2, 3)

13. Choose one part of exercise 12. Write to explain how you determined the coordinates of C.

14. A line segment has endpoints C(6, 2) and D(8, 5). Point P is such that PC is perpendicular to CD. Determine the coordinates of P if P is on:

   a) the x-axis   b) the y-axis

15. a) Graph the quadrilateral with vertices A(−2, 7), B(−3, −1), C(4, −5), and D(5, 3). Calculate the lengths of the four sides.

   b) What kind of quadrilateral is ABCD?

   c) Determine the coordinates of the midpoint of each diagonal.

   d) Determine the slope of each diagonal.

   e) What can you conclude from parts c and d?

16. Points A, B, and C are three vertices of a rectangle. Plot the points on a grid. Determine the coordinates of the fourth vertex.

   a) A(2, −1), B(5, −3), C(7, 0)   b) A(1, 8), B(−3, −2), C(6, 6)

   c) A(−4, 7), B(−6, 4), C(3, −2)   d) A(2, 4), B(−2, 2), C(1, −4)

17. Choose one part of exercise 16. Write to explain how you determined the coordinates of the vertex.

18. a) The coordinates of the endpoints of a line segment are given. For each line segment, write the coordinates of the endpoints of a perpendicular line segment.

   i) A(7, 6), B(−6, 3)   ii) C(−3, 7), D(1, −5)

   iii) E(2, 3), F(−2, −7)   iv) G(−4, 2), H(6, −4)

   b) Compare your answers for part a with those of a classmate. Write to explain any differences.

19. In this diagram, ABCD is a square.

Determine the coordinates of C and D for each pair of coordinates of A and B.

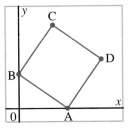

   a) A(1, 0), B(0, 4)   b) A(3, 0), B(0, 7)

   c) A(9, 0), B(0, 1)   d) A(a, 0), B(0, b)

20. The line segment joining R(8, 6) and S(4, 8) is the shortest side of a right △RST. Point T is on the x-axis. Determine the possible coordinates of T.

C 21. Two vertices of an isosceles triangle are A(−5, 4) and B(3, 8). The third vertex is on the x-axis. Determine the possible coordinates of the third vertex.

22. The coordinates of two vertices of a square are given. Determine the possible coordinates of the other two vertices.

   a) O(0, 0), A(2, 5)   b) B(3, 1), C(0, 5)

23. This diagram shows right △OAB with squares drawn on the three sides. Points P, Q, and R are the centres of the three squares.

   a) Draw the diagram on a grid.

   b) Show that each pair of segments is perpendicular and equal in length.

   i) AP and QR   ii) BQ and PR   iii) OR and PQ

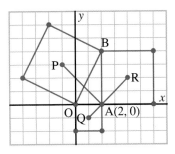

Suppose your friend asks you for help with tonight's homework. How would you explain how to tell if two line segments are perpendicular? Use examples to illustrate your explanations. Record your answer in your journal.

Square Patterns on a Grid

Young children may recognize the first figure below as a square, but not the other three. All four figures are squares, because they all have four equal sides and four right angles. These characteristic properties are not affected by orientation.

Keep this in mind as you solve the following problems.

1. The pattern (below left) was drawn on a 5 by 5 grid of dots. The dots are 1 cm apart.

   a) How many squares are there on the diagram?

   b) The pattern was created by drawing as many squares as possible with every square having one vertex at the centre of the grid. How many squares were drawn?

   c) Your answers to parts a and b should be different. Explain why.

2. Although some of the squares on the pattern (below left) have the same area, there are squares with several different areas.

   a) How many different areas of squares are there?

   b) What are the areas of these squares? What are the side lengths?

   c) What kind of sequence do the areas of these squares form? What kind of sequence do the side lengths of these squares form? Explain your answers.

3. a) Create another pattern similar to that below left, in which all the squares have vertices on the same dot of the grid.

   b) Depending on which dot you choose, your diagram may or may not be symmetrical. Explain why your diagram is or is not symmetrical.

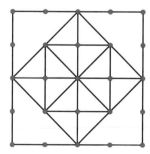 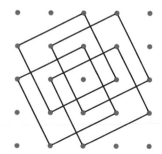

4. Four squares were drawn to create the pattern (below right on page 190).

a) What common property do these squares have?

b) How many other squares also have this property?

c) Create another pattern in which all the squares have a similar property.

5. Squares are drawn with vertices on the dots of a 5 by 5 grid.

a) How many different areas of squares are there? Support your answer with appropriate drawings and calculations.

b) Organize your drawings and your calculations in a systematic way so that you could convince someone else that you have covered all the possibilities.

6. Squares that have the same area are congruent. Squares that do not have the same area are *non-congruent*.

a) How many non-congruent squares can be drawn on a 5 by 5 grid?

b) Use your results from part a. Predict how many non-congruent squares can be drawn on a 7 by 7 grid, and on a 9 by 9 grid.

c) State a rule you could use to determine how many non-congruent squares can be drawn on an *n* by *n* grid, where *n* is any odd number.

7. a) Verify that the diagram (below left) can be used to illustrate the sum $1 + 3 + 5 + 7 + 9$.

b) Explain why similar diagrams can be used to illustrate the sum of the arithmetic series $1 + 3 + 5 + \cdots + n$, where *n* is any odd number.

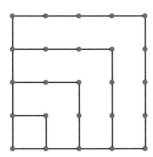 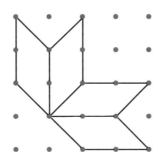

8. The diagram (above right) shows four congruent parallelograms, each with an area of 2 square units. Each parallelogram has one vertex at a common point.

a) Identify this point on the diagram.

b) What are the side lengths of the parallelograms?

c) Each parallelogram has two possible heights, measured between a pair of parallel sides. Determine the two possible heights for each parallelogram.

d) Draw a diagram like that above right. Include all other parallelograms that have an area of 2 square units, and one vertex at the same common point.

1. Determine the length of each line segment with each pair of endpoints.

   a) A(−6, 3), B(9, 3)
   b) C(4, 11), D(4, −3)
   c) E(3, 7), F(−7, −3)

   d) G(−8, 9), H(−3, 4)
   e) J(5, 2), K(1, −6)
   f) L(−5, 3), M(3, −5)

2. A rectangle has vertices A(−2, −4), B(6, −4), C(6, 2), and D(−2, 2).

   a) Determine the lengths of the sides.
   b) Determine the perimeter.

   c) Determine the lengths of the diagonals.
   d) Determine the area.

3. On a grid, draw the triangle with each set of vertices. Name the type of triangle.

   a) A(−3, 2), B(0, −2), C(4, 1)
   b) P(−3, 7), Q(5, 3), R(7, 7)

4. Choose one part of exercise 3. Write to explain how you identified the triangle.

5. Determine the coordinates of the midpoint of the line segment with each pair of endpoints.

   a) A(3, −8), B(3, 12)
   b) C(5, −1), D(−11, −1)
   c) E(2, 6), F(−10, −12)

   d) G(5, −3), H(13, −7)
   e) J(−2, 8), K(5, 4)
   f) L(3, 7), M(4, −2)

6. A triangle has vertices K(3, 7), L(−5, 1), and M(7, −5).

   a) Determine the coordinates of the midpoint of each side.

   b) Determine the length of the median from L to KM.

7. Determine the slope of the line segment with each pair of endpoints.

   a) A(3, 8), B(−1, −2)
   b) C(8, −6), D(−3, −6)
   c) E(−3, 8), F(4, −5)

   d) G(−7, 2), H(5, 10)
   e) J(6, −3), K(−8, 7)
   f) L(2, −6), M(−3, 9)

8. A triangle has vertices P(−3, 4), Q(−2, 2), and R(6, 2). Points S and T are the midpoints of PR and QR respectively.

   a) Determine the slopes of ST and PQ.
   b) Determine the lengths of ST and PQ.

9. The vertices of a quadrilateral are P(1, 4), Q(7, 6), R(11, 2), and S(3, −6). Show that the midpoints of the sides of the quadrilateral are the vertices of a parallelogram.

10. A triangle has vertices A(5, 7), B(−3, 4), and C(2, −5).

   a) Determine the slopes of AB and AC.

   b) Determine the slope of a line segment through point A parallel to BC.

   c) Determine the slope of the median from point B to AC.

11. A quadrilateral has vertices A(-1, 4), B(-3, -2), C(3, -1), and D(4, 5). Determine whether the quadrilateral is a parallelogram.

12. A triangle has vertices P(-4, -2), Q(6, 4), and R(-7, 3). Show that $\angle$QPR is a right angle. Write to explain how you completed the exercise.

13. State the slope of a line segment parallel to a segment with each given slope.

  a) $-\frac{3}{5}$ **b)** 4   **c)** $-\frac{7}{4}$ **d)** 0.3   **e)** -8

14. Determine the slope of a line segment perpendicular to each segment in exercise 13.

15. The slopes of parallel line segments are given. Determine each value of k.

  a) $\frac{2}{3}$, $\frac{10}{k}$ **b)** $\frac{-1}{2}$, $\frac{4}{k}$ **c)** $-\frac{1}{3}$, $\frac{k}{4}$ **d)** $\frac{k}{5}$, -0.6

16. Suppose the line segments in exercise 15 are perpendicular. Determine each value of k.

17. The coordinates of the endpoints of line segments are given. Determine the slope of line segments that are parallel to each line segment.

  a) A(2, -3), B(-5, 3)   **b)** P(5, 2), Q(-1, 6)   **c)** M(-3, 7), N(-3, -5)

18. Determine the slope of line segments that are perpendicular to each line segment in exercise 17.

19. A quadrilateral has vertices A(-2, 2), B(-1, 3), C(5, -2), and D(4, 3). Determine whether it is a parallelogram or a rectangle. Write to explain how you identified the quadrilateral.

20. The points P(1, 4), Q(-1, -2), and R(4, -3) are given. Determine the coordinates of a point S so that RS is parallel to PQ and S is on:

  a) the x-axis   **b)** the y-axis

21. Use the points in exercise 20. Determine the coordinates of a point S so that RS is perpendicular to PQ and point S is on:

  a) the x-axis   **b)** the y-axis

22. In a game of "Flags," one team's flag is at A(8, 3), and the other team's is at B(-4, -3). The playing field is a rectangle with vertices P(-7, 2), Q(-5, -6), R(11, -2), and S(9, 6).

  a) Determine the distance between the flags.

  b) Determine the coordinates of the midpoint of the segment joining the flags.

  c) Determine the perimeter of the playing field.

  d) Determine the lengths of the diagonals of the field.

  e) Determine the coordinates of the midpoint of each diagonal.

1. Determine the indicated term of each arithmetic sequence.

   a) 1, 9, 17, … t_4 **b)** 4, 10, 16, … t_7 **c)** 5, 3, 1, … t_9

   d) −5, −3, −1, … t_{10} **e)** −7, 0, 7, … t_{12} **f)** −4, 1, 6, … t_8

2. **a)** Insert two numbers between 7 and 55 so that the four numbers form an arithmetic sequence.

   b) Insert five numbers between 7 and 55 so that the seven numbers form an arithmetic sequence.

3. Copy and complete each geometric sequence.

   a) 1.75, ■, ■, 112, ■   **b)** ■, 10, ■, ■, 80   **c)** ■, ■, ■, 189, 567

   d) 6, ■, ■, −6, ■   **e)** 3, ■, 12, ■, ■   **f)** 4.5, ■, ■, ■, 72

4. Suppose you buy a sweater in a province where the GST is 7% and PST is 8%. The regular price for the sweater is $49.99.

   a) Calculate.   **i)** the GST   **ii)** the PST   **iii)** the total cost of the sweater

   b) The sweater is then put on sale for 20% off. Repeat the calculations in part a when the discount is applied before the GST and PST are calculated.

   c) Suppose the discount is applied *after* the GST and PST are calculated. What is the total cost of the sweater?

   d) Compare the results of parts b and c. Explain.

5. Emile invests $1000 in a term deposit that pays interest at 12% per year compounded annually. Julianna invests $1000 in a term deposit that pays interest at 1% per month, compounded monthly.

   a) Estimate which term deposit is worth more at the end of one year. Calculate the value of each term deposit at the end of one year. Compare your estimate with your calculation.

   b) Repeat part a if the term deposits are invested for 20 years.

6. Simplify without using a calculator.

   a) $\sqrt{4}$ **b)** $\sqrt[3]{8}$ **c)** $\sqrt[4]{16}$ **d)** $\sqrt[5]{32}$ **e)** $\sqrt[6]{64}$

   f) $\sqrt[7]{128}$ **g)** $\sqrt[8]{256}$ **h)** $\sqrt[9]{512}$ **i)** $\sqrt[10]{1024}$

7. One can of floor paint covers an area of 33 m².

   a) A basement floor measures 10 m by 5 m. The furnace and water heater together occupy an area of 2 m², and cannot be moved. How many cans of paint must be purchased to paint the floor?

   b) Suppose two coats of paint are used. How many cans of paint are needed?

c) Determine the dimensions of a basement that has a square floor with the same area as the basement in part a.

8. An equilateral triangle has sides of length 10 cm.

   a) Calculate the height of the triangle.

   b) Calculate the area of the triangle.

9. Use the *Olympic Summer Games* database.

   a) Find the records for the women's 400-m track event. Each time there was a change in the world record, record the year and the world record time on paper or in a spreadsheet. Graph the data, using line segments to join the points.

   b) Describe how the slopes of the line segments change over time. If you wish, confirm your observations by calculating the slope of each segment.

   c) Use the pattern in your graph. What do you think is the fastest possible time for this event? Explain your thinking.

   d) Suppose you connected the first and last points on the graph with a line segment. How would your answer to part c change? Which answer do you think is more realistic? Explain.

   e) Repeat parts a to d for the men's 200-m track event.

   f) Compare your results with those of other students.

10. On a grid, draw the triangle with vertices D(−4, 4), E(2, −2), and F(4, 6).

   a) Determine the length of each side.

   b) Determine the coordinates of the midpoint of each side.

   c) Determine the slope of each side.

11. Point M(3, 2) is the midpoint of AB. Determine the coordinates of B for each point A.

   a) A(5, −2)   **b)** A(−1, 4)   **c)** A(−2, −3)   **d)** A(0, 0)

12. A triangle has vertices A(3, 8), B(2, 1), and C(10, 7). Determine the coordinates of the midpoint of the longest side. How far is this point from each vertex?

13. Line segment JK has slope $-\frac{2}{3}$. It joins J(−3, 6) to a point K on the *x*-axis. Determine the coordinates of point K and the point where JK intersects the *y*-axis.

14. The line segment joining each pair of points has the given slope. Determine each value of *a*.

   a) P(−2, −1), Q(*a*, 2); slope $\frac{3}{4}$ **b)** C(3, −2), D(−5, *a*); slope $-\frac{3}{8}$

It's All in the Packaging

Many manufacturers work with standard sizes of shipping containers. They might design their boxes to fit a certain product. Or they might work backward to design a product so the boxes fit on the warehouse skids and shelves, and in transport trucks or railway containers.

 CONSIDER THIS SITUATION

A book publisher produces two books with the same cover size, but different page counts. The printer uses only one size of carton. Each carton holds 30 thin books or 20 thick books.

Kayla manages a bookstore. She phones the publisher to order these two books. The publisher will pay for shipping if Kayla orders full cartons of books. Kayla does not need two full cartons. She could fill one carton with some of both books. How many of each book should Kayla order to fill one carton?

- Why do you think the books come in same-sized cartons?
- Why do you think the publisher would be willing to pay the shipping costs if Kayla orders full cartons of books?
- What is your answer to Kayla's question?

On pages 238 and 239, you will develop a model to explore the question concerning the numbers of books to order. Your model will involve a straight line graph.

 FYI Visit www.awl.com/canada/school/connections

For information related to the above problem, click on <u>MATHLINKS</u> followed by <u>AWMath</u>. Then select a topic under It's All in the Packaging.

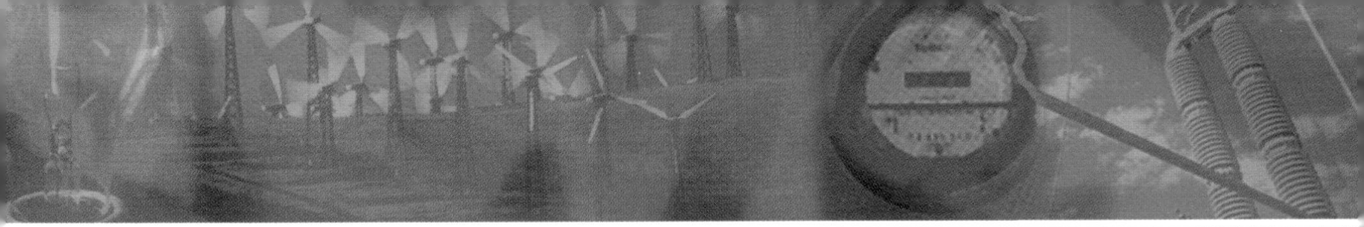
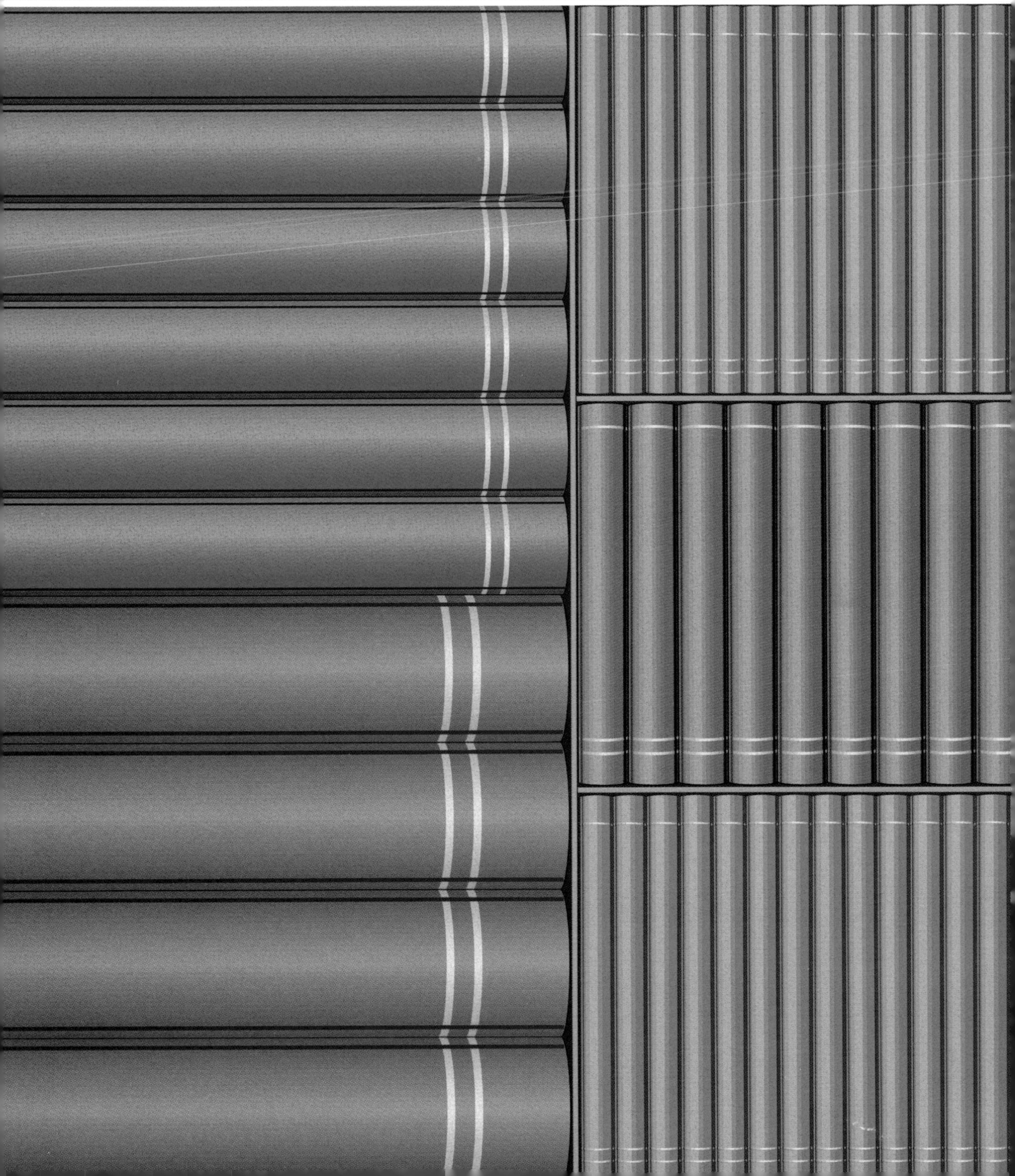

4.1 Using an Equation to Draw a Graph

The cost, C dollars, to rent a hall for a party is given by the formula
$C = 100 + 5n$, where n is the number of people who plan to attend.
The hall can accommodate a maximum of 100 people. To calculate
the cost for 20 people, we substitute 20 for n in the formula:

$$C = 100 + 5n$$
$$= 100 + 5(20)$$
$$= 100 + 100$$
$$= 200$$

For 20 people, it would cost $200 to rent the hall.

In the same way, we can calculate the cost for other numbers of people.
Substitute in the formula each value of n from the table below. The
corresponding value of C is shown.

Number of people, n	Cost, C dollars
20	200
40	300
60	400
80	500
100	600

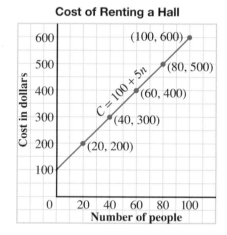

Cost of Renting a Hall

We can use these values to plot points on a grid. The result is a graph that
shows the cost for different numbers of people. The points appear to lie on a
straight line, but the variable n represents only whole numbers. However, if we
were to plot all 100 points, the graph would look like a solid line, so that is how
we draw it.

The graph represents ordered pairs that relate the cost to the number of people.
That is, if you know the number of people, you can determine the cost.
For example, you can see from the graph that the cost for 30 people is $250, the
cost for 50 people is $350, and so on.

Observe that the equation $C = 100 + 5n$ contains two variables, C and n.
We can use a table of values to draw a graph for any equation containing
two variables.

Example 1

Draw a graph of the equation $y = 8 - 2x$.

Solution

$y = 8 - 2x$
Choose several values of x.
When $x = -2$, $y = 8 - 2(-2)$
$\qquad\qquad\quad = 8 + 4$
$\qquad\qquad\quad = 12$
When $x = 0$, $y = 8 - 2(0)$
$\qquad\qquad\quad = 8$
Continue in this way to determine values of y for other values of x. Make a table of values, then plot the coordinates (x, y) on a grid. Join the points with a straight line, and label it with its equation.

x	y
−2	12
0	8
2	4
4	0
6	−4

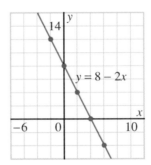

In *Example 1*, the equation $y = 8 - 2x$ acts like a formula that relates the values of y to the values of x. That is, if you know a value of x, the value of y is determined by the equation. All the pairs of values of x and y that can be obtained in this way can be plotted on a graph. These points all lie on a straight line. We say that the line is the graph of the equation.

The graph of an equation contains all the points, and only those points, whose coordinates satisfy the equation.

Equation of a Line Property

The coordinates of every point on the line satisfy the equation of the line.

Every point whose coordinates satisfy the equation of the line is on the line.

An equation that contains two variables can be written in different forms. Whichever form is used, the equation is still graphed the same way. That is, we make a table of values. To make a table of values for an equation such as $3x - 2y = 6$, substitute values of one variable and solve for the other variable.

Example 2

a) Graph the equation $3x - 2y = 6$.

b) Use the graph to verify the Equation of a Line Property.

Solution

a) Substitute $x = 0$ into the equation $3x - 2y = 6$.

$$3(0) - 2y = 6$$
$$-2y = 6$$
$$y = -3$$

One point on the graph is $(0, -3)$.

Substitute $x = 2$ into the equation $3x - 2y = 6$.

$$3(2) - 2y = 6$$
$$-2y = 0$$
$$y = 0$$

Another point on the graph is $(2, 0)$.

Substitute $x = 4$ into the equation $3x - 2y = 6$.

$$3(4) - 2y = 6$$
$$12 - 2y = 6$$
$$-2y = -6$$
$$y = 3$$

A third point on the graph is $(4, 3)$.

Record the results in a table, then plot the coordinates on a grid. Draw a straight line through the points.

x	y
0	-3
2	0
4	3

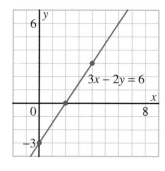

b) The Equation of a Line Property has two parts. To verify the first part, choose some other point on the line and verify that its coordinates satisfy the equation. From the graph, another point on the line is $(3, 1.5)$. Substitute 3 for x and 1.5 for y into both sides of the equation $3x - 2y = 6$.

L.S. $= 3x - 2y$ R.S. $= 6$
$\quad = 3(3) - 2(1.5)$
$\quad = 9 - 3$
$\quad = 6$

The coordinates $(3, 1.5)$ satisfy the equation. This verifies the first part of the property.

To verify the second part, find another point whose coordinates satisfy the equation. Substitute $x = 6$ into $3x - 2y = 6$.

$3(6) - 2y = 6$
$18 - 2y = 6$
$2y = 12$
$y = 6$

Another point whose coordinates satisfy the equation is $(6, 6)$. Observe that this point lies on the graph. This verifies the second part of the property.

DISCUSSING THE IDEAS

1. Examine the tables in the examples. Observe that the x-coordinates are in arithmetic sequence in each table.

   a) Why are the x-coordinates in arithmetic sequence? Did they have to be in arithmetic sequence?

   b) What kind of sequence do the y-coordinates form? Explain why they form this type of sequence.

   c) Suppose we had not chosen the x-coordinates to be in arithmetic sequence. Would the y-coordinates be in arithmetic sequence? Explain.

2. Explain what it means to say that the coordinates of a point satisfy an equation.

3. What other way is there to obtain the values of x and y that satisfy the equation in *Example 2*?

A **1. a)** Copy and complete each table of values.

i) $y = 2x + 3$

x	y
0	■
■	0
■	■
■	■

ii) $y = 5 - 3x$

x	y
0	■
■	0
■	■
■	■

iii) $4x + y = 12$

x	y
0	■
■	0
■	■
■	■

iv) $2x - 3y = 12$

x	y
0	■
■	0
■	■
■	■

v) $5x + 2y = 20$

x	y
0	■
■	0
■	■
■	■

vi) $3x - 2y = 15$

x	y
0	■
■	0
■	■
■	■

b) Choose one table from part a. Write to explain how you completed it.

2. a) For each equation, make a table of values and draw a graph.

i) $y = 2x - 4$ **ii)** $y = 3 - 2x$ **iii)** $x + y = 6$

iv) $3x + y = 9$ **v)** $x - 2y = -10$ **vi)** $4x - 3y - 24 = 0$

b) Choose one equation from part a. Use your graph to verify the Equation of a Line Property.

3. The cost, C cents, of printing and binding n copies of a manual is given by the formula $C = 70 + 20n$.

a) Make a table of values to show the costs for up to 100 copies.

b) Use your table to draw a graph.

c) Use the graph to estimate the cost of 75 copies.

d) Use the graph to estimate how many copies can be made for $10.

B **4.** The cost, C dollars, for a school basketball team to play in a tournament is given by the formula $C = 300 + 20n$, where n is the number of players.

a) Make a table of values using appropriate values of n.

b) Use your table to draw a graph.

c) Suppose $550 is available for travel costs. How many players can play in the tournament?

In exercise 4, your graph represents a model of the cost to play in the tournament.

- Does the model consist of the line joining the points, or only the points? Explain.
- Which points are reasonable ones to use for the model? Why?

5. A car travels at a constant speed from Edmonton to Calgary. The distance d kilometres from Calgary after t hours of driving is given by the formula $d = 280 - 100t$.

a) Make a table of values and draw a graph of d against t.

b) Explain how the graph shows the distance between these two cities. What is the distance?

c) Explain how the graph shows the total travelling time. What is the total travelling time?

d) After 2 h, how far is the car from Calgary? How far is it from Edmonton?

e) Explain why the formula is correct.

6. a) Draw the graph of each equation.

    i) $2x - y = 10$ **ii)** $4x - y = -8$ **iii)** $-3x + y = 15$

    iv) $6x + y = 18$ **v)** $x + 3y = 0$ **vi)** $2x = 5y$

b) Choose one equation from part a. Write to explain how you graphed the equation.

7. a) Draw the graph of each equation.

    i) $x + 2y = 7$ **ii)** $3x + 2y = 12$ **iii)** $5x - 3y = -15$

    iv) $3x - y - 10 = 0$ **v)** $2x + 4y + 9 = 0$ **vi)** $5x + 7y = 1$

b) Choose one equation from part a. Use your graph to verify the Equation of a Line Property.

8. The thermometer of an old oven is calibrated in Fahrenheit degrees. This formula converts Fahrenheit temperatures to Celsius temperatures.

$$C = \tfrac{5}{9}(F - 32)$$

a) Make a table of values, then draw a graph of Celsius temperatures against Fahrenheit temperatures.

b) Use your graph to determine the Fahrenheit temperature for each Celsius temperature.

    i) 90°C   **ii)** 120°C   **iii)** 200°C

c) Use your graph to determine the Celsius temperature for each Fahrenheit temperature.

    i) 90°F   **ii)** 120°F   **iii)** 200°F

d) Extend the graph to determine the Celsius temperature for each Fahrenheit temperature.

    i) 20°F   **ii)** 0°F   **iii)** −10°F   **iv)** −20°F

e) There is only one temperature that is the same in both scales. What temperature is it?

9. Two numbers are related by the following rule. If you double the first number and add the second number you always get 12.

a) Make a table of values of pairs of numbers that obey this rule.

b) Use your table to draw a graph.

c) Write the coordinates of one other point on the graph. Verify that those two numbers obey the rule.

d) Write the coordinates of any point that is not on the graph. Verify that those two numbers do not obey the rule.

10. In *Example 1* on page 11, the general term of the arithmetic sequence 2, 5, 8, … was expressed as $t_n = 3n - 1$. Use this equation. Draw a graph that represents the terms of the sequence. Decide whether you should join the points on your graph. Explain the reason for your decision.

C **11.** Find an example of an equation whose graph is not a straight line. Make a table of values for your equation, then draw the graph to show that it is not a straight line.

12. The graph of a line divides the coordinate plane into three different regions — points on the line, points on one side of the line, and points on the other side of the line. You know that the coordinates of points on the line satisfy the equation of the line.

a) Explain why the coordinates of points that are not on the line do not satisfy its equation.

b) Find out if there is any property that relates to the equation and distinguishes the points on one side of a line from those on the other side.

Use examples to illustrate your explanations.

COMMUNICATING THE IDEAS

In your journal, write an explanation of what it means to say that a line is the graph of an equation. Use an example to illustrate your explanation.

Investigating Hall Rental Costs

On page 198, the cost C dollars, to rent a hall for a party was represented by the formula $C = 100 + 5n$, where n is the number of people planning to attend. For another hall, the formula might be $C = 125 + 6.85n$. A graph and a table of values for this equation, obtained with a graphing calculator, are shown below. These screen images are from the TI-83 calculator. Other graphing calculators will produce similar results.

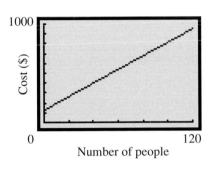

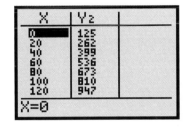

Displaying a graph

To produce this graph on your graphing calculator, follow these steps. The keystrokes required depend on your calculator.

- *Enter the equation.*
  You will need to use x and y for the variables. Enter this equation: $y = 125 + 6.85x$.

- *Define the viewing window.*
  On the graph above, observe that x runs from 0 to 120 with a tick mark every 20 units, and y runs from 0 to 1000 with a tick mark every 100 units. Enter these numbers in your calculator.

- *Graph the equation.*
  Press the appropriate key to display the graph. The result should be similar to the graph above.

Displaying a table of values

Your calculator may have a tables feature. To use it, follow these steps. Again, the keystrokes required will depend on your calculator.

- *Set up the table.*
  In the table above, x starts at 0 and increases in steps of 20. Set up your table to match the table above.

- *Display the table.*
  Press the appropriate key to display the table. The result should be similar to the one above.

LINKING IDEAS

Mathematics & Technology

Displaying the coordinates of a point on the graph

If your calculator has a trace feature, use it along with arrow keys to obtain coordinates of points on the graph. In the example (below left), the x-coordinates are not integers. We cannot tell the cost for 57 people from the graph. In this example, we would prefer the x-coordinates be integers.

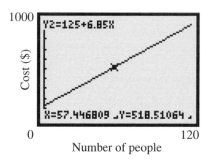

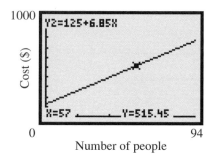

Adjusting the window

The values of x used during tracing depend on the window settings and the number of pixels in the calculator display. On the TI-83 calculator, the display is 95 pixels wide. Hence, if we adjust the window so that x runs from 0 to 94 we will obtain integral values of x when we use the trace feature (above right). This graph shows that the cost for 57 people is $515.45.

Displaying the graph and the table together

Your calculator may have a split screen feature that allows you to view the graph and the table of values together.

If you have completed the above steps successfully, you are ready to complete the following problems. Use the same equation for exercises 1 and 2.

1. Use your calculator to determine the cost for each number of people.

   a) 17 people   **b)** 37 people   **c)** 77 people   **d)** 97 people

2. What is the greatest number of people who can attend for each cost?

   a) $250   **b)** $500   **c)** $750   **d)** $1000

When you completed exercises 1d and 2d, you found what happened when you tried to trace past the edge of the screen. You may not have noticed that that also changed the window setting. You should change it back to continue.

3. Use your calculator to determine the cost for 57 people when the cost formula changes to $C = 425 + 3.25n$.

4. Which of the two cost formulas is better? Does your decision depend on the number of people who attend? Support your answer with some results you obtained from your calculator.

4.2 The Slope of a Line

In Chapter 3, we determined the slope of a line segment using the coordinates of its endpoints.

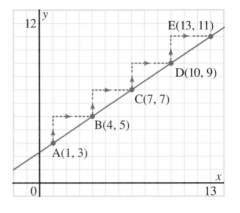

We now extend the concept of the slope of a segment to the slope of a line.

Start at A(1, 3), then move 2 units up and 3 units right to point B. Then move 2 units up and 3 units right to C. Continue in this way to D and E. Observe that A, B, C, D, and E all lie on a line.

Choose any two segments of this line; for example, AB and BD, or AD and CE. Determine their slopes.

Slope of AB $= \frac{5-3}{4-1}$ Slope of BD $= \frac{9-5}{10-4}$

        $= \frac{2}{3}$ $= \frac{4}{6}$

                                $= \frac{2}{3}$

Slope of AD $= \frac{9-3}{10-1}$ Slope of CE $= \frac{11-7}{13-7}$

        $= \frac{6}{9}$ $= \frac{4}{6}$

        $= \frac{2}{3}$ $= \frac{2}{3}$

The fact that these slopes are all $\frac{2}{3}$ suggests that the slope of every segment of the line is $\frac{2}{3}$. Similar results apply for other lines.

Constant Slope Property

The slopes of all segments of a line are equal.

The Constant Slope Property allows us to define the slope of a line to be the slope of any segment of the line. For example, the slope of the line shown above is $\frac{2}{3}$.

Points A and B are any two points on a line. Visualize a right triangle below AB. Visualize sliding this triangle so that segment AB is always on the line. If it helps, draw the line on a grid, and cut out a right triangle. Move the triangle as described.

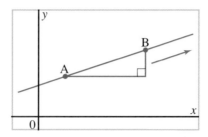

1. As the triangle slides along the line, what happens to:

   a) the rise from A to B?   b) the run from A to B?   c) the slope of AB?

2. How do your answers to exercise 1 illustrate the Constant Slope Property of a line?

Example 1

Determine the slope of this line.

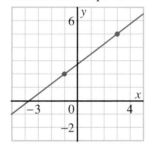

Solution

Two points on the line are R(−1, 2) and S(3, 5). Determine the slope of the segment joining them.

$$\text{Slope of RS} = \frac{5-2}{3-(-1)}$$
$$= \frac{3}{4}$$

The slope of the line is $\frac{3}{4}$.

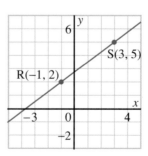

We can use the Constant Slope Property to draw a line passing through a given point and with a given slope.

Example 2

On a grid, draw a line through K(4, −2) with slope $-\frac{1}{3}$.

Solution

Start at K(4, −2). Move 1 unit up and 3 units left (or, 1 unit down and 3 units right). Repeat two or three times to obtain several points on the line. Then draw the line through these points.

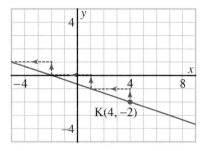

In the previous chapter, we found the relations between the slopes of parallel line segments and between the slopes of perpendicular line segments. Since the slope of a line is equal to the slope of any segment of the line, the same relations must be true for lines.

> If the slopes of two lines are equal, the lines are parallel.
>
> Conversely, if two non-vertical lines are parallel, their slopes are equal.
>
> If the slopes of two lines are negative reciprocals, the lines are perpendicular.
>
> Conversely, if two lines are perpendicular (and neither one is vertical), their slopes are negative reciprocals.

Example 3

On the same axes as in *Example 2*
a) Draw a line through L(2, 2) parallel to the line drawn in *Example 2*.
b) Draw a line through L(2, 2) perpendicular to the line drawn in *Example 2*.

Solution

a) The line in *Example 2* has slope $-\frac{1}{3}$.
A line parallel to this line also has slope $-\frac{1}{3}$. Start at L(2, 2).
Repeat the steps of *Example 2* to locate points on the line.

b) The negative reciprocal of $-\frac{1}{3}$ is 3.
A line perpendicular to the line in *Example 2* has slope 3. Start at L. Move 3 units up and 1 unit right (or, 3 units down and 1 unit left), and repeat. Draw the line through these points.

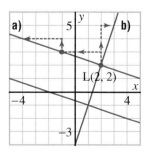

1. What do you think would happen to a line if the Constant Slope Property did not apply? Discuss your ideas with a partner.

2. In *Example 1*, how could you determine the coordinates of some other points on the line? Use your idea to determine the coordinates of two other points on the line.

4.2 EXERCISES

A 1. State the slope of each line.

a)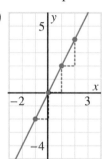

b)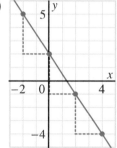

c)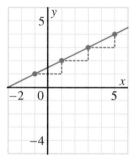

2. a) On the same axes, draw a line through E(0, 4) with each given slope. Then determine the coordinates of two more points on each line.

   i) 3   ii) 2   iii) 1   iv) $\frac{1}{2}$ v) $\frac{1}{3}$

   vi) 0   vii) $-\frac{1}{2}$ viii) -1 ix) -2 x) $-\frac{1}{4}$

   xi) -3 xii) not defined by a real number

   b) Choose one slope from part a. Write to explain how you determined the coordinates of two points.

3. a) On a grid, draw a line through A(0, −3) with slope $\frac{2}{3}$.

   b) On the same axes, draw three lines parallel to the line in part a: one through O(0, 0); one through C(0, 3); and one through D(0, 6).

   c) On the same axes, draw lines perpendicular to the line in part a through A, O, C, and D.

4. On a grid, draw a line through each point with each given slope. Then determine the coordinates of two more points on each line.

   a) A(0, 3) with slope   i) 1   ii) −1

   b) R(2, 1) with slope   i) $\frac{2}{3}$ ii) $-\frac{3}{2}$

   c) L(1, −3) with slope   i) $-\frac{1}{2}$ ii) 2

   d) C(5, 4) with slope   i) 0   ii) not defined by a real number

5. On a grid, draw a line with each given slope. Determine the coordinates of two points on the line.

   a) 3   **b)** $\frac{4}{3}$ **c)** −2   **d)** $-\frac{2}{5}$ **e)** $-\frac{3}{4}$

6. Choose one slope from exercise 5. Compare the coordinates of the two points you determined with those of a classmate. Do your points lie on your classmate's line? Do your classmate's points lie on your line? If your answer to either of these questions is no, explain why the points lie on different lines.

B **7.** On a grid, draw a line through each point to satisfy each condition. Then determine the coordinates of two more points on each line.

   a) W(3, 5)   **i)** with slope $-\frac{2}{3}$ **ii)** perpendicular to the line in part i
   b) G(0, 2)   **i)** with slope 1   **ii)** perpendicular to the line in part i
   c) T(−2, −1)   **i)** with slope 3   **ii)** perpendicular to the line in part i
   d) Q(3, 1)   **i)** undefined slope   **ii)** perpendicular to the line in part i

8. a) Draw a line through E(4, 1) with slope $\frac{3}{2}$.

   b) On the same axes, draw two lines through F(0, 3): one parallel to the line in part a, the other perpendicular to the line in part a.

9. a) Draw a line through J(−2, 5) with slope −2.

   b) On the same axes, draw two lines through K(5, 1): one parallel to the line in part a, the other perpendicular to the line in part a.

10. a) A line has slope 4. It passes through the points A(3, 8) and B(2, k). What is the value of k?

   b) Write to explain how you determined the value of k.

11. A line has slope −1. It passes through the points C(−q, 3) and D(4, −2). What is the value of q?

12. Graph each set of three points. Determine the slopes of segments AB, BC, and AC. Compare their slopes. What do you notice?

   a) A(0, 1), B(3, 3), C(9, 7)
   b) A(−6, 1), B(−2, −1), C(4, −4)
   c) A(8, 5), B(−4, 1), C(3, 4)

13. Points M(1, −1), N(3, −5), and Q(d, 5) lie on the same line. What is the value of d?

14. Each pair of points lies on a line. Determine the coordinates of two more points that lie on each line.

   a) E(2, 3) and F(1, 7)   **b)** G(−4, 7) and H(1, 0)
   c) J(−6, −2) and K(5, 8)   **d)** L(−3, −7) and M(−4, −6)

15. Points that lie on the same line are *collinear* points. In the diagram, three points, A, B, and C appear to be collinear.

    a) Determine the slopes of AB, BC, and AC.

    b) Determine whether the three points are collinear.

    c) Find another way to determine if the points are collinear.

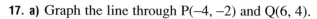

16. Are the points in each set collinear?

    a) D(−4, 4), E(0, 2), F(6, −1)

    b) J(5, 9), K(1, 4), L(−2, 1)

    c) R(−3, −3), S(2, 1), T(11, 8)

17. **a)** Graph the line through P(−4, −2) and Q(6, 4).

    b) Draw the lines perpendicular to PQ through P and through Q.

    c) How are the lines you drew in part b related?

 18. On April 14, 1981, the first American space shuttle, *Columbia*, returned to Earth. At one point in its reentry, it was travelling at approximately 1080 km/h and dropping at 4200 m/min. What was the slope of the reentry path to two decimal places?

MODELLING the Space Shuttle's Reentry Path

When you determined the slope of the reentry path in exercise 18, you were assuming that the space shuttle is following a path that is a straight line.

- Give a reason why the reentry path may not be a straight line.
- As the space shuttle approaches the ground, how would you expect the slope of the reentry path to change?

---

19. The points A(−2, 3) and B(4, 5) are two vertices of a right △ABC. Point C is on the *y*-axis. Determine the possible coordinates of C for each angle.

    a) ∠A = 90°   **b)** ∠B = 90°   **c)** ∠C = 90°

COMMUNICATING THE IDEAS

In your journal, explain the difference between the slope of a line segment and the slope of a line. Illustrate your explanation with some examples, including diagrams.

Investigating y = mx + b

1. The graphs below were drawn using the same equations but different window settings.

   a) Which screen shows a more accurate graph?

   b) What are the advantages and disadvantages of using these two window settings?

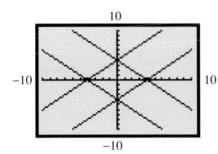

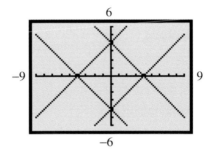

2. a) On the same screen, graph each equation in the lists below left.

   b) Describe the pattern in the graphs. Explain how the coefficients in the equations account for the pattern.

   c) Visualize how the graph of $y = mx + 1$ changes as m varies. What special case occurs when $m = 0$?

$y = 2x + 1$	$y = -0.5x + 1$	$y = 0.5x + 5$	$y = 0.5x - 1$
$y = x + 1$	$y = -x + 1$	$y = 0.5x + 3$	$y = 0.5x - 3$
$y = 0.5x + 1$	$y = -2x + 1$	$y = 0.5x + 1$	$y = 0.5x - 5$

3. a) On the same screen, graph each equation in the lists above right.

   b) Repeat exercise 2b.

   c) Visualize how the graph of $y = 0.5x + b$ changes as b varies. What special case occurs when $b = 0$?

4. Based on your investigations, describe how the values of m and b affect the graph of the equation $y = mx + b$.

5. Predict what the graph of each equation would look like. Sketch each graph on paper, then use your calculator to check your answer.

   a) $y = 2x + 3$ b) $y = -2x + 3$ c) $y = -x - 4$

6. What equations were used to make the graphs on the screens above?

INVESTIGATE

One way to graph the equation of a straight line is to make a table of values. Another way to graph an equation depends on obtaining information about the graph from the numbers in the equation.

Equations of the Form y = mx

1. a) Use a table of values to graph each equation on the same axes.

$$y = x \qquad y = 2x \qquad y = \frac{1}{2}x \qquad y = 0x$$

$$y = -x \qquad y = -2x \qquad y = -\frac{1}{2}x$$

b) Compare the graphs in part a. How are they the same? How are they different?

2. a) Determine the slope of each line in exercise 1.

b) Compare each slope with the corresponding equation. What do you notice?

3. Each equation you graphed in exercise 1 has the form $y = mx$.

> What does this number tell you about the graph?

Equations of the Form y = mx + b

4. a) Use a table of values to graph each set of equations on the same axes.

i) $y = 2x + 5 \qquad y = -\frac{1}{2}x + 5 \qquad$ **ii)** $y = 2x + 5 \qquad y = 2x - 1$

$\quad y = x + 5 \qquad\quad y = -x + 5 \qquad\qquad\quad y = 2x + 3 \qquad y = 2x - 3$

$\quad y = \frac{1}{2}x + 5 \qquad y = -2x + 5 \qquad\qquad\quad y = 2x + 1 \qquad y = 2x - 5$

$\quad y = 0x + 5$

b) Compare the graphs in part a. How are they the same? How are they different?

5. a) Determine the slope of each line in exercise 4.

b) Compare each slope with the corresponding equation. What do you notice?

6. Each equation you graphed in exercise 4 has the form $y = mx + b$.

> What does this number tell you about the graph?

> What does this number tell you about the graph?

Consider the equation $y = 2x + 3$ and the table of values below. Plot the points on a grid. Join the points with a straight line.

x	y
−1	1
0	3
1	5
2	7
3	9

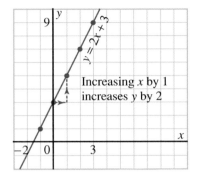

Increasing x by 1 increases y by 2

The table and graph suggest another method for graphing a linear equation. This method is based on two numbers in the equation:

The slope

This is the coefficient of x when the equation has the form $y = mx + b$. In the case of $y = 2x + 3$, the slope is 2.

The y-intercept

It is the value of b when the equation has the form $y = mx + b$. It is also the value of y when $x = 0$. In the case of $y = 2x + 3$, the y-intercept is 3.

$$y = 2x + 3$$

slope | y-intercept

The graph of the equation $y = mx + b$ is a straight line with slope m and y-intercept b.

The equation $y = mx + b$ is called the *slope y-intercept form* of the equation of a line. We can draw the graph of an equation in this form without making a table of values.

Example 1

Graph the line represented by each equation.

a) $y = \frac{2}{3}x - 5$ **b)** $y = -2x + 4$

Solution

a) $y = \frac{2}{3}x - 5$

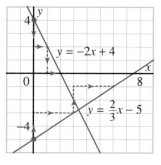

The slope is $\frac{2}{3}$. The y-intercept is -5.

The corresponding point has coordinates $(0, -5)$.

Begin at $(0, -5)$. Move 2 up and 3 right. This
is a point on the line. Other points on the line
can be obtained by continuing in this way, or
by moving 2 down and 3 left.

b) $y = -2x + 4$

The slope is -2, or $\frac{-2}{1}$. The y-intercept is 4. The corresponding point
has coordinates $(0, 4)$. Begin at $(0, 4)$. Move 2 down and 1 right. This
is a point on the line. Other points on the line can be obtained by
continuing in this way, or by moving 2 up and 1 left.

VISUALIZING

$P(x, y)$ is any point on the line in *Example 1a*. Visualize what
happens if P begins to move along the line in the direction shown.

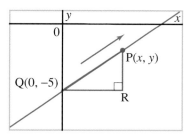

1. As P moves along the line, what happens to:

 a) the size and shape of $\triangle$PQR? **b)** the rise from Q to P?

 c) the run from Q to P? **d)** the slope of QP?

2. Do the answers to exercise 1 change if P moves in the opposite direction?

3. Determine how x and y are related for any position of P on the line.

We can also determine the equation of a line in the slope y-intercept form when its graph is given.

Example 2

Determine the equation of each line on this grid.

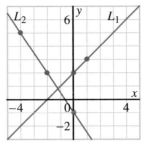

Solution

The slope and the y-intercept of each line can be read from its graph.

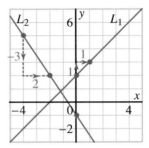

L_1 has a slope of 1 and a y-intercept of 2.

Its equation is $y = 1x + 2$, or $y = x + 2$.

L_2 has a slope of $-\frac{3}{2}$ and a y-intercept of -1.

Its equation is $y = -\frac{3}{2}x - 1$.

DISCUSSING THE IDEAS

1. Describe the line you think each equation represents.

   a) $y = x$
   b) $y = 5$, $y = 0$, $y = -3$
   c) $x = 5$, $x = 0$, $x = -3$

2. Does every line have an equation that can be written in the form $y = mx + b$? Discuss your ideas with a partner.

4.3 EXERCISES

A 1. State the slope and the y-intercept for the line represented by each equation.

   a) $y = 3x + 5$
   b) $y = -2x + 3$
   c) $y = \frac{2}{5}x - 4$
   d) $y = -\frac{1}{2}x + 6$

   e) $y = -4x - 7$
   f) $y = \frac{3}{8}x - \frac{5}{2}$
   g) $y = \frac{4}{3}x - 2$
   h) $y = \frac{9}{5}x + 1$

2. Write the equation of each line with the given slope and y-intercept.

   a) $m = 2$, $b = 3$
   b) $m = -1$, $b = 4$
   c) $m = \frac{2}{3}$, $b = -1$

   d) $m = -\frac{4}{5}$, $b = 8$
   e) $m = -3$, $b = \frac{5}{2}$
   f) $m = 0$, $b = 3$

3. For each line, state the slope, the y-intercept, and the equation.

a)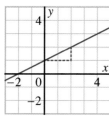

b)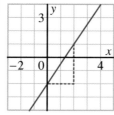

c)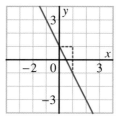

4. a) Determine the equation of each line.

i)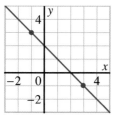

ii)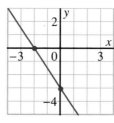

iii)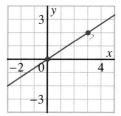

b) Choose one line from part a. Write to explain how you determined its equation.

5. a) Graph the line represented by each equation.

    i) $y = \frac{2}{5}x + 3$ **ii)** $y = \frac{3}{4}x - 2$ **iii)** $y = -\frac{1}{2}x + 1$ **iv)** $y = -\frac{3}{2}x - 1$

    v) $y = 2x - 3$ **vi)** $y = -x + 5$ **vii)** $y = -3x + 2$ **viii)** $y = 0x - 4$

b) Choose one equation from part a. Write to explain how you graphed the line.

6. a) Graph the line represented by $y = -\frac{1}{2}x + 3$.

b) What are the coordinates of the point where the line in part a intersects the x-axis?

c) Determine the length of the segment between the x- and y-axes.

d) Shade in the triangle formed by the line and the x- and y-axes. Determine its area and its perimeter.

7. The equations $y = 2x + 4$ and $y = -x + 7$ are given.

a) Graph the line represented by each equation. Determine the coordinates of the point of intersection.

b) Shade in the triangle formed by the two lines and the x-axis. Determine the area of this triangle.

B **8.** The equations of the three sides of a triangle are $y = 2x - 4$, $y = -\frac{1}{2}x + 6$, and $y = -3x + 1$. Graph these lines on the same axes. Determine the coordinates of the vertices of the triangle.

9. a) Identify the patterns in the values of m and b in each list of equations.

   i) $y = 2x + 4$
   $y = x + 3$
   $y = 0x + 2$
   $y = -x + 1$
   $y = -2x + 0$

   ii) $y = 2x - 6$
   $y = x - 3$
   $y = 0.5x - 1.5$
   $y = -0.5x + 1.5$
   $y = -x + 3$
   $y = -2x + 6$

   b) Plot the graphs in each list on the same grid. Describe what you see.

   c) Explain how the patterns in the equations account for the patterns on the graphs.

10. a) The equation of a line is $y = 3x + b$. Determine the value of b when the line passes through each point.

    i) R(2, 1)   ii) K(−1, 4)   iii) A(3, −2)   iv) B(−2, 2)

    b) Choose one point from part a. Write to explain how you determined the value of b.

11. The equation of a line is $y = mx + 2$. Determine the value of m when the line passes through each point.

    a) D(12, 5)   b) S(1, −3)   c) E(−2, 6)   d) A(−5, 1)

12. a) Identify the pattern in the values of m and b in these equations.

    $y = x + 1$ $y = -x - 1$
    $y = 2x + 0.5$ $y = -2x - 0.5$
    $y = 0.5x + 2$ $y = -0.5x - 2$

    b) Plot the graphs of all six equations on the same screen. Describe what you see.

    c) Explain how the patterns in the equations account for the patterns on the graphs.

13. In exercise 12, m and b were reciprocals. Suppose similar equations were listed in which m and b were negative reciprocals. Predict the pattern formed by the graphs of these equations. Use your calculator to check your prediction.

14. Graph the line represented by $y = 2x + 1$. Draw three lines perpendicular to $y = 2x + 1$. Determine the slope of each line.

15. a) Compare the perpendicular lines you drew in exercise 14 with those of a classmate. Did you draw the same lines?

    b) Compare the slopes of the perpendicular lines you drew with the slopes of the lines drawn by a classmate. Are the slopes of your lines equal to those of your classmate?

c) Write to explain why your answer to part a could be no, while your answer to part b should be yes.

16. a) Graph the line represented by $y = -3x - 2$. Draw three lines perpendicular to $y = -3x - 2$. Determine the slope of each line.

b) Use the results of part a and exercise 15. Write a rule to determine the slope of a line that is perpendicular to a given line.

17. Write the equations of the lines in each pattern.

a)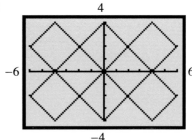

b)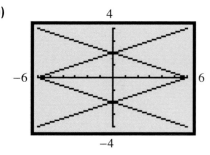

18. Predict what the graphs of the equations in each list would look like. Use your calculator to check your prediction.

a) $y = x + 3$
$y = 10x + 3$
$y = 100x + 3$
$y = 1000x + 3$

b) $y = x + 3$
$y = 10x + 30$
$y = 100x + 300$
$y = 1000x + 3000$

19. The line represented by $y = 3x - 2$ and a line perpendicular to it intersect at R(1, 1). Determine the equation of the perpendicular line.

20. Two perpendicular lines intersect on the y-axis. One line has equation $y = -x + 5$. What is the equation of the other line?

21. Two perpendicular lines intersect on the x-axis. One line has equation $y = 3x + 6$. What is the equation of the other line?

C 22. Determine the equation of the line that passes through the point of intersection of the lines $y = 2x - 5$ and $y = -x + 4$, and is also:

a) parallel to the line $y = \frac{2}{3}x + 4$

b) perpendicular to the line $y = \frac{3}{4}x - 1$

COMMUNICATING THE IDEAS

Suppose your friend telephones you to discuss tonight's homework. How would you explain, over the telephone, how to graph an equation of the form $y = mx + b$? How would you explain what the graph represents? Use examples to illustrate your explanations.

Patterns in Equations and Lines

Some computer programs are designed to graph equations in the coordinate plane. This graph was produced with *Graphmatica*.

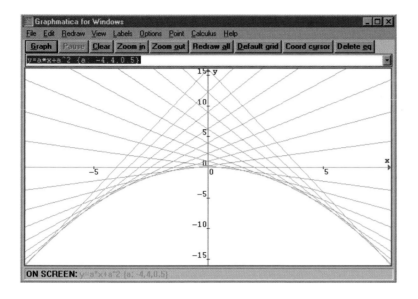

The graphing window shows values of *x* from −8 to 8 and values of *y* from −16 to 16. You can complete these exercises using only this graph and your knowledge of the graphs of equations of the form $y = mx + b$.

1. One line passes through the points (0, 16) and (4, 0). What is the equation of this line? Write the equation in the form $y = mx + b$.

2. Other lines pass through the following pairs of points. Determine the equation of each line, then write it in the form $y = mx + b$.

   a) (0, 9) and (3, 0)   **b)** (0, 4) and (2, 0)   **c)** (0, 1) and (1, 0)

3. Examine the equations you found in exercises 1 and 2.

   a) Identify any pattern you see in the equations.

   b) Use the pattern to determine the equations of other lines on the graph. Try to determine the equations of all the lines.

4. The computer graphed the lines one at a time, starting with the line in exercise 1. Try to visualize the screen as the computer graphed these lines. Write to describe what you would see.

5. Suppose the pattern in the equations were extended to include more equations. Describe how the graph would change.

How Slope Applies to Speed and Acceleration

The speed and acceleration of an airplane

The table shows the time taken and the distance travelled by an airplane after it has reached its cruising altitude.

Time (h)	0	0.1	0.2	0.3	0.4	0.5
Distance (km)	0	60	120	180	240	300

We know that average speed $= \frac{\text{distance}}{\text{time}}$.

1. Use the information in the table and the equation above. Determine the average speed of the plane.

2. On grid paper, plot the ordered pairs from the table above. Draw a graph of distance against time, with time along the horizontal axis.

   a) Calculate the slope of the line.

   b) How does the slope of the line compare with the average speed of the plane?

   c) Write the equation of the line in the form $d = mt + b$. Write to explain how you did this.

3. Recall the Constant Slope Property of a line. Copy and complete the table below, for the plane in exercise 1.

Time (h)	0	0.1	0.2	0.3	0.4	0.5
Speed (km/h)						

Acceleration tells us how speed changes.

When the speed is increasing, the acceleration is positive.

When the speed is constant, the acceleration is zero.

When the speed is decreasing, the acceleration is negative.

4. What is the acceleration of the plane? Explain your answer.

5. On grid paper, plot the ordered pairs from the table in exercise 3. Draw a graph of speed against time, with time along the horizontal axis.

   a) Calculate the slope of the line.

b) How does the slope of the line compare with the acceleration of the plane?

c) Write the equation of the line in the form $v = mt + b$. Explain how you did it.

The acceleration of a skydiver

Suppose we ignore air resistance. This table shows how the speed of a skydiver changes with time. The skydiver's motion is affected by the force of gravity and the constant acceleration due to this force.

Time (s)	0	1	2	3	4	5
Speed (m/s)	0	9.8	19.6	29.4	39.2	49.0

6. We can use the information in the table to calculate the acceleration of the skydiver.

Use the formula: $\text{Acceleration} = \dfrac{\text{difference in speed}}{\text{difference in time}}$

a) From the table, choose two values of speed. Calculate their difference.

b) Use the two values of time that correspond to the two values of speed you chose in part a. Calculate the difference in time.

c) Substitute the differences in parts a and b in the formula for acceleration.

d) What is the acceleration of the skydiver? Use the units of time and speed. What are the units for acceleration?

7. On grid paper, plot the ordered pairs from the table above. Draw a graph of speed against time, with time along the horizontal axis.

a) Calculate the slope of the line.

b) How does the slope of the line compare with the acceleration of the skydiver?

c) Write the equation of the line in the form $v = mt + b$. Explain your equation.

8. Write to explain how slope can be used to determine speed and acceleration.

9. Sketch a graph of speed against time for each situation.

a) The speed is increasing at a constant rate.

b) The acceleration is zero.

c) The acceleration is negative and constant.

Mathematics & Science

4.4 The Equation of a Line: Part II

In Section 4.3, we determined the equation of a line when its slope and
y-intercept were known. If other information about a line is known, we can
find its equation using the Constant Slope Property.

Given the slope and a point on the line

With a given slope, there is only one line that passes through a given point.
The next example shows how we can determine its equation.

Example 1

Determine the equation of the line, with slope $\frac{2}{3}$, that passes through the
point A(−1, 3).

Solution

Graph the line. Start at A(−1, 3).

Move 2 up and 3 right. Mark a point.
Draw a line through this point and A.

Let P(x, y) be any point on the line.

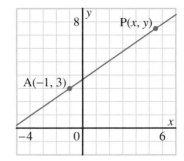

$$\text{Slope of AP} = \frac{y-3}{x-(-1)}$$

$$= \frac{y-3}{x+1}$$

The slope of the line is $\frac{2}{3}$.

Since these slopes are equal, the equation is:

$$\frac{y-3}{x+1} = \frac{2}{3}$$
$$3(y-3) = 2(x+1)$$
$$3y - 9 = 2x + 2$$
$$2x - 3y + 11 = 0$$

The equation of the line is $2x - 3y + 11 = 0$.

In the solution of *Example 1*, all the terms were collected on the left side of the
equation. We say that the equation is written in the form $Ax + By + C = 0$. This
is the *standard form* of the equation of a line.

Given two points on the line

There is only one line that passes through two given points. We determine its equation by first calculating its slope, then using the Constant Slope Property.

Example 2

Determine the equation of the line that passes through the points R(1, 4) and S(4, −2).

Solution

Graph the line.

Slope of RS $= \frac{-2-4}{4-1}$

$= \frac{-6}{3}$

$= -2$

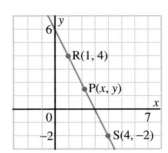

Let P(x, y) be any point on the line.

Slope of RP $= \frac{y-4}{x-1}$

Since these slopes are equal:

$$\frac{y-4}{x-1} = -2$$
$$y - 4 = -2(x - 1)$$
$$y - 4 = -2x + 2$$
$$2x + y - 6 = 0$$

The equation of the line is $2x + y - 6 = 0$.

---

1. How could you check that the equation of the line in *Example 1* is correct?

2. Can you think of other ways to determine the equation of the line in *Example 1*? Discuss your ideas with a partner.

3. How could you check that the equation of the line in *Example 2* is correct?

4. Can you think of other ways to determine the equation of the line in *Example 2*? Discuss your ideas with a partner.

A **1.** For each line shown, P(x, y) is any point on the line.

   a) Determine an expression for the slope of the segment AP.

   b) Determine the slope of the line.

   c) Write the equation of the line.

   i) 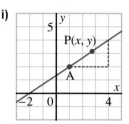   **ii)** 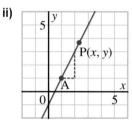   **iii)**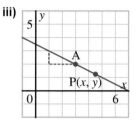

2. a) The coordinates of a point and the slope of a line are given. Graph the line through the point with the given slope. Determine the equation of the line.

   i) A(2, 5), 3   **ii)** R(−4, 2), 7   **iii)** K(6, −8), −4   **iv)** G(−4, −5), 2

   v) M(−10, 3), $-\frac{3}{5}$ **vi)** B(5, −2), $\frac{2}{3}$ **vii)** W(−3, 7), $-\frac{7}{2}$ **viii)** C$\left(\frac{1}{2}, \frac{3}{4}\right)$, 0

   b) Choose one point and corresponding slope from part a. Write to explain how you determined the equation of the line.

3. Which of the given points lie on the line with equation $3x - 2y + 12 = 0$?

   a) A(3, −2)   **b)** B(−4, 0)   **c)** C(0, −6)   **d)** D(0, 6)

   e) E(2, 9)   **f)** F(−2, 3)   **g)** G(4, 12)   **h)** H(3, 10)

4. a) The equations of lines are given. Which lines pass through the point (−4, 2)?

   i) $x + y = 0$ **ii)** $x - y + 6 = 0$ **iii)** $3x - y + 14 = 0$

   iv) $x + 2y = 0$ **v)** $x - 2y - 8 = 0$ **vi)** $2x + 5y - 2 = 0$

   b) Choose one equation from part a. Write to explain how you determined whether the line passed through the point.

5. Determine the equation of each line.

   a) 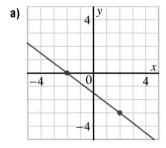   **b)** 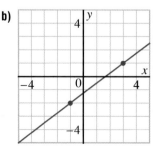   **c)**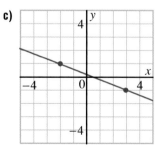

6. The coordinates of two points are given. Draw the line through each pair of points. Determine the equation of the line. Check that the coordinates of the points satisfy the equation.

   a) A(2, 2), B(4, 3)   **b)** R(3, 4), S(−3, 0)   **c)** F(−1, 5), G(2, −1)

   d) M(3, −3), N(−6, 0)   **e)** C(0, 2), D(4, 1)   **f)** W(−4, 7), V(3, 5)

7. Determine the equation of the line through each pair of points.

   a) E(2, 1) and F(5, 7)   **b)** S(−3, 2) and T(1, −10)

   c) A(5, −2) and B(7, 5)   **d)** L(−1, −3) and M(4, 7)

   e) Q(4, −1) and R(−2, −5)   **f)** C(−7, −12) and D(−4, −4)

B 8. A triangle has vertices R(0, 5), S(3, 0), and T(5, 6). Draw the triangle on a grid. Determine the equations of the three sides.

9. **a)** Determine the equation of each line.

   i) slope 3, through M(2, 1)   **ii)** slope $-\frac{2}{5}$, through R(−1, 4)

   iii) slope 2, x-intercept 3   **iv)** x-intercept 3, y-intercept 4

   v) slope $\frac{3}{4}$, through F(4, 0)   **vi)** slope −2, through N(0, 5)

   vii) x-intercept $-\frac{2}{3}$, slope $\frac{5}{6}$ **viii)** x-intercept −5, y-intercept 2

   b) Choose one equation from part a. Write to explain how you determined the equation.

10. A square has vertices A(0, 4), B(−6, 0), C(−2, −6), and D(4, −2). Draw the square on a grid.

    a) Determine the equations of the four sides.

    b) Determine the equations of the two diagonals.

11. **a)** The equation of a line is $3x + 2y + k = 0$. Determine the value of k when the line passes through each point.

    i) A(2, −1)   **ii)** B(−4, 3)   **iii)** C(5, 2)   **iv)** D(0, 4)   **v)** E$\left(\frac{1}{3}, \frac{1}{2}\right)$

    b) Choose one point from part a. Write to explain how you determined the value of k.

12. In △ABC, M is the midpoint of side AB. Determine the equation of each line.

    a) the line BC

    b) the line through A parallel to BC

    c) the line through M parallel to BC

    d) the line through M perpendicular to AB

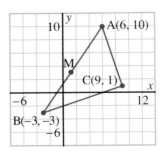

13. The equation of a line is $2x - 5y + k = 0$. Determine the value of k when the line passes through each point.

a) F(3, 0) **b)** G(−1, 4) **c)** H(−7, −2)

C **14. a)** Draw this design on a grid, as shown. Start with the black square with vertices (1, 1), B(1, 2), (2, 2), and D(2, 1).

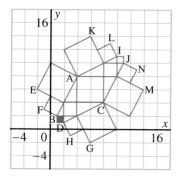

b) Determine the equation of each line: AB, CD, EF, and GH. What do the lines have in common?

c) Determine the equation of each line: AI, CJ, KL, and MN. Show that they all pass through the same point. What are the coordinates of that point?

15. a) On a grid, draw the line segment joining S(−3, 3) and T(7, 7).

b) Draw the perpendicular bisector of segment ST.

c) Determine the equation of the perpendicular bisector.

16. a) Graph △ABC with vertices A(−3, 1), B(5, −1), and C(9, 9).

i) Draw the median from A to the midpoint of BC.

ii) Draw the altitude from C to AB.

iii) Draw the perpendicular bisector of AC.

b) Determine the equation of each line drawn in part a.

17. Points P(−6, −4), Q(6, 2), and R(−2, 8) are the vertices of a triangle.

a) Determine the equations of the three sides.

b) Determine the equation of the perpendicular bisector of each side.

c) Determine the point of intersection, A, of the perpendicular bisectors.

d) Determine the lengths of AP, AQ, and AR.

COMMUNICATING THE IDEAS

To determine the equation of a line, you need to know two facts about it. For example, two facts that determine a line are the slope and y-intercept. In your journal, list as many pairs of facts as you can that determine a line. Refer to the examples and exercises in this chapter for ideas.

4.5 Interpreting the Equation $Ax + By + C = 0$

In previous sections, equations of lines satisfying certain conditions were found, and the results were written in the standard form, $Ax + By + C = 0$. When the equation of a line is given in this form, certain information about its graph can be obtained from it.

Determining the intercepts

The x-intercept of a line is the x-coordinate of the point where the line intersects the x-axis.

Similarly, the y-intercept is the y-coordinate of the point where the line intersects the y-axis. The diagram suggests a method for finding these intercepts.

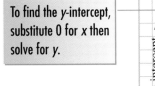

To find the y-intercept, substitute 0 for x then solve for y.

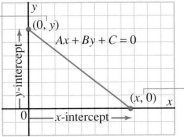

To find the x-intercept, substitute 0 for y then solve for x.

Example 1

The equation $2x + 3y - 12 = 0$ defines a line.

a) Determine the x-intercept.

b) Determine the y-intercept.

c) Graph the line.

Solution

a) Substitute 0 for y.

$$2x + 3(0) - 12 = 0$$

Solve for x.

$$2x = 12$$
$$x = 6$$

The x-intercept is 6.

b) Substitute 0 for x.

$$2(0) + 3y - 12 = 0$$

Solve for y.

$$3y = 12$$
$$y = 4$$

The y-intercept is 4.

c) Use the intercepts to graph the line.

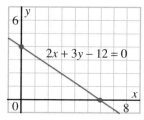

2x + 3y − 12 = 0

$P(x, y)$ is any point on the line in *Example 1*. Visualize what happens when P begins to move along the line in the direction shown.

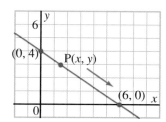

1. For every increase of 3 units in the x-coordinate of P, what happens to the y-coordinate?

2. How are the changes in the coordinates in exercise 1 related to the equation $2x + 3y − 12 = 0$?

3. How would your answers to exercises 1 and 2 be affected if P moves in the opposite direction?

Determining the slope and the y-intercept

When the equation of a line is given, its slope may be determined by solving the equation for y. Recall that this means writing the equation in the form $y = mx + b$.

Example 2

The equation $4x − 3y + 15 = 0$ defines a line.

a) Determine the slope and the y-intercept.

b) Graph the line.

Solution

a) Solve the equation for y.

$$4x - 3y + 15 = 0$$
$$-3y = -4x - 15$$
$$\frac{-3y}{-3} = \frac{-4x}{-3} - \frac{15}{-3}$$
$$y = \frac{4}{3}x + 5$$

From the equation in this form, the slope of the line is $\frac{4}{3}$, and the y-intercept is 5.

b) Use the slope and y-intercept to graph the line.

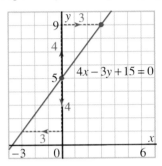

Special cases of the equation Ax + By + C = 0

When one or more of A, B, or C are 0, the line has certain special properties.

Example 3

Graph the line with each equation. Describe each line.

a) $x - 3y = 0$

b) $2y + 4 = 0$

c) $3x - 12 = 0$

Solution

a) $x - 3y = 0$

Solve the equation for y.

$$3y = x$$
$$y = \frac{1}{3}x$$

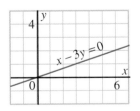

This line has slope $\frac{1}{3}$ and y-intercept 0. Use these to graph the line.

From its graph, we see that the line passes through the origin.

b) $2y + 4 = 0$

Solve the equation for y.

$$2y = -4$$
$$y = -2$$

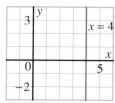

This equation may be written as $y = 0x - 2$. From this slope y-intercept form, the line has a y-intercept of -2 and a slope of 0. It is a horizontal line 2 units below the x-axis. The equation $y = -2$ indicates that every point on the line has a y-coordinate of -2.

c) The equation $3x - 12 = 0$ cannot be solved for y. This means that it cannot be written in slope y-intercept form.

Solve for x.

$$3x - 12 = 0$$
$$3x = 12$$
$$x = 4$$

The form $x = 4$ indicates that every point on the line has an x-coordinate of 4. The line is a vertical line 4 units to the right of the y-axis.

Example 3 illustrates these general properties:

The equation $Ax + By = 0$ represents a line through the origin.

The equation $By + C = 0$, or $y = k$, represents a horizontal line.

The equation $Ax + C = 0$, or $x = k$, represents a vertical line.

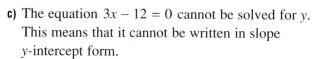

DISCUSSING THE IDEAS

1. In *Example 1*, we graphed a line by determining its intercepts from the equation. Are there any disadvantages to this method of graphing a line?

2. In *Examples 1* and *2*, how could we check that each graph is correct?

3. In an earlier grade when you solved an equation for x and obtained $x = 4$, you may have graphed the solution by locating the point, 4, on a number line. Why is the graph in *Example 3c* different? Discuss your ideas with a partner.

4.5 EXERCISES

A **1.** For each line, state the *x*-intercept, the *y*-intercept, and the slope.

a)
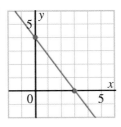

b)
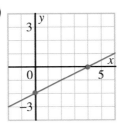

c)
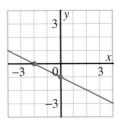

2. Which lines have an *x*-intercept of 3?

a) $x + 2y - 3 = 0$ b) $2x - 3y - 4 = 0$ c) $2x + 5y - 6 = 0$

3. Which lines have a *y*-intercept of −2?

a) $x + y + 2 = 0$ b) $3x + 4y + 8 = 0$ c) $2x + 3y - 6 = 0$

4. a) Determine the *x*- and *y*-intercepts for each line. Graph each line.

    i) $x - y - 5 = 0$ **ii)** $3x + y - 9 = 0$ **iii)** $x - 2y + 4 = 0$

    iv) $2x - 3y + 12 = 0$ **v)** $4x - 3y - 24 = 0$ **vi)** $2x + 5y - 6 = 0$

  b) Choose one equation from part a. Write to explain how you determined
the intercepts.

5. Determine the slope of each line.

  a) $2x + y - 6 = 0$ **b)** $x - y + 3 = 0$ **c)** $3x - 2y + 8 = 0$

  d) $x - 3y - 7 = 0$ **e)** $4x - 2y - 9 = 0$ **f)** $2x - 8y - 3 = 0$

6. a) For each line, determine the slope and *y*-intercept. Graph each line.

    i) $3x - 4y - 12 = 0$ **ii)** $5x - 2y - 10 = 0$ **iii)** $2x + y - 3 = 0$

    iv) $2x + 5y + 20 = 0$ **v)** $x + 2y - 5 = 0$ **vi)** $4x - 7y + 15 = 0$

  b) Choose one equation from part a. Write to explain how you determined
the slope and *y*-intercept.

7. a) What is the same in these equations? What is different?

    $6x + 2y - 4 = 0$; $3x + 2y - 4 = 0$; $x + 2y - 4 = 0$; $-0.25x + 2y - 4 = 0$;
    $-x + 2y - 4 = 0$; $-6x + 2y - 4 = 0$

  b) Plot the graphs on the same screen. Describe what you see. Explain why
the pattern occurs.

  c) Write the equations of 6 different lines that have the same values for *B*
and for *C*, but different values for *A*. Predict the pattern the graphs will
display on the screen. Plot the graphs on the same screen to check your
prediction.

B **8.** Graph each line.

a) $3x - 2y = 6$ b) $2x + y = 4$ c) $5x - 2y = 10$

d) $4x - 3y + 12 = 0$ e) $x - 2y - 8 = 0$ f) $5x + 3y - 6 = 0$

9. Graph each line.

a) $x - 2y = 0$ b) $2x + y = 0$ c) $3x - y = 0$ d) $3x + 2y = 0$

e) $2y - 6 = 0$ f) $3x + 6 = 0$ g) $x = 4$ h) $y = -5$

10. Compare each equation below with the graph of $2x - 3y + 6 = 0$.

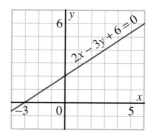

$2x - 3y + 12 = 0$ $x + y + 6 = 0$

$3x + 2y - 8 = 0$ $3x - 2y + 18 = 0$

$4x - 6y - 9 = 0$ $2x - y + 4 = 0$

a) Which lines have the same x-intercept?

b) Which lines have the same y-intercept?

c) Which lines have the same slope?

11. Two perpendicular lines intersect on the y-axis. The equation of one line is $2x - y + 8 = 0$. Determine the equation of the other line.

12. The intersection point of two perpendicular lines lies on the x-axis. The equation of one line is $2x + y + 6 = 0$. Determine the equation of the other line.

13. Choose either exercise 11 or 12. Write to explain how you determined the equation.

14. For the line $3x + 2y - k = 0$, determine the value of k for each condition.

a) x-intercept 3   b) x-intercept -2 c) y-intercept -3 d) y-intercept 7

15. The graph (below left) shows four lines containing the sides of a rhombus. The equation of one of the lines is $2x + 3y - 12 = 0$.

a) Determine the equations of the other three lines.

b) Calculate the area and the perimeter of the rhombus.

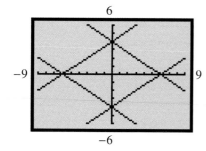

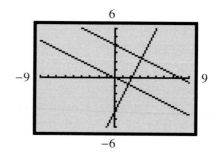

16. A certain rectangle is twice as long as it is wide. The graph (at the right on page 234) shows the lines containing three sides of the rectangle.

 a) The equation of one line on the graph is $x + 2y - 8 = 0$. Determine the equations of the other two lines on the graph.

 b) Determine as many possible equations for the line containing the fourth side of the rectangle as you can. Illustrate your results on a graph.

17. The intersection point of two perpendicular lines is N$(-1, -4)$. The equation of one line is $2x + y + 6 = 0$. Determine the equation of the other line.

18. Compare each equation below with the equation $x + 2y - 4 = 0$.

 $x - y + 2 = 0$ $\qquad$ $3x + 6y + 5 = 0$ $\qquad$ $2x - y - 8 = 0$

 $2x + 4y - 9 = 0$ $\qquad$ $4x - 3y + 6 = 0$ $\qquad$ $5x - 7y - 20 = 0$

 a) Which lines have the same x-intercept?

 b) Which lines have the same y-intercept?

 c) Which lines have the same slope?

19. Choose either exercise 10 or 18. Write to explain how you identified lines with:

 a) the same x-intercept $\qquad$ **b)** the same y-intercept $\qquad$ **c)** the same slope

20. The equation of a line is given in the form $Ax + By + C = 0$. Determine an expression for:

 a) the slope $\qquad\qquad$ **b)** the y-intercept $\qquad\qquad$ **c)** the x-intercept

C **21.** The equations of the sides of a triangle are $2x + 3y = 18$, $5x + y + 7 = 0$, and $3x - 2y = 14$.

 a) Graph the lines on the same axes. Determine the coordinates of the vertices of the triangle.

 b) Determine the lengths of the sides of the triangle.

 c) What are the slopes of the sides?

 d) What kind of triangle is it?

 e) Determine the area and the perimeter of the triangle.

COMMUNICATING THE IDEAS

In your journal, write a clear explanation of the nature of the lines represented by each equation.

 a) $y = mx + b$ $\qquad$ **b)** $y = x$ $\qquad$ **c)** $y = b$ $\qquad$ **d)** $x = a$

The Vanishing Square Puzzle

This right triangle has been divided into two triangles, two L-shaped pieces, and a unit square.

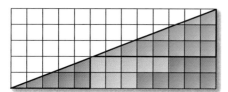

When the pieces are rearranged, there is no room for the square!

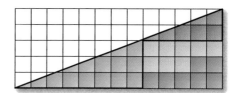

What happened to the square?

If you are not sure how to explain the puzzle, try to think of some things you could do to explain it. Here are some ideas.

- You could draw large, accurate diagrams on graph paper.
- You could carry out some calculations. What kinds of calculations do you think these should be?

Can you explain what happened to the square, and why?

After you have explained the vanishing square puzzle, solve these problems.

1. ***The Vanishing Rectangle Puzzle***
   The puzzle below contains three beige rectangular pieces. When they are arranged one way (below left), you can see ten mauve rectangles. When the top two beige pieces are interchanged (below right), there are only nine mauve rectangles. What happened to the other mauve rectangle?

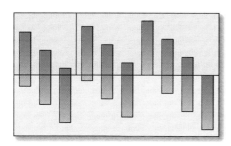 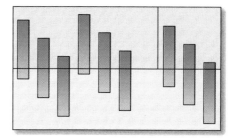

2. *The Vanishing Plane Puzzle*
In this puzzle, 78 planes are
enclosed in a 13 by
6 rectangle. The rectangle is
cut apart along the solid lines
and the pieces rearranged.
Only 77 planes remain. What
happened to the other plane?

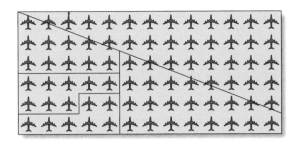

3. Compare the puzzles.

   a) How is the vanishing plane
   puzzle similar to the
   vanishing square puzzle?

   b) How is the vanishing plane
   puzzle similar to the
   vanishing rectangle puzzle?

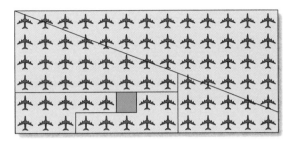

4. In the vanishing square puzzle, you should have discovered that the
endpoints of the hypotenuses of the three right triangles are not
collinear on either diagram. These points form a very slender obtuse
triangle. Try to visualize this triangle on both diagrams. Then choose
one of the triangles.

   a) What is the area of this triangle? How do you know?

   b) Sketch the triangle on plain paper, drawing a distorted view so you
   can see all three sides clearly.

   c) Calculate the lengths of the three sides of the triangle, and its
   perimeter. Express your answers in exact form and in decimal form.

   d) Calculate the height of the triangle using its longest side as the base.

   e) Calculate the height of the triangle using its shortest side as the base.

5. Create a vanishing puzzle of your own.

6. When you studied geometry in an earlier grade, you may have cut the
corners from a triangle and arranged them along a line.

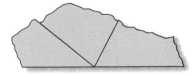

   a) What geometrical property does this demonstration illustrate?

   b) Based on your experience with the problems above, can you be sure
   this demonstration verifies the geometrical property?

It's All in the Packaging

Page 196 presented a problem to determine how many of two different books will fit in one carton. The books have the same cover size, but one is thicker than the other. The carton holds 30 thin books or 20 thick books. On these pages, we will develop a model to solve this problem and similar problems.

DEVELOP A MODEL

Assume the length and width of the carton are the same as the length and width of the books packed in it. That is, the books are packed in a single pile.

A full carton of thin books holds 30 books. Each book fills $\frac{1}{30}$ of the carton.

A full carton of thick books holds 20 books. Each book fills $\frac{1}{20}$ of the carton.

Visualize a single carton filled with both books.

Let x and y represent the number of thin and thick books, respectively, in the carton. Since each thin book fills $\frac{1}{30}$ of the carton, the thin books fill $x \times \frac{1}{30}$, or $\frac{x}{30}$ of the carton. Similarly, the thick books fill $\frac{y}{20}$ of the carton.

Thick books

Thin books

$\frac{y}{20}$ of the carton

$\frac{x}{30}$ of the carton

1. What number represents the full carton of books? Use your answer to write an equation relating x and y in the form $Ax + By + C = 0$.

2. **a)** On grid paper, graph the equation you wrote in exercise 1.

   b) Your graph should be a straight line. On your graph, mark the points on the line that represent solutions to the problem.

   c) Why is there more than one solution? Why should the points representing solutions to the problem lie on a line?

3. Make a table to show the solutions. How many solutions are there?

 LOOK AT THE IMPLICATIONS

4. Write the equation from exercise 1 in slope y-intercept form. Explain what the y-intercept and the slope represent.

5. The first equation you wrote in exercise 1 had the form $\frac{x}{a} + \frac{y}{b} = 1$. Explain what the numbers a and b represent.

6. What changes would there be in the results of exercises 1, 2, and 3 in each case?

   a) The carton holds only 18 thick books instead of 20.

   b) The carton holds 19 thick books.

 REVISIT THE SITUATION

7. This model assumes the books are packed in a single pile in the carton. It is more common to pack books in two piles.

   a) Why do you think it is more common to pack books this way?

   b) What changes are there, if any, in the results? Explain your answer.

8. Explain why we did not need to know the thickness of a book or the height of the carton to solve this problem.

9. **a)** Describe how the model for solving this problem could be used for similar problems in which the numbers of thin and thick books that fill a carton are represented by m and n, respectively.

   b) How could you tell from the values of m and n how many solutions there are? Illustrate your answer with some examples.

Sequences and $Ax + By + C = 0$

You will graph several equations of the form $Ax + By + C = 0$ in which the coefficients are consecutive terms of a sequence.

Coefficients in arithmetic sequence	Coefficients in geometric sequence
$2x + 5y + 8 = 0$	$x + 3y + 9 = 0$
$3x - y - 5 = 0$	$x + 2y + 4 = 0$
$x + y + 1 = 0$	$x + y + 1 = 0$
$x - 3y - 7 = 0$	$x - 3y + 9 = 0$

Since most graphing calculators can only graph equations in the form $y = mx + b$, these equations must be solved for y. Instead of doing this with every equation, we will solve the equation $Ax + By + C = 0$ once, then use the calculator. The method described below applies to the TI-82 and TI-83 calculators. Other calculators may have similar features.

In exercise 20, page 235, you solved $Ax + By + C = 0$ for y and obtained $y = -\frac{A}{B}x - \frac{C}{B}$.

To graph the equations in the list (above left), enter the coefficients in the list editor. Enter the equation in the screen (below right). Press the ⟨GRAPH⟩ key. The calculator will graph the equation corresponding to each row in the list editor. List L_1 contains the values of A, list L_2 contains the values of B, and list L_3 contains the values of C.

▪L1	L2	L3	1
2	5	8	
3	-1	-5	
1	1	1	
1	-3	-7	
------	------	------	

$L1 = \{2, 3, 1, 1\}$

```
Plot1 Plot2 Plot3
\Y1冒-(L1/L2)X-(L
3/L2)
\Y2=
\Y3=
\Y4=
\Y5=
\Y6=
```

1. Explain how $y = -\frac{A}{B}x - \frac{C}{B}$ corresponds to the equation in the screen (above right).

2. **a)** Graph several equations in which the coefficients A, B, and C are consecutive terms of an arithmetic sequence.

   b) What can you conclude about equations with this property?

3. Repeat exercise 2 for geometric sequences.

4 Review

1. For each equation, make a table of values then draw a graph.
   a) $3x - y = 9$ b) $x + 2y = 10$ c) $4x - 3y = -12$

2. A person estimates that the annual cost, C dollars, of operating her car is given by the formula $C = 1000 + 0.2n$, where n kilometres is the distance driven in one year.
   a) Make a table for values of n up to 10 000.
   b) Draw a graph.
   c) Use the graph. Estimate how far the person drove when her annual expenses were $2850.
   d) Use the graph. Estimate how far the person drove when her annual expenses were $3350.

3. State the slope of each line.

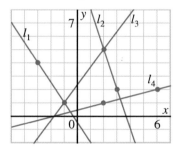

4. a) On a grid, draw a line with slope $\frac{2}{5}$ through A(−3, 2).
   b) Draw a line through C(−2, −1) that is parallel to the line in part a.
   c) Draw a line through C(−2, −1) that is perpendicular to the line in part a.

5. a) On the same axes, draw a line through P(5, 2) with each slope.
   i) $\frac{2}{3}$ ii) $-\frac{1}{4}$ iii) $-\frac{5}{7}$ iv) 0
   v) not defined by a real number
   b) Choose one slope from part a. Write to explain how you drew the line.

6. Determine whether the points R(8, 5), S(1, −2), and T(5, 2) are collinear.

7. State the slope and the y-intercept for the line represented by each equation.
   a) $y = 4x - 3$ b) $y = -\frac{5}{3}x + 7$ c) $y = -\frac{9}{4}x - 3$

8. Write the equation of the line that has each slope and y-intercept.
   a) $m = -\frac{1}{2}$, $b = -4$ b) $m = \frac{4}{3}$, $b = -6$ c) $m = -\frac{3}{2}$, $b = \frac{3}{4}$

9. a) Graph the line represented by each equation.

  i) $y = \frac{2}{3}x - 1$ **ii)** $y = -2x + 3$ **iii)** $y = -\frac{4}{3}x + 1$ **iv)** $y = x - 5$

  b) Choose one equation from part a. Write to explain how you graphed the line.

10. Determine the value of b when the line $y = \frac{3}{2}x + b$ passes through each point.

  a) C(4, 7)   **b)** D(−6, −4)

11. Determine the value of m when the line $y = mx - 3$ passes through each point.

  a) A(1, 5)   **b)** B(4, 0)

12. Which of the following points lie on the line $2x - 5y + 4 = 0$?

  a) C(3, 2)   **b)** D(7, 4)   **c)** E(−5, 1)   **d)** F(8, 4)   **e)** G$\left(\frac{1}{2}, 1\right)$

13. Which of the following lines pass through the point H(−3, 4)?

  a) $2x + 3y - 5 = 0$ **b)** $3x - 2y + 17 = 0$ **c)** $5x + 2y + 7 = 0$

14. Determine the equation of the line that passes through each pair of points.

  a) A(3, 5) and B(−5, −3)   **b)** C(−4, 7) and D(5, −4)
  c) E(2, 10) and F(−2, −6)   **d)** G(−8, 3) and H(6, −5)

15. Determine the equation of the line that passes through J(5, 5) and is:

  a) parallel to the line $3x + 4y = -16$
  b) perpendicular to the line $5x + 2y = 10$

16. A triangle has vertices P(−3, −1), Q(9, 3), and R(3, 7).

  a) Determine the equation of each side of the triangle.
  b) Determine the equation of the line through R parallel to PQ.
  c) Determine the equation of the line through R perpendicular to PQ.
  d) Determine the equation of the median from Q to the midpoint of PR.

17. For each line, determine the x-intercept, the y-intercept, and the slope.

  a) $2x - 5y + 10 = 0$ **b)** $4x + y - 12 = 0$ **c)** $3x - 7y - 14 = 0$

18. Graph each line without making a table of values.

  a) $2x - 3y + 12 = 0$ **b)** $3x + 5y = 0$ **c)** $5x + 2y - 15 = 0$
  d) $2y - 8 = 0$ **e)** $4x - 3y - 16 = 0$ **f)** $3x + 12 = 0$

19. For what value of k are the lines $3kx - 7y - 10 = 0$ and $2x + y - 7 = 0$

  a) parallel?   **b)** perpendicular?

1. Copy and complete each arithmetic sequence.

a) 1, 7, ▩, ▩, ▩   **b)** 3, ▩, 11, ▩, ▩

c) ▩, ▩, 6, 14, ▩   **d)** ▩, 9, ▩, 23, ▩

e) ▩, ▩, 4, ▩, 23   **f)** 5, ▩, ▩, ▩, 73

2. Determine the indicated term of each geometric sequence.

a) $\frac{1}{4}, \frac{1}{2}, 1, \ldots t_6$ **b)** $-\frac{1}{4}, -\frac{1}{2}, -1, \ldots t_7$

c) $1, \frac{1}{2}, \frac{1}{4}, \ldots t_7$ **d)** $2.25, 1.5, 1, \ldots t_5$

e) $2, 6, 18, \ldots t_8$ **f)** $4, 4.8, 5.76, \ldots t_8$

3. Calculate each exact value without using a calculator.

a) $\left(\frac{4}{9}\right)^{-\frac{1}{2}}$ **b)** $64^{\frac{5}{6}}$ **c)** $(-64)^{-\frac{2}{3}}$ **d)** $2.25^{1.5}$ **e)** $\left(\frac{27}{64}\right)^{-\frac{1}{3}}$

f) $\left(-\frac{1}{8}\right)^{-\frac{1}{3}}$ **g)** $\left(\frac{16}{81}\right)^{\frac{1}{4}}$ **h)** $100\,000\,000^{-\frac{3}{4}}$ **i)** $0.0001^{-\frac{1}{4}}$ **j)** $\left(\frac{49}{81}\right)^{-\frac{1}{2}}$

4. Determine x and y to 2 decimal places.

a)
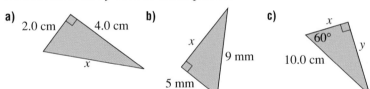
2.0 cm   4.0 cm x

b) x 9 mm   5 mm

c) x 60° y 10.0 cm

d) x 3 m   6 m y

5. Simplify.

a) $\frac{3\sqrt{6}}{2\sqrt{5}} \times \frac{\sqrt{20}}{\sqrt{27}}$ **b)** $3\sqrt{3} + 2\sqrt{12} + 5\sqrt{18} + 2\sqrt{8}$ **c)** $\frac{\sqrt{3} - \sqrt{2}}{\sqrt{3} + \sqrt{2}} + \frac{\sqrt{3} + \sqrt{2}}{\sqrt{3} - \sqrt{2}}$

6. Determine whether points P(−4, 3), Q(2, 1), and R(11, −2) are collinear.

7. Determine the slope of the line segment with each pair of endpoints.

a) C(−3, 7), D(7, −3)   **b)** E(2, 8), F(−6, 4)   **c)** G(8, −2), H(−1, 6)

8. The points A(2, 6), B(5, 2), and C(2, 2) are given. Determine the coordinates of point D on the x-axis so that CD is perpendicular to AB.

9. The equations of the sides of a triangle are $4x − 3y + 9 = 0$, $x − 7y + 21 = 0$, and $3x + 4y − 37 = 0$. What kind of triangle is it?

10. a) Draw two perpendicular lines through the point P(−3, 5).

b) Determine the equation of each line you drew in part a.

c) Compare your equations with those of a classmate. If the equations are different, write to explain why this is possible.

5 FUNCTIONS

Pushing Your Physical Limits

CONSIDER THIS SITUATION

Have you ever watched a marathon or a long-distance bicycle race? One athlete may burst through the finish line, while another collapses at or near the finish. How can two people respond so differently to the same race?

Perhaps you participate in an aerobic sport. Think about the training you undergo, and how it influences your level of fitness. How do you recognize your own physical limits? How do you try to extend those limits?

Think about it.

• What are some factors that influence fitness? How does training affect those different factors?

• Describe some ways to measure fitness.

On pages 302 and 303, you will examine a model related to measuring your body's efficiency during exercise. You will use functions to investigate how fitness is measured.

 FYI Visit www.awl.com/canada/school/connections

For information related to the above problem, click on
<u>MATHLINKS</u> followed by <u>AWMath</u>. Then select Personal Fitness.

5.1 What Is a Function?

Many calculators have an $\boxed{x^2}$ key.

You enter a number, press $\boxed{x^2}$, and the calculator displays the result.

Input
5
→
Press
$\boxed{x^2}$
→
Output
25

The number you enter is the *input number*.

The calculator display is the *output number*.

The $\boxed{x^2}$ key illustrates the idea of a function.

A *function* is a rule that gives a single output number for every valid input number.

The rule that defines a function can be expressed in different ways.

For the above function we can express the rule as follows.

- In words:

  Multiply the number by itself.

- As an equation:

  $y = x^2$

  In an equation, we use x for each input number and y for each output number.

- As a table of values:

Input number x	Output number y
−4	16
−3	9
−2	4
−1	1
0	0
1	1
2	4
3	9
4	16

- As a set of ordered pairs:

  The numbers in the table can be written
  $\{(-4, 16), (-3, 9), (-2, 4), (-1, 1), (0, 0), (1, 1), (2, 4), (3, 9), (4, 16)\}$

- As a graph:

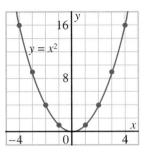

In the graph, the plotted points appear to lie on a curve. If we included more points corresponding to input numbers such as $-4.5, -3.5, \ldots, 2.5, 3.5, 4.5$, we would see that the points do lie on a curve.

INVESTIGATE

The rules defining several functions are given below. For each function:

a) Make a table of values.

b) Draw a graph of the function.

c) Decide if there are any input numbers that cannot be used.

d) Write an equation for the function.

Rule 1: Add 3 to the number.

Rule 2: Multiply the number by 3.

Rule 3: Double the number and add 3.

Rule 4: Subtract the number from 10.

Rule 5: Determine the positive square root of the number.

We can describe a function in various ways. Some simple functions can be described in words. We can also use a table of values, a set of ordered pairs, an equation, or a graph to describe a function.

Example 1

Monique has a part-time job at a garden centre. She earns $6.50 per hour. Employees are paid for whole numbers of hours worked – not for parts of hours worked. Let h hours represent the time Monique works in a week. Let p dollars represent her pay.

a) Make a table of values. Draw a graph of p against h.

b) Write an equation for the function.

c) Suppose Monique gets a raise to $7.00 per hour. How would the graph and the equation change?

Solution

a)

Hours worked, h	Pay, p dollars
0	0.00
1	6.50
2	13.00
3	19.50
4	26.00
5	32.50

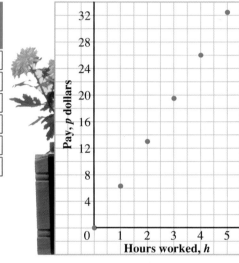

b) Each value of h was multiplied by 6.5 to get the corresponding value of p. An equation for the function is $p = 6.5h$.

c) The points on the graph lie along a straight line. We do not join the points because employees are paid for whole numbers of hours only. If Monique gets a raise to $7.00 per hour, the points would lie along a slightly steeper line. The equation would be $p = 7h$.

In *Example 1*, the function was represented by a table of values. These values could have been written as the set of ordered pairs:
$\{(0, 0.00), (1, 6.50), (2, 13.00), (3, 19.50), (4, 26.00), (5, 32.50)\}$.

In *Example 1b*, an equation for the function was $p = 6.5h$. We say that this equation *expresses p as a function of h*.

Example 2

A sunscreen's effectiveness is indicated by a number called the sunscreen protection factor. The percent p of the sun's ultraviolet light that passes through a sunscreen is expressed as a function of its protection factor s by the formula $p = \frac{100}{s}$.

a) Some sunscreens have protection factors of 2, 8, 15, 25, and 45. Determine the percent of the sun's ultraviolet light that passes through each sunscreen.

b) Draw a graph to show how the percent of the sun's ultraviolet light that passes through a sunscreen depends on the protection factor.

Solution

a) Substitute 2, 8, 15, 25, and 45, in turn, for s in the formula $p = \frac{100}{s}$ to obtain the following table.

b)

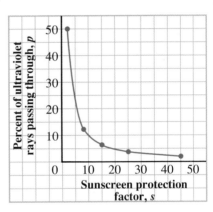

Sunscreen protection factor, s	Percent of ultraviolet light passing through, p
2	50.0
8	12.5
15	6.7
25	4.0
45	2.2

DISCUSSING THE IDEAS

1. What is a function?

2. What ways are there to express a function? How are these ways similar? How are they different?

A 1. Write a rule to describe each function.

a)

x	y
1	4
2	5
3	6
4	7

b)

x	y
−2	−6
0	0
1	3
2	6

c) {(1, 3), (2, 5), (3, 7), (5, 11)}

2. Below are two statements about functions. Decide if each statement is true or false. Explain your answers.

a) A table of values always shows all possible input numbers.

b) When you graph a function, you always join the plotted points.

3. a) Your calculator should have the keys $\boxed{\sqrt{x}}$ and $\boxed{1/x}$. How does each key illustrate the concept of a function?

b) Are there any values of x that cannot be used as input numbers for the functions defined by $y = \sqrt{x}$ and $y = \frac{1}{x}$? Explain your answer.

4. Three functions are graphed below.

a) Describe how the graphs are similar.

b) Account for the differences among the graphs.

i)

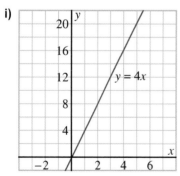

ii)

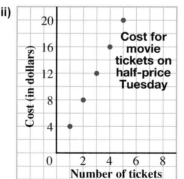

iii)

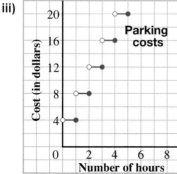

B **5.** The rules defining three functions are given below. For each function:

   a) Make a table of values.

   b) Draw a graph of the function.

   c) Decide if there are any input numbers that cannot be used.

      i) *Rule 1*: Add 2.5 to the number.

      ii) *Rule 2*: Multiply the number by 0.5.

      iii) *Rule 3*: $y = \sqrt{x} + 2$

   d) Choose one function from parts a to c. Write to explain how you decided if there were any input numbers that could not be used.

6. List 5 ordered pairs for each function in exercise 5.

7. In a family, the children receive a weekly allowance from their sixth birthday until they are eighteen. The amount depends on the child's age. At 6 years of age, each child receives $1 a week. Each year, the amount is increased by $1. Let A years represent the child's age. Let D dollars represent the allowance received.

   a) Make a table of values for this function.

   b) Write an equation that expresses D as a function of A.

   c) Graph this function with A on the horizontal axis.

8. Graph each function on a separate grid.

   a) *Rule 1*: The output number is the same as the input number.

   b) *Rule 2*: The output number is always 3.

   c) *Rule 3*: $y = (x + 1)(x - 1)$

9. List 5 ordered pairs for each function in exercise 8.

 10. In *Example 2*, you learned that the percent p of the sun's ultraviolet light that penetrates a sunscreen with protection factor s is $p = \frac{100}{s}$.

   a) On your graphing calculator, define the function $Y = \frac{100}{X}$. Graph the function using window settings: $0 \le X \le 47$ and $-10 \le Y \le 100$.

   b) Use the trace feature. Determine the approximate percent penetration for protection factors 4 and 35.

   c) Use the trace feature. Estimate the protection factor that provides 15% penetration.

   d) Change the window settings to $0 \le X \le 50$ and $0 \le Y \le 100$. Try to use the trace feature. What are the advantages of using the window settings in part a?

   e) Does doubling the protection factor have the greatest effect when s is small or when s is large? Explain.

f) Write to describe how the percent penetration changes as the protection factor increases.

11. Lo Cost Car Rentals charges $25.00 per day to rent a mid-sized car. This includes 100 km of free driving. If the customer drives more than 100 km, she is charged an additional 15¢ per kilometre. Let T dollars represent the total rental charge. Let D represent the number of kilometres travelled in one day.

   a) Make a table of values for this function. Include $D = 500$ km.

   b) Graph this function with D on the horizontal axis.

   c) Write an equation that expresses T as a function of D.

   d) Suppose the first 200 km of driving was free. How would the graph in part b change?

12. a) Graph these functions on the same grid.

   Rule 1: Double the number and add 3 to the result.
   Rule 2: Add 3 to the number and double the result.
   Rule 3: $y = 2x - 3$
   Rule 4: $y = 2(x - 3)$

   b) Account for the similarities and the differences in the graphs.

13. Money invested in Guaranteed Investment Certificates (GICs) earns compound interest. The table shows the amounts to which a principal of $1250 grows after various years, at an interest rate of 3.75%.

Number of years	Amount ($)
0	1250.00
1	1296.88
2	1345.51
3	1395.96
4	1448.31
5	1502.62

   a) Graph the data and draw a smooth curve through the points.

   b) Estimate the amount after three and a half years.

   c) Suppose the amount after 5 years was reinvested at the same interest rate. Estimate the amount after 7 years.

   d) Compare your answer to part c with that of a classmate. If your answers differ, explain why.

14. a) Graph each function on a separate grid.

   i) *Rule 1*: The output number is the opposite of the input number.
   ii) *Rule 2*: $y = -1$
   iii) *Rule 3*: $y = \dfrac{x(x - 1)}{2}$

   b) Choose one function from part a. Write to explain how you graphed it.

15. Usually, light does not penetrate below 100 m into an ocean. The table shows the percent of surface light present at various depths.

Depth (m)	Percent of light present
0	100
20	63
40	40
60	25
80	16
100	10

a) Graph the data and join the points with a smooth curve.

b) Use the graph. Estimate the percent of surface light present at a depth of 15 m.

c) Estimate the depth at which 30% of the light is present.

16. a) Graph these functions on the same grid.

Rule 1: Square the number and subtract 2 from the result.
Rule 2: Subtract 2 from the number and square the result.
Rule 3: $y = x^2 + 2$
Rule 4: $y = (x + 2)^2$

b) Account for the similarities and the differences in the graphs.

17. A graphing calculator produced these graphs. The equations of the functions are shown. Identify the equation that corresponds to each graph.

i)
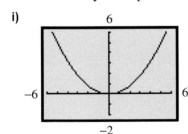

ii)
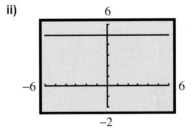

iii)
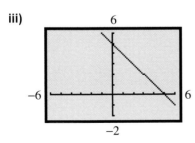

iv)
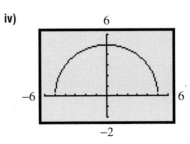

a) $y = 5 - x$ b) $y = \frac{x^2}{5}$ c) $y = 5$ d) $y = \sqrt{25 - x^2}$

C **18.** The Body Mass Index (BMI) provides an indication of a person's physical fitness. This is derived from the formula $\text{BMI} = \dfrac{\text{mass (kg)}}{(\text{height (m)})^2}$

The desirable range for the BMI is between 20 and 25 for men and women.

a) Consider people whose mass is 70 kg. Substitute 70 for mass in the formula. This expresses the BMI for these people as a function of their height.

b) Graph the function in part a for some reasonable heights.

c) Describe how the BMI changes as height increases.

d) How would the graph be different for people with a mass less than 70 kg? How would it be different for people with a mass greater than 70 kg?

e) Make a sketch to show BMI functions corresponding to different masses.

MODELLING Physical Fitness

The Body Mass Index is only a crude measure of physical fitness. You can think of it as replacing height and weight tables that you may have seen. In exercise 18, you substituted for mass to express the BMI as a function of height for 70-kg people. You could have substituted for height to express the BMI as a function of mass.

- Consider people who are 150 cm tall. In the formula, substitute 1.5 for the height to express the BMI for these people as a function of their mass. Graph this function. Why is this different from the functions you graphed in exercise 18?

- Make a sketch to show BMI functions corresponding to different heights.

---

19. In *Example 1*, suppose the garden centre changes its policy and pays employees parts of hours.

a) How would the graph change if employees are paid for:

i) half-hour time periods worked?

ii) quarter-hour time periods worked?

b) Would it be possible for the graph to be drawn as a straight line? Explain your answer.

COMMUNICATING THE IDEAS

In your journal, write a few sentences to explain what is meant by a function. Illustrate your definition with some examples.

5.2 Interpreting Graphs of Functions

The graph below shows how the median age of Canadians has changed over the past 75 years. It also predicts how the median age will continue to change over the next 25 years. The ages of half the population are above the median, and the ages of the other half are below it. So, this measure gives a good indication of the aging of our society.

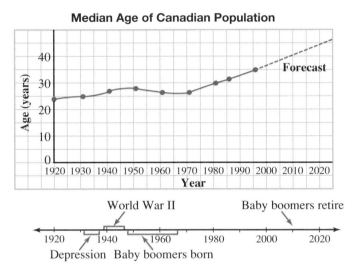

From the graph, we see that the median age peaked in the 1940s, then declined until the 1960s. Since the 1960s, the median age of Canadians has been steadily increasing.

The time line below the graph illustrates some of the major events from 1930 to 1960. During the depression in the 1930s, few families were having children. This resulted in a rise in the median age until the late 1940s, when the baby boom occurred after World War II. This brought the median age down. Since this time, the birth rate has moderated and the median age has steadily increased as the baby boom generation ages. Factors such as better nutrition and health care could also explain why the median age is increasing.

Example 1

Use the graph at the right.

a) What does the graph illustrate?

b) What was the death rate in each year?

    i) 1900   ii) 1930   iii) 1960

c) What general trend does the graph show?

d) Suggest some reasons why this trend is occurring.

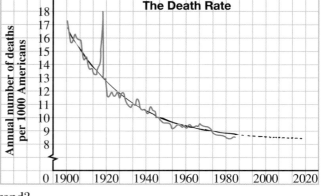

e) Are there any exceptions to this trend? Suggest reasons why these may have occurred.

f) What does the graph suggest will happen to the trend in future years?

Solution

a) The broken-line graph shows how the death rate in the United States has changed since 1900. The curve illustrates the trend in the death rate.

b) From the graph, the death rate:

    i) in 1900 was about 17 people per 1000 people.

    ii) in 1930 was between 11 and 12 people per 1000 people.

    iii) in 1960 was between 9 and 10 people per 1000 people.

c) The death has generally been decreasing since 1900.

d) The death rate may be decreasing because of better health care and living conditions.

e) There was one major exception to this trend. It occurred just prior to 1920, which coincided with World War I. Notice that there was no large peak during World War II.

f) From the graph, the death rate appears to stabilize and stop decreasing.

In *Example 1*, the input numbers for the function are the years from 1900 to 1985 plotted on the horizontal axis. This set of numbers is called the domain of the function. The output numbers are the number of deaths per thousand plotted on the vertical axis. These vary from about 8.5 to 18. This set of numbers is called the range of the function.

For any function, we define two sets as follows.

- The domain is the set of all possible input numbers.
- The range is the set of all possible output numbers.

The *domain* of a function is represented by the shadow of its graph on the *x*-axis.

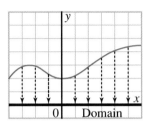

The *range* of a function is represented by the shadow of its graph on the *y*-axis.

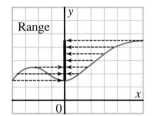

Given the graph of a function:

The domain is the set of *x*-values represented by the graph.

The range is the set of *y*-values represented by the graph.

Example 2

Triangle ABC is a right triangle. Suppose $x°$ represents the measure of the smaller acute angle in the triangle. Suppose $y°$ represents the measure of the larger acute angle. The graph illustrates the measure of the larger acute angle as a function of the smaller acute angle.

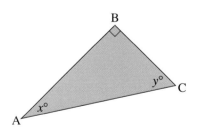

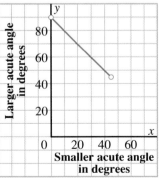

a) What is the equation of the function?

b) What happens to *y* as *x* increases? Explain, using the diagram, the graph, and the equation.

c) State the domain and range of the function.

Solution

a) The sum of the angles in $\triangle ABC$ is 180°. So, the two acute angles have a sum of 90°.

That is, $x + y = 90$

Hence, the equation of the function is $y = 90 - x$.

b) As x increases, y decreases. On the diagram, if C moves farther from B, x becomes larger and y becomes smaller. On the graph, as you move to the right along the line, you also move down. In the equation, as larger numbers are subtracted from 90, the differences are smaller.

c) The domain is the set of all possible angle measures for the smaller acute angle; that is, $0 < x < 45$. Notice that when $x = 45$, there is no smaller angle.

The range is the set of all possible measures for the larger acute angle; that is, $45 < y < 90$.

DISCUSSING THE IDEAS

1. For each graph in the preceding examples, explain why it is a function.

2. What is the difference between the domain and the range of a function?

5.2 EXERCISES

Ⓐ 1. A person rides a bike up and down a steep hill. Which graph best represents the speed of the bike as a function of time? Briefly explain your choice.

a) 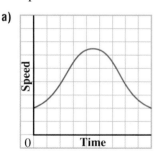 **b)** 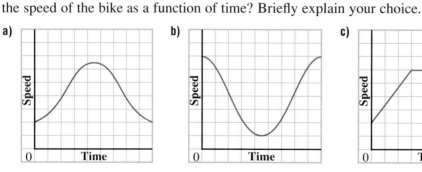 **c)**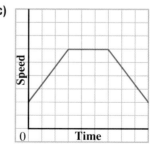

2. a) Give an example of a function that can be defined by listing all its ordered pairs.

b) Give an example of a function that cannot be defined by listing all its ordered pairs.

3. The school volleyball team is raising money for team jackets. Each day the team buys a dozen doughnuts, which costs $5. The team then sells the doughnuts for $1 each. Which graph best represents the team's profit for one day as a function of the number of doughnuts sold? Justify your choice.

a)
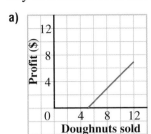

b)
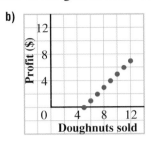

c)
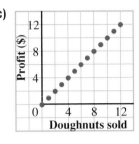

4. A child steadily climbs a slide, stops at the top, then slides down.

i)
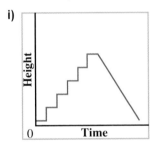

ii)
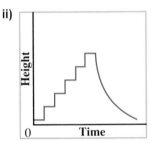

iii)
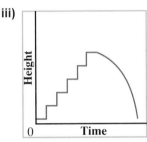

   a) Which graph best represents the height of the child as a function of time?

   b) What does the slope of each part of the graph represent?

5. State the domain and the range of each function. Then choose one function and write to explain how you determined the domain and range.

a)
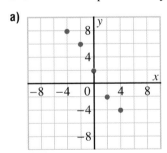

b)
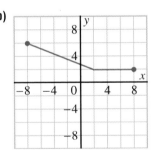

c)
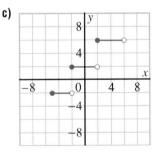

B 6. a) A car is travelling at a constant speed on a highway. It increases its speed to pass a second car, then returns to its original speed. Which graph on the next page best describes this motion? Explain.

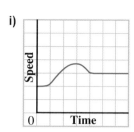 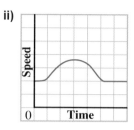 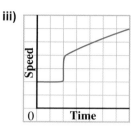

i)   ii)   iii)

b) Choose one graph from part a that does not describe the motion. Write to explain why it does not describe the motion. Describe a situation that could be represented by that graph.

7. Habiba walks in her neighbourhood for exercise. Here is a graph that describes her walk.

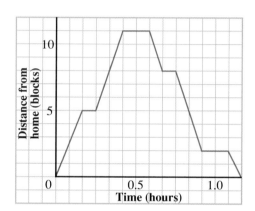

a) How long did it take Habiba to complete her walk?

b) What percent of the time in part a was Habiba walking?

c) Write to describe Habiba's walk.

d) What is the domain of this function? What does the domain represent?

e) What is the range of this function? What does the range represent?

8. For each graph, describe a practical situation it could represent. When you describe the situation, state the meaning of any intercepts, slopes, and maximum and/or minimum points, if possible. For each graph, describe an appropriate domain and range.

a) 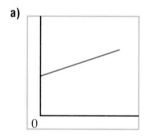   **b)** 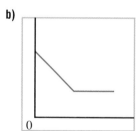   **c)**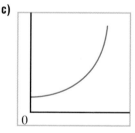

d) 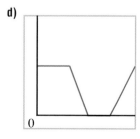   **e)** 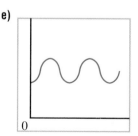   **f)**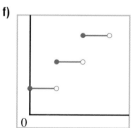

9. The graph shows the price of publishing a two-page colour newsletter.

a) What does the y-intercept represent?

b) Calculate the slope of each line segment.

c) What does the slope of each line segment represent?

d) How much would it cost to publish 200 newsletters?

e) Describe the publishing company's pricing scheme. Provide as much detail as possible.

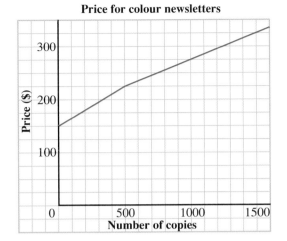

Price for colour newsletters

10. The graph describes the amount of fuel in pounds in a helicopter's fuel tank. To begin, the tank is full.

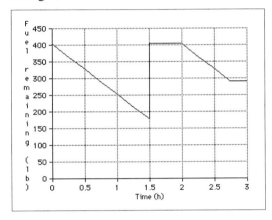

a) What is the fuel capacity of this helicopter?

b) What happens to the amount of fuel in the tank during the first 1.5 h? Why do you think this occurs? What happens after 1.5 h?

c) Describe what happens from 1.5 h to 3.0 h.

d) Calculate the slope of the first line segment. What does this represent? Why is the slope negative? The opposite of the slope represents the rate at which the engine uses fuel, and is called fuel flow.

e) Use the *Helicopters* database and your answers to parts a and d. Identify the model of helicopter. For each line segment in the graph that represents time in flight, use the airspeed for maximum range to estimate the distance the helicopter flew.

f) Choose another helicopter from the database. Create a similar exercise with a different graph. Ask a classmate to describe your graph and identify the helicopter.

11. The graph illustrates the number of similar pages *n* printed by an ink jet printer as a function of time *T* minutes.

a) Describe how the graph illustrates the function.

b) Determine the domain and range for this function.

c) Determine the slope and explain what it represents.

d) Determine any intercepts and explain what each represents.

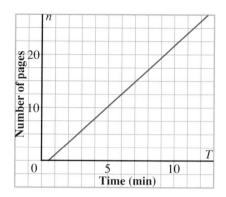

 MODELLING the Printing of Pages

The function in exercise 11 models the action of a printer. In practice, some time is lost between printing one page and the next. That is, some time passes but no page is printed.

- Draw a graph to show the action of a printer, with a short delay between the printing of consecutive pages.

- Some print jobs are more complex and take longer to print because the printer requires more time to process complex diagrams or photographs. Suggest how each graph might change for each complex print job.

  a) the graph in exercise 11

  b) the graph allowing for a short delay in printing consecutive pages

12. The graph shows how the orbital velocities of the planets relate to their mean distances from the sun. The distances are measured in astronomical units (A.U.). One astronomical unit is the mean distance between Earth and the sun, which is approximately 150 000 000 km.

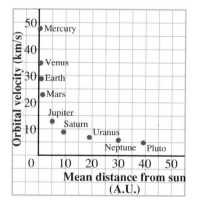

a) Which planet is closest to the sun? Which is farthest from the sun?

b) Which planet has the greatest orbital velocity? Which has the least?

c) Which planet is approximately 6 times as far from the sun as Jupiter? How are the velocities of these two planets related?

d) The points on this graph appear to lie on a curve. Why was the curve not drawn on the graph?

e) Write to explain how the orbital velocities of the planets are related to their distances from the sun.

13. Right triangle PQR has $\angle P = 90°$. Let $x°$ represent the measure of the larger acute angle and $y°$ the measure of the smaller acute angle.

a) Draw a graph to represent the measure of the smaller acute angle as a function of the larger acute angle.

b) What is the equation of the function?

c) What happens to y as x decreases? Explain, using the graph and equation.

d) State the domain and range of the function.

14. The graph shows how a car's volume of fuel varies with time during a trip. The graph consists of six line segments.

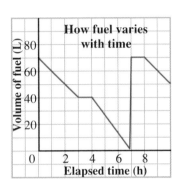

a) Write to explain what each line segment tells about the trip.

b) Suppose the car were driven at an average speed of 100 km/h. Determine its rate of fuel consumption in litres per 100 km.

15. In a power test, a car accelerated from a standing start to a top speed in less than 10 s. The distance, s metres, travelled by the car as a function of time, t seconds, is given by the equation $s = -0.07t^3 + 3.15t^2 + 1.2t$.

a) Graph the function $Y = -0.07X^3 + 3.15X^2 + 1.2X$ for $0 \le X \le 10$ and $0 \le Y \le 300$. Compare your graph with those below. Which graph represents s as a function of t?

i) 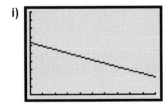 **ii)** 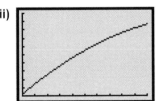 **iii)**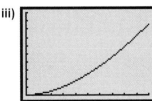

b) The two other graphs in part a show the velocity v as a function of time, and the acceleration a as a function of time. Identify the graph that represents each function. Write to explain how you did this.

c) The equation of v as a function of t is $v = -0.21t^2 + 6.3t + 1.2$. Graph this equation on your graphing calculator with $0 \le X \le 10$ and $0 \le Y \le 500$. Check your answers to part b.

16. Sketch a graph to illustrate each situation. If there is sufficient information, label your sketch and write an equation to represent the function.

　a) the circumference of a circle as a function of its radius

　b) your height as function of your age in years

　c) the height of a basketball during a free-throw as a function of time

　d) the cost of a speeding ticket as a function of the speed of the car

　e) the area of a square as a function of the length of one side

　f) the average daily temperature as a function of the day of the year

　g) the final cost of an item as a function of its price before taxes

　h) the speed of a soccer ball after it as been kicked as a function of time

　Choose one part from a to h. Write to explain how you determined the shape of the graph you sketched.

C 17. The graph shows the value of a computer over time.

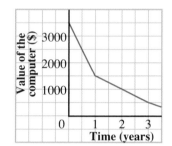

　a) What happens to the value of a computer after one year?

　b) What is the value of a $3500 computer after two years?

　c) What does the *y*-intercept of this graph represent?

　d) Suppose the axes were extended. Will this graph ever intersect the *x*-axis? Explain your answer.

　e) Calculate the slope of each line segment. Write to explain what each slope represents.

COMMUNICATING THE IDEAS

In this section, some graphs were drawn using a continuous curve or line, while others were drawn using points or line segments. In your journal, write to explain why this is. In your explanation, use examples from this text.

Some Examples of Functions

The graphs of these four functions were drawn using *Graphmatica*.

Graph 1

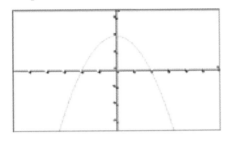

Graph 2

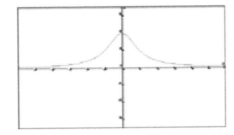

Graph 3

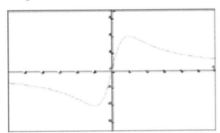

Graph 4

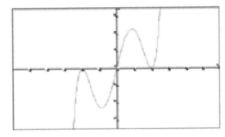

These are the equations of the functions, but they are not in the same order as the graphs:

$$y = \frac{x(x^2 - 4)^2}{4} \qquad y = \frac{4x}{x^2 + 1} \qquad y = \frac{4 - x^2}{2} \qquad y = \frac{2}{x^2 + 1}$$

Recall the Equation of a Line Property from Chapter 4. A similar property applies to functions.

Equation of a Function Property

- The coordinates of every point on the graph of a function satisfy its equation.
- Every point whose coordinates satisfy the equation of a function is on its graph.

Use this property.

1. Determine the intercepts of each function from its graph.

2. Use your answer to exercise 1. Identify the equation of each function.

3. Determine the domain and the range of each function.

Mathematics & Technology

LINKING IDEAS

5.3 Finding Relationships in Data

A group of students was curious about how a fine for exceeding the speed limit was determined. The students contacted the local police, who gave them these data.

Table A

Number of kilometres per hour over the speed limit	Fine ($)
5	85
10	90
15	95
20	100
25	105

Table B

Number of kilometres per hour over the speed limit	Fine ($)
30	150
35	160
40	170
45	180
50	190
55	200

1. Graph the data.

2. For the data in each table, use your graph to write an equation relating the fine, f dollars, to the speed, s kilometres per hour, by which a person exceeds the speed limit.

3. Use the appropriate equation. Determine the fine for a person who exceeds the speed limit by each amount.

 a) 12 km/h **b)** 28 km/h **c)** 39 km/h **d)** 51 km/h

4. What do you think the fines should be for exceeding the speed limit by 60 km/h or more? Prepare a Table C showing your recommended fines for exceeding the speed limit by amounts greater than those in Table B.

For Table A in *Investigate*, you may have noticed that the numbers in the second column are 80 greater than those in the first column: $f = 80 + s$. For table B, the numbers in the second column are 90 greater than double those in the first column: $f = 90 + 2s$. In the following example, we will look for similar equations relating two variables.

Example 1

For each table of values, determine a rule that expresses y as a function of x.

a)

x	y
0	0
1	4
2	8
3	12
4	16
5	20

b)

x	y
0	−1
1	2
2	5
3	8
4	11
5	14

c)

x	y
0	1
1	2
2	5
3	10
4	17
5	26

Solution

a) Each y-value is 4 times the corresponding x-value.

A rule is $y = 4x$.

b) Each y-value is 1 less than 3 times the corresponding x-value.

A rule is $y = 3x - 1$.

c) Each y-value is 1 more than the square of the corresponding x-value.

A rule is $y = x^2 + 1$.

Sometimes we can use a pattern in experimental data to determine unknown measurements.

Example 2

Ethanol is the intoxicating substance in alcoholic drinks. It is produced by the yeast fermentation of sugar. For a chemistry experiment, different volumes of ethanol are poured into a beaker. After each volume is poured into the beaker, the beaker is weighed. Here are the results.

Volume of ethanol (mL)	Mass of beaker and ethanol (g)
0	90
50	129
100	168
150	207
200	246

a) Identify a pattern in the data. State the effect this pattern will have on a graph of the data.

b) Plot the data to confirm your statement in part a.

c) Determine an equation expressing the mass of the beaker and ethanol as a function of the volume of ethanol.

d) What do the numbers in your equation represent?

e) Determine the mass of the beaker and 250 mL of ethanol.

Solution

a) The difference between consecutive volumes in column 1 is 50 mL, and the difference between consecutive masses in column 2 is 39 g. Hence, for every 50 mL increase in volume, there is a 39 g increase in mass. This means that a graph of the data will be a straight line.

b) Plot the points and draw a straight line through them.

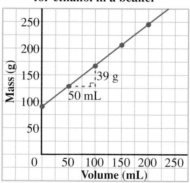

Graph of mass against volume for ethanol in a beaker

c) The vertical intercept of the line is 90.

The slope of the line is $\frac{39}{50} = 0.78$.

Let m represent mass in grams, and let v represent volume in millilitres.

The equation of the line is $m = 0.78v + 90$.

d) The vertical intercept represents the mass of the empty beaker, 90 g.

Written in the form $\frac{39}{50}$, the slope tells us that 50 mL of ethanol has a mass of 39 g. Hence, 1 mL of ethanol has a mass of $\frac{39}{50}$ g, or 0.78 g.

The slope represents the mass of 1 mL of ethanol.

e) Substitute 250 for v in $m = 0.78v + 90$.

$$m = 0.78 \times 250 + 90$$
$$= 285$$

The mass of the beaker and 250 mL of ethanol is 285 g.

In *Example 2d*, the mass of 1 mL of ethanol is called its density. The density of a substance is defined as the mass of a unit volume of the substance.

DISCUSSING THE IDEAS

1. In *Example 2*, how could we have used the patterns in the data to establish the equation $m = 90 + 0.78v$ without drawing the graph? Do you think it would be a good idea to do this? Discuss your ideas with a partner.

2. Could the method in the solution of *Example 2* be used to determine the rules corresponding to the tables in *Example 1*? Explain why or why not for each table in *Example 1*.

5.3 EXERCISES

 1. Suppose the input number is 2. Determine three different rules that give an output number 4.

2. Describe how you find a rule for each table of values.

a)

x	y
1	6
2	7
3	8
4	9
5	10

b)

x	y
1	1
2	3
3	5
4	7
5	9

c)

x	y
1	3
2	6
3	11
4	18
5	27

3. Determine a rule for each table of values.

a)

x	y
1	2
2	4
3	6
4	8
5	10

b)

x	y
1	11
2	21
3	31
4	41
5	51

B 4. Determine a rule for each table of values.

a)

x	y
−3	−9
−1	−3
0	0
2	6
7	21

b)

x	y
−4	2
0	0
3	−1.5
6	−3
8	−4

5. Choose one part of exercise 4. Write to explain how you determined the rule.

6. On each graph below, four points are labelled.

i)

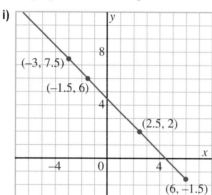

ii)

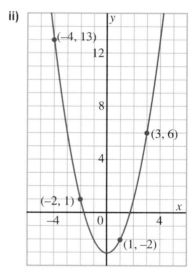

a) Make a table of values for each graph.

b) Determine a rule for each table of values.

c) Use the rule to obtain three other entries in each table. Do the points corresponding to these entries lie on the graph?

7. This table shows the mass of sulfur for each of several different volumes.

Mass (g)	Volume (cm^3)
23.5	11.4
60.8	29.2
115	55.5
168	81.1

a) Graph these data, with volume along the horizontal axis. Draw a line of best fit through the points.

b) Determine the slope of the graph.

c) Recall that density is defined as mass per unit volume. This means that the density of a solid is the mass in grams of 1 cm^3 of the solid. What is the density of sulfur?

d) Compare your answers for parts b and c. What do you notice?

e) When you drew the best straight line in part a, the line did not pass through all the plotted points. Explain why this happened.

8. Write a rule for each table of values.

a)

x	y
0	2
1	3
2	4
3	5
4	6

b)

x	y
−5	−9
−2	−3
0	1
3	7
7	15

c)

x	y
−4	1
−1	−2
0	−3
2	−5
8	−11

d)

x	y
0	0
1	2
2	6
3	12
4	20

e)

x	y
5	125
3	27
1	1
0	0
−2	−8

f)

x	y
24	0.5
3	4
2	6
−1	−12
−4	−3

9. Choose one part of exercise 8. Write to explain how you determined the rule.

10. Write a rule for each table of values.

a)

x	y
0	5
1	3
2	1
3	−1
4	−3
5	−5
6	−7

b)

x	y
−2	7
−1	4
0	3
1	4
2	7
3	12
4	19

c)

x	y
1	12
2	6
3	4
4	3
5	2.4
6	2

d)

x	y
4	1
5	4
6	9
7	16
8	25
9	36

11. A number of rectangular tables are arranged in a room. Eight people can sit at each table: 3 on each side and 1 on each end.

a) Suppose the tables are arranged end-to-end. How many people could be seated for each number of tables arranged this way?

    i) 2   ii) 3   iii) 4   iv) 5   v) 6

b) Write an equation to express the number of people who can be seated as a function of the number of tables placed end-to-end.

c) Suppose the tables are arranged side-by-side. Use the numbers of tables in part a. How many people could be seated for each number of tables?

d) Write an equation to express the number of people who can be seated as a function of the number of tables placed side-by-side.

e) Suppose the tables were square and could seat two people on each side. How would the equations in part b and d change?

12. Different volumes of the same liquid were added to a flask on a balance. After each addition of liquid, the mass of the flask with liquid was measured. The results are shown.

Volume (mL)	Mass (g)
14	103.0
27	120.4
41	139.1
55	157.9
82	194.1

a) Graph these data, with volume along the horizontal axis.

b) What is the mass of the flask?

c) What is the density of the liquid?

d) What other way could you have used to find the density of the liquid?

13. A pizza restaurant advertises these prices.

Number of pieces	Cost ($)
12	9.77
16	12.77
20	14.77
24	15.77
28	19.77

a) Plot these data.

b) Describe the pattern in the data.

c) Which pizza is the best buy?

14. Computer memory is expressed in bytes. One megabyte is approximately 1 000 000 bytes. The exact number of bytes in one megabyte is 2^x, where x is the exponent of the least power of 2 that exceeds 1 000 000.

a) In your graphing calculator, define $Y = 2^X$. Create a table of values. Move through the table to find the exact number of bytes in one megabyte.

b) One gigabyte, which is approximately one billion bytes, is defined in a way similar to one megabyte. Write a definition for one gigabyte.

c) Use the table in part a to find the exact number of bytes in one gigabyte.

C **15.** A function is defined by this equation:
$$y = 2x + 0.1(x - 1)(x - 2)(x - 3)(x - 4)(x - 5)$$

a) Use the equation. Determine the values of y for $x = 1, 2, 3, 4, 5,$ and 6. Show the results in a table.

b) How does your table compare with the table in exercise 3a?

c) Are the rules for the two tables the same? Explain.

16. Sam entered an equation into his TI-83 graphing calculator. He used the tables feature to produce the table (below left).

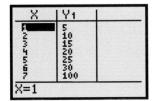

a) Describe the pattern in the first 6 rows of the table.

b) Why do you think the pattern does not extend to the 7th row?

c) What equation did Sam enter in his calculator?

d) Sam changed the equation and obtained the table (above right). What equation did Sam use this time?

e) Sam claims that he can change the equation so that the values of y for the first 6 values of x are the ones shown above right, and the value of y for $x = 7$ is any number you want. Explain how Sam can do this. Illustrate your explanation by choosing a number and modifying the equation.

COMMUNICATING THE IDEAS

Make up a rule of your own and use it to generate a table of values. Give your table to a friend. Challenge your friend to write a rule for the table and explain how he or she found the rule. If possible, program your rule into a computer spreadsheet or a programmable calculator. Challenge a friend to discover the rule by inputting numbers and looking for patterns in the output numbers.

Recall that a function is defined as a rule that gives a single output for a given input. In this section we will examine linear functions.

These are linear functions	These are non-linear functions
$y = 3x - 4$	$y = x^2$
$2x + y = 6$	$y = 0.1x^3 - 2$
$x - 2y + 8 = 0$	$2x^2 + y = 8$
$3x = 2y$	$xy = 10$

1. Use a graphing calculator. Graph the first function. Sketch the graph on a grid and write its equation. Clear the screen.

2. Repeat exercise 1 for each function.

3. Look at the graphs of the linear functions and the non-linear functions. What do all the graphs of linear functions have in common?

4. Look at the equations of the linear functions and the non-linear functions. What do all the equations of linear functions have in common?

A *linear function* has a defining equation that can be written in the form $y = mx + b$, where m and b are constants. Its graph is a straight line with slope m and y-intercept b.

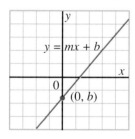

By examining the linear functions in *Investigate*, we see that each one can be rewritten in the form $y = mx + b$.

$y = 3x - 4$ is already in this form.	$2x + y = 6$ $y = -2x + 6$	$x - 2y + 8 = 0$ $-2y = -x - 8$ $y = \frac{1}{2}x + 4$	$3x = 2y$ $y = \frac{3}{2}x$ $m = \frac{3}{2}$ and $b = 0$

The non-linear functions in *Investigate* cannot be written in the $y = mx + b$ form.

Example 1

The student council is organizing a dance. The profit from the dance is a function of the number of tickets sold. The function relating the profit P dollars and the number of tickets sold T is represented by the equation: $P = 5T - 1000$, where $T \leq 400$.

a) Plot a graph of P against T.

b) Determine the slope of the graph. Explain what it represents.

c) Determine the P- and T-intercepts. Explain what each represents.

d) State the domain and range of this function.

e) What is the maximum profit that can be made at the dance?

Solution

a) The equation $P = 5T - 1000$ defines a linear function whose graph has slope 5 and P-intercept -1000. Choose a suitable scale for each axis. Plot a point at the P-intercept, move horizontally 200 units and vertically 1000 units for a slope of 5. Plot another point. Join the two plotted points and extend the line to $T = 400$.

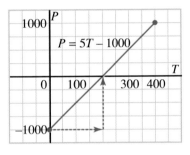

b) The slope of the graph is 5. This represents the price of a ticket, $5. The price is positive because it represents money received.

c) The P-intercept is -1000. This represents the cost to organize the dance, $1000. The cost is negative because it represents money paid out. The T-intercept is 200. This represents the number of tickets that must be sold to recover the cost to organize the dance.

d) The domain consists of all possible values of T: {0, 1, 2, 3, ..., 400} The range consists of all the values of P that correspond to the values of T: {-1000, -995, -990, ..., 995, 1000}

e) From the graph, the maximum profit is the value of P when $T = 400$; that is, $1000.

In *Example 1*, since only a whole number of tickets can be sold, and it is given that $T \leq 400$, the variable T represents whole numbers up to 400. However, if we were to plot all 400 points, the graph would look like a solid line, so that is how we draw it.

Example 2

a) Graph $2x - y = 6$ for $x = 0, 1, 2, 3, 4,$ and 5.

b) What are the intercepts and the slope of the graph?

c) The graph in part a defines a function. What are its domain and range?

Solution

a) Make a table of values. Use the table of values to plot the ordered pairs.

x	y
0	−6
1	−4
2	−2
3	0
4	2
5	4

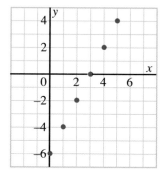

b) The y-intercept is −6. The x-intercept is 3. The slope is 2.

c) The domain is the set of x-values: $\{0, 1, 2, 3, 4, 5\}$.
The range is the corresponding set of y-values: $\{-6, -4, -2, 0, 2, 4\}$.

Example 3

Consider the linear function defined by this graph.

a) What are its domain and range?

b) Write an equation that describes the function.

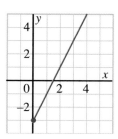

Solution

a) The domain consists of all non-negative real numbers: $x \geq 0$. The range consists of all real numbers greater than or equal to −3: $y \geq -3$

b) The slope of the graph is $\frac{6}{3}$, or 2. The y-intercept
is −3. Use the equation $y = mx + b$.

$b = -3$ $m = 2$

Substitute these values in the equation.

An equation defining the function is $y = 2x - 3$.

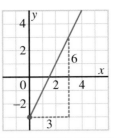

Frequently, the defining equation of a linear function can only be approximated
from a graph.

Example 4

The temperature at which water boils is a function of altitude.
The table shows the boiling points at different altitudes.

Location	Altitude, h metres	Boiling point of water, $t°C$
Halifax, N.S.	0	100
Banff, Alberta	1383	95
Quito, Ecuador	2850	90
Mount Logan, B.C.	5951	80

a) Draw a graph of boiling point against altitude.

b) Use the graph. Estimate the boiling point of water at
each location.

  i) Lhasa, Tibet, altitude 3680 m

  ii) the summit of Earth's highest mountain,
     Mount Everest, altitude 8848 m.

c) Write an equation that relates boiling point to altitude.
Use the equation to check the answers in part b.

d) Use the equation. Calculate the boiling point of water
at the deepest exposed depression on Earth's surface,
the shore of the Dead Sea at an altitude of −399 m.

e) Write the domain and range of the function.

Solution

a) Plot the ordered pairs from the table. Ensure the axes are long enough
to include the point that will represent Mount Everest in part b. Draw a
line of best fit through the points and extend the line.

b) i) From the graph, when $h = 3680$, t is about 87.5. The boiling point at Lhasa is about 88°C.

ii) From the graph, when $h = 8848$, t is about 70. The boiling point at the summit of Mount Everest is about 70°C.

c) Determine the vertical intercept and the slope. From the graph, the vertical intercept is 100. For the slope, choose any two points on the line; for example, (5951, 80) and (0, 100). The slope of the line through these points is:

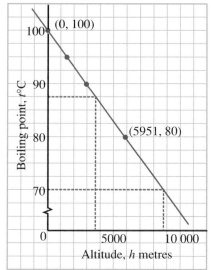

Boiling point of water

$$\frac{100 - 80}{0 - 5951} = \frac{20}{-5951}$$
$$\doteq -0.0034$$

Use the form $y = mx + b$. Substitute $m = -0.0034$ and $b = 100$.

An equation is $t = -0.0034h + 100$.

To check the answers in part b:

Substitute $h = 3680$, then $t = -0.0034(3680) + 100$
$$\doteq 87$$

Substitute $h = 8848$, then $t = -0.0034(8848) + 100$
$$\doteq 70$$

The estimates in part b were reasonable.

d) Substitute $h = -399$ in the equation $t = -0.0034h + 100$.

$t = -0.0034(-399) + 100$
$\doteq 101.36$

The boiling point of water is approximately 101°C.

e) The domain is the set of all possible altitudes. Since Mount Everest is the highest point on Earth, it is reasonable to have its height as the upper limit. Since the shore of the Dead Sea is the lowest point on Earth, we take its altitude as the lower limit.

The domain is $-399 \leq h \leq 8848$.

The range is the set of all possible boiling points. These correspond to the heights in the domain. The range is $70 \leq t \leq 101$.

In *Example 4*, when we estimated the boiling point at Lhasa, we were estimating the value of a variable *between* known values. This is called *interpolating*. When we estimated the boiling point at Mount Everest, we were estimating the value *beyond* known values. This is called *extrapolating*.

 MODELLING the Dependence of Boiling Point of Water on Altitude

- Why do you think the temperature at which water boils depends on altitude?
- According to *The Guinness Book of Records*, the world's deepest mine is 3581 m deep in South Africa. The deepest penetration into Earth's surface is 15 000 m at a geological drilling in Russia. Do you think the domain of the function in *Example 4* should be extended to include these depths below sea level? Explain your answer.

DISCUSSING THE IDEAS

1. Explain how you can determine if a function is linear by examining its graph.

2. Explain how you can determine if a function is linear by examining its equation.

3. Describe a linear function that has a restricted domain.

4. In *Example 1a*, describe another method you could have used to draw the graph.

5.4 EXERCISES

If you have a graphing calculator, use it to draw the graphs in exercises 9, 13, 14, 15, and 21.

 1. State whether each function is linear. If the function is linear, identify its slope and *y*-intercept.

a) $y = 4x$

b) $y = 2 - x$

c) $y = x^2$

d) $y = \frac{2x + 5}{10}$

e) $y = \frac{1}{x}$

f) $x = 3y + 4$

g) $y = (x - 2)(x + 1)$

h) $\frac{1}{y} = 2x + 1$

i) $x = 2y$

2. Graph the function represented by each equation. Identify the intercepts, slope, domain, and range. If no restrictions are stated, assume that the domain is the set of all real numbers.

a) $y = 2x - 5$

b) $y = -\frac{3x}{2}$; $x = -4, -2, 0, 2, 4, 6$

c) $y = \frac{3x}{4} - 2$

d) $y = 5$

e) $y = 4 - x$; $x \geq 0$

f) $y = -3$

B 3. Franc is a salesperson at a furniture store. His monthly income I dollars depends on his monthly sales S dollars. His income and sales are related by the function $I = 0.05S + 800$.

a) Plot a graph with S on the horizontal axis and I on the vertical axis.

b) Determine the slope of this graph. Explain what it represents.

c) Determine the intercepts. Explain what each represents.

d) What are the domain and range of this function?

e) Describe this function using a rule.

4. Graph each function individually using a graphing calculator. Explain how you ensured the graph appeared on the screen.

a) $y = x + 50$

b) $y = -2(x + 16)$

c) $y = 0.05x - 11$

5. a) For each linear function, state its domain and range; then write an equation to describe the function.

i)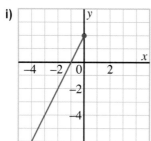

ii)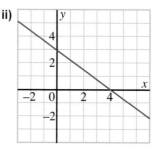

iii)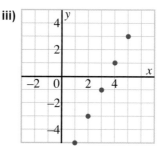

b) Choose one function from part a. Write to explain how you completed the exercise.

6. A rectangle has a perimeter of 24 cm.

a) Write the length of the rectangle as a function of its width.

b) What is the domain of the function in part a?

7. For each diagram, express y as a function of x; then state the domain of each function.

a)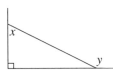

b)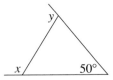

c)

8. This graph shows how the distance travelled by a car changes with time.

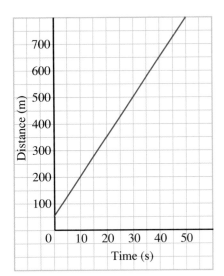

   a) At what time will the distance travelled be 600 m?

   b) For part a, did you interpolate or extrapolate?

   c) After 45 s, how far has the car travelled?

   d) Determine the slope of the line. Explain what it represents.

   e) Write an equation to describe this function.

   f) At what time will the distance travelled be 1200 m?

   g) After 100 s, how far has the car travelled?

   h) To answer parts f and g, what assumption did you make?

9. A taxi company sets its fares using the following scheme. It charges $2.50 when you enter the cab and an additional 10¢ for every 12 s after the meter has started.

Time in cab (minutes)	0	1	2	3	4	5	6
Total fare ($)							

   a) Copy and complete the table, which relates the total fare to the time in the cab.

   b) Which type of sequence do the fares in the table form?

   c) Graph the function defined by the table.

   d) What is the vertical intercept? What does it represent?

   e) Write an equation to represent the function.

   f) Use the graph. Determine the fare for a 15-min taxi ride. Use the equation to check your answer.

10. To attend an indoor playground, a child pays a $2 admission fee plus an additional fee for each half-hour of play. It costs $5 for a child to play at the indoor play ground for one hour.

a) Graph this relationship.

b) Determine the fee for each half-hour of play.

c) Determine the cost for a child to play for 3 h.

11. To attend a skateboard park, a person pays a $2 admission fee plus an additional fee for each half-hour of use. Suppose you know the total fee and the number of hours the park was used. Write to explain how you could determine the fee per half-hour.

12. An outfitter rents canoes by the day, charging an additional fixed cost that includes an insurance premium. The total cost to rent a canoe for 3 days is $61. The total cost to rent a canoe for 5 days is $85.

a) Graph the total cost as a function of the number of days of rental.

b) Determine the fixed cost and the daily rate to rent a canoe.

c) Write to explain how you completed part b.

13. A car is travelling at a speed of 20 m/s. It begins to accelerate at a rate of 0.5 m/s^2 and continues to do so for 30 s. The speed of the car v metres per second at any time t seconds is described by the equation $v = 20 + 0.5t$.

a) Plot a graph with v on the vertical axis and t on the horizontal axis.

b) What are the domain and range of this function?

14. The approximate temperatures of Earth's atmosphere at different altitudes up to 10 km are shown.

Altitude (km)	Temperature (°C)
0	15
2	2
4	−11
6	−24
8	−37
10	−50

a) Draw a graph to show temperature as a function of altitude.

b) Write an equation that relates temperature and altitude.

c) i) Determine the temperature at an altitude of 7 km.

ii) Determine the altitude at which the temperature is 0°C.

d) Above 11 km, the temperature remains fairly constant at −56°C. Show this on the graph you drew in part a.

15. A helicopter is travelling at its maximum sea level speed. The approximate distances of the helicopter from its destination are given in nautical miles.

Time (min)	Distance from destination (nm)
0	235.0
20	194.0
45	142.8
60	112.0
72	87.4

a) Draw a graph to show the distance from the destination as a function of time.

b) At about what time will the helicopter arrive at its destination? What assumption are you making?

c) Calculate the slope of the line.

d) Calculate the speed of the helicopter in nautical miles per minute. How does your answer compare with the slope in part c?

e) Write an equation to describe the distance of the helicopter from its destination. Use your equation to check your extrapolation in part b.

f) The standard measure for a helicopter's speed is knots, or nautical miles per hour. Express the speed from part d in knots. Find a helicopter in the *Helicopters* database with this maximum sea level speed.

g) Use the hourly operating cost from the database. Calculate the operating cost of this trip.

h) How would your equation in part e change for a different length of trip?

16. A youth conference is in Saskatoon. The table shows the approximate cost of a round trip to Saskatoon by bus from various cities.

City	Distance (km)	Cost ($)
Regina	258	52
Edmonton	546	121
Calgary	613	137
Winnipeg	793	161

a) Plot the cost of travelling to Saskatoon as a function of the distance. Draw a line of best fit.

b) Is this a linear relationship? Justify your answer.

c) Use your line of best fit. Write an equation relating the cost of travelling to the conference and the distance from Saskatoon.

d) Estimate the cost of travelling to the conference for a person living in Lethbridge, which is 658 km from Saskatoon.

e) Estimate the cost of travelling to the conference for a person living in Vancouver, which is 1089 km from Saskatoon.

f) Of the estimates in parts d and e, which one is more likely to be incorrect? Explain.

g) Compare your equation in part c with those of your classmates. Did all of you write the same equation? If your equation differs, explain the differences.

17. The following data on used vans were collected from the classified section of a newspaper.

Age of van (years)	Asking price ($)
6	7 000
1	23 500
9	5 500
3	15 500
5	11 900

a) Create a scatterplot of the age of the van against the asking price.

b) Draw a line of best fit, then write an equation to describe the line.

c) How well does a linear function describe these data? Explain.

d) A person is asking $6000 for a van that is six years old. Is this asking price reasonable? Explain.

e) What factors, besides the age of the van, may affect the asking price?

18. Each description below represents the graph of a function with equation $y = mx + b$. Determine m and b for each graph.

a) a graph with slope $\frac{2}{5}$ and y-intercept 7

b) a graph that passes through P(6, 2) and has slope $-\frac{4}{3}$

c) a graph that passes through C(−3, 10) and has y-intercept 4

d) a graph that passes through A(5, 9) and B(0, −6)

19. For each function in exercise 18, state its domain, range, and intercepts.

20. The sign below was on the back of a truck.

To Stop at:	I Need:
50 km/h	50 m
70 km/h	80 m
90 km/h	110 m

a) Draw a graph to show how stopping distance is related to the truck's speed.

b) Determine an equation that expresses the stopping distance as a function of the truck's speed.

c) Predict the stopping distance for a truck travelling at 100 km/h.

d) Suppose it took a truck 145 m to stop. Calculate how fast the truck was travelling before the driver applied the brakes.

21. A person who evaluates computer programs charges a fee based on the number of days required to evaluate a given program. The maximum number of days that the evaluator will devote to one job is 60. The fee structure for the first 5 days is given below. The fees for additional days follow the same pattern.

Day	1	2	3	4	5
Fee ($)	500	800	1100	1400	1700

a) Graph this function.

b) Write an equation to describe this function.

c) How many days would the evaluator have to work to earn $5000?

d) Write to explain how you could have completed part c by using your knowledge of arithmetic sequences.

C 22. ABCD is a trapezoid, with AB = 10 cm and DC = 6 cm. P is a point on AB so that AP = x centimetres. Q is a point on DC so that DQ = y centimetres.

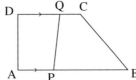

a) Suppose trapezoids APQD and PBCQ have equal areas. Determine an equation that relates x and y.

b) For what positions of P on AB is it possible to find a point Q on DC so that trapezoids APQD and PBCQ have equal areas?

COMMUNICATING THE IDEAS

Explain what a linear function is. In your explanation, use at least one example of a linear function and one example of a non-linear function.

Explain how to write the equation of a linear function if you are given its graph.

The Line of Best Fit

You can use a graphing calculator to draw a scatterplot for a given set of data, to calculate the equation of the line of best fit, and to graph it with the data. We shall use the data in *Example 4* on page 277. The screens on these pages are from the TI-83 calculator.

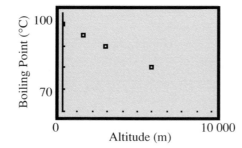

- *Enter the data in the list editor*
  Enter the altitudes in the first list and the boiling points in the second list.

- *Define the viewing window*
  Choose values so that the data will appear on the screen.

- *Define a statistical plot*
  You must identify the list for each axis and the type of graph you want. Refer to your manual to determine how to input this information.

- *Graph the data*
  Make sure that the calculator screen is clear, and press GRAPH.

Using the statistical features of your calculator, you can determine much information about the data in the two lists. This information includes the coefficients m and b in the equation of the line of best fit. Refer to your manual to determine how to do this.

This type of calculation is called a *linear regression*. This screen shows the result for these data. The equation of the line has the form $y = ax + b$, where $a \doteq -0.003\ 34$ and $b \doteq 99.8$. Hence, the equation of the line of best fit is $y \doteq -0.003\ 34x + 99.8$. Compare this with the equation in the solution of *Example 4* on page 278.

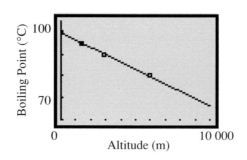

Graphing the line of best fit

You can enter the equation $y = -0.003\,34x + 99.8$ manually. You can also use the calculator to enter the calculated equation directly. Refer to your manual to find how this is done.

Using the line of best fit

You can trace along the line of best fit to view the coordinates of many points on the line. You may need to press the ▼ key to move the cursor from the data points to the line. To determine the boiling point at any given altitude, simply enter the altitude and press ENTER. The cursor will jump to that location on the line and display the y-coordinate.

1. Students determined the resistance (R ohms) in a electrical circuit by measuring the voltage (V volts) when a current (I amperes) was passing through the circuit. According to Ohm's law, these electrical measurements are related by the equation $V = RI$. These data were collected.

Current (A)	Voltage (V)
0.5	5.0
1.0	10.3
1.5	15.2
2.0	20.4
2.5	25.5
3.0	30.5
3.5	35.8
4.0	4.7
4.5	46.0

   a) Construct a scatterplot of voltage against current.

   b) Determine the equation of the line of best fit. What is the resistance of the circuit?

   c) Graph the line of best fit on the scatterplot.

   d) Use the equation to predict the voltage for a current of 5.9 A.

   e) Use the equation to predict the current for a voltage of 17.8 V.

2. Use the data from exercise 16, page 283.

   a) Construct a scatterplot for these data.

   b) Draw the line of best fit.

   c) Determine an equation for the line of best fit. How is this equation different from the equation you determined in exercise 16?

   d) Compare the equation of the line of best fit from your calculator with those of your classmates. Are they different? Explain.

Use a Graph

Problems like the one below are frequently encountered by investors.

Jim's grandmother has $130 000 she would like to invest in fixed-income investments. From this amount, she would like to receive $9000 annual income, before taxes. Her financial advisor suggests two different bonds: one yielding 6.1% and the other yielding 8.3%. How much should Jim's grandmother invest in each type of bond?

We use a graph to model then solve a problem like this. Follow these steps.

Step 1	**Step 2**	**Step 3**
Begin a graph. Use units of thousands of dollars.	Sketch the line $y = 0.061x$ to represent the yield on the 6.1% investment. Stop at point A.	Join A to B(130, 9). Segment AB has a greater slope to represent the additional yield on the 8.3% investment.

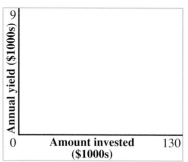
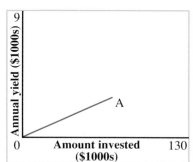
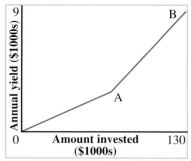

Step 4

Point A has coordinates $(x, 0.061x)$, where x thousand dollars is the amount invested at 6.1%.

Point B has coordinates (130, 9).

Slope of AB: $\dfrac{9 - 0.061x}{130 - x} = 0.083$

Solve the equation. Obtain the amount that should be invested at 6.1%, in thousands of dollars. Subtract that amount from $130 000 to obtain the amount that should be invested at 8.3%.

DISCUSSING THE IDEAS

1. **a)** What do the coordinates of A represent? What does the slope of OA represent?

   b) What do the coordinates of B represent? What does the slope of OB represent?

Complete these exercises.

1. Check that your solution to the problem on the preceding page is correct.

2. Some people might think that Jim's grandmother should invest all the money at 8.3%. Suggest a reason why this might not be a good idea.

3. You should have found that Jim's grandmother would have to invest $81 363.64 at 6.1% and $48 636.36 at 8.3% to receive exactly $9000 annual income.

   a) Do you think it is realistic for her to invest these amounts to the nearest cent?

   b) Do you think she expects to receive an annual income from this investment that is exactly $9000, or would she be satisfied with an income that is close to $9000?

   c) Suppose you were her financial advisor. How much would you suggest investing at each rate? Calculate the total annual income for your proposal.

Use a graph to solve each problem.

4. Brenda plans to cycle at 30 km/h and jog at 8 km/h. She hopes to cover 50 km in 2 h. How much time should she spend jogging?

5. A butcher has supplies of lean beef containing 15% fat, and fat trim containing 100% fat. How many kilograms of lean beef and fat trim does she need to make 50 kg of hamburger, which is 25% fat?

6. A car averages 12.5 L/100 km in city driving and 7.5 L/100 km on the highway. In a week of mixed driving, the car used 35 L of fuel and travelled 400 km. Estimate the distance travelled in highway driving.

7. A laboratory technician has acid solution in two concentrations, 50% and 100%. He wants to mix the right amount of each to make 400 mL of 60% acid solution. How many millilitres of each solution does he need?

8. A water tank has two taps, A and B. Line A on the graph shows how fast the tank drains if only tap A is open. Line B shows how fast the tank drains if only tap B is open. Use the graph. Estimate the time to drain the tank if both taps were open.

Graph of percent of volume against time

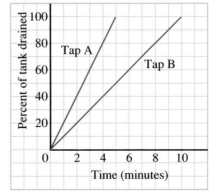

INVESTIGATE

In its advertising, a car dealership states that leasing and driving a new compact car will cost about $400/month.

1. Create a table of values relating the cost of leasing and driving a compact car to the number of months it is driven.

2. Graph the cost against the number of months.

3. a) What happens to the cost of driving the car when the number of months is doubled? What happens when the number of months is tripled?

   b) Explain how your answers in part a can be determined from both the table and the graph.

4. Use the graph in exercise 2.

   a) What type of function does this graph represent?

   b) What is the vertical intercept?

   c) What is the slope? What does it represent?

5. Write an equation that represents this function.

Marcie works as a commission salesperson at a car dealership. At the end of every month, she is paid 1.5% of her total sales. The table and the graph illustrate how her earnings are related to her total sales.

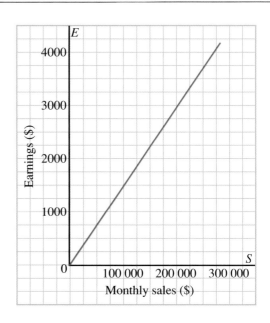

Monthly sales, S ($)	Earnings, E ($)
0	0
40 000	600
80 000	1200
120 000	1800
160 000	2400
200 000	3000
240 000	3600

Marcie's pay is a product of her total sales and 1.5% expressed as a decimal.

This means that the function could also be described by the equation $E = 0.015S$.

The table, the graph, and the equation show that when S is doubled, E is doubled.

Similarly, when S is tripled, E is also tripled.

We say that E *varies directly as* S.

The graph of this direct variation is linear and passes through the origin.

Hence, it can be described by a function with an equation of the form $y = mx$, or in this case, $E = kS$.

The constant k is called the *constant of proportionality*. We can see from the equation that the constant of proportionality is also the slope of the line.

In the example above, the constant of proportionality is 0.015, which represents Marcie's rate of commission.

In general, when y varies directly as x:

- The graph of y against x is a straight line passing through the origin.
- The equation relating the variables has the form $y = mx$, where m is the constant of proportionality.

Example 1

In the table, y varies directly as x.

x	y
2	4
5	
6	
	16
	17

a) By what number is each x-value multiplied to give the corresponding y-value?

b) Copy and complete the table.

c) Write the equation relating x and y.

d) What is the constant of proportionality?

e) Graph y against x.

Solution

a) Since $y = 4$ when $x = 2$, each x-value is multiplied by 2 to give the corresponding y-value.

b) To complete the table, multiply each x-value by 2 to get the corresponding y-value. Divide each y-value by 2 to get the corresponding x-value.

x	y
2	4
5	10
6	12
8	16
8.5	17

c) The equation relating x and y is $y = 2x$.

d) The constant of proportionality is 2.

e)

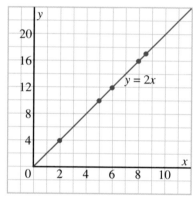

Example 2

Have you noticed that during a thunderstorm there is sometimes a delay between seeing the flash of lightning and hearing the thunderclap? You can use this time delay to estimate the distance to the point where the lightning struck the ground. The distance d kilometres is expressed as a function of the time delay t seconds by the equation $d = 0.35t$.

a) Suppose the time delay is 2.5 s. How far are you from the lightning?

b) Suppose you are 4.0 km from the lightning. What is the time delay?

c) Graph the function.

Solution

a) Substitute 2.5 for t in the formula.

$$d = 0.35t$$
$$= 0.35 \times 2.5$$
$$= 0.875$$

When the time delay is 2.5 s, you are about 0.88 km from the lightning.

b) Substitute 4.0 for d in the formula.

$$d = 0.35t$$
$$4.0 = 0.35t$$
$$t = \frac{4.0}{0.35}$$
$$\doteq 11.4$$

When you are 4.0 km from the lightning, the time delay is approximately 11.4 s.

c)

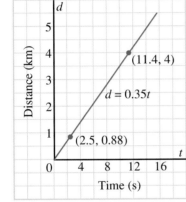

 MODELLING the Distance to a Thunderstorm

The speed of light is approximately 300 000 000 m/s. The speed of sound in air is approximately 350 m/s.

- Use this information. Explain how the equation $d = 0.35t$ is derived.

- Suggest some reasons why it might be important for a person to be able to estimate one's distance to a thunderstorm.

Example 3

The amount of simple interest earned on an investment varies directly as the amount of money invested. For one particular investment, a person who invested $2250 earned $258.75 interest. How much interest would an investment of $1700 earn?

Solution

Let I dollars represent the interest earned. Let D dollars represent the amount of money invested.

Since I varies directly as D, then $I = kD$

Using the given information, calculate the constant of proportionality.

$I = kD$

Substitute $I = 258.75$ and $D = 2250$, then solve for k.

$$258.75 = k(2250)$$
$$k = \frac{258.75}{2250}$$
$$= 0.115$$

Substitute $k = 0.115$ in the equation $I = kD$.

$I = 0.115D$

Use this equation to calculate the interest earned on a $1700 investment. Substitute $D = 1700$.

$$I = 0.115(1700)$$
$$= 195.50$$

An investment of $1700 would earn $195.50 in interest.

DISCUSSING THE IDEAS

1. a) Are all examples of direct variation also examples of linear functions?

   b) Are all examples of linear functions also examples of direct variation?

   Explain your answers with examples.

A 1. Suppose *y* varies directly as *x*.

   a) How is *y* affected when *x* is doubled?

   b) How is *y* affected when *x* is halved?

2. For each table of values below, *y* varies directly as *x*. For each table:

   a) State by what number each *x*-value is multiplied to give the corresponding *y*-value.

   b) Copy and complete the table.

   c) State the relationship between *x* and *y*.

i)

x	y
1	
2	10
3	
4	
5	

ii)

x	y
1	
2	
3	
4	−12
5	

iii)

x	y
2	
4	
6	
8	
10	5

iv)

x	y
2	
4	
6	9
8	
10	

3. Suppose *D* varies directly as *T*, and *D* is equal to 28 when *T* is 4. Describe how you would graph this relationship, knowing only one value of *D* and *T*.

B 4. **a)** State whether the relationship between each pair of variables is an example of a direct variation. If it is, write an equation to describe the variation.

    i) the circumference of a circle and its diameter

    ii) the tax paid when purchasing an item and the price of the item

    iii) the area of a square and the length of a side of the square

    iv) the area of a circle and its radius

    v) the cost of purchasing gas and the number of litres of gas purchased

   b) Choose one relationship in part a. Write to explain how you determined whether the situation represented a direct variation.

5. Refer to the graph in *Example 2*, page 6.

   a) Does this graph represent direct variation? Explain.

   b) Write an equation that expresses the salesperson's weekly salary as a function of the number of cars sold.

6. For each table of values below, y varies directly as x. For each table:

   a) Copy and complete the table.

   b) Write an equation relating x and y.

   c) Graph the function.

   i)

x	y
–6	
	–2
0	
3	2
9	

   ii)

x	y
	2
	1
	0
4	
6	–3

   iii)

x	y
0	
3	4
6	
9	
12	

   iv)

x	y
	0
2	
4	10
8	
	30

7. For a bike-a-thon, Steven had pledges totalling $1.40 per kilometre. The maximum distance he can cycle is 100 km.

   a) How much would he collect if he cycled 20 km?

   b) How far would he have to ride to collect $40.00?

   c) Write the equation relating the amount collected and the distance covered.

   d) Graph the function.

8. y varies directly as x. When x is 12, y is 8.

   a) Write y as a function of x.

   b) Determine y when x is 21.

   c) Determine x when y is 15.

9. It is estimated that the volume of blood in the human body varies directly as the body mass. An 80-kg person has a blood volume of about 6 L.

   a) Write an equation to express the blood volume as a function of the body mass.

b) Calculate the blood volume of a 60-kg person.

c) Sketch a graph of the function.

10. When a driver decides to apply the brakes, there is a short delay due to reaction time. During this time, the distance the car travels varies directly as the speed of the car. Suppose a driver's reaction time is one second. Then, a car travelling at 80 km/h will travel 22 m before the brakes are applied.

   a) Write an equation that expresses the distance the car travels as a function of its speed.

   b) Determine how far a car will travel if its initial speed is 100 km/h.

   c) Write to explain how reaction time is related to the distance the car travels.

11. The World Wildlife Federation has stated that every hour Canada loses approximately 240 acres of wilderness.

   a) Copy and complete this table.

Time in hours	0	1	2	3	4	5	6	7	8
Total area lost (acres)									

   b) Is this relationship an example of direct variation? Explain.

   c) Look at the numbers in the second row of the table. What kind of sequence do they form? Explain.

   d) Write an equation that expresses the total number of acres lost as a function of time.

   e) Calculate how many days it will take for Canada to lose over 1 million acres of wilderness.

12. A super tanker travelling at 25 km/h needs 5 km to come to a complete stop. Suppose stopping distance varies directly as speed.

   a) Write an equation to express the stopping distance as a function of the speed.

   b) What distance will the super tanker need to stop from a speed of 15 km/h?

   c) Sketch a graph of the function.

   d) Use the graph. Determine the stopping distance for a super tanker travelling at 12 km/h.

   e) Use the graph. The stopping distance for a super tanker is 3.5 km. At what speed was it travelling before it began to slow down?

   f) Use the equation to check your answers to parts d and e.

13. The mass of garbage collected in a city varies directly as the population. A city of 200 000 people generates 125 t of garbage per day. What mass of garbage would be generated per day in a city with each population?

a) 50 000 b) 500 000 c) 3 200 000

14. The universe contains billions of galaxies. These galaxies are moving away from Earth at very high speeds, called *recession velocities*. The farther away the galaxy, the faster it is moving. This means that the recession velocities vary directly as the distances from Earth. Look at the table below.

Constellation containing the galaxy	Distance (light years)	Recession velocity (km/s)
Virgo	7.8×10^7	1 200
Ursa Major	1.0×10^9	15 000
Corona Borealis	1.4×10^9	22 000
Bootes	2.5×10^9	39 000
Hydra	4.0×10^9	61 000

a) Graph these data. Draw a line of best fit.

b) Through which point do you know your line must pass? Explain.

c) Write an equation for your line of best fit.

d) What does the slope of the line represent?

e) Use the graph. Estimate each distance from Earth.

   i) the 300 galaxies in the Hercules cluster that are receding at about 10 000 km/s

   ii) a galaxy that is receding at about 75% of the speed of light (the speed of light is 300 000 km/s)

f) Use the equation to check your answers to part e.

15. In each ordered pair below, the first coordinate is the age of a randomly selected woman. The second coordinate is the age of her husband.

{(29, 34), (37, 38), (19, 20), (57, 57), (34, 32), (23, 25), (51, 54), (72, 81), (29, 23), (70, 70), (45, 54), (39, 37), (58, 56), (64, 71), (35, 35), (42, 50), (25, 24), (37, 48), (36, 36), (27, 32), (56, 42), (17, 24), (28, 28), (57, 26), (41, 39), (47, 47), (16, 18), (24, 27), (84, 87), (55, 59)}

a) Should the age of a husband vary directly as the age of his wife? Explain.

b) Use a graphing calculator. Create a scatterplot of the age of the husband against the age of the wife. Is the relationship linear? Explain.

c) Use the method explained on page 286. Draw the line of best fit.

d) Does the line of best fit represent direct variation? Explain.

e) Use the line of best fit. Estimate the age of the husband for each wife with the following age.

   i) 20   **ii)** 40   **iii)** 60

f) The line of best fit illustrates that the ages of women and their husbands approximate a linear function. What is the slope of the line? What does it represent?

C **16.** In the diagram (below left), the inner circle and the shaded region have equal areas.

   a) Write an equation relating x and y.

   b) Does y vary directly as x? Explain.

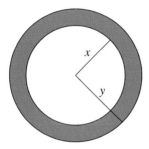

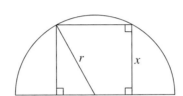

17. In the diagram (above right), a square with side length x is inscribed in a semicircle with radius r. Show that r varies directly as x. State the constant of proportionality.

18. When an object falls freely, the distance it falls varies directly as the square of the time for which it is falling. An object that falls for 10 s will travel about 490 m.

   a) Write an equation relating the distance travelled and the time the object falls.

   b) An object falls freely for 5 s. What distance will it travel?

   c) An object falls freely for 15 s. What distance will it travel?

COMMUNICATING THE IDEAS

Find a real-world example of direct variation. In your journal, explain how you obtained your data for this direct variation.

Arithmetic Sequences and Linear Functions

Recall *Example 1* on page 11 in Chapter 1, which concerned the arithmetic sequence 2, 5, 8, …. The first term is $a = 2$ and the common difference is $d = 3$. In the solution of the example, we used the formula for t_n to determine the general term of the sequence:

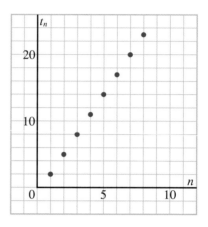

$$t_n = a + (n - 1)d$$
$$= 2 + (n - 1) \times 3$$
$$= 3n - 1$$

The general term is $t_n = 3n - 1$.

When the terms of the sequence are plotted on a graph, the points lie on a straight line with slope 3 and vertical intercept −1. The graph and the equation for t_n show that the arithmetic sequence is a linear function. Its domain is the set of natural numbers.

The equation of the function has the form $y = mx + b$, where $m = 3$ and $b = -1$. This equation is $y = 3x - 1$. We see that m is the common difference, 3, but what about b? Visualize extending the sequence backward one term. This term would be $2 - 3$, or −1. We can call this the *zeroth term*.

Compare the equation of the function with the equation for t_n.

Linear function	**Arithmetic sequence**
$y = 3x - 1$	$t_n = 3n - 1$
↑ ↑	↑ ↑
slope y-intercept	common zeroth difference term

An arithmetic sequence is a linear function whose domain is the set of natural numbers. The general term of the sequence may be written in this form:

$y = mx + b$ (x is a natural number)

where m is the common difference and b is the zeroth term.

We can use this idea to write the general term of any arithmetic sequence. For example, consider the sequence 7, 11, 15, 19, …

Think: 3   7   11   15   19

-4 $+4$ $+4$ $+4$

The common difference is 4. If the sequence were extended backward one term, the zeroth term would be 3.
Hence, the general term of the sequence is $t_n = 4n + 3$.

1. Complete these exercises from page 14, Chapter 1, using the above method for writing t_n.

   a) Exercise 4   b) Exercise 5   c) Exercise 6   d) Exercise 9   e) Exercise 10

2. Refer to *Example 1* on page 248 in Section 5.1. In that example, the pay for a part-time job is expressed as a linear function of the number of hours worked.

   a) Copy and complete this table.

Number of hours worked	1	2	3	4	5
Pay ($)					

   b) The numbers in the second row are terms of an arithmetic sequence. Write the first term a and the common difference d. Use the formula for t_n to determine the pay for working n hours.

   c) Compare the equation you wrote in part b with the equation of the function in *Example 1* on page 248. What do you notice about the two equations?

3. Refer to *Example 1* on page 275 in Section 5.4. In that example, the profit from a dance is expressed as a linear function of the number of tickets sold.

   a) Copy and complete this table.

Number of tickets sold	1	2	3	4	5
Profit ($)					

   b) The numbers in the second row are terms of an arithmetic sequence. Write the first term a and the common difference d. Use the formula for t_n to determine the profit when n tickets are sold.

   c) Compare the equation you wrote in part b with the equation of the function in *Example 1* on page 275. What do you notice about the two equations?

MATHEMATICAL MODELLING

Pushing Your Physical Limits

On page 244, you listed different ways to measure personal fitness. On these pages, you will examine one measure of fitness.

Your muscles use more oxygen during aerobic exercise than they do at rest. One measure of fitness is the maximum volume of oxygen (VO_2 max) that your body can transfer from the bloodstream to its muscle cells during exercise. In general, the higher your VO_2 max, the higher your level of fitness.

VO_2 max depends on both body mass and time. It is the volume of oxygen transferred by 1 kg of body mass in 1 min. For example, a female swimmer may have a VO_2 max of 100 mL/kg/min. This means that each kilogram of her body mass transfers 100 mL of oxygen to her muscle tissue every minute. If her mass is 50 kg, then 5000 mL of oxygen is transferred every minute.

Physiologists can measure VO_2 max in a laboratory using a treadmill and specialized measuring equipment. Since these precise measurements can be costly and time-consuming, physiologists have developed mathematical models to help predict VO_2 max using heart rate.

 DEVELOP A MODEL

One model for predicting VO_2 max involves the step test. In this 3-min test, a person repeats a four-step cycle (up-up-down-down) using gymnasium bleachers with height about 36 cm. Women complete 22 cycles per minute; men complete 24 cycles per minute, with time regulated by a metronome.

After stepping for 3 min, subjects remain standing and take their pulse rate about 15 s into recovery (this is called the recovery heart rate). When the step test was conducted with large numbers of college students, researchers arrived at these equations:

Men: Women:
$$v = 113.33 - 0.42r$$ $$v = 65.81 - 0.18r$$

where v represents the VO_2 max (in millilitres per kilogram per minute) and r represents the recovery heart rate (beats/min)

This pair of equations provides a model for predicting VO_2 max based on recovery heart rate. You can use graphing to explore the model.

1. **a)** Graph the two equations on the same axes. Plot VO_2 max (mL/kg/min) on the vertical axis and recovery heart rate (beats/min) on the horizontal axis. Make your axes long enough to fit v up to 120 and r up to 200.

   b) What do you think would be reasonable values of r? What does this tell you about possible values of v?

2. **a)** A female skier took the step test and had a recovery heart rate of 145 beats/min. Estimate her VO_2 max.

   b) Canadian rower Silken Laumann estimates that her recovery heart rate, after maximum energy output, is 130 beats/min. Try to make a reliable estimate of her VO_2 max, knowing that the step test was designed to reflect moderate exercise. Explain.

3. **a)** Spanish cyclist Miguel Indurain registered a VO_2 max of 88 mL/kg/min in the season when he won his fifth consecutive Tour de France. Estimate his recovery heart rate at the time.

   b) A male cyclist has a VO_2 max of about 55 mL/kg/min. Estimate his recovery heart rate after a step test.

 LOOK AT THE IMPLICATIONS

4. Look at your graph for exercise 1. What is the relationship between VO_2 max and recovery heart rate? Describe a VO_2 max and recovery heart rate for an exceptionally fit person.

5. The graph indicates significantly different results for men and women. Explain why you think this is so.

6. Where do you fit in? To find out, conduct the step test. Have someone time 3 min of stepping, then 15 s before pulse rates are taken. Take your pulse for 15 s, then multiply by 4 to determine the number of beats per minute. Work with your classmates to construct a class graph to display your results.

 REVISIT THE SITUATION

7. Your body uses only a percent of the oxygen available. That percent reflects how close you are to your physical limit. Someone on the point of exhaustion is using a very high percent of the available oxygen. How does this explain why some athletes finish a race without being winded, whereas another athlete may be near collapse?

5.6 Function Notation

In algebra, symbols such as x and y are used to represent numbers.

To represent functions, we use symbols such as $f(x)$ and $g(x)$.

For example, we write $f(x) = x^2 - 3x - 4$.

The symbol $f(x)$ is read "f of x," and means that the expression that follows contains x as a variable. This notation is useful because it simplifies recording the values of the function for several values of x.

For example, $f(6)$ means substitute 6 for every x in the expression.

$$f(x) = x^2 - 3x - 4$$

$$f(6) = 6^2 - 3(6) - 4$$
$$= 36 - 18 - 4$$
$$= 14$$

Example 1

Given $f(x) = 3x^2 - x - 6$, determine each value.

a) $f(2)$ **b)** $f(-1)$ **c)** $f(\sqrt{2})$

Solution

a) For $f(2)$, substitute 2 for x in:
$$f(x) = 3x^2 - x - 6$$
$$f(2) = 3(2)^2 - 2 - 6$$
$$= 12 - 8$$
$$= 4$$

b) For $f(-1)$, substitute -1 for x in:
$$f(x) = 3x^2 - x - 6$$
$$f(-1) = 3(-1)^2 - (-1) - 6$$
$$= 3 + 1 - 6$$
$$= -2$$

c) For $f(\sqrt{2})$, substitute $\sqrt{2}$ for x in:
$$f(x) = 3x^2 - x - 6$$
$$f(\sqrt{2}) = 3(\sqrt{2})^2 - \sqrt{2} - 6$$
$$= 3(2) - \sqrt{2} - 6$$
$$= 6 - \sqrt{2} - 6$$
$$= -\sqrt{2}$$

Example 2

A function is defined by $g(x) = \sqrt{x + 2}$.

a) What are the domain and range of the function?

b) Graph the function for $-2 \leq x \leq 7$.

Solution

a) Only non-negative numbers have real square roots.

So, the function $g(x)$ is only defined when $x + 2 \geq 0$, or $x \geq -2$.

The domain is the set of all real numbers greater than or equal to -2.

The range is the set of non-negative real numbers.

b) Make a table of values. Then plot the ordered pairs from the table.

x	g(x)
−2	0.00
−1	1.00
0	1.41
1	1.73
2	2.00
3	2.24
4	2.45
5	2.65
6	2.83
7	3.00

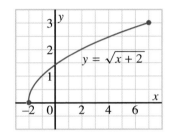

Function notation can be used even when there is no known equation relating the variables.

Example 3

From the graph of $y = f(x)$, determine each value.

a) $f(5)$ **b)** $f(0)$ **c)** $f(-4)$

d) an ordered pair describing the point on the graph with y-coordinate $f(2)$

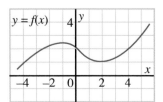

Solution

a) For $f(5)$, determine the value of y when $x = 5$. Start at 5 on the x-axis. Move vertically to the graph, then horizontally to the y-axis, to meet it at 3.
Hence, $f(5) = 3$

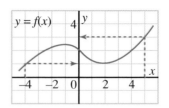

b) For $f(0)$, determine the value of y when $x = 0$. This is the y-intercept.
$$f(0) = 2$$

c) For $f(-4)$, determine the value of y when $x = -4$. From the graph,
$f(-4) = 1$

d) For $f(2)$, determine the value of y when $x = 2$. From the graph, $y = 1$
The point has coordinates $(2, 1)$.

Other algebraic expressions may be substituted for the variable in the equation of a function.

Example 4

Given the function $f(x) = 3x - 1$, determine each expression.

a) $f(c)$ **b)** $f(2x)$ **c)** $f\left(\frac{1}{x}\right),\ x \neq 0$ **d)** $f(x - 2)$

Solution

a) $f(x) = 3x - 1$ Substitute c for x.

$$f(c) = 3c - 1$$

b) $f(x) = 3x - 1$ Substitute $2x$ for x.

$$f(2x) = 3(2x) - 1$$
$$= 6x - 1$$

c) $f(x) = 3x - 1$ Substitute $\frac{1}{x}$ for x.

$$f\left(\frac{1}{x}\right) = 3\left(\frac{1}{x}\right) - 1$$
$$= \frac{3}{x} - 1$$
$$= \frac{3 - x}{x} \qquad x \neq 0$$

d) $f(x) = 3x - 1$

Substitute $x - 2$ for x.

$$f(x - 2) = 3(x - 2) - 1$$
$$= 3x - 6 - 1$$
$$= 3x - 7$$

DISCUSSING THE IDEAS

In earlier grades, you learned that $a(x - 2)$ can be expanded using the Distributive Law to obtain $ax - 2a$. In *Example 4d*, why did we not write $fx - 2f$? How can you tell that $a(x - 2)$ means $ax - 2a$, but $f(x - 2)$ does not mean $fx - 2f$?

5.6 EXERCISES

A **1.** For $f(x) = 1 - x^2$, evaluate each expression.

    a) $f(2)$ **b)** $f(3)$ **c)** $f(0.5)$

2. For $g(x) = 3x - 1$, evaluate each expression.

    a) $g(1)$ **b)** $g(5)$ **c)** $g\left(\frac{1}{2}\right)$

3. a) For each function, determine $f(-1)$, $f(\sqrt{2})$, and $f(0.5)$.

    i) $f(x) = 3x^2 - 2x + 1$ **ii)** $f(x) = 2x^3 + 5x^2 + 3x - 4$

    b) Choose one function in part a. Write to explain how you evaluated the expression.

B **4. a)** Graph each function.

    i) $f(x) = \frac{1}{2}x - 1$ **ii)** $f(x) = \sqrt{x - 1}$

    b) State the domain and range of each function in part a.

5. For each graph of $y = f(x)$, determine $f(-2)$, $f(1)$, and $f(3)$.

a)

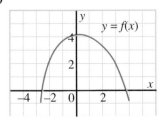

b)

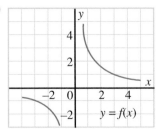

c)

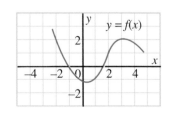

6. Write the ordered pair for each point whose y-coordinate you found in exercise 5.

7. For $f(x) = 3x - 5$, determine each expression.

   a) $f(m)$ b) $f(4x)$ c) $2f(x)$ d) $f\left(\frac{2}{x}\right)$ e) $f(2x + 1)$

8. For $g(x) = 5x + 1$, determine each expression.

   a) $g(k)$ b) $g(x - 1)$ c) $g(2x + 1)$ d) $g(4 - 3x)$

9. For each function, determine $f(2)$, $f(-5)$, $f(\sqrt{3})$, and $f(0.5)$.

   a) $f(x) = 4x - 7$ b) $f(x) = 8x^2 + x - 9$ c) $f(x) = \sqrt{6x - 1}$

   d) $f(x) = x^2 + \frac{1}{x}$ e) $f(x) = x^3 - x^2$ f) $f(x) = \frac{4x}{2x + 1}$

10. a) Graph each function. State its domain and range.

    i) $f(x) = \frac{1}{2}x + 3$ ii) $f(x) = x^2 + 1$ iii) $f(x) = x(x - 3)$

    b) Choose one function in part a. Write to explain how you drew the graph and determined the domain and range.

11. Graph each function for $-3 \le x \le 3$.

    a) $f(x) = x^3$ b) $f(x) = x^3 - 4x$ c) $f(x) = 2^x$ d) $f(x) = \frac{6}{x^2 + 1}$

12. For $f(x) = 2x^2 + 3x - 5$, determine each expression.

    a) $f(x + 1)$ b) $f(x + 2)$ c) $f(x + 3)$

    d) $f(2x)$ e) $f(3x)$ f) $f(-x)$

13. a) Write an expression, $f(x)$, for which $f(3) = 6$.

    b) Write an expression, $f(x)$, for which $f(-1) = 4$.

    c) Write an expression, $f(x)$, for which $f(\sqrt{3}) = 1$.

    d) Choose one of parts a to c. Write to explain how you determined the expression given its value for a particular value of x.

14. Compare your answers to exercise 13 with those of a classmate. If your answers are different, explain the differences.

15. If $f(x) = mx + b$, determine m and b if the graph of the function:

    a) has slope $\frac{2}{5}$ and $f(5) = -1$

    b) has y-intercept 2 and $f(2) = 8$

    c) has y-intercept -3 and $f(-2) = -3$

    d) has $f(-1) = -7$ and $f(5) = 9$

    e) has $f(2) = 2$ and $f(-4) = 11$

16. Given $g(x) = 2x + 3$, determine x for each value of $g(x)$.

   a) $g(x) = 5$ **b)** $g(x) = -9$ **c)** $g(x) = 0$

17. On your graphing calculator, define $Y = 2X + 3$. Display a table of values from which you can read values of X for $Y = 5, -9$, and 0. Verify your answers for exercise 16.

18. You are given $f(x) = \frac{x}{1+x}$.

   a) Evaluate.   **i)** $f(2) + f\left(\frac{1}{2}\right)$ **ii)** $f(3) + f\left(\frac{1}{3}\right)$

   b) Predict the value of $f(n) + f\left(\frac{1}{n}\right)$, where n is any real number. Use algebra to prove your prediction is correct.

   c) For what values of n does the result in part b hold?

19. a) To verify your answer to exercise 18a part i, enter $f(x)$ in a graphing calculator as Y_1. Return to the home screen. Use the VARS function to enter $Y_1(2) + Y_1\left(\frac{1}{2}\right)$.

   b) Verify your answer to exercise 18a part ii in a similar way.

   c) Verify your answers to exercise 18b and c by defining Y_2 as $Y_1(X) + Y_1\left(\frac{1}{X}\right)$, then looking through the table of values.

20. You are given $g(x) = 3^x$.

   a) i) Show that $g(2x) = [g(x)]^2$.   **ii)** Show that $g(3x) = [g(x)]^3$.

   b) To what is $g(nx)$ equal?

21. The numbers $7^1, 7^2, 7^3, \dots$ are the positive integral powers of 7.

   a) On your graphing calculator, define $Y = 7^X$. Display a table of values for this function.

   b) Move the cursor to the y-values. Scroll through the table. Record all the powers of 7 whose last digit is 1. What do all these powers have in common?

   c) Write an expression for the values of x for which 7^x has the last digit 1.

COMMUNICATING THE IDEAS

In your journal, write to describe what function notation is; then explain how it is useful when you work with functions.

Absolute Value

On the number line, the numbers 3 and −3 are each located 3 units from 0.

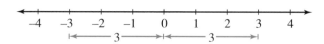

Each number is said to have an absolute value of 3.

We write: $|3| = 3$ and $|−3| = 3$

Read "absolute value of −3"

> The *absolute value* of any real number x is written $|x|$, and it represents the distance from x to 0 on the number line.
>
> • The absolute value of any number other than 0 is positive.
> • The absolute value of 0 is 0.

1. Write each expression without the absolute value symbol.

   a) $|−29|$ **b)** $|12|$ **c)** $|107|$

   d) $|−15|$ **e)** $|6.7|$ **f)** $|−1.8|$

2. Write each expression without the absolute value symbol.

   a) $|3 − 7|$ **b)** $|−5 + 11|$ **c)** $|9 − 1|$ **d)** $|−2 − 5|$

   e) $|14| − |−9|$ **f)** $|−6| + |−21|$ **g)** $|−7| + |−3|$ **h)** $|−2| − |−6|$

3. Write each expression without the absolute value symbol.

   a) $5|1 − 7|$ **b)** $−4|5 − 8|$ **c)** $−9|−1 − 3|$

   d) $5|−2| + 3|4|$ **e)** $3|−1| − 2|−6|$ **f)** $7|−3| − 5|−4|$

4. Determine all the values of x that satisfy each equation. Which equations have no solution?

   a) $|x + 1| = 7$ **b)** $|2x − 1| = 9$ **c)** $|5x − 7| = −3$

   d) $|x + |x|| = 10$ **e)** $|x + |x|| = 0$ **f)** $x + |x| = −2$

5. a) Graph the function $y = |x|$.

  b) How is the graph of $y = |x|$ related to the graph of $y = x$?

6. Graph each function.

  a) $y = |x - 1|$ **b)** $y = |x + 1|$ **c)** $y = |-x|$

  d) $y = |2x - 3|$ **e)** $y = |0.5x + 3|$ **f)** $y = |x| - 3$

7. A graphing calculator was used to graph the four functions below. Three graphs involve absolute value.

  a) Which graph does not involve absolute value? What is the equation of this graph?

  b) Determine the equations of the other three graphs.

  c) Explain why each graph appears the way it does.

Graph 1

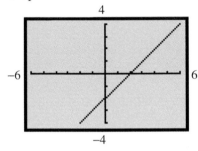

Graph 2

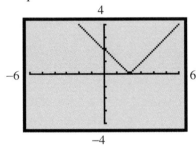

Graph 3

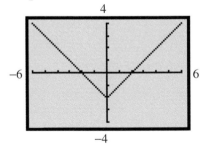

Graph 4

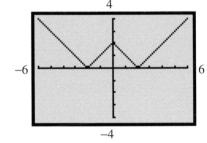

5.7 Relations

Consider the following rule:

If you square the output number y, the result is the input number x.

We generate a table of values for this rule.

x (input)	0	1	1	4	4	9	9
y (output)	0	1	−1	2	−2	3	−3

Recall that a function is a rule that gives a single output number for every valid input number.

The table of values above does not describe a function because several input numbers result in two possible output numbers. For example, the input number $x = 1$ results in output values of $y = 1$ and $−1$.

The above rule and corresponding table describe a set of ordered pairs that is called a relation.

> A *relation* is a rule that produces one or more output numbers for every valid input number.

As with a function, we can express the rule that defines a relation in different ways. We illustrate these ways with a different relation.

- In words:

    The sum of the squares of the input number and output number is 25.

- As an equation:

    $$x^2 + y^2 = 25$$

    In an equation, we use x for each input number and y for each output number.

- As a table of values:

x	y
−5	0
−4	±3
−3	±4
0	±5
3	±4
4	±3
5	0

- As a set of ordered pairs:

    The numbers in the table can be written

    $\{(-5, 0), (-4, 3), (-4, -3), (-3, 4), (-3, -4), (0, 5), (0, -5), (3, 4), (3, -4),$
    $(4, 3), (4, -3), (5, 0)\}$

- As a graph:

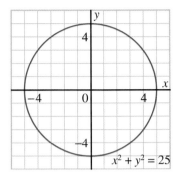

In the graph, the plotted points appear to lie on a circle. If we included more points corresponding to input numbers such as −4.5, −3.5, …, 2.5, 3.5, we would see that the points do lie on a circle. Through the plotted points, we drew a smooth circle. The domain and range of the circle include all real numbers from −5 to 5 inclusive.

The graph shows that this relation is not a function.

Consider the input value of 4. From the graph, there are two output values: 3 and −3.

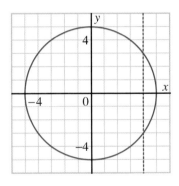

When we examine the graph of a function, such as $y = 2x - 2$, each input value results in only one output value.

For example, when $x = 2$, $y = 2$; when $x = -1$, $y = -4$

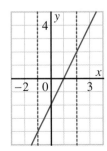

To locate an output value for any input value, in each graph above, we drew a vertical line through the input value and found where it intersected the graph.

We can use this method to determine if the graph of a relation represents a function.

If a graph is given, visualize a vertical line moving across the graph.

| If the vertical line never meets more than one point on the graph, the graph represents a function. | If there is any place where the vertical line meets more than one point on the graph at the same time, the graph does not represent a function. |

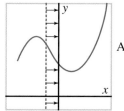 A function

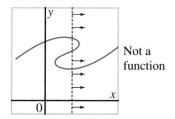 Not a function

We call this the vertical line test.

Vertical line test

If no two points on a graph can be joined by a vertical line, then the graph represents a function.

Example 1

State the domain and range of each relation.

a) {(2, 3), (4, −1), (6, 3), (8, −1)}

b)

x	y
3	4
5	4
7	8
9	8
11	12

c)

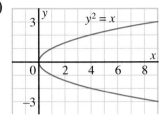

d) $y = |x|$

Solution

Recall that the domain is the set of all possible x-values; and the range is the set of all possible y-values.

a) To determine the domain, list the x-values. The domain is $\{2, 4, 6, 8\}$.
To determine the range, list the y-values. The range is $\{-1, 3\}$.

b) The domain is $\{3, 5, 7, 9, 11\}$. The range is $\{4, 8, 12\}$.

c) The absence of dots at the ends of the graph indicates that the relation continues indefinitely in the positive x-direction. The domain is the set of real numbers greater than or equal to 0. From the graph, y can be any real number. The range is the set of real numbers.

d) For the equation, $y = |x|$, x can be any real number. The domain is the set of real numbers. Since $|x|$ is never negative, y is never negative. The range is the set of real numbers greater than or equal to 0.

Example 2

Determine whether each relation in *Example 1* is a function.

Solution

a) $\{(2, 3), (4, -1), (6, 3), (8, -1)\}$
For each x-value, there is only one y-value. Hence, the relation is a function.

b)

x	y
3	4
5	4
7	8
9	8
11	12

The table shows that for each *x*-value, there is only one *y*-value. The relation is a function.

c)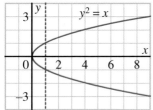

Use the vertical line test. The points $(1, 1)$ and $(1, -1)$, for example, can be joined with a vertical line. The relation is not a function.

d) For $y = |x|$, sketch its graph. Use the vertical line test.

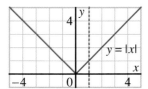

No two points on this graph can be joined using a vertical line. This relation is a function.

DISCUSSING THE IDEAS

1. We know that the graph of a linear function is a straight line. Consider all possible positions of a straight line on a grid. Does each line describe a function? Explain.

2. Explain the difference between a function and relation.

5.7 EXERCISES

 1. Are all functions relations? Explain.

2. Are all relations functions? Explain.

3. Is it possible to have a function whose graph is a circle? Explain.

4. If the same *y*-value occurs twice in a relation, is the relation a function? Explain.

B **5.** For each relation below

    i) State its domain and range.

    ii) State whether it is a function. If necessary, draw a graph to find out.

a) {(1, 1), (2, 4), (3, 9), (4, 16)}   **b)** {(4, 3), (7, 6), (2, 3), (10, 9)}

c)

x	y
3	3
3	−3
6	−6
6	6
0	0

d)

x	y
1	2
2	4
3	6
5	6
8	10

6. For each relation below

    i) State its domain and range.   **ii)** State whether it is a function.

a)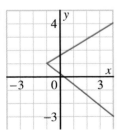

b)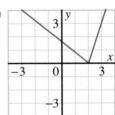

c)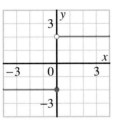

d)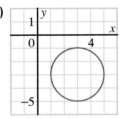

e)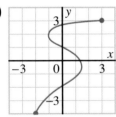

f)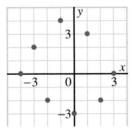

g)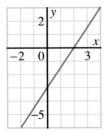

h)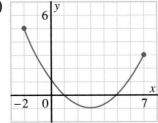

7. For each relation in exercise 6, write to explain how you identified it as a function or a relation.

8. For each relation defined below

    i) State its domain and range.

    ii) State whether it is a function. If necessary, draw a graph to find out.

    a) $y = x^2$ b) $y + 1 = |x|$

    c) $y = 3x + 7$ for $x = 1, 3, 5, 7, 9$ d) $x = 7$

9. Graph each relation. State its domain and range. Is it a function? Explain.

    a) $y = -2x + 3$ b) $x - y^2 = 1$

    c) $2x + 3y = 12$ for $x \geq 0$ d) $y = x^3$

    e) $y = |2x - 1|$ f) $y = |x + 3|$ for $x \geq -3$

C 10. Some geography textbooks contain a graph similar to this. The graph illustrates how the temperature of the atmosphere varies with increasing altitude.

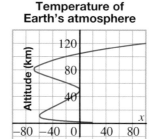

Temperature of Earth's atmosphere

a) Is the relation a function? That is, is altitude a function of temperature?

b) Is temperature a function of altitude? How might the graph be drawn to show this? Give a reason why the graph is not drawn this way in geography textbooks.

11. Is this statement true or false?
A graph represents a function if no vertical line can be drawn that intersects the graph at more than one point.
Explain your answer.

12. a) Determine whether the relation defined by $x = |y|$ is a function.

    b) Graph the relation $x = |y|$.

COMMUNICATING THE IDEAS

You are given an equation defining a relation that is not a function. How would you use a calculator to determine the coordinates of points you could use to graph the relation? Illustrate your answer using a specific example. Write your ideas in your journal.

Classifying Functions

In this chapter, you learned what a function is, and studied linear functions. Here are some other ways that functions can be classified.

Increasing	**Decreasing**	**Periodic**
As x increases, $f(x)$ increases.	As x increases, $f(x)$ decreases.	There is a repeating pattern in the graph.

Piecewise linear	**Step**	**Discontinuous**
The graph has parts, all of which are linear.	The graph comprises a set of unconnected horizontal lines.	The graph cannot be drawn without lifting the pencil.
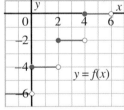	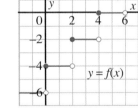	

Read this text.

1. Try to find at least two examples of each kind of function illustrated above. Are there any for which you could not find an example?

2. Try to find a function that has none of the above properties.

1. The rules defining three functions are given. For each function:

   a) Make a table of values.

   b) Graph the function.

   c) Decide if there are any input numbers that cannot be used.

   d) Write an equation to define the function.

       i) *Rule 1*: Double the number, then subtract 1.

       ii) *Rule 2*: Square the number, then subtract the answer from 8.

       iii) *Rule 3*: Add 3 to the square root of the number.

2. a) Graph these functions on the same grid.

       i) $y = x^2 - 3$ **ii)** $y = (x - 3)^2$ **iii)** $y = x^2 + 3$ **iv)** $y = (x + 3)^2$

   b) Write to explain the similarities and differences among the graphs in part a.

3. The size of a car engine is related to the car's fuel consumption. Here is a table listing the fuel consumptions for different sizes of car engines.

Engine size (L)	Consumption (L/100 km)
2.2	6.4
3.0	7.5
3.8	8.1
4.1	8.6

   a) Plot the data to show the fuel consumption as a function of engine size. Draw a line of best fit.

   b) Write an equation that relates fuel consumption and engine size.

   c) Estimate the fuel consumption for a 2.5-L engine.

   d) Estimate the fuel consumption for a 6.0-L engine.

   e) Of the estimates in parts c and d, which one is more likely to be incorrect? Explain.

   f) Compare the equation you wrote in part b with those of your classmates. Did all of you write the same equation? If your equations differ, explain the differences.

4. Michel works as a lifeguard and earns $9.50/h for a 32-h week. He is paid time and a half for any additional hours.

   a) Make a table of values to show Michel's weekly earnings for up to 50 h.

   b) Graph the data in part a. Should the plotted points be joined? Explain.

5. A graduation dance will cost $5000 plus $20 for each person attending.

a) Write an equation to express the cost as a function of the number of people attending.

b) Determine the cost for 400 people.

c) Graph the function.

d) Suppose the dance committee wants the dance to break even. What should be the cost per person for each number of people attending?

    i) 200   **ii)** 250   **iii)** 270   **iv)** 320

6. The height to which a ball bounces varies directly as the distance from which it was dropped. When it is dropped from a height of 2.4 m, it bounces to a height of 1.6 m.

a) Express the height the ball bounces as a function of the height from which it was dropped.

b) The ball is dropped from a height of 3.0 m. How high will it bounce?

c) Suppose the ball bounces 80 cm. From what height was it dropped?

d) Sketch a graph of the function.

7. Given the function $f(x) = 3x^2 - 4x + 5$, evaluate each expression.

    a) $f(3)$ **b)** $f(-2)$ **c)** $f(\sqrt{2})$

8. Given the function $f(x) = -2x^2 + 6x - 1$, write the ordered pair describing the point on the graph with each y-coordinate.

    a) $f(1)$ **b)** $f(-10)$ **c)** $f(\sqrt{2})$ **d)** $f(0.5)$

9. State the domain and range of each relation.

a)
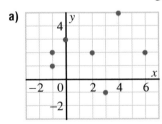

b)
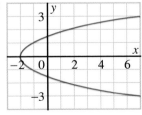

c)
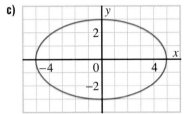

d)
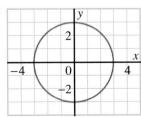

e)

f)
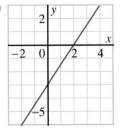

10. For each relation in exercise 9, state whether it is a function. Write to explain how you identified each function.

1. Determine the sum of the first 10 terms of each arithmetic series.

   a) $1 + 2 + 3 + \cdots$ b) $1 + 3 + 5 + \cdots$ c) $1 + 4 + 7 + \cdots$

   d) $1 + 5 + 9 + \cdots$ e) $1 + 6 + 11 + \cdots$ f) $1 + 7 + 13 + \cdots$

2. Look at the results of exercise 1. What pattern do you notice?
   Write to explain why the pattern exists.

3. A certain type of sports cars depreciates by 10% as you drive it off the car
   lot, and by 20% each year after that. A new sports car costs $90 000.

   a) Construct a table. Show the value of the sports car each year for 10 years
   after it is bought.

   b) How long does it take for the car to decrease to each value?

       i) $\frac{1}{2}$ of its original value

       ii) $\frac{1}{4}$ of its original value

       iii) $\frac{1}{10}$ of its original value

4. Use a calculator to determine each value, to 3 decimal places.

   a) $40^{1.3}$ b) $26^{2.1}$ c) $15^{\frac{2}{3}}$ d) $(-7)^{-\frac{2}{3}}$ e) $14^{-\frac{5}{4}}$

   f) $\sqrt[5]{31.9}$ g) $\sqrt[5]{(-1.7)^2}$ h) $4.32^{\frac{3}{2}}$ i) $0.004^{-\frac{1}{7}}$ j) $\sqrt[6]{972^2}$

5. a) Determine the ratio of the
   perimeter of the large square
   to the perimeter of the
   small square.

   b) Determine the ratio of the
   area of the large square to the
   area of the small square.

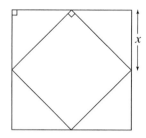

6. Determine the slopes of two line segments; one parallel to, the other
   perpendicular to the line segment with each pair of endpoints.

   a) B(2, 7), C(−3, 1)   b) D(−2, 4), E(4, −3)   c) F(−3, −2), G(2, 5)

7. Graph each line.

   a) $3x + 2y - 18 = 0$ b) $5x - 3y - 15 = 0$ c) $2x + 5y + 10 = 0$

8. The equations of three sides of a square are $y = x + 4$, $y = -x + 6$, and
   $y = -x + 10$.

   a) Graph these lines on the same grid.

   b) Determine a possible equation for the fourth side.

9. Select the graph that best illustrates each given function.

a) how a person's mass changes with age

b) how the temperature of a cup of coffee changes with time

c) how the number of cars in the school parking lot changes during the day

d) how the braking distance of a car changes with speed

e) how the cost of parking depends on the time a car is parked

f) how the height of a kicked football changes with time

i)

ii)

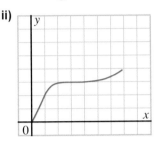

iii)

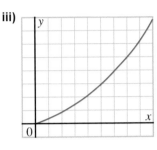

iv)

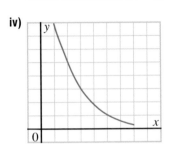

v)

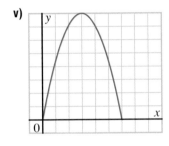

vi)

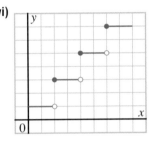

10. a) For each graph in exercise 9

i) Describe an appropriate domain and range.

ii) State the meaning of any intercepts, and maximum and minimum points.

b) For one graph in exercise 9, write to explain how you identified the graph, and how you determined the domain and range.

11. Write a rule for each table of values.

a)

x	y
0	10
1	9
2	8
3	7
4	6

b)

x	y
−2	1
−1	2
0	3
1	4
2	5

Could a Giant Survive?

CONSIDER THIS SITUATION

People have always been interested in the gigantic. For
example, in *Gulliver's Travels*, written in the 1700s,
Gulliver discovered the Brobdingnag giants. The
Brobdingnagians were 12 times as tall as Gulliver.

The Guinness Book of Records reports some real-life
examples of exceptionally tall people. The tallest
person ever, Robert Wadlow, was 272 cm tall. Wadlow
died at 22. The tallest woman, Zeng Jinlian, was 248 cm
tall. She died at 18.

- Do you think a fictional giant could survive? Explain
  your thinking.

- What special needs do you think exceptionally tall
  people have?

On pages 338 and 339, you will develop a mathematical
model to explore this situation. You will use your skills
with evaluating measurement formulas and manipulating
polynomial expressions.

 FYI Visit www.awl.com/canada/school/connections

For information related to the above problem, click on
<u>MATHLINKS</u> followed by <u>AWMath</u>. Then select a topic
under GIANTS! Could a Giant Survive?

6.1 Measurement Formulas and Monomials

Many measurement formulas involve monomials.

Expressions such as $6s^2$, and $\frac{4}{3}\pi r^3$ are *monomials*. A monomial is the product of a coefficient and one or more variables. For example, consider the monomials $5x^2$ and $-9a^3b$.

$$\underset{\substack{\nwarrow\qquad\nwarrow\\ \text{coefficient}\quad\text{variable}}}{5x^2}\qquad\qquad\underset{\substack{\nwarrow\qquad\nwarrow\\ \text{coefficient}\quad\text{variables}}}{-9a^3b}$$

To evaluate a monomial, substitute a number for each variable.

Cube

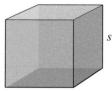

Surface area, $A = 6s^2$
Volume, $V = s^3$

Sphere

Surface area, $A = 4\pi r^2$
Volume, $V = \frac{4}{3}\pi r^3$

Example 1

The mean radius of Earth is approximately 6365 km. Calculate the surface area and the volume of Earth.

Solution

Visualize Earth as a sphere. Substitute 6365 for r in the formulas for the surface area and volume of a sphere:

$A = 4\pi r^2$

$\quad = 4 \times \pi \times 6365^2$

$\quad \doteq 5.1 \times 10^8$

$V = \frac{4}{3}\pi r^3$

$\quad = \frac{4}{3} \times \pi \times 6365^3$

$\quad \doteq 1.1 \times 10^{12}$

The surface area of Earth is approximately 5.1×10^8 km². Its volume is approximately 1.1×10^{12} km³.

MODELLING Earth

Earth is not spherical. It is flattened at the poles. Its polar radius is 8 km less than its mean radius, while its equatorial radius is 13 km more than its mean radius.

1. Calculate the surface area and volume using:

   a) the polar radius   b) the equatorial radius

2. Would the actual surface area and volume be closer to the results obtained using the polar radius or the results obtained using the equatorial radius? Explain.

Example 2

A spherical hot air balloon has a diameter of 3.6 m. We pump 20 additional cubic metres of air into the balloon. What are the new diameter, surface area, and volume? Give the answers to 2 significant digits.

Solution

The radius of the balloon is 1.8 m. To determine its volume, substitute 1.8 for r in the formula for the volume of a sphere.

$$V = \frac{4}{3}\pi r^3$$
$$= \frac{4}{3} \times \pi \times 1.8^3$$
$$\doteq 24.4290$$

The volume of the balloon is approximately 24 m^3.

After pumping in 20 additional cubic metres, the new volume of the balloon is approximately 44.4290 m^3.

To determine the new diameter of the balloon, first find the new radius. Substitute 44.4290 for V in the formula for the volume of a sphere. Then solve for r:

$$44.4290 = \frac{4}{3}\pi r^3$$
$$3 \times 44.4290 = 4\pi r^3$$
$$\frac{3 \times 44.4290}{4\pi} = r^3$$
$$r^3 \doteq 10.6066$$
$$r \doteq \sqrt[3]{10.6066}$$
$$r \doteq 2.1971$$

Use the cube-root function on your calculator.

The new radius is approximately 2.2 m, and the new diameter is approximately 4.4 m.

To determine the new surface area, substitute 2.1971 for r in the formula for the surface area of a sphere:

$$A = 4\pi r^2$$
$$= 4 \times \pi \times 2.1971^2$$
$$\doteq 60.6610$$

The new surface area is approximately 61 m^2.

1. In *Examples 1* and *2*, suppose you use the decimal approximation for π, 3.14, instead of the $\boxed{\pi}$ key on your calculator. How would this affect the final answers?

2. In the solution of *Example 2*, intermediate values were rounded to 4 decimal places. Why do you think this was done? What would have happened to the final answer if the values had been rounded to 1 decimal place?

6.1 EXERCISES

 1. Calculate the surface area and volume of each sphere. The radius is given.

a)

10 cm

b)

5 cm

c)

15.2 cm

2. Calculate the surface area and the volume of each ball listed.

	Sport	Diameter of ball (cm)
a)	Baseball	7.4
b)	Golf	4.3
c)	Table tennis	3.7
d)	Volleyball	20.9

3. In this self-portrait, the Dutch artist, M.C. Escher, is holding a reflecting sphere.

a) Estimate the diameter of the sphere in this picture.

b) Use your estimate. Calculate the surface area and the volume of the sphere.

4. The radius of Uranus is almost 4 times as great as the radius of Earth. The radius of the moon is approximately $\frac{1}{4}$ that of Earth. Use the information in *Example 1*. Determine the surface area and volume of:

a) Uranus   **b)** the moon

5. Calculate the radius of the sphere with each volume.

a) 45.7 m³   **b)** 23.8 cm³   **c)** 1356 mm³

6. Calculate the diameter of the sphere with each surface area.

a) 2367 mm² **b)** 325.6 cm² **c)** 3.9 m²

7. Choose one part of exercise 6. Write to explain how you calculated the diameter.

8. A basketball has a circumference of 75 cm. Calculate each measurement.

a) radius **b)** surface area **c)** volume

9. A squash ball has a radius of 2.0 cm. It fits in a box in the shape of a cube. Visualize a sphere in a cube.

a) Calculate the surface area of the ball and the surface area of the box. What assumptions are you making about the dimensions of the box?

b) Calculate the ratio of the surface area of the ball to the surface area of the box.

c) Calculate the volume of the ball and the volume of the box.

d) Calculate the ratio of the volume of the ball to the volume of the box. Compare the result with part b. What do you notice?

10. Calculate each surface area and volume.

a) Pluto, with radius 1139 km **b)** the sun, with radius 696 260 km

11. A cylinder has radius r and height h. Its volume is given by the formula $V = \pi r^2 h$.

Calculate the volume of each cylinder.

a) radius 3.5 cm; height 5.7 cm

b) height 5.5 m; radius 1.9 m

c) diameter 44 mm; height 156 mm

12. a) Use a graphing calculator or a computer graphing program. Graph the equation for the surface area of a sphere, $A = 4\pi r^2$.

b) Use the trace and zoom features. Verify the surface areas you calculated for the balls in exercises 2, 8, and 9.

c) Describe what happens to the surface area as the radius increases.

13. a) Use a graphing calculator or a computer graphing program. Graph the equation for the volume of a sphere, $V = \frac{4}{3}\pi r^3$.

b) Use the trace and zoom features. Verify the volumes you calculated for the balls in exercises 2, 8, and 9.

c) Describe the graph.

14. Recall the hot air balloon in *Example 2*. It has a volume of 24.43 m³.

    a) Calculate the new diameter, surface area, and volume for each additional volume of air pumped in: 10 m³; 30 m³; 40 m³; 50 m³; 60 m³; 70 m³. Summarize your results in a table.

    b) Draw a graph of the new surface area against the volume of air pumped in. Write to describe the graph.

    c) Draw a graph of the new volume against the volume of air pumped in. Write to describe the graph.

15. Twenty cubic metres of air are pumped into a spherical hot air balloon with each given diameter: 1.0 m; 2.0 m, 3.0 m; 4.0 m; 5.0 m; 6.0 m.

    a) Calculate the new diameter, new surface area, and new volume of each balloon. Summarize your results in a table.

    b) Draw a graph of the new surface area against the original diameter of the balloon. Describe the graph.

    c) Draw a graph of the new volume against the original diameter of the balloon. Describe the graph.

16. Which has a greater volume: a sphere with radius r, or a cube with edges of length r? Support your answer with a written explanation or some examples.

17. A spherical balloon is blown up from a diameter of 20 cm to a diameter of 60 cm. By how many times has its:

    a) surface area increased?   **b)** volume increased?

18. What happens to the surface area and the volume of a sphere when its radius is:

    a) doubled?   **b)** tripled?   **c)** multiplied by n?

19. Determine expressions for the surface area and the volume of a sphere with:

    a) diameter d **b)** circumference C

20. Choose one part of exercise 19. Write to explain how you calculated the expressions.

21. According to Bergmann's Law, an animal generates heat in proportion to its volume. It loses heat in proportion to its surface area. Suppose we visualize an animal as spherical, as it would appear if it were curled up. Use the formulas for the volume V and surface area A of a sphere. The larger the value of $\frac{V}{A}$, the better the animal's ability to survive cold temperatures.

    a) Draw a graph of V against A for $0 \leq r \leq 2$, where the radius of the sphere is r metres.

b) Describe how the $\frac{\text{volume}}{\text{surface area}}$ ratio changes as the size of the animal increases.

c) Are larger or smaller animals better able to maintain their body heat in cold climates? Explain.

C **22.** A spherical soap bubble contains a fixed volume of air. It lands on a flat surface and its shape changes to that of a hemisphere. Visualize a hemisphere on a table.

a) Is the volume of air in the hemisphere greater than, less than, or the same as the volume of air in the sphere? Explain your answer.

b) Is the radius of the hemisphere greater than, less than, or the same as the radius of the sphere? Explain your answer.

c) Determine the ratio of the radius of the hemisphere to the radius of the sphere.

d) Determine the ratio of the surface area of the soap film of the hemisphere to the surface area of the soap film of the sphere.

23. The Hubble Space Telescope was designed to operate in orbit, high above Earth's atmosphere. The diagram shows that the Space Telescope can see seven times as far into space as ground observatories.

Limit of Ground Observatories

Limit of Space Telescope

2 billion light-years

14 billion light-years

a) Compare the volume of space that can be studied by the Space Telescope with the volume of space that can be studied by ground observatories.

b) The unaided eye can see objects up to 600 000 light-years away. Compare the volume of space that can be studied by ground observatories with the volume that can be studied by the unaided eye.

COMMUNICATING THE IDEAS

All the measurement formulas in this section are monomials. Find a measurement formula that is not a monomial. In your journal, write to explain why it is not a monomial.

Archimedes of Syracuse

Archimedes of Syracuse was the greatest mathematician of antiquity. His many accomplishments include the invention of a pump for raising water, a planetarium to show the motions of the sun, moon, and planets, and the use of levers and pulleys to move heavy objects.

In his work, *On the Sphere and the Cylinder*, Archimedes described many formulas he had derived for the areas and volumes of geometric solids. One of his results involves the relationships among the volumes and surface areas of a sphere and a cylinder that touches it.

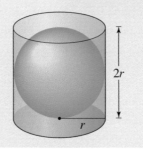

When a sphere is inscribed in a cylinder:

The volume of the sphere is $\frac{2}{3}$ the volume of the cylinder.

The surface area of the sphere is $\frac{2}{3}$ the surface area of the cylinder.

1. Verify that Archimedes' relationships between the surface areas and volumes of a sphere and the touching cylinder are correct.

2. In each diagram below, the cylinder has a height equal to the radius of its base. The second and third diagrams show a hemisphere and a cone inscribed in the cylinder. Find monomials to represent the volumes of the cylinder, the hemisphere, and the cone. How are these volumes related?

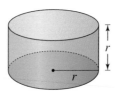

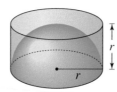

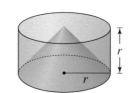

6.2 Multiplying and Dividing Monomials

Recall that to multiply or divide monomials, we use the exponent laws.

Using the law for multiplying powers	Using the law for dividing powers
Add the exponents.	Subtract the exponents.
$(2x^6y^3)(-5x^2y) = -10x^8y^4$	$\dfrac{20x^6y^3}{-5x^2y} = -4x^4y^2$
Multiply the coefficients.	Divide the coefficients.

Example 1

Simplify.

a) $(5x^3y^4)(2xy^2)$ 　　 b) $(5ab^2)(-2a^2b)^3$ 　　 c) $\dfrac{9x^5y^4}{6x^3y}$ 　　 d) $\dfrac{(6m^2n^2)(8m^4n^3)}{(-4mn^2)^2}$

Solution

Use mental math to multiply the coefficients and add the exponents.

a) $(5x^3y^4)(2xy^2) = 10x^4y^6$ 　　　　 b) $(5ab^2)(-2a^2b)^3 = (5ab^2)(-8a^6b^3)$
$$= -40a^7b^5$$

Use mental math to divide the coefficients and subtract the exponents.

c) $\dfrac{9x^5y^4}{6x^3y} = \dfrac{3}{2}x^2y^3$ 　　　　 d) $\dfrac{(6m^2n^2)(8m^4n^3)}{(-4mn^2)^2} = \dfrac{48m^6n^5}{16m^2n^4}$
$$= 3m^4n$$

Example 2

The area of a rectangle is 10 cm^2. The area is to increase by 20 cm^2 so that the new rectangle is similar to the original rectangle. By what factor must each dimension of the rectangle be multiplied?

Solution

Recall that for two figures to be similar, the ratio of the lengths of two sides of one figure must equal the ratio of the lengths of the corresponding sides of the other figure. For this to be true for the rectangles, the length and width of the original rectangle must be multiplied by the same factor.

The length and width of the rectangle are not known. Let them be represented by l and w, respectively.

The original area is 10 cm^2; so we know that $lw = 10$.

Let x represent the factor by which the length and width are multiplied.

The new length is then lx and the new width is wx.

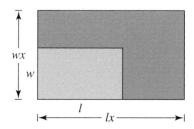

Since the area is *increased* by 20 cm^2, the area of the larger rectangle is 30 cm^2. Hence, we write:

$$(lx)(wx) = 30$$
$$lwx^2 = 30$$

Since we know that $lw = 10$, we replace lw with 10 in this equation to obtain:

$$10x^2 = 30$$
$$x^2 = 3$$
$$x = \sqrt{3}$$

To increase the area of the rectangle by 20 cm^2, both the length and the width must be multiplied by $\sqrt{3}$.

DISCUSSING THE IDEAS

1. Use the meaning of *exponents* to explain each result in *Example 1*.

2. In *Example 2*, recall that the factor $\sqrt{3}$ is called the *scale factor*. By what scale factor must the length and the width of the original rectangle be multiplied to increase the area by 30 cm^2? Justify your answer.

3. In *Example 2*, are the two rectangles congruent? Justify your answer.

4. In *Example 2*, how many times as great as the original perimeter is the perimeter of the larger rectangle?

6.2 EXERCISES

A 1. Simplify.

    a) $(3a)(2a)$
    b) $(-3a)(2a)$
    c) $(-3a^2)(2a)$
    d) $(-3a^2)(2a^2)$

    e) $(-3a^2)(-2a^2)$
    f) $(-3a^3)(-2a^2)$
    g) $(-3a^3)(2a^3)$
    h) $(-6a^3)(-2a^4)$

2. Simplify.

    a) $(7x^2)(5x^4)$
    b) $(3a)(4a)$
    c) $(-6m^3)(9m^2)$
    d) $(5n^2)(2n)$

    e) $(-8y^5)(-7y^4)$
    f) $(-7b^3)(4b)$
    g) $-(t^3)(-6t^5)$
    h) $(12p^5)(-3p^3)(-2p)$

3. Simplify.

    a) $(6x)(3y)$
    b) $(8p^2)(4q)$
    c) $(-5m^3)(-7n)$
    d) $(2ab)(7a)$

    e) $(-9r)(5rs^3)$
    f) $(4cd)(-11c)$
    g) $(-2st^2)(-3st)$
    h) $(2b^2c^3)(bc^2)$

4. Simplify.

    a) $(-4ab)(2b^2)$
    b) $(6m^2n^3)(-2m^3n^4)$
    c) $(7a^2b^2)(3ab^3)$

    d) $(-8p^2q^5)(-3pq^2)$
    e) $(2xy)(3x^2y)(4x^3y)$
    f) $(-3ab)(2ab^2)(a^2b)$

5. Simplify.

    a) $\dfrac{12x^5}{3x^2}$
    b) $\dfrac{36m^6}{4m^2}$
    c) $\dfrac{54a^8}{-6a^2}$
    d) $\dfrac{-52b^3}{-13b^2}$
    e) $\dfrac{-24cd^3}{-6cd}$
    f) $\dfrac{28m^2n^5}{-7m^2n^3}$

6. Write each monomial as a product of two monomials, in three different ways. Which monomials can be written as a product of two equal factors?

    a) $36a^4b^2$
    b) $-15xyz$
    c) $7p^2qr$
    d) $-48a$
    e) $25c^2d^2$
    f) $4m^4$

7. Write each monomial in exercise 6 as a quotient of two monomials, in three different ways.

8. Choose one part of exercise 7. Write to explain how you found each quotient.

B 9. Simplify.

    a) $\dfrac{36r^6s^4}{8r^2s^2}$
    b) $\dfrac{-45x^8y^5}{15x^4y^2}$
    c) $\dfrac{(4m^4n^2)(6m^8n^4)}{-3m^6n^6}$
    d) $\dfrac{(3a^4b^2)(6ab^3)}{9a^3b^4}$
    e) $\dfrac{(-9x^3y^6)(8x^7y^4)}{(2x^2y^3)(-6x^3y)}$

10. Simplify.

    a) $(2x^5)(3x^2)^3$
    b) $(-3a^2b)(2a^4)^2$
    c) $(5s^2t^5)(3s^4)^2$

    d) $(-4pq)^3(-5p^3)$
    e) $(3m^2n^3)(-2mn^2)^3$
    f) $(5x^2y)^2(3x^2y^4)^3$

11. Simplify.

    a) $\dfrac{(2xy^2)(3x^5y^4)}{4x^2y^5}$
    b) $\dfrac{(-3p^2q^5)(-4pq^3)^2}{8p^4q^4}$
    c) $\dfrac{(-7a^2b^4)(4a^3b^5)}{(-2ab^2)^3}$

    d) $\dfrac{(-5x^4y^5)(2xy^2)^3}{(10x^3y^8)^2}$
    e) $\dfrac{(-4m^2n^4)^3(-3m^3n)^2}{(-6m^2n^3)^2}$
    f) $\dfrac{(3a^6b^2)^2(-2a^2b^4)^3}{(6ab)^2(-2a^4b)}$

Use the following information to complete exercises 12 and 13.

A packaging company makes rectangular juice boxes. The regular size box has a capacity of 250 mL.

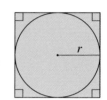

12. By what factor must each dimension of this box be multiplied to increase the capacity to:

   a) 500 mL?   **b)** 1 L?

13. In each part of exercise 12, how many times as great as the original surface area is the surface area of each larger box?

14. A circle with radius r is inscribed in a square.

   a) Determine the ratio of the perimeter of the circle to the perimeter of the square. Does the result depend on the radius of the circle?

   b) Determine the ratio of the area of the circle to the area of the square. What do you notice about the result?

15. A ball with radius r fits snugly in a cubical box.

   a) Determine the ratio of the surface area of the box to the surface area of the ball. Does the result depend on the radius of the ball?

   b) Determine the ratio of the volume of the box to the volume of the ball. What do you notice about the result?

16. Three golf balls fit snugly in a rectangular box. What fraction of the space in the box do they occupy?

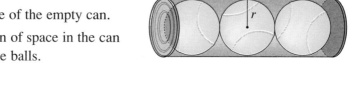

17. A model train is built to a scale of 1 : 60.

   a) What is the scale factor for the model train?

   b) The length of the model engine is 25 cm. What is the length of the engine?

   c) The area of metal used to cover the outside surface of the model engine is 240 cm². What is the area of the metal used to cover the engine?

   d) The volume of the model engine is 180 cm³. What is the volume of the engine?

18. Tennis balls are packed in cans of 3. A tennis ball has a radius of 3.6 cm.

   a) Determine the total volume of the three balls.

   b) Determine the volume of the empty can.

   c) Determine the fraction of space in the can that is occupied by the balls.

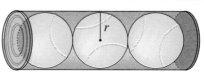

In exercise 18, you modelled the containers in which tennis balls are sold.

- In what ways might the design of a real can of tennis balls differ from the model?
- Modify the model to account for one of the differences you identified.
- Would the results from your modified model differ from the results you obtained in exercise 18? Support your answer with some calculations.

19. a) Measure the base radius and the height of a pop can. Calculate its total surface area in square centimetres and its volume in cubic centimetres.

b) The balloon in the photograph is similar to a pop can. The balloon was made to a scale of 250 : 1. Use the scale and your answers to part a to help you answer these questions.

   i) What is the scale factor for the balloon?

   ii) How many square metres of material were used to make the balloon?

   iii) What is the volume of the balloon, in cubic metres?

C 20. The base of a hemisphere is also the base of an inscribed cone. Determine the ratio of the volume of the hemisphere to the volume of the cone.

COMMUNICATING THE IDEAS

In your journal, write an answer to each question. Explain each answer.

Suppose each dimension of a figure is multiplied by a constant k. By what factor is the area multiplied? By what factor is the perimeter multiplied?

Suppose each dimension of a solid is multiplied by a constant k. By what factor is the volume multiplied? By what factor is the surface area multiplied?

Suppose k has a value between 0 and 1. How does the new figure compare with the original? How does the new solid compare with the original?

MATHEMATICAL MODELLING

Could a Giant Survive?

See page 324 for some examples of fantasy giants and exceptionally tall people. Could the giants in the land of Brobdingnag survive?

 **DEVELOP A MODEL**

Visualize a continuing sequence of cubes, like this.

   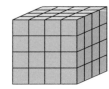

1. Make and complete a table for larger and larger cubes.

Edge length (cm)	Volume (cm³)	Surface area (cm²)	Volume / Surface area
1			
2			
3			
⋮			
n			

2. **a)** As n increases, which grows more rapidly, surface area or volume?

   b) As n increases, what happens to the ratio of volume to surface area?

   c) Write a formula for the ratio of volume to surface area for any edge length n.

3. Suppose you multiply each dimension of a person by 12.

   a) Approximately how many times as great would the surface area and the volume be for the giant than for the person?

   b) How would the giant's ratio of volume to surface area compare with this ratio for a person?

4. In the movie *Honey, I Blew Up the Kid*, an inventor accidentally enlarges his two-year-old. The child eventually grows from 1 m to 32 m tall, terrorizing Las Vegas until the change is reversed. Suppose you multiply each dimension of a person by 32. Repeat parts a and b of exercise 3.

338 CHAPTER 6 POLYNOMIALS

LOOK AT THE IMPLICATIONS

5. Choose one or more of the biological systems described below. What health problems are implied for the giants in the land of Brobdingnag, or for the giant child in the movie *Honey, I Blew Up the Kid*?

   Transpiration system Your body generates heat according to its volume. But the rate of cooling depends on the area of your skin, since an important cooling mechanism is perspiration. As the water on your skin evaporates, it helps to keep you cool.

   Respiratory system All the cells in your body require oxygen. The number of cells depends on volume, but the amount of oxygen processed by your respiratory system depends on the surface area of your lungs.

   Skeletal system Your mass depends on volume, but the strength of your bones depends on the area of their cross-section.

6. See page 324 for two real-life examples of the gigantic. Jinlian suffered from severe scoliosis, or curvature of the spine. Wadlow died of an infected blister on his ankle, caused by a poorly fitting brace.

   a) Do you think that Jinlian's and Wadlow's physical ailments, indicated above, might have been related to their physical size? Explain.

   b) How do you think their exceptional heights may have contributed to their early deaths?

 REVISIT THE SITUATION

7. On the preceding page you modelled a giant using a cube. A cube may not be the best model because it does not represent the shape of a person well.

   a) Try using models that represent the human shape more closely. Would your answers to exercise 5 change? Explain.

   b) Would your answers be different if you used a complex model that more closely resembled the shape of a person? Explain.

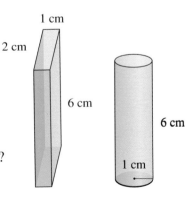

8. Create a new problem related to the situation on these pages. Solve it by applying or modifying the model you developed above.

6.3 Adding and Subtracting Polynomials

Recall that a monomial, and an expression formed by adding or subtracting monomials, are called *polynomials*. Examples are:

$$y^2 \qquad 3x^3 \qquad 5x - 6y^2 \qquad\qquad 3a^2b^3 - 5ab - 6b^3 + 7ab^2$$

To add or subtract polynomials, combine like terms. These like terms have the same variables raised to the same exponents. The term with the greatest exponent, or exponent sum, determines the *degree* of the polynomial.

Example 1

Simplify, and state the degree of each polynomial.

a) $6a - 4b - 2c - 5a + 5b + 3c$

b) $(2x - 5z + y) - (7x + 4y - 2z)$

c) $(2x^3 - 4xy^2 + 5x^2y^2) + (3x^3 + 2x^2y - 6x^2y^2)$

Solution

Use mental math to add or subtract the coefficients.

a) $\quad 6a - 4b - 2c - 5a + 5b + 3c$
$\quad = 6a - 5a - 4b + 5b - 2c + 3c$
$\quad = a + b + c$

Each term has an exponent of 1.
This is a first-degree polynomial.

b) $\quad (2x - 5z + y) - (7x + 4y - 2z)$
$\quad = (2x - 5z + y) + (-7x - 4y + 2z)$ ⟵ Adding the opposite
$\quad = 2x - 5z + y - 7x - 4y + 2z$ ⟵ Removing brackets
$\quad = -5x - 3y - 3z$

Each term has an exponent of 1.
This is a first-degree polynomial.

c) $\quad (2x^3 - 4xy^2 + 5x^2y^2) + (3x^3 + 2x^2y - 6x^2y^2)$
$\quad = 2x^3 - 4xy^2 + 5x^2y^2 + 3x^3 + 2x^2y - 6x^2y^2$
$\quad = 5x^3 - x^2y^2 + 2x^2y - 4xy^2$

The second term, $-x^2y^2$, has the greatest exponent sum, 4.
This is a fourth-degree polynomial.

Example 2

The cost, in dollars, of producing n video cassettes is $1.5n + 16\ 000$. The income, in dollars, from sales to a retailer is $6n$.

a) Write a formula for the profit from producing and selling n cassettes.

b) Calculate the profit from the production and sale of 10 000 cassettes.

Solution

a) Profit, $P = $ Income $-$ Cost

$$P = 6n - (1.5n + 16\ 000)$$
$$= 6n - 1.5n - 16\ 000$$
$$= 4.5n - 16\ 000$$

A formula for the profit from producing and selling n cassettes is $P = 4.5n - 16\ 000$.

b) Substitute $n = 10\ 000$ in the formula for P.

$$P = 4.5(10\ 000) - 16\ 000$$
$$= 45\ 000 - 16\ 000$$
$$= 29\ 000$$

The profit from the production and sale of 10 000 cassettes is $29 000.

DISCUSSING THE IDEAS

1. In the cost expression in *Example 2*, what do you think the coefficient 1.5 represents? What do you think the constant term 16 000 represents?

2. In the profit expression in *Example 2*, what do you think the coefficient 4.5 represents? What do you think the constant term $-16\ 000$ represents?

6.3 EXERCISES

A 1. a) Choose any two numbers. Calculate their sum and their difference, then add the results. How is the answer related to the numbers with which you started?

b) Try this with other numbers, such as fractions and negative numbers.

c) Will this always happen? Use algebra to explain your answer in writing.

2. a) In exercise 1, predict what would happen if you subtract the sum and the difference of the numbers instead of adding them.

  b) Repeat exercise 1 but, after finding the sum and the difference, subtract the results.

3. Simplify.

  a) $(6a + 9) + (4a - 5)$ **b)** $(2x - 7y) + (8x - 3y)$

  c) $(3m^2 + 4) - (7 + 9m^2)$ **d)** $(5p + 11) - (2p + 4)$

  e) $(17x^2 - 9x) - (6x^2 - 5x)$ **f)** $(7k + 3l) - (12k - 8l)$

4. The sum of two polynomials is $15a + 4$. One polynomial is $3a - 6$. What is the other polynomial? Write to explain how you found it.

5. Write each polynomial as the sum of two polynomials, in three different ways.

  a) $8a + 10$ **b)** $-6b + 15$ **c)** $-4m - 11n$ **d)** $16 - 17x$

6. Write each polynomial in exercise 5 as the difference of two polynomials, in three different ways.

7. The sum of two polynomials is $4x^2 - 7x + 3$. One polynomial is $-5x^2 - 8x + 5$. What is the other polynomial? Write to explain how you found it.

B **8.** Simplify.

  a) $(3m^2 - 5m + 9) + (8m^2 + 2m - 7)$

  b) $(7a^3 - 2a^2 + 5a) + (-3a^3 + 6a^2 - 2a)$

  c) $(x^2 - 5x + 6y) - (4x^2 + 15x - 11y)$

  d) $(5t^2 - 13t + 17) + (9t^3 - 7t^2 + 3t - 26) - (16t^2 + 5t - 8)$

  e) $(2 + 8a^2 - a^3) + (2a^3 - 3a^2 - 8) - (12 - 7a^3 + 5a^2)$

  f) $(4x^2 - 7x + 3) - (x^2 - 5x + 9) - (8x^2 + 6x - 11)$

9. The cost, in dollars, of producing n board games is $4.5n + 30\ 000$. The income, in dollars, from sales is $20n$.

  a) Write a formula for the profit from producing and selling n board games.

  b) What is the profit from the production and sale of each number of games?

    i) 5000 games   **ii)** 10 000 games   **iii)** 20 000 games

10. Suppose you needed to determine the value of each expression below when $x = -3$ and when $x = 3$. Would it be easier to substitute each value for x in the given expression, or would it be better to simplify the expression first then substitute for x in the result? Write to explain your thinking.

  Evaluate each expression for each value of x.

  a) $(2x^2 + 3x - 5) + (x^2 - 5x + 3)$ **b)** $(17x^2 - 7x + 9) - (7x^2 + 3x - 3)$

11. Simplify, then state the degree of each polynomial.

    a) $(5x^2 - 3x^2y + 2y^2x - y) - (2xy^2 - y - 5x^2 + 3yx^2)$

    b) $(p^3q^2 + 7q^2p^2 - 3p) + (2q - 6p^2q^2 - q^2p^3)$

    c) $(3m^2 - 3m + 7) - (2n^2 - 2n + 7)$

    d) $(2ab - 2ac - 2bc) + (2ca + 2cb - 2ba + 3)$

    e) $(xy + 3yx + 2x^2y - 3xy^2) - (2xy + 2yx^2)$

12. Add.

    a) $3x^2 - 6x + 7$
       $\underline{7x^2 - 2x + 9}$

    b) $7z^3 - 6z^2 + z - 3$
       $\underline{2z^3 + 4z^2 - z + 4}$

13. Subtract.

    a) $5x^2y - 3xy + 2xy^2$
       $\underline{3x^2y + 4xy - 3xy^2}$

    b) $6m^2n^2 - 4mn + 3mn^3$
       $\underline{2m^2n^2 + 2mn - 4mn^3}$

14. The sum of three polynomials is $-5x^2 - 8x + 5$. One polynomial is $4x^2 - 7x + 3$. What could the other two polynomials be? Write to explain how you found them.

15. Simplify, then state the degree of each polynomial.

    a) $(3x^2 - 2y^2) + (y^2 - 2x^2) - (4x^2 + 2)$

    b) $(m^2 - n^2) - (n^2 - m^2) + (2n^2 + m^2)$

    c) $(4x^2y - 2yx) + (3yx^2 - 6xy^2) - (3x^2y^2 + 2y^2x^2 - xy)$

    d) $(a^2b^2 - b^2) - (3b^2a + 2a^2b) - (b^2 - b^2a^2 - b^2a)$

16. Give an example of each polynomial.

    a) degree 4 with 3 terms   **b)** degree 3 with 4 terms

    c) degree 2 with 4 terms   **d)** degree 5 with 2 terms

17. Choose one part of exercise 16. Write to explain how you found the polynomial.

COMMUNICATING THE IDEAS

Imagine you have a younger brother who thinks that $2x + 3y = 5xy$. In your journal, write an argument to convince him that his reasoning is not correct.

INVESTIGATE **Area Models for Multiplication**

1. In the diagram (below left)

   a) Write the area of the rectangle as the product of its length and width.

   b) Write the area of the rectangle as the sum of the areas of the squares and rectangles inside it.

   c) Write to explain why the area model shows that $2(x + 3) = 2x + 6$.

2. In the diagram (above right)

   a) Write the area of the rectangle as the product of its length and width.

   b) Write the area of the rectangle as the sum of the areas of the square and the rectangles inside it.

   c) Write to explain why the area model shows that $x(x + 3) = x^2 + 3x$.

3. In each diagram above, imagine that x varies, but the other lengths remain the same. Visualize how the diagram changes if x becomes smaller, and if x becomes larger. Draw diagrams to illustrate these situations. Do the results apply in all cases?

4. Use an area model to illustrate then determine each product.

   a) $3(x + 2)$ **b)** $x(x + 4)$ **c)** $x(2x + 5)$

5. Use an area model to illustrate each polynomial then write it as a product.

   a) $4x + 8$ **b)** $x^2 + 6x$ **c)** $3x^2 + 12x$

The area model suggests an algebraic method for multiplying a monomial by a polynomial: multiply the monomial by each term of the polynomial. Recall that this process is called *expanding*.

Example 1

Expand.

a) $4x^2(2x - 5)$ **b)** $-5a^3(3a - 2a^2 + 1)$

Solution

Use mental math to multiply the coefficients and add the exponents.

a) $4x^2(2x - 5) = 8x^3 - 20x^2$ **b)** $-5a^3(3a - 2a^2 + 1) = -15a^4 + 10a^5 - 5a^3$

Recall that to *factor* a polynomial means to express it as a product. Any product of a monomial and a polynomial can always be factored by reversing the procedure.

The polynomial $10x^2 + 35x$ can be factored in different ways:

$$10x^2 + 35x = 5(2x^2 + 7x) \qquad 10x^2 + 35x = x(10x + 35) \qquad 10x^2 + 35x = 5x(2x + 7)$$

5 is a common factor of $10x^2$ and $35x$. x is a common factor of $10x^2$ and $35x$. $5x$ is the *greatest common factor* of $10x^2$ and $35x$.

We could also write $10x^2 + 35x = 10(x^2 + 3.5x)$. Although this is correct, we will not consider $10(x^2 + 3.5x)$ to be a factored form of $10x^2 + 35x$. When we factor polynomials in this chapter, it will be understood the polynomials have integer coefficients, and the factored forms also have integer coefficients.

When factoring a polynomial, always look for the greatest common factor.

Example 2

Factor.

a) $6x^2 - 9x + 15$ **b)** $12a^2b - 8ab^2$

Solution

Use mental math to find the greatest common factor.

a) $6x^2 - 9x + 15 = 3(2x^2 - 3x + 5)$ **b)** $12a^2b - 8ab^2 = 4ab(3a - 2b)$

Recall that a polynomial with two terms is a *binomial* and a polynomial with three terms is a *trinomial*.

Some expressions have binomials or trinomials as common factors.

Example 3

Factor.   **a)** $3a(4a + 5) - 2(4a + 5)$ **b)** $(x^2 + 3x - 2)x + 5(x^2 + 3x - 2)$

Solution

a) $3a(4a + 5) - 2(4a + 5)$

The binomial $4a + 5$ is the common factor.

$= (4a + 5)(3a - 2)$

b) $(x^2 + 3x - 2)x + 5(x^2 + 3x - 2)$

The trinomial $x^2 + 3x - 2$ is the common factor.

$= (x^2 + 3x - 2)(x + 5)$

Example 4

Expand, then simplify.   **a)** $4(2a - 3b) - 3(a + 5b)$ **b)** $2x(3x^2 - 5xy) - 5y(x^2 - 2y^2)$

Solution

Use mental math to multiply the coefficients and add the exponents.
Then use mental math to combine like terms.

a) $4(2a - 3b) - 3(a + 5b)$

$= 8a - 12b - 3a - 15b$

$= 5a - 27b$

b) $2x(3x^2 - 5xy) - 5y(x^2 - 2y^2)$

$= 6x^3 - 10x^2y - 5x^2y + 10y^3$

$= 6x^3 - 15x^2y + 10y^3$

Example 5

In the 200-m sprint, competitors must stay in their own lanes, which are 1.22 m wide. Since the track is curved, staggered start positions are needed to equalize the distances run. How far apart should the start positions be?

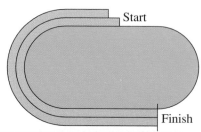

Solution

Assume the curved part of the inside lane is a semicircle with radius r. Then, when going around the semicircle from A to B, the runner in the inside lane runs a distance of πr, and the runner in the next lane runs a distance of $\pi(r + 1.22)$. The difference in these distances is equal to the distance between the start positions. This is:

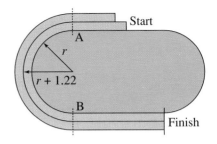

$$\pi(r + 1.22) - \pi r = \pi r + 1.22\pi - \pi r$$
$$= \pi r - \pi r + 1.22\pi$$
$$= 1.22\pi$$
$$\doteq 3.83$$

The start positions should be approximately 3.83 m apart.

DISCUSSING THE IDEAS

1. In *Example 5*, does the answer depend on the radius of the semicircle? Does the answer depend on the width of the lanes? Why would the answers to these questions be useful information for designers of tracks?

2. In *Example 5*, how far apart should the staggered start positions be for the 400-m race? What assumptions do you need to make?

6.4 EXERCISES

1. Expand.

   a) $4(5x^2 + 10)$

   b) $7(2a - 5)$

   c) $-8(3k^2 - 2k)$

   d) $12(2b^2 - 3b + 9)$

   e) $-9(-5m^2 + 7m - 3)$

   f) $3(8p^2 - 5p + 7)$

   g) $12x(5x - 4)$

   h) $3a(-7a + 2)$

   i) $6p(2p - q)$

   j) $-15n^2(6 - 9n)$

   k) $7m^3(3mn + 6)$

   l) $-8x^2(5x + 7y)$

2. Factor.

   a) $5y - 10$

   b) $8m + 24$

   c) $6x + 7x^2$

   d) $35a + 10a^2$

   e) $45d^3 - 36d$

   f) $49b^2 - 7b$

   g) $3x^2 + 6x$

   h) $8y^2 - 4y$

   i) $5p^3 - 15p^2$

   j) $24m^2n + 16mn^2$

   k) $12ab + 18a^2b$

   l) $-28xy^2 - 35x^2y$

B 3. Choose one part of exercise 2. Write to explain how you completed the factoring.

4. Expand, then simplify.

   a) $3(x + 4) + 7$ b) $-8(2a - 3) + 11a$ c) $5(y + 2) - 7y$

   d) $4(7m - 5) - 13$ e) $-6(3p^2 + 2p) + 5p^2$ f) $7(5x - 3y) - 43x$

5. Expand, then simplify.

   a) $3(x + 2) + 2(x - 6)$ b) $2(x + 9) - 3(x + 7)$

   c) $3(2a + 10b - 2c) - 6(a - 2b + 5c)$ d) $3(2m - 4n + 3) - 5(-2m + 5n - 1)$

6. Factor.

   a) $3x^2 + 12x - 6$ b) $3x^2 + 5x^3 + x$ c) $a^3 + 9a^2 - 3a$

   d) $3x^2 + 6x^3 - 12x$ e) $16y^2 - 32y + 24y^3$ f) $8x^2y - 32xy + 16xy^2$

Use the following information to complete exercises 7 to 11.

A packaging company makes boxes with no tops. One style of box is made from a piece of cardboard 20 cm long and 10 cm wide. Equal squares are cut from each corner and the sides are folded up.

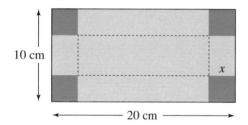
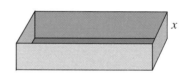

7. Let x centimetres represent the side length of each square cut out. Write a polynomial to represent the surface area of the box.

8. a) What values of x are possible in your polynomial in exercise 7? Visualize, then describe how the size and the shape of the box change as x varies.

   b) Do you think all the boxes would have the same surface area? Visualize, then describe how the surface area changes as x varies.

9. Complete a table like this for at least three different values of x.

Side length of square cut out (cm)	Total surface area (cm²)

10. Compare your results from exercise 9 with those of other students. Are any of the areas close to 150 cm²? What value of x do you think is needed to make a box with a surface area of 150 cm²?

Two other ways to make the box are shown below. In both cases, three rectangular pieces are needed.

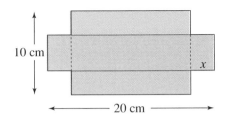

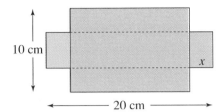

11. For each of these ways:

    a) Write a polynomial for the area of each of the three rectangular pieces.

    b) Use the polynomials in part a to determine a polynomial for the total surface area of the box. Compare the results with exercise 7. What do you notice?

MODELLING the Surface Area of a Box

In exercises 7 to 11, you modelled the surface area of a box.

- Describe the model.
- In what ways might the design of a real box differ from our model?
- How might you modify the model to account for the differences?
- How might the results from your modified model differ from the results you obtained in exercises 7 to 11?

---

12. Factor.

    a) $6b^2 - 3b + 12$ **b)** $5y^3 + 6y^2 + 3y$ **c)** $16x + 32x^2 + 48x^4$

    d) $12y^4 - 12y^3 + 24y^2$ **e)** $9a^3 + 7a^2 + 18a$ **f)** $10z^3 - 15z^2 + 30z$

13. Factor.

    a) $25xy + 15x^2$ **b)** $14m^2n - 21mn^2$ **c)** $9a^2b^3 - 12a^2b^2$

    d) $4x^2y - 16xy^2$ **e)** $12p^2q + 18pq^2$ **f)** $27m^3n - 15m^2n^3$

14. Factor.

    a) $3x(a + b) + 7(a + b)$ **b)** $m(2x - y) - 5(2x - y)$

    c) $(x + 4)x^2 + (x + 4)y^2$ **d)** $5x(a + 3b) - 9y(a + 3b)$

    e) $10y(x - 3) + 7(x - 3)$ **f)** $7w(x + w) - 10(w + x)$

15. Choose one part of exercise 14. Write to explain how you completed the factoring.

16. Factor.

    a) $3x^2(x - 7) + 2x(x - 7) + 5(x - 7)$ **b)** $4m^2(2x + y) - 3m(2x + y) + 7(2x + y)$

    c) $5a^2(x^2 + y) - 7a(x^2 + y) + 8(x^2 + y)$ **d)** $2m(a - b) - 3n(a - b) - 7(a - b)$

    e) $2x^2(3a - 2b) + 5x(3a - 2b) - 9(3a - 2b)$ **f)** $6a(b - a) + 4b(b - a) - 7(b - a)$

17. Expand, then simplify.

    a) $3x^2(x + y) + 2x^2(3x + 5y)$ **b)** $3a^3(2a - 5b) - 4a^3(2a + 3b)$

    c) $5p^2(4p - q) - 8p^2(2p - 7q)$ **d)** $6a^3(-3a + 7b - 4) - 8a^3(2a - 3b + 7)$

    e) $-2ab^2(ab - a^2 + b)$ **f)** $3x^2y(xy^2 + xy - y)$

    g) $-5m^2n(3mn + mn^2 - n^2)$ **h)** $5x(x - y) - 2y(x + y - 1) + y^2$

    i) $2b(b^2 - bc) - 2c(b - c) + (7bc - 4c^2)$ **j)** $7x(x^2 - y^2) - 2xy - 2y(x^2 + y^2)$

18. A tin can has base radius r and height h.

    a) The label of the can is rectangular. What are its dimensions?

    b) Write a formula for the total surface area of the can, including the top and the bottom. Express the formula in two different ways.

    c) Use the results of part b. Determine the dimensions of a rectangle that has the same area as the total surface area of the can. How does this rectangle compare with the rectangle in part a?

19. Determine an expression for the surface area of a closed cylinder when:

    a) The radius is 5 cm and the height is 24 cm.

    b) The height is three times the radius.

    c) The height is 5 cm less than three times the radius.

    d) The height is 7 cm more than half the radius.

20. Choose one part of exercise 19. Write to explain how you found the expression.

21. Expand, then simplify. State the degree of the polynomial.

    a) $3a^2b - 7a(a - 4) + (5ba^2 - 13a)$

    b) $6m(2mn - n^2) - 3(m^2n - 19mn^2) + 6n(-5mn - 7m^2)$

    c) $4xy(x + 2y) - 5x(2xy - 3y^2) - 3y(3x^2 + 7xy)$

    d) $5s(3s^2 - 7s + 2) + 2s(5s^2 + s - 6) + 9s(2s^2 + 6s - 4)$

    e) $4x(5x^2 - 2xy + y^2) - 9x(2x^2 + 6y^2 - xy) + 3x(x^2 - 5xy - y^2)$

    f) $5xy - 3y(2x - y) + 7x(8y - x) + 9(x^2 - 2y^2)$

Express your answers to exercises 22, 23, and 24 in factored form.

22. Determine an expression for the area of each shaded region.

a)
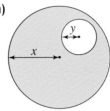

b)
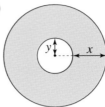

23. Four congruent circular arcs are drawn. Their centres are at the vertices of a square with side length x. Determine an expression for the total area of the figure in terms of x.

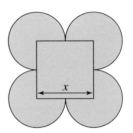

24. A cylindrical silo has a hemispherical top. Determine an expression in terms of r and h to represent each measurement.

a) the volume of the silo

b) the total surface area of the silo, including its base

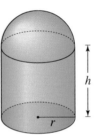

C **25.** Factor.

a) $2m(a - b) - 3n(b - a) - 7(a - b)$

b) $2x^2(3a - 2b) + 5x(3a - 2b) - 9(2b - 3a)$

c) $6a(b - a) + 4b(a - b) - 7(a - b)$

26. Factor.

a) $x^2 + 3x + xy + 3y$ **b)** $x^3 + x^2 + x + 1$ **c)** $5am + a + 10bm + 2b$

d) $3x^2 - 6xy + 5x - 10y$ **e)** $5m^2 + 10mn - 3m - 6n$ **f)** $2a^2 - 6ab - 3a + 9b$

COMMUNICATING THE IDEAS

According to the rules for the order of operations, you must evaluate expressions in brackets first.

Explain why this rule is not violated by writing something like $2(a + 3b) = 2a + 6b$.

Suppose that $a = 5$ and $b = 2$.

Is the rule violated by evaluating the expression as follows?

$$2(a + 3b) = 2(5 + 6)$$
$$= 2 \times 5 + 2 \times 6$$
$$= 10 + 12$$
$$= 22$$

Write your ideas in your journal.

Round Robin Scheduling

Have you heard the term "round robin"? This is a system of scheduling a number of games in sports. Every player or team is matched with every other one. For example, suppose there are 4 teams in a round robin baseball tournament. You can determine the total number of games to be played in two different ways.

Method 1 — *Count the games each team plays.*

Team A plays teams B, C, D.   3 games
This is true for every team.   12 games

But this counts every game twice.
(B against A is the same game as A against B.)

Hence, there must be 6 games.

Method 2 — *Visualize a schedule.*

The schedule has 16 spaces, but 4 of these cannot be used because a team cannot play itself. There are $16 - 4 = 12$ spaces for games. But this counts each game twice. Hence, there must be 6 games.

1. Use both methods. Determine the number of games in a round robin hockey tournament with each number of teams.

   a) 5 teams   **b)** 6 teams

2. Suppose there are *n* teams in a round robin tournament.

   a) Use Method 1. Determine a formula for the number of games.

   b) Use Method 2. Determine a formula for the number of games.

   c) Compare the two formulas. Explain why they are equivalent.

3.

Use graph paper.	or	Use a graphing calculator.
a) Use the formula you found in exercise 2. Make a table showing the number of games in round robin tournaments for up to 10 teams. **b)** Draw a graph. Show how the number of games increases as the number of teams increases.		**a)** Graph the formula you found in exercise 2. Show how the number of games increases as the number of teams increases. **b)** Use the tables feature. Determine the number of games required for up to 10 teams.

Solve each problem below.

4. The designs below were made by marking equally spaced points on a circle, and joining them in all possible ways with line segments.

   a) How many of these line segments are there on each diagram?

   b) How many such line segments would there be on a similar diagram with n equally spaced points on a circle?

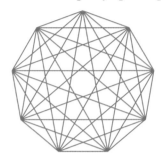 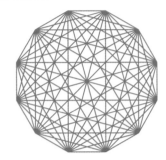

5. a) Count the diagonals in each polygon. Complete the table.

Number of sides	3	4	5	6	7
Number of diagonals					

   b) Determine a formula for the number of diagonals in a polygon with n sides in two ways:

   i) How many diagonals pass through each vertex of the polygon? How many diagonals are there in all?

   ii) How many different line segments (including the sides) can be drawn to join the vertices of the polygon? How many of these are not diagonals of the polygon?

6. A 20-cm pizza serves two people. How many people should a 40-cm pizza serve? Explain.

7. Write down as many second-degree polynomials in x and y as you can find, in which the coefficient of each term is 1, and there is no constant term. How many such polynomials are there?

8. Suppose x has a value between 1 and 5, and y has a value between 5 and 10. What are the possible values for each expression?

   a) $x + y$ b) $x - y$ c) xy d) $\frac{x}{y}$

Area Models of Polynomial Products

INVESTIGATE

1. Use algebra tiles to make a rectangle with length $x + 4$ and width $x + 1$.

2. Determine the area of each small rectangle in the diagram at the right.

   a) Write the area of the rectangle as the product of its length and width.

   b) Write the area of the rectangle as the sum of the areas of the individual rectangles.

   c) Explain why your area model shows that $(x + 4)(x + 1) = x^2 + 5x + 4$.

3. In the diagram above, imagine that x varies, but the lengths marked 1 remain the same. Visualize how the diagram changes if x becomes smaller, and if x becomes larger. Draw diagrams to illustrate these situations. Does the result $(x + 4)(x + 1) = x^2 + 5x + 4$ apply in all cases?

4. Substitute 1, 2, 3, and 4 for x, in turn, in both sides of $(x + 4)(x + 1) = x^2 + 5x + 4$. What patterns can you find in the results?

5. Use an area model to illustrate then determine each product. Record your results.

   a) $(x + 3)(x + 5)$ **b)** $(a + 3)(2a + 1)$ **c)** $(2t + 1)^2$

   d) $(t + 2)(t + v + 1)$ **e)** $(m + 1)(2m + 2n + 3)$ **f)** $(3x + y)(x + y + 3)$

The area model suggests an algebraic method for multiplying binomials: multiply each term of one binomial by each term of the other binomial.

VISUALIZING

When you multiply polynomials, visualize patterns like the ones below. The patterns help you make sure that you include the product of each possible pair of terms.

$$(a + b)(x + y + z) = ax + ay + az + bx + by + bz$$

Alternatively, we can think: $(a + b)(x + y + z) = a(x + y + z) + b(x + y + z)$
$$= ax + ay + az + bx + by + bz$$

Example 1

Expand.

a) $(3x + 2)(x + 7)$ **b)** $(x - 5)^2$

Solution

Use mental math to multiply the coefficients and constant terms, then combine like terms.

a) $(3x + 2)(x + 7) = 3x^2 + 21x + 2x + 14$
$$= 3x^2 + 23x + 14$$

b) $(x - 5)^2 = (x - 5)(x - 5)$
$$= x^2 - 5x - 5x + 25$$
$$= x^2 - 10x + 25$$

Example 2

Expand, then simplify. $(2x - 1)(3x + 4) - (2x - 3)^2$

Solution

Use mental math to multiply the coefficients and constant terms, then combine like terms.

$$(2x - 1)(3x + 4) - (2x - 3)^2 = (6x^2 + 5x - 4) - (4x^2 - 12x + 9)$$
$$= 6x^2 + 5x - 4 - 4x^2 + 12x - 9$$
$$= 2x^2 + 17x - 13$$

Example 3

Write an expression for the area of the shaded region.

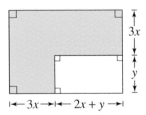

Solution

Shaded area = Area 1 − Area 2
$= (5x + y)(3x + y) - (2x + y)y$
$= 15x^2 + 8xy + y^2 - 2xy - y^2$
$= 15x^2 + 6xy$

Rectangle	°Length	Width	Area	
Large	$5x + y$	$3x + y$	$(5x + y)(3x + y)$	← Area 1
Small	$2x + y$	y	$(2x + y)y$	← Area 2

The shaded region has area $15x^2 + 6xy$.

Example 4

Expand. $(a - 2b)^3$

Solution

Use mental math to multiply the coefficients and constant terms, then combine like terms.

$(a - 2b)^3 = (a - 2b)(a - 2b)(a - 2b)$

$\qquad = (a - 2b)(a^2 - 4ab + 4b^2)$

$\qquad = a^3 - 4a^2b + 4ab^2 - 2a^2b + 8ab^2 - 8b^3$

$\qquad = a^3 - 6a^2b + 12ab^2 - 8b^3$

DISCUSSING THE IDEAS

1. a) Describe how you could check the solution to *Example 1a* by substituting a number for *x*.

   b) Describe a similar check for *Example 1b*.

2. *Example 3* could be solved by adding the areas of two rectangles.

   a) Identify the two rectangles to be added.

   b) Which polynomial products result when using these two rectangles? Solve *Example 3* using these rectangles.

   c) Discuss which method of solution you prefer, and why.

3. In *Example 1b*, the expression $(x - 5)^2$ is called a *binomial square*. Why do you think it has this name?

6.5 EXERCISES

A 1. Draw a rectangle to illustrate each product. Write the area of the rectangle in two different ways.

   a) $(x + 3)(x + 2)$ **b)** $(x + 1)(2x + 6)$ **c)** $(x + y)(x + y + 3)$

   d) $(x + 2)(x + y + 1)$ **e)** $(a + b)(a + b + c)$ **f)** $(3 + t)(2 + t + s)$

2. Expand each binomial product.

a) $(x + 4)(x + 3)$　　　　**b)** $(2a - 4)(a + 2)$　　　　**c)** $(3 - 3b)(2 + b)$

d) $(t - 1)(t - 9)$　　　　**e)** $(x + 3)(2x - 5)$　　　　**f)** $(4 + k)(6 - k)$

g) $(y + 7)(4y - 5)$　　　**h)** $(k - 8)(k - 3)$　　　　**i)** $(m + 6)(m - 5)$

3. Expand each binomial square.

a) $(k + 2)^2$　**b)** $(3m + 3)^2$　**c)** $(a - 6)^2$　**d)** $(2b + 7)^2$

e) $(y - 1)^2$　**f)** $(z + 6)^2$　**g)** $(8 - 4x)^2$　**h)** $(a - 3b)^2$

4. Choose one part of exercise 3. Write to explain how you expanded the binomial square.

5. Here is a multiplication in arithmetic and the corresponding multiplication in algebra:

$$\begin{array}{r} 13 \\ \times\ 27 \end{array} \qquad\qquad \begin{array}{r} x + 3 \\ \times\ 2x + 7 \end{array}$$

a) Compare the two multiplications. In what ways are they similar? In what ways are they different?

b) Complete the two multiplications. In what ways are the steps the same? In what ways are the steps different?

c) How many individual multiplications do you do when you multiply two 2-digit numbers? How many do you do when you multiply two binomials?

6. Expand the products in each list. Describe in writing any patterns you observe.

a) $(x + 2)(x + 3)$　**b)** $(x + 1)(x + 4)$　**c)** $(x + 5)(x + 2)$　**d)** $(x + 6)(x + 3)$
$(x + 2)(x - 3)$　　　$(x + 1)(x - 4)$　　　$(x + 5)(x - 2)$　　　$(x + 6)(x - 3)$
$(x - 2)(x + 3)$　　　$(x - 1)(x + 4)$　　　$(x - 5)(x + 2)$　　　$(x - 6)(x + 3)$
$(x - 2)(x - 3)$　　　$(x - 1)(x - 4)$　　　$(x - 5)(x - 2)$　　　$(x - 6)(x - 3)$

B **7.** Expand the products in each list.

a) $(x + 1)(x + 1)$　　**b)** $(x + 1)(x - 2)$　　**c)** $(x + 2)(x + 1)$
$(x + 1)(x + 2)$　　　　$(x + 2)(x - 2)$　　　　$(x + 2)(x + 2)$
$(x + 1)(x + 3)$　　　　$(x + 3)(x - 2)$　　　　$(x + 2)(x + 3)$
　⋮　　　　　　　　　　⋮　　　　　　　　　　⋮

8. a) Describe in writing any patterns you found in exercise 7. Explain why you think those patterns occur.

b) Predict the next three products for each part of exercise 7.

c) Suppose you extended the lists in exercise 7 upward. Predict the preceding three products in each list.

9. Repeat the process of exercises 7 and 8, using the lists below.

a) $(t + 1)(t - 1)$
$(t + 2)(t - 2)$
$(t + 3)(t - 3)$
$\vdots$

b) $(t + v)(t - v)$
$(t + 2v)(t - 2v)$
$(t + 3v)(t - 3v)$
$\vdots$

10. What is the area of each shaded region? In each case, write to explain how you could calculate the area in two different ways.

a)

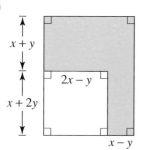

$x + y$

$2x - y$

$x + 2y$

$x - y$

b)

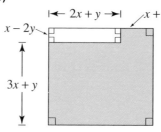

$x - 2y$

$\mid\!\leftarrow 2x + y \rightarrow\!\mid$

$x + y$

$3x + y$

c)

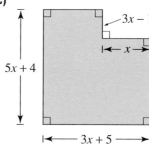

$3x - 1$

x

$5x + 4$

$\mid\!\leftarrow\!\!\longrightarrow\!\!\mid\ 3x + 5$

Use the following information to answer exercises 11 to 15.

A packaging company makes boxes with no tops. One style of box is made from a piece of cardboard 20 cm long and 10 cm wide. Equal squares are cut from each corner and the sides are folded up.

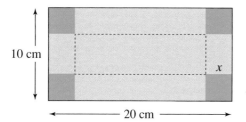

10 cm

x

20 cm

x

11. Let x centimetres represent the side length of each square cut out. Write a polynomial to represent each measurement.

a) the length of the box

b) the width of the box

c) the area of the base of the box

d) the volume of the box

12. What values of x are possible in your polynomials in exercise 11? Visualize how the size and the shape of the box change as x varies. Do you think all the boxes would have the same volume? Explain.

13. Complete a table like this for at least three different values of x.

Side length of square cut out (cm)	Volume (cm³)

14. Compare your results from exercise 13 with those of other students.

a) Are any of the volumes close to 160 cm³? What value(s) of x do you think are needed to make a box with a volume of 160 cm³?

b) What do you think is the maximum volume? What value of x is needed to make a box with the maximum volume?

The company referred to above also makes boxes with tops. One style is made from the same size of cardboard as before, 20 cm long and 10 cm wide. The diagrams below show that there are two different ways to make the box.

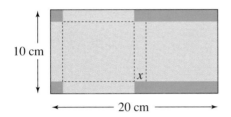 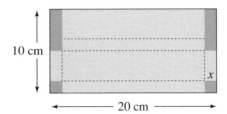

15. Let x centimetres represent the side length of each square cut out. Visualize how each box is formed using the above diagrams as a guide. Then, for each box, write a polynomial to represent each measurement.

a) the length of the box   **b)** the width of the box

c) the area of the base of the box   **d)** the volume of the box

 MODELLING the Volume of a Box

In exercises 11 to 15, you modelled the volume of a box made from a piece of cardboard.

- In what ways might the design of a real box differ from our models?
- How might you modify the models to account for the differences?
- How might the results from your modified models differ from the results you obtained in exercises 11 to 15?

16. Determine each product. Investigate patterns in these products. Describe each pattern in writing. Explain why you think it occurs. Predict the next three lines in each list.

a) $(x + 1)(x + 1)$
$(x + 1)(x^2 + x + 1)$
$(x + 1)(x^3 + x^2 + x + 1)$
$\vdots$

b) $(x + 1)(x - 1)$
$(x + 1)(x^2 - x + 1)$
$(x + 1)(x^3 - x^2 + x - 1)$
$\vdots$

17. Determine each product.

a) $(2x + y)(3x + y)$ **b)** $(3a + 2)^2$ **c)** $(2x - y)(3x + 4y)$

d) $(x + 3y)(x + 4y)$ **e)** $(5m - 2n)(7m - n)$ **f)** $(6x - 2y)(3x - 7y)$

18. Expand each product of polynomials.

   a) $(3x - 1)(2x^2 + 3x - 4)$ **b)** $(n - 3)(4n^2 - 7n + 12)$ **c)** $(2a - 5)(3a^2 + 8a - 9)$

   d) $(3p + 2)(5p^2 - 6p + 2)$ **e)** $(3m + 7)(2m^2 - 3m + 3)$ **f)** $(2y - 3z)(4y^2 - y - 3)$

19. Expand.

   a) $(x + 1)^3$ **b)** $(x + y)^3$ **c)** $(2x + y)^3$ **d)** $(2x + 3)^3$ **e)** $(3x - 2y)^3$

20. A rectangle has length $(2x - 4)$ centimetres and width $(5 - x)$ centimetres.

   a) Write a polynomial for the area of the rectangle.

   b) Each dimension of the rectangle is increased by 1 cm. Determine the increase in area.

   c) The value of x increases by 1 cm. Write a polynomial to represent the change in area.

21. Expand, then simplify.

   a) $(3x + 4)(x - 5)(2x + 8)$ **b)** $(b - 7)(b + 8)(3b - 4)$ **c)** $(2x - 5)(3x + 4)^2$

   d) $(5a - 3)^2(2a - 7)$ **e)** $(5m - 2)^3$ **f)** $(2k - 3)(2k + 3)^2$

C **22.** Determine each product. Investigate patterns in these products. Describe each pattern. Explain why you think it occurs. Predict the next three lines in each list.

   a) $(x + 1)^2$ **b)** $(x - 1)^2$ **c)** $(x + 1)(x - 1)$
   $(x + 1)^3$ $(x - 1)^3$ $(x + 1)^2(x - 1)^2$
   $(x + 1)^4$ $(x - 1)^4$ $(x + 1)^3(x - 1)^3$
   $\vdots$ $\vdots$ $\vdots$

23. An artist has 120 cm of frame for a picture. Let w, l, and A represent its width, length, and area. Write a polynomial in terms of w for each measurement.

   a) the length   **b)** the area

   c) the change in the area when the width is increased by 1 cm

24. Suppose you invest $100 at an annual interest rate of $r\%$, compounded annually.

   a) Write an algebraic expression for the value of your investment after 2 years.

   b) Expand your expression. Explain what each term represents.

COMMUNICATING THE IDEAS

Suppose your friend telephones you to discuss tonight's homework. How would you explain how to multiply two polynomials? How would you explain how to multiply three polynomials? Use examples to illustrate your explanations. Record your answer in your journal.

Graphing the surface area and volume of a box

1. a) Use a graphing calculator. Graph the surface area polynomial you wrote in exercise 7 on page 348. Your graph should look similar to this:

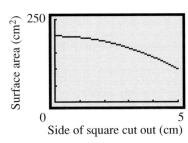

b) Visualize the points on the graph that correspond to the surface areas you calculated in exercise 9 on page 348.

c) Use the trace and zoom features of your calculator. Determine the value of x for a surface area of 150 cm^2.

2. a) Use a graphing calculator. Graph the volume polynomial you wrote in exercise 11d on page 358. Your graph should look similar to this:

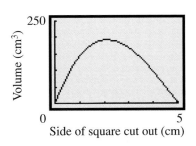

b) Visualize the points on the graph that correspond to the volumes you calculated in exercise 13 on page 358.

c) Use the trace and zoom features of your calculator. Determine the value(s) of x for a volume of 160 cm^3.

d) Use the trace and zoom features. Determine the value of x for maximum volume.

3. a) Use a graphing calculator. Graph the polynomial for the volume of a box from exercise 15d on page 359.

b) Compare your graph in part a with the graph in exercise 2a. Explain the similarities and the differences.

6.6 Factoring Trinomials of the Form $x^2 + bx + c$

Multiplying a pair of binomials often results in a trinomial.

To express a trinomial such as $x^2 + 5x + 6$ as a product, recall how you used algebra tiles to create a rectangular array.

The polynomial is represented by the area of the rectangle: $x^2 + 5x + 6$

The factors of the trinomial are represented by the length $(x + 3)$ and the width $(x + 2)$ of the rectangle.

$x^2 + 5x + 6 = (x + 3)(x + 2)$ — The integers 3 and 2 have a product of 6 and a sum of 5.

Observe how the numbers in the product are related to the coefficients of the trinomial. You can use this pattern to factor other trinomials.

Example 1

Factor.

a) $x^2 + 11x + 24$ **b)** $a^2 + a - 12$ **c)** $m^2 - 6m + 9$

Solution

a)
$x^2 + 11x + 24$ — Use mental math to find which two integers have a product of 24 and a sum of 11.

The integers are 8 and 3.

$x^2 + 11x + 24 = (x + 8)(x + 3)$ ◁ Check by expanding.

b)
$a^2 + a - 12$ — Use mental math to find which two integers have a product of -12 and a sum of 1.

The integers are 4 and -3.

$a^2 + a - 12 = (a + 4)(a - 3)$ ◁ Check by expanding.

c)

$m^2 - 6m + 9$ ——————— Use mental math to find which two integers have a product of 9 and a sum of −6.

The integers are −3 and −3.

$m^2 - 6m + 9 = (m - 3)(m - 3)$ ← Check by expanding.

When the terms of a trinomial have a common factor, remove the common factor first.

Example 2

Factor.

a) $5x^2 - 35x + 60$ **b)** $3x^3 + 9x^2 - 12x$

Solution

Use mental math to find the common factor.

a) $5x^2 - 35x + 60 = 5(x^2 - 7x + 12)$

Find two integers that have a product of 12 and a sum of −7.

$$= 5(x - 4)(x - 3)$$

b) $3x^3 + 9x^2 - 12x = 3x(x^2 + 3x - 4)$

Find two integers that have a product of −4 and a sum of 3.

$$= 3x(x + 4)(x - 1)$$

Example 3

Factor.

a) $x^4 + 11x^2 + 24$ **b)** $2a^4 + 18a^2 + 36$

Solution

Use mental math to find the product and sum of two appropriate integers.

a) $x^4 + 11x^2 + 24$
$= (x^2 + 8)(x^2 + 3)$

b) $2a^4 + 18a^2 + 36$
$= 2(a^4 + 9a^2 + 18)$
$= 2(a^2 + 3)(a^2 + 6)$

Example 4

Factor. $(x + b)^2 + 6(x + b) + 8$

Solution

Replace $x + b$ with a single variable, z.
$(x + b)^2 + 6(x + b) + 8 = z^2 + 6z + 8$

Factor $z^2 + 6z + 8$ in the usual way.
$z^2 + 6z + 8 = (z + 4)(z + 2)$

Replace z with $x + b$ to obtain:
$(x + b)^2 + 6(x + b) + 8 = (x + b + 4)(x + b + 2)$

DISCUSSING THE IDEAS

1. Describe some mental math strategies you can use to find two integers that satisfy each set of conditions.

   a) product of 15, sum of 8   **b)** product of −12, sum of 1
   c) product of 20, sum of 21   **d)** product of −7, sum of −6

2. Compare *Example 3a* with *Example 1a*. In what ways are the trinomials and their factored forms similar? In what ways are they different?

3. How could you check the results in *Examples 3* and *4*?

4. In *Example 1c*, the trinomial $m^2 - 6m + 9$ is called a *perfect square*. Why do you think this name is appropriate?

6.6 EXERCISES

A 1. In each list, factor the trinomials then extend the pattern for three more trinomials.

a) $x^2 + 3x + 2$	**b)** $x^2 + 3x + 2$	**c)** $x^2 + 3x + 2$
$x^2 + 4x + 3$	$x^2 + 4x + 4$	$x^2 + 5x + 6$
$x^2 + 5x + 4$	$x^2 + 5x + 6$	$x^2 + 7x + 12$
$\vdots$	$\vdots$	$\vdots$

2. Make up other lists of trinomials like those in exercise 1, in which there is a pattern in the given trinomials and a pattern in their factors.

3. Factor.

a) $x^2 + 10x + 24$

b) $m^2 + 5m + 6$

c) $a^2 + 10a + 16$

d) $p^2 + 8p + 16$

e) $y^2 + 13y + 42$

f) $d^2 + 4d + 4$

g) $x^2 - 7x + 12$

h) $c^2 - 17c + 72$

i) $a^2 - 7a + 6$

j) $x^2 - 9x + 8$

k) $s^2 - 12s + 20$

l) $x^2 - 14x + 49$

4. Choose one part of exercise 3. Write to explain how you factored the trinomial.

B 5. Factor.

a) $y^2 + 8y - 9$

b) $b^2 + 19b - 20$

c) $p^2 + 15p - 54$

d) $x^2 + 12x - 28$

e) $a^2 + 2a - 24$

f) $k^2 + 3k - 18$

g) $x^2 - 2x - 8$

h) $n^2 - 5n - 24$

i) $a^2 - a - 20$

j) $d^2 - 4d - 45$

k) $m^2 - 9m - 90$

l) $y^2 - 2y - 48$

6. Factor. Check by expanding.

a) $x^2 + 14x + 24$

b) $m^2 - 15m + 50$

c) $a^2 - 11a + 30$

d) $x^2 + 6x + 9$

e) $x^2 + 17x + 72$

f) $a^2 - 12a + 36$

g) $x^2 + 8x - 20$

h) $d^2 - 6d - 16$

i) $b^2 + 7b - 18$

j) $x^4 + x^2 - 20$

k) $a^4 + 14a^2 + 45$

l) $m^4 - 7m^2 - 60$

7. Factor.

a) $p^2 + 15p + 26$

b) $x^2 + 40x - 41$

c) $m^2 - m - 72$

d) $t^2 - 13t + 12$

e) $y^2 + 13y + 36$

f) $m^2 + 4m - 96$

g) $s^2 + 21s - 100$

h) $c^2 - 8c + 16$

i) $p^2 - p - 90$

j) $k^4 + 11k^2 + 30$

k) $r^4 - 10r^2 + 9$

l) $c^4 - 16c^2 - 80$

8. Factor, if possible.

a) $x^2 + 4x + 4$

b) $x^2 + 5x + 5$

c) $x^2 + 5x + 4$

d) $x^2 + 4x + 5$

e) $x^2 + 4x - 5$

f) $x^2 - 5x + 4$

9. Choose one trinomial from exercise 8 that you could not factor. Write to explain why you could not factor the trinomial.

10. a) Evaluate the trinomial $x^2 + 3x - 4$ for $x = 0, 1, 2, 3,$ and 4.

b) Factor the trinomial in part a. Evaluate the factored trinomial for $x = 0, 1, 2, 3,$ and 4.

c) Compare the results of parts a and b. What do you notice?

11. In exercises 1 and 2, suppose you extended the lists upward. Predict the preceding four trinomials in each list, then factor them.

12. a) Factor each trinomial in the two lists below.

$$x^2 + 7x + 12 \qquad x^2 - 7x + 12$$
$$x^2 + 8x + 12 \qquad x^2 - 8x + 12$$
$$x^2 + 13x + 12 \qquad x^2 - 13x + 12$$

b) What are the integral values of k for which the trinomial $x^2 + kx + 12$ can be factored?

13. Determine the integral values of k so that each trinomial can be factored.

a) $x^2 + kx + 10$ **b)** $a^2 + ka - 9$ **c)** $m^2 + km + 8$

d) $y^2 + ky - 12$ **e)** $n^2 + kn + 18$ **f)** $b^2 + kb - 16$

14. For which integral values of k can each trinomial be factored?

a) $x^2 + x + k$ **b)** $x^2 - x + k$ **c)** $x^2 + 2x + k$

d) $x^2 - 2x + k$ **e)** $x^2 + 3x + k$ **f)** $x^2 - 3x + k$

15. Choose one part of exercise 14. Write to explain how you found the values of k.

16. Factor.

a) $2x^2 + 8x + 6$ **b)** $5a^2 + 15a - 20$ **c)** $3y^2 - 12y - 36$

d) $5y^2 - 40y + 80$ **e)** $2m^2 + 2m - 112$ **f)** $9x^2 + 54x + 45$

17. Factor.

a) $4x^2 + 28x + 48$ **b)** $5a^2 - 20a - 60$ **c)** $3n^2 - 27n + 60$

d) $3p^2 + 15p - 108$ **e)** $7x^2 - 84x - 196$ **f)** $3a^2 - 18a + 15$

18. In all the previous examples and exercises in this section, each trinomial contains only one variable, and the terms are written in order of decreasing powers. Use the result of the example at the top of page 362 as a guide. Factor the following trinomials. Check your results by expanding.

a) $x^2 + 5xy + 6y^2$ **b)** $a^2b^2 + 5ab + 6$ **c)** $1 + 5c + 6c^2$

19. Factor.

a) $4y^2 - 20y - 56$ **b)** $3m^2 + 18m + 24$ **c)** $4x^2 + 4x - 48$

d) $10x^2 + 80x + 120$ **e)** $5am^2 - 40am + 35a$ **f)** $7c^2d - 35cd^2 + 42d^3$

20. Factor.

a) $x^4 + 7x^2 + 10$ **b)** $a^4 + 9a^2 + 14$ **c)** $m^4 + 13m^2 + 36$

d) $2b^4 + 16b^2 + 30$ **e)** $3c^4 + 24c^2 + 21$ **f)** $5x^4 + 25x^2 + 30$

21. Factor.

a) $(x + y)^2 + 9(x + y) - 10$ b) $(p - 2q)^2 - 11(p - 2q) + 24$

c) $(3y - 4)^2 - 2(3y - 4) - 63$ d) $(x^2 + 4x)^2 + 8(x^2 + 4x) + 15$

e) $(2m - n)^2 - (2m - n)p - 20p^2$ f) $3(2x + 4)^2 + 12(2x + 4)y - 36y^2$

22. Choose one part of exercise 21. Write to explain how you factored the expression.

23. When $x^2 + bx + c$ is a perfect square, how are b and c related?

C 24. Substitute $x = a + b$ and $y = a - b$. Write each trinomial in terms of a and b, then simplify.

a) $x^2 + 2xy + y^2$ b) $x^2 - 5xy + 6y^2$

c) $x^2 + 4xy - 12y^2$ d) $x^2y^2 - xy - 2$

25. a) Factor each trinomial.

$x^2 + 10x + 24$
$x^2 - 10x + 24$
$x^2 + 10x - 24$
$x^2 - 10x - 24$

b) In part a, the trinomial $x^2 + 10x + 24$ factors, and the trinomials obtained by replacing either or both $+$ signs with $-$ signs also factor. Find other examples like this, in which the four trinomials $x^2 \pm bx \pm c$ can all be factored.

26. a) Factor each trinomial.

$x^2 + 5x + 6$
$x^2 + 6x + 5$

b) In part a, the trinomial $x^2 + 5x + 6$ can be factored, and if the 5 and 6 are interchanged, the resulting trinomial can also be factored. Find other examples like this in which a trinomial of the form $x^2 + bx + c$ can be factored, and the trinomial formed by interchanging b and c can also be factored.

COMMUNICATING THE IDEAS

Think about the different factoring strategies you have learned so far in this chapter. In your journal, write a description of the steps you would use to factor $2a^2 - 10a + 8$.

Verifying Trinomial Factorizations

1. Use a graphing calculator or a computer graphing program.

   a) Display the graph of $y = x^2 + 5x + 4$. Have your partner display the graph of $y = (x + 4)(x + 1)$.

   b) Compare your displays. Explain why your results confirm that $x^2 + 5x + 4 = (x + 4)(x + 1)$.

   c) Use the tables feature of your calculator. Display a table of values for $y = x^2 + 5x + 4$. Have your partner display a table of values for $y = (x + 4)(x + 1)$. Compare your tables.

   d) Use your calculator to display the graphs and tables of values for other trinomials and their factored forms. Select two or more examples from 6.6 Exercises.

2. The spreadsheet started below will display the values of $x^2 - 7x + 10$ and its factored form, $(x - 2)(x - 5)$, for several values of the variable x.

	A	B	C
1	x	x^2 - 7x + 10	(x - 2)(x - 5)
2	1	=A2^2-7*A2+10	=(A2-2)*(A2-5)
3	=A2+1	=A3^2-7*A3+10	=(A3-2)*(A3-5)

   a) Start a new spreadsheet document. Enter the text and formulas above.

   b) Explain the purpose of the formulas in cells B2, C2, and A3.

   c) Extend the spreadsheet down to 10 values of x, by selecting a block of cells starting with row 3, then using the Fill Down option.

   d) Examine the numbers in columns B and C. Try different starting values of x. Examine columns B and C again.

   e) Explain why your spreadsheet confirms $x^2 - 7x + 10 = (x - 2)(x - 5)$.

   f) Use the graphing feature of your spreadsheet program. Display the graph of $y = x^2 - 7x + 10$. Then display the graph of $y = (x - 2)(x - 5)$.

   g) Modify the spreadsheet to display tables of values and graphs for other trinomials and their factored forms. Select two or more examples from 6.6 Exercises.

3. a) Could you use a graphing calculator or a spreadsheet to verify the factorization of any trinomial? Explain.

   b) Could you use a graphing calculator or a spreadsheet to factor a trinomial? Explain.

INVESTIGATE ## Algebra Tile Models of Trinomials

In Section 6.6, you factored trinomials where the x^2 term had coefficient 1. However, other trinomials might result from a product of binomials. One example is $3x^2 + 17x + 10$, which results from $(3x + 2)(x + 5)$.

Work with a partner.

1. Select algebra tiles to represent $2x^2 + 7x + 3$. Arrange the tiles in a rectangle. Explain why the arrangement shows that
   $2x^2 + 7x + 3 = (2x + 1)(x + 3)$.

2. Use algebra tiles to factor each trinomial. Sketch each result.

   a) $2x^2 + 5x + 3$ b) $3x^2 + 5x + 2$

   c) $4x^2 + 12x + 9$ d) $4x^2 + 11x + 6$

3. Examine your results from exercise 2. Describe any patterns you could apply to factor a trinomial algebraically.

4. Use algebra tiles to factor the trinomial $2x^2 + 6x + 4$. In how many different ways can you do this? Why can you factor this trinomial in more than one way, when there is only one way to factor each trinomial in exercises 1 and 2?

You can learn how to factor a trinomial of the form $ax^2 + bx + c$ by examining how a trinomial is formed from the product of two binomials:

$$(3x + 2)(x + 5) = 3x(x + 5) + 2(x + 5)$$
$$= 3x^2 + 15x + 2x + 10$$

> 15 and 2 have a sum of 17 and a product of 30, which is the same as the product of 3 and 10.

$$= 3x^2 + 17x + 10$$

To factor a trinomial of the form $ax^2 + bx + c$, look for two integers with a sum of b and a product of ac.

Example 1

Factor. $3x^2 - 10x + 8$

Solution

$$3x^2 - 10x + 8$$

> Use mental math to find which two integers have a product of 24 and a sum of −10.

The integers are −6 and −4. The trinomial can be factored after writing the second term as $-6x - 4x$:

$$3x^2 - 10x + 8 = 3x^2 - 6x - 4x + 8$$
$$= 3x(x - 2) - 4(x - 2)$$
$$= (x - 2)(3x - 4)$$

> Check by expanding.

Example 2

Factor. $8a^2 + 18a - 5$

Solution

$$8a^2 + 18a - 5$$

> Use mental math to find which two integers have a product of −40 and a sum of 18.

$$8a^2 + 18a - 5 = 8a^2 + 20a - 2a - 5$$
$$= 4a(2a + 5) - 1(2a + 5)$$
$$= (2a + 5)(4a - 1)$$

> Check by expanding.

Example 3

Factor. $6a^4 + 7a^2 - 10$

Solution

$$6a^4 + 7a^2 - 10$$

> Use mental math to find which two integers have a product of −60 and a sum of 7.

$$6a^4 + 7a^2 - 10 = 6a^4 + 12a^2 - 5a^2 - 10$$
$$= 6a^2(a^2 + 2) - 5(a^2 + 2)$$
$$= (a^2 + 2)(6a^2 - 5)$$

> Check by expanding.

As always, look for a common monomial factor before factoring any trinomial.

Example 4

Factor. $10y^2 - 22y + 4$

Solution

Use mental math to find the common factor, then the product and sum of two appropriate integers.

$$\begin{aligned} 10y^2 - 22y + 4 &= 2(5y^2 - 11y + 2) \\ &= 2(5y^2 - 10y - y + 2) \\ &= 2[(5y(y - 2) - 1(y - 2)] \\ &= 2(y - 2)(5y - 1) \end{aligned}$$

Check by expanding.

DISCUSSING THE IDEAS

1. The method of *Example 1* should work regardless of the order used to rewrite the second term. Complete *Example 1* using $-4x - 6x$ to replace $-10x$. Discuss your results.

2. In *Example 4*, suppose we had not noticed that 2 is a common factor. Factor the trinomial $10y^2 - 22y + 4$ directly by finding two integers with a product of 40 and a sum of -22. Compare the result with the solution for *Example 4*. Are the results the same?

6.7 EXERCISES

A 1. In each list, factor the trinomials, and extend the pattern for three more trinomials.

   a) $2x^2 + 5x + 2$
   $2x^2 + 7x + 3$
   $2x^2 + 9x + 4$
   $\vdots$

   b) $2x^2 + 5x + 2$
   $2x^2 + 7x + 6$
   $2x^2 + 9x + 10$
   $\vdots$

   c) $2x^2 + 5x + 2$
   $3x^2 + 7x + 2$
   $4x^2 + 9x + 2$
   $\vdots$

2. Make up other lists of trinomials like those in exercise 1, in which there is a pattern in the trinomials and in the factored results.

3. Factor. Use algebra tiles if you wish.

   a) $2x^2 + 7x + 6$
   b) $2a^2 + 7a + 6$
   c) $2d^2 + 3d + 1$
   d) $3s^2 + 4s + 1$
   e) $6y^2 + 11y + 3$
   f) $8x^4 + 10x^2 + 3$

B 4. Factor. Check by expanding.

   a) $5x^2 - 7x + 2$
   b) $3n^2 - 11n + 6$
   c) $14c^2 - 13c + 3$
   d) $2x^2 - 11x + 15$
   e) $3x^2 - 22x + 7$
   f) $4a^4 - 4a^2 + 1$

5. Factor and check.

   a) $3t^2 + 7t - 6$ b) $6k^2 + 5k - 4$ c) $8r^2 + 2r - 3$

   d) $4m^2 + 3m - 10$ e) $5y^2 + 19y - 4$ f) $4d^4 + 4d^2 - 15$

   g) $5a^2 - 7a - 6$ h) $3x^2 - 13x - 10$ i) $2m^2 - m - 21$

   j) $4k^2 - 9k - 9$ k) $6x^2 - x - 12$ l) $15a^4 - a^2 - 2$

6. Examine your results for each of exercises 3, 4, and 5. Write to describe how the trinomials and their factored forms in each exercise are similar.

7. Factor.

   a) $3x^2 + 13x + 4$ b) $2m^2 - 11m + 12$ c) $4s^2 - 20s + 25$

   d) $5x^2 + 15x + 10$ e) $6a^2 + 17ab + 12b^2$ f) $8x^4 - 14x^2y^2 + 3y^4$

   g) $10x^2 - x - 3$ h) $6k^2 + k - 5$ i) $15g^2 - 7g - 2$

   j) $6x^2 - 9x + 3$ k) $8c^2 + 18cd - 5d^2$ l) $15x^2 - 4xy - 4y^2$

8. a) Factor $4x^2 + 20x + 25$. Compare the two factors.

   b) For the product in part a, how are the coefficients of the terms of the factors related to the coefficients of the terms of the trinomial? What makes this product special?

   c) Is the pattern you found in part b just a coincidence, or is there a relationship? Use some examples to explain your answer in writing.

9. Factor.

   a) $4x^2 - 4x + 1$ b) $2h^2 + 5h + 2$ c) $9q^2 - 30q + 25$

   d) $10u^4 - 29u^2 + 10$ e) $10m^2 - 17mn - 6n^2$ f) $8 + 14d - 15d^2$

10. Factor, if possible.

   a) $16h^2 - 24h + 9$ b) $10r^4 + 13r^2 - 3$ c) $2w^2 + 13w + 15$

   d) $9t^2 + 12t - 4$ e) $10x^2 - 33xy - 7y^2$ f) $9a^2 - 24ab + 16b^2$

11. In each list, factor the trinomials then extend the pattern for three more trinomials.

   a) $2x^2 + x - 1$ b) $2x^2 + 3x - 2$ c) $4x^2 + 4x + 1$
      $2x^2 + 3x - 2$ $3x^2 + 4x - 4$ $9x^2 + 12x + 4$
      $2x^2 + 5x - 3$ $4x^2 + 5x - 6$ $16x^2 + 24x + 9$
      $2x^2 + 7x - 4$ $5x^2 + 6x - 8$ $25x^2 + 40x + 16$

      $\vdots$ $\vdots$ $\vdots$

12. Make up another list of trinomials like those in exercise 11, in which there is a pattern in the trinomials and in the factored results.

13. In exercises 1, 2, and 11, suppose you extended the lists upward. Predict the preceding four trinomials in each list, and factor them.

14. a) Factor each trinomial in the two lists below.

$$5x^2 + 11x + 6; \ 5x^2 + 13x + 6; \ 5x^2 + 17x + 6; \ 5x^2 + 31x + 6$$
$$5x^2 - 11x + 6; \ 5x^2 - 13x + 6; \ 5x^2 - 17x + 6; \ 5x^2 - 31x + 6$$

b) For which integral values of k can the trinomial $5x^2 + kx + 6$ be factored?

c) In part a, all the trinomials begin with $5x^2$ and end with $+6$. Make up other examples like these, in which several trinomials with two coefficients in common can be factored.

15. For which integral values of k can each trinomial be factored?

a) $4x^2 + kx + 3$ **b)** $4x^2 + kx + 25$ **c)** $6x^2 + kx - 9$

d) $12x^2 + kx + 10$ **e)** $9x^2 - kx + 1$ **f)** $kx^2 + 2x + k$

16. Factor.

a) $20x^2 + 70x + 60$ **b)** $15a^2 - 65a + 20$ **c)** $18a^2 + 15a - 18$

d) $16 - 12r + 10r^2$ **e)** $24x^2 - 72xy + 54y^2$ **f)** $12a^4 - 52a^2b^2 - 40b^4$

Ⓒ **17.** Check that when you factor $6x^2 + 13x + 6$, the result is $(2x + 3)(3x + 2)$. The coefficients in the original trinomial form a symmetrical pattern: 6, 13, 6. Also, the coefficients are reversed in the two binomial factors: 2, 3 in the first factor and 3, 2 in the second factor. Find out if this always happens. That is, when a trinomial whose coefficients form a symmetrical pattern is factored, will the coefficients always be reversed in the two binomial factors?

18. Factor, if possible.

a) $3xy^2 - 22xy + 6x$ **b)** $3m^2n - 13mn + 12n$ **c)** $4x^2y - 17xy - 15y$

d) $2x^3y + 7x^2y - 15xy$ **e)** $2m^3n - m^2n - 21mn$ **f)** $6x^3y - 7x^2y^2 - 3xy^3$

19. Factor, if possible.

a) $6x^3 + 33x^2 + 45x$ **b)** $6a^3 + 26a^2 - 20a$ **c)** $18x^2y - 3xy^2 - 45y$

d) $10m^3 - 25m^2 - 60m$ **e)** $9a^3 - 39a^2 + 42a$ **f)** $42ab^2 + 49ab - 28a$

20. Factor.

a) $32x^2 - 20x + 3$ **b)** $24s^2 - 13s - 2$ **c)** $4a^2 + 19a + 21$

d) $4x^2 - 21xy - 18y^2$ **e)** $10a^2 - 19ab - 15b^2$ **f)** $21x^2 + 25xy - 4y^2$

g) $21x^2 + 17x - 30$ **h)** $72x^2 + 11x - 6$ **i)** $15x^2 - 28x - 32$

21. When $ax^2 + bx + c$ is a perfect square, how are a, b, and c related?

COMMUNICATING THE IDEAS

Is it possible to make two different rectangular arrays using the same set of algebra tiles? In your journal, draw diagrams, then explain your answer and its significance.

INVESTIGATE **Removing a Square from a Square**

1. From a piece of stiff paper or cardboard, cut out a square. Cut a smaller square from one corner. Cut the L-shaped piece that remains into two congruent parts.

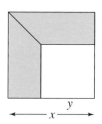

2. Let x represent the length of a side of the original square. Let y represent the length of the side of the square cut out. Write an expression for the area covered by the two congruent parts when they are arranged in each of the two ways shown. What do you notice about the results?

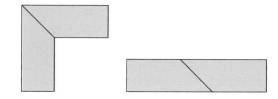

3. **a)** Factor the binomial $a^2 - 25$. Compare the two factors.

   b) For the product in part a, how are the coefficients of the terms of the factors related to the coefficients of the terms of the binomial? What makes this product special?

   c) Is the pattern you found in part b just a coincidence, or is there an underlying relationship? Justify your answer with an explanation or with more examples.

A polynomial that can be expressed in the form $x^2 - y^2$ is called a *difference of squares*.

A difference of squares results when you multiply two binomials that are the sum and the difference of the same two quantities.

$$(x + y)(x - y) = x^2 - xy + yx - y^2$$
$$= x^2 - y^2$$

| Product of the sum and the difference of two quantities x and y | Difference of the squares of x and y |

Using this pattern, you can always factor a difference of squares.

Example 1

Factor.

a) $36x^2 - 49$
b) $16m^2 - 121n^2$

Solution

Use mental math to write each monomial as a square.

a) $\quad 36x^2 - 49$
$\quad = (6x)^2 - (7)^2$
$\quad = (6x + 7)(6x - 7)$

b) $\quad 16m^2 - 121n^2$
$\quad = (4m)^2 - (11n)^2$
$\quad = (4m + 11n)(4m - 11n)$

A difference of squares may occur after a common factor is removed.

Example 2

Factor.

a) $8m^2 - 2n^2$
b) $3a^3 - 12ab^2$

Solution

Use mental math to find the common factor, then to write each monomial as a square.

a) $\quad 8m^2 - 2n^2$
$\quad = 2(4m^2 - n^2)$
$\quad = 2(2m - n)(2m + n)$

b) $\quad 3a^3 - 12ab^2$
$\quad = 3a(a^2 - 4b^2)$
$\quad = 3a(a + 2b)(a - 2b)$

Sometimes one of the factors is a difference of squares. An additional step is required to factor the expression fully.

Example 3

Factor. $81a^4 - 16$

Solution

$81a^4 - 16 = (9a^2 + 4)(9a^2 - 4)$
$\quad\quad\quad\quad = (9a^2 + 4)(3a + 2)(3a - 2)$

Sometimes a difference of squares may have terms that are not monomials.

Example 4

Factor.

a) $(x + 3)^2 - y^2$ **b)** $x^4 - (2x - 1)^2$

Solution

a) $(x + 3)^2 - y^2$
     $= [(x + 3) + y][(x + 3) - y]$
     $= (x + 3 + y)(x + 3 - y)$

b) $x^4 - (2x - 1)^2$
     $= [x^2 - (2x - 1)][(x^2 + (2x - 1)]$
     $= (x^2 - 2x + 1)(x^2 + 2x - 1)$
     $= (x - 1)^2(x^2 + 2x - 1)$

DISCUSSING THE IDEAS

1. In *Investigate*, visualize how the shapes of the cardboard pieces change as x remains fixed and y varies, or as y remains fixed and x varies. Does the difference of squares pattern, $(x + y)(x - y) = x^2 - y^2$, depend on the values of x and y, or is it true for all values of x and y? Explain.

2. Explain the meaning of *difference of squares* if the "squares" are geometric figures, and if the squares are numbers or algebraic expressions.

3. Explain why there are only two terms in the expansion of a product of the form $(x + y)(x - y)$.

4. Can a sum of two squares be factored? Explain your answer.

6.8 EXERCISES

Ⓐ **1. a)** Choose any two natural numbers that differ by 2. Multiply the numbers. Calculate the mean of the numbers with which you started, then square it. How do the two results compare?

    b) Try this with other natural numbers. Based on these results, state a probable conclusion.

2. a) In exercise 1, suppose the two chosen numbers that differ by 2 are not natural numbers. Investigate whether the conclusion in exercise 1b is still true.

    b) For what numbers is the conclusion true? Explain your answer in writing.

3. Factor.

a) $x^2 - 49$ b) $4b^2 - 121$ c) $9m^2 - 64$ d) $81f^2 - 16$

e) $25y^2 - 144$ f) $49x^2 - 36$ g) $16 - 81y^2$ h) $169 - 16t^2$

i) $100m^2 - 49$ j) $64b^2 - 1$ k) $121a^2 - 400$ l) $36b^2 - 25$

m) $25p^2 - 81$ n) $144m^2 - 49$ o) $36 - 121x^2$ p) $1 - 25q^2$

B   **4.** Factor.

a) $4s^2 - 9t^2$ b) $16x^2 - 49y^2$ c) $81a^2 - 64b^2$ d) $121c^2 - 100d^2$

e) $p^2 - 36q^2$ f) $144y^2 - 81z^2$ g) $25m^2 - 169n^2$ h) $4e^2 - 225f^2$

i) $m^4 - 1$ j) $x^4 - 16$ k) $1 - 16y^4$ l) $16a^4 - 81b^4$

5. Factor.

a) $8m^2 - 72$ b) $6x^2 - 150$ c) $20x^2 - 5y^2$ d) $18b^2 - 128$

e) $12a^2 - 75$ f) $18p^2 - 98$ g) $80s^2 - 405$ h) $12p^2 - 363$

i) $12x^3 - 27x$ j) $32m^3 - 98m$ k) $63a^2b - 28b$ l) $75s^2t^2 - 27t^2$

m) $(x - y)^2 - z^2$ n) $(2a + b)^2 - 81$ o) $81a^2 - (3a + b)^2$ p) $4(2x - y)^2 - 25z^2$

6. Choose one part of exercise 5. Write to explain how you factored the expression.

7. Factor.

a) $(x + 2)^2 - (x + 7)^2$ b) $(5m - 2)^2 - (3m - 4)^2$ c) $(2a + 3)^2 - (2a - 3)^2$

d) $(3y + 8z)^2 - (3y - 8z)^2$ e) $(3p - 7)^2 - (8p + 2)^2$ f) $(2x - 1)^2 - (7x + 4)^2$

g) $x^4 - 13x^2 + 36$ h) $a^4 - 17a^2 + 16$ i) $y^4 - 5y^2 - 36$

8. a) Factor each trinomial in the list below. Find as many other polynomials that belong in the list as you can.

$x^2 + 5xy - 36y^2$
$x^2 + 9xy - 36y^2$
$x^2 + 16xy - 36y^2$

b) What are the integral values of k for which the polynomial $x^2 + kxy - 36y^2$ can be factored?

c) In part a, all the polynomials begin with a perfect square, x^2, and end with the opposite of a perfect square, $-36y^2$. Make up another list of examples like these. Each polynomial in your list should begin with a perfect square and end with the opposite of a perfect square.

9. Find some examples of consecutive integers a, b, and c such that the trinomial $ax^2 + bx + c$ can be factored. How many possibilities can you find?

10. A circular fountain is 150 cm in diameter. It is surrounded by a circular flower bed 325 cm in diameter. Calculate the area of the flower bed.

11. The formula for the volume, V, of a square-based pyramid is $V = \frac{1}{3}l^2h$, where l is the length of a side of the base and h is the height.

   a) Express V in terms of s and h, where s is the slant height (that is, the height measured from the top of the pyramid to the midpoint of a side of the base). Write this formula in two different ways.

   b) Calculate the volume of the Great Pyramid of Khufu (Cheops), for which $h = 146.6$ m and $s = 186.4$ m.

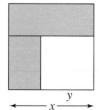 **MODELLING the Great Pyramid**

The ancient engineers built the Great Pyramid with incredible accuracy. Although the four sides are approximately 228.6 m long, they differ in length by less than 20 cm.

- Suppose the dimensions given in exercise 11 differ from the true dimensions by 20 cm. Carry out calculations to estimate the difference this could make to the volume.

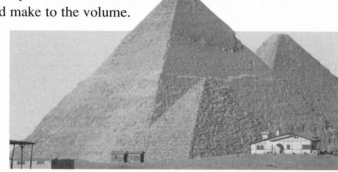

- You may have noticed from photographs that a tiny portion is missing from the top. The height given in exercise 11 is the original height, but it is now 9.1 m shorter. Calculate the volume of the missing portion.

- You may also have noticed from photographs that the surface of the Great Pyramid is very rough. Over the centuries, the original smooth limestone surface has been removed. Carry out calculations to estimate the volume of limestone that may have been removed.

- Combine the above results to estimate the difference between your calculated volume in exercise 11 and the actual volume. Express this difference as a percent.

12. In *Investigate* on page 374, you could have cut the L-shaped piece into parts in other ways, as shown below. For each way, determine expressions for the areas of the rectangles indicated, and verify that their sum can be written in the form $x^2 - y^2$.

   a) divided into two rectangles

   b) divided into three rectangles

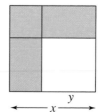

13. In *Investigate*, you could have cut the smaller square from the middle of the larger square, and formed four congruent trapezoids, as shown. Determine an expression for the area of each trapezoid. Verify that the sum of the areas can be written in the form $x^2 - y^2$.

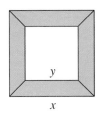

C 14. Factor, if possible.

a) $8d^2 - 32e^2$

b) $25m^2 - \frac{1}{4}n^2$

c) $18x^2y^2 - 50y^4$

d) $10a^2 - 7b^2$

e) $25s^2 + 49t^2$

f) $p^2 - \frac{1}{9}q^2$

g) $5x^4 - 80$

h) $\frac{x^2}{16} - \frac{y^2}{49}$

15. Choose one part of exercise 14 that could not be factored. Write to explain why it could not be factored.

16. a) Find as many prime numbers, p, as you can so that $5p + 1$ is a perfect square.

b) How many prime numbers like this do you think there are?

c) Prove that your answer in part b is correct.

17. A metal strip has length l. It is secured at both ends to fixed points. When the strip is heated, its length increases by a factor x. This causes the strip to buckle.

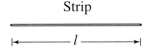

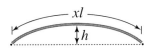

a) Determine an expression for the approximate distance, h, through which the middle point of the strip moves.

b) Estimate h to two decimal places for each strip in the table.

Metal	Length, l	Expansion factor, x
Steel	250 cm	1.0012
Brass	250 cm	1.0020
Aluminum	250 cm	1.0024

c) Would the actual distance h be greater than, less than, or the same distance as your estimate? Explain your answer.

COMMUNICATING THE IDEAS

How would you explain to someone, over the telephone, how to recognize a difference of squares, and how to factor a difference of squares? Record your answer in your journal.

6.9 | Solving Quadratic Equations

In the equations you have studied until now, the variable has never been squared. Examples of equations in which the variable is squared are:

$$x^2 + 3x - 28 = 0 \qquad 3x^2 + 10x = 8 \qquad 4x^2 = 25$$

These are examples of *quadratic equations*.

In quadratic equations such as $4x^2 = 25$, there is no first-degree term. These equations can be solved by isolating the variable, then taking the square root of both sides. You can use mental math to find the square root of a perfect square.

Example 1

Solve.

a) $4x^2 = 25$ **b)** $9a^2 - 2 = 8$

Solution

a) $4x^2 = 25$

$$x^2 = \frac{25}{4}$$

$$x = \pm\sqrt{\frac{25}{4}}$$

Since $\left(\frac{5}{2}\right)^2 = \frac{25}{4}$ and $\left(\frac{-5}{2}\right)^2 = \frac{25}{4}$,

x could be $\frac{5}{2}$ or $-\frac{5}{2}$.

We write $x = \pm\frac{5}{2}$, or ± 2.5.

b) $9a^2 - 2 = 8$

$$9a^2 = 10$$

$$a^2 = \frac{10}{9}$$

$$a = \pm\sqrt{\frac{10}{9}}$$

$$= \frac{\sqrt{10}}{3} \text{ or } -\frac{\sqrt{10}}{3}$$

We write $a = \pm\frac{\sqrt{10}}{3}$

$$\doteq \pm 1.054$$

Many quadratic equations have first-degree terms. You can solve some of these equations by factoring. The solution by factoring depends on the following important property:

> If two numbers have a product of 0, then one or both of them must be 0.
>
> That is, if $m \times n = 0$, then either $m = 0$, or $n = 0$, or both
>
> This is also true for algebraic expressions.
>
> If $(a + b)(c + d) = 0$, then either $a + b = 0$, or $c + d = 0$, or both

Example 2

Solve and check.

a) $x^2 - x - 6 = 0$ **b)** $x^2 + 10x + 25 = 0$

Solution

a) $x^2 - x - 6 = 0$

Use mental math to factor.

$(x - 3)(x + 2) = 0$

Either $x - 3 = 0$ or $x + 2 = 0$
          $x = 3$ $x = -2$

b) $x^2 + 10x + 25 = 0$

Use mental math to factor.

$(x + 5)(x + 5) = 0$

That is, $x + 5 = 0$
                $x = -5$

Check: Always check by substituting in the original equation.
Then use mental math for the indicated operations.

a) If $x = 3$,   If $x = -2$,

$\ x^2 - x - 6$ $\ x^2 - x - 6$
$= 3^2 - 3 - 6$ $= (-2)^2 - (-2) - 6$
$= 9 - 3 - 6$ $= 4 + 2 - 6$
$= 0$ $= 0$

Both solutions are correct.

b) If $x = -5$,

$\ x^2 + 10x + 25$
$= (-5)^2 + 10(-5) + 25$
$= 25 - 50 + 25$
$= 0$

The solution is correct.

Quadratic equations often arise in problems involving projectiles. When an object is projected into the air, its speed changes. The speed decreases as the object goes up, then increases as it falls to the ground.

Example 3

When a football is kicked with a vertical speed of 20 m/s, its height, h metres, after t seconds is given by the formula:

$$h = 20t - 5t^2$$

How long after the kick is the football at a height of 15 m?

Solution

Substitute 15 for h in the formula:

$$15 = 20t - 5t^2$$

This produces a quadratic equation. The solution of the equation gives the times when the height is 15 m.

Collect all the terms on one side of the equation.

$5t^2 - 20t + 15 = 0$
$5(t^2 - 4t + 3) = 0$ ⟵ Removing 5 as a common factor
$t^2 - 4t + 3 = 0$
$(t - 1)(t - 3) = 0$
Either $t - 1 = 0$ or $t - 3 = 0$
$t = 1$ $t = 3$

The football is at a height of 15 m twice: first on the way up, 1 s after the kick; then on the way down, 3 s after the kick.

MODELLING the Height of a Projectile

In *Example 3*, the height of a football is modelled using the formula $h = 20t - 5t^2$. This formula is not realistic because it has been constructed to lead to a quadratic equation in which the trinomial can be factored. We can improve the model by making some changes to the equation to make it more realistic. From physics, we know that if an object is projected vertically upwards, its height, h metres, after time t seconds is given by the formula:

$$h = ut - 0.5gt^2$$

where u is the initial speed and g is the acceleration due to gravity. On Earth, the acceleration due to gravity is $g \doteq 9.8$ m/s^2.

- Substitute 19.2 for u and 9.8 for g in this formula to obtain a formula that gives the height of a football after t seconds when it is kicked with a vertical speed of 19.2 m/s.

- Substitute 15 for h to obtain a quadratic equation whose solution gives the time after the kick when the football is at a height of 15 m.

- Suggest some strategies you could use to solve the equation. Use one of these strategies. Attempt to solve the equation to the nearest tenth of a second.

- Suggest some reasons why the actual height of the football might differ from the height predicted by the model.

Some problems can be solved by using the information given to write a quadratic equation.

Example 4

The ones digit of a two-digit number is 1 less than the tens digit. The sum of the squares of the digits is 85. Find the number.

Solution

Let n represent the tens digit.

Use the fact that the ones digit is 1 less than the tens digit to write an expression for the ones digit in terms of n. The ones digit is $n - 1$.

Use the fact that the sum of the squares of the digits is 85 to write an equation.

$$n^2 + (n - 1)^2 = 85$$
$$n^2 + n^2 - 2n + 1 = 85$$
$$2n^2 - 2n - 84 = 0 \quad\quad \boxed{\text{Divide each term by 2.}}$$
$$n^2 - n - 42 = 0 \quad\quad \boxed{\text{Factor.}}$$
$$(n - 7)(n + 6) = 0$$
$$\text{Either } n - 7 = 0 \quad \text{or} \quad n + 6 = 0$$
$$n = 7 \quad\quad\quad\quad n = -6$$

Since n represents the tens digit in a two-digit number, the negative value for n is not permissible.

So, the tens digit is 7 and the ones digit is 1 less, so it is 6. The two-digit number is 76.

DISCUSSING THE IDEAS

1. The numbers that satisfy an equation are called its *roots*.

   a) Why do you think some quadratic equations have two different roots?

   b) Do all quadratic equations have two different roots?

2. In the solution of *Example 3*, explain why the equation $5(t^2 - 4t + 3) = 0$ was written as $t^2 - 4t + 3 = 0$.

3. In *Example 3*, does it matter if we use x instead of t for the variable? Explain.

4. In *Example 4*, how could we find the number without using a quadratic equation? Explain.

6.9 EXERCISES

A **1.** Solve.

a) $x^2 - 2 = 7$ b) $3x^2 = 75$ c) $2x^2 - 3 = 5$ d) $4p^2 - 5 = 11$

e) $3t^2 + 7 = 10$ f) $2a^2 = 12$ g) $2n^2 - 49 = n^2$ h) $8b^2 = 49 + b^2$

B **2.** Solve and check.

a) $x^2 + 8x + 15 = 0$ b) $x^2 - 7x + 12 = 0$ c) $x^2 - x - 20 = 0$

d) $x^2 + 5x - 24 = 0$ e) $x^2 + 8x + 12 = 0$ f) $x^2 - 5x - 36 = 0$

g) $x^2 - 10x + 24 = 0$ h) $x^2 + 15x + 56 = 0$ i) $x^2 - x - 42 = 0$

3. Determine the roots of each equation.

a) $x^2 - 9x + 25 = 5$ b) $x^2 - 16x + 50 = -13$ c) $x^2 - 6x - 20 = -4$

d) $x^2 + 10x + 25 = 4$ e) $x^2 - 5x - 20 = -6$ f) $x^2 + 6x - 15 = 4x$

g) $x^2 - 5x + 16 = 3x$ h) $x^2 - 10x + 16 = 4 - 2x$ i) $x^2 - 8x - 40 = 4 - x$

4. Two numbers differ by 6. The sum of their squares is 90. Find the numbers.

5. The height, h metres, of an infield fly ball t seconds after being hit is given by the simplified formula $h = 30t - 5t^2$. How long after being hit is the ball at a height of 25 m?

6. A baseball is hit with a vertical speed of 31.3 m/s. Use the formula in the modelling box following *Example 3*. How long after being hit is the ball at a height of 25 m? Give your answer to 1 decimal place.

7. A packaging company makes tin cans with no tops. One type of can has a height of 4.0 cm, and a capacity of 1000 mL. Let r centimetres represent the radius of the base of the can.

a) Write an expression for the volume of the can in terms of r.

b) Determine the radius and the diameter of the base of the can.

Use the following information to answer exercises 8 and 9.

The company mentioned in exercise 7 also makes boxes with no tops. One type of box is made from a square piece of tin by cutting equal squares from each corner and folding up the sides. The box has a height of 4.0 cm, and a volume of 1000 cm^3.

8. Do you think the length of the base of the box will be greater than, less than, or the same as the diameter of the can in exercise 7? Explain your thinking.

9. Let x centimetres represent the length of a side of the square piece of tin from which the box is made.

a) Write an expression for the volume of the box in terms of x.

b) Determine the dimensions of the square piece of tin that is needed to make the box.

c) What are the dimensions of the box? Does the result agree with your prediction in exercise 8?

10. The sum, S, of the first n terms of the series $2 + 4 + 6 + 8 + \cdots$ is given by the formula $S = n(n + 1)$.

a) Determine the sum of the first 20 terms.

b) Suppose the sum of the first n terms is 110. Determine the value of n.

11. The sum, S, of the first n terms of the series $10 + 8 + 6 + 4 + \cdots$ is given by the formula $S = n(11 - n)$.

a) Determine the sum of the first 20 terms.

b) Suppose the sum of the first n terms is 28. Determine the value of n.

c) Why are two values of n possible in part b, but only one is possible in exercise 10b? Explain your answer in writing.

12. An object falls d metres in t seconds when dropped from rest. The quantities d and t are related by the formula $d = 4.9t^2$.

a) Solve the formula for t.

b) How long would it take an object to hit the ground when dropped from each height?

   i) 10 m   ii) 20 m   iii) 30 m

13. Recall from Chapter 2, page 110, the area of an equilateral triangle $A = \frac{\sqrt{3}}{4}x^2$, where x represents the length of its sides.

a) Solve the formula for x.

b) What is the side length of an equilateral triangle with each area?

   i) 10 cm^2 ii) 20 cm^2 iii) 40 cm^2

COMMUNICATING THE IDEAS

What are some of the advantages and disadvantages of solving quadratic equations by:

a) factoring?   b) using guess and check?

Which method do you think is better? Write your ideas in your journal.

6.10 Dividing a Polynomial by a Binomial

Dividing a polynomial by a binomial is similar to long division in arithmetic. Compare the steps in these two examples.

divisor ────── quotient
        ┌──── dividend

$$
\begin{array}{r}
32 \\
21\overline{)679} \\
63 \\
\hline
49 \\
42 \\
\hline
7
\end{array}
$$

| Divide 6 by 2 to get 3. |
| Multiply 3 by 21 to get 63. |
| Subtract 63 from 67 to get 4. Bring down 9. |
| Divide 4 by 2 to get 2. Multiply 2 by 21 to get 42. |
| Subtract 42 from 49 to get 7. |

remainder ┘

divisor ────── quotient
        ┌──── dividend

$$
\begin{array}{r}
3x + 2 \\
2x + 1\overline{)6x^2 + 7x + 9} \\
6x^2 + 3x \\
\hline
4x + 9 \\
4x + 2 \\
\hline
7
\end{array}
$$

| Divide $6x^2$ by $2x$ to get $3x$. |
| Multiply $3x$ by $2x + 1$ to get $6x^2 + 3x$. |
| Subtract $6x^2 + 3x$ from $6x^2 + 7x$ to get $4x$. Bring down 9. |
| Divide $4x$ by $2x$ to get 2. Multiply 2 by $2x + 1$ to get $4x + 2$. |
| Subtract $4x + 2$ from $4x + 9$ to get 7. |

remainder ┘

The results of these divisions may be written as follows:

$$\frac{679}{21} = 32 + \frac{7}{21}$$

or

$$679 = 21 \times 32 + 7$$

$$\frac{6x^2 + 7x + 9}{2x + 1} = 3x + 2 + \frac{7}{2x + 1}$$

or

$$6x^2 + 7x + 9 = (2x + 1)(3x + 2) + 7$$

These are called *division statements*.

Example 1

Divide $3x^2 + 8x + 11$ by $x + 2$. Express the result as a division statement in two ways.

Solution

Use mental math for each operation.

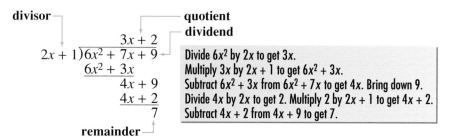

$$
\begin{array}{r}
3x \\
x + 2\overline{)3x^2 + 8x + 11}
\end{array}
$$

Divide $3x^2$ by x to get $3x$.

$$
\begin{array}{r}
3x\phantom{{}+8x+11} \\
x+2\overline{)3x^2+8x+11} \\
\underline{3x^2+6x\phantom{{}+11}} \\
2x\phantom{{}+11}
\end{array}
$$

Multiply $3x$ by $x+2$ to get $3x^2+6x$.
Subtract $3x^2+6x$ from $3x^2+8x$ to get $2x$.

$$
\begin{array}{r}
3x+2 \\
x+2\overline{)3x^2+8x+11} \\
\underline{3x^2+6x}\phantom{{}+11} \\
2x+11
\end{array}
$$

Bring down 11.
Divide $2x$ by x to get 2.

$$
\begin{array}{r}
3x+2 \\
x+2\overline{)3x^2+8x+11} \\
\underline{3x^2+6x}\phantom{{}+11} \\
2x+11 \\
\underline{2x+4} \\
7
\end{array}
$$

Multiply 2 by $x+2$ to get $2x+4$.
Subtract $2x+4$ from $2x+11$ to get 7.

Since the remainder has a lower degree than the divisor, the division is now complete. Expressed as a division statement, the result is:

$$
\frac{3x^2+8x+11}{x+2} = 3x+2+\frac{7}{x+2}
$$

or

$$
3x^2+8x+11 = (x+2)(3x+2)+7
$$

Example 2

Divide $-4x^3+6x^2+4x-7$ by $2x-3$, and write the division statement.

Solution

Use mental math for each operation.

$$
\begin{array}{r}
-2x^2\phantom{{}+6x^2+}+2 \\
2x-3\overline{)-4x^3+6x^2+4x-7} \\
\underline{-4x^3+6x^2}\phantom{{}+4x-7} \\
0+4x-7 \\
\underline{+4x-6} \\
-1
\end{array}
$$

Since the remainder is 0 at this stage, bring down the next *two* terms.

The quotient is $-2x^2+2$ with a remainder of -1. The division statement is:

$$
\frac{-4x^3+6x^2+4x-7}{2x-3} = -2x^2+2+\frac{-1}{2x-3}
$$

or

$$
-4x^3+6x^2+4x-7 = (2x-3)(-2x^2+2)-1
$$

If a power is missing in the dividend, you must still include it, using 0 as the coefficient.

Example 3

Divide $4t^4 + 25t^2 - 33t + 10$ by $2t - 1$.

Solution

Use mental math for each operation. Include a t^3 term as $0t^3$ in the dividend.

$$
\begin{array}{r}
2t^3 + t^2 + 13t - 10 \\
2t - 1 \overline{)\, 4t^4 + 0t^3 + 25t^2 - 33t + 10} \\
\underline{4t^4 - 2t^3} \\
+ 2t^3 + 25t^2 \\
\underline{+ 2t^3 - t^2} \\
26t^2 - 33t \\
\underline{26t^2 - 13t} \\
- 20t + 10 \\
\underline{- 20t + 10} \\
0
\end{array}
$$

The quotient is $2t^3 + t^2 + 13t - 10$, with a remainder of 0. This means that $2t - 1$ is a factor of $4t^4 + 25t^2 - 33t + 10$.

If a power is missing in the divisor, make sure that you write like terms in the same column.

Example 4

Divide $4x^3 + x^2 - 2x + 1$ by $2x^2 - 3$.

Solution

Use mental math for each operation.

$$
\begin{array}{r}
2x + 0.5 \\
2x^2 - 3 \overline{)\, 4x^3 + x^2 - 2x + 1} \\
\underline{4x^3 - 6x} \\
x^2 + 4x + 1 \\
\underline{x^2 - 1.5} \\
4x + 2.5
\end{array}
$$

Divide $1x^2$ by $2x^2$ to get 0.5.

The quotient is $2x + 0.5$, with a remainder of $4x + 2.5$.

In *Example 4*, the remainder is not a constant term as it was in the other examples. The remainder in *Example 4* is a polynomial with degree 1, which is less than the degree of the divisor (otherwise we could have continued to divide).

Division Statement

In any division problem, the divisor D, quotient Q, dividend P, and remainder R are related as follows:

$$\frac{P}{D} = Q + \frac{R}{D}$$

or $P = DQ + R$

In arithmetic, D, Q, P, and R are whole numbers, where $D \neq 0$ and $0 \leq R < D$.

In algebra, D, Q, P, and R are polynomials, where $D \neq 0$ and the degree of R is less than the degree of D.

DISCUSSING THE IDEAS

1. After you have divided a polynomial by a binomial, how could you check the result? Test your idea using one or more of the above examples.

2. When a polynomial is divided by a binomial, which of these statements is true: the result is always a polynomial; the result is never a polynomial; the result is sometimes a polynomial? Give examples to support your answer.

3. Give examples of divisions in arithmetic that correspond to *Example 3* and *Example 4*.

4. *Example 4* shows that you may encounter a fraction at some stage in the division. Do you think it would be better to use decimal form or fraction form when this occurs? Explain your answer.

5. In *Example 4*, the remainder is a polynomial with degree 1. Could the remainders in the other examples be considered as polynomials? If so, what is their degree?

6.10 EXERCISES

 1. These questions refer to *Example 1*.

   a) Suppose the dividend were $3x^2 + 8x + 9$. How would the quotient and remainder change?

   b) For a remainder of 0, what should the dividend be? How are the dividend and the divisor related in this case?

   c) Suppose the divisor were $x + 3$. How would the quotient and remainder change?

2. These questions refer to *Example 2*.

   a) Suppose the constant term in the dividend were changed to 0. How would the quotient and remainder change?

   b) For a remainder of 0, what should the dividend be?

   c) Suppose the divisor were $2x - 4$. How would the quotient and remainder change?

3. Write the division statements in *Example 3* and *Example 4* in the form $P = DQ + R$.

4. Divide, and write the division statement in the form $P = DQ + R$.

   a) $(x^2 + 7x + 14) \div (x + 3)$ b) $(x^2 - 3x + 5) \div (x - 2)$

   c) $(x^2 + x - 2) \div (x + 3)$ d) $(n^2 - 11n + 6) \div (n + 5)$

5. Choose one part of exercise 4. Write to explain how you determined the division statement.

6. Divide, and write the division statement in the form $\frac{P}{D} = Q + \frac{R}{D}$.

   a) $(x^3 - 5x^2 + 10x - 15)$ by $(x - 3)$ b) $(x^3 - 5x^2 - x - 10)$ by $(x - 2)$

   c) $(3x^3 + 11x^2 - 6x - 10)$ by $(x + 4)$ d) $(2x^3 + x^2 - 27x - 36)$ by $(x + 3)$

B 7. Divide, and write the division statement in either form.

   a) $(2x^2 + 5x - 1)$ by $(x + 1)$ b) $(3x^2 + 2x - 5)$ by $(x - 2)$

   c) $(25u^2 + 1)$ by $(5u + 3)$ d) $(6x^2 - 3)$ by $(2x + 4)$

   e) $(8x^2 - 6x + 11)$ by $(2x - 3)$ f) $(9m^2 - 5)$ by $(3m + 2)$

8. Divide. In each case, state whether the divisor is a factor of the dividend.

   a) $(c^3 + 13c^2 + 39c + 20)$ by $(c + 9)$

   b) $(x^3 - 8x^2 + x + 37)$ by $(x - 2)$

   c) $(-2n^3 - 11n^2 + 7n + 6)$ by $(n + 6)$

   d) $(x^3 - 12x - 20)$ by $(x + 2)$

   e) $(m^3 - 19m - 24)$ by $(m - 3)$

   f) $(4x^4 + x^2 - 3x - 1)$ by $(2x + 1)$

9. Divide, and write the division statement in the form $P = DQ + R$.

   a) $\dfrac{x^2 - 9}{x - 3}$ b) $\dfrac{x^2 - 9}{x + 3}$ c) $\dfrac{x^2 + 9}{x + 3}$ d) $\dfrac{x^2 + 9}{x - 3}$

10. a) Divide. Explain the results in parts i to iv.

   i) $\dfrac{x^2 + 7x + 12}{x + 3}$ ii) $\dfrac{x^2 + 7x + 12}{x + 4}$ iii) $\dfrac{x^2 + 7x + 12}{x + 7}$ iv) $\dfrac{x^2 + 7x + 12}{x}$

   v) $\dfrac{x^2 + 7x + 12}{x + 2}$ vi) $\dfrac{x^2 + 7x + 12}{x + 5}$ vii) $\dfrac{x^2 + 7x + 12}{x + 1}$ viii) $\dfrac{x^2 + 7x + 12}{x + 6}$

b) Summarize the results of part a in a table. Describe the patterns you see. Explain why these patterns occur.

$$x^2 + 7x + 12$$

Divisor	Quotient	Remainder
x		
$x + 1$		
$x + 2$		
$x + 3$		
$x + 4$		
$x + 5$		
$x + 6$		
$x + 7$		

c) Use the patterns in part b. Predict each quotient and remainder.

i) $\dfrac{x^2 + 7x + 12}{x + 8}$ **ii)** $\dfrac{x^2 + 7x + 12}{x - 1}$ **iii)** $\dfrac{x^2 + 7x + 12}{x + 9}$ **iv)** $\dfrac{x^2 + 7x + 12}{x - 2}$

11. For each polynomial, complete a table similar to the one in exercise 10. Compare each table with the one in exercise 10. Describe the similarities and differences.

a) $x^2 + 5x + 6$ **b)** $x^2 + 8x + 12$ **c)** $x^2 + 6x + 9$

12. Determine the quotient and remainder for each expression in each list.

a) $\dfrac{2x^2 + 7x + 6}{x + 3}$

$\dfrac{2x^2 + 8x + 9}{x + 3}$

$\dfrac{2x^2 + 9x + 12}{x + 3}$

b) $\dfrac{x^3 + 6x^2 + 9x + 5}{x + 2}$

$\dfrac{x^3 + 6x^2 + 10x + 7}{x + 2}$

$\dfrac{x^3 + 6x^2 + 11x + 9}{x + 2}$

13. a) Describe any patterns you found in exercise 12. Explain why you think those patterns occur.

b) Predict the next three quotients and remainders for each part of exercise 12.

c) Suppose you extended the lists in exercise 12 upward. Predict the previous three quotients and remainders in each list.

14. Divide. Investigate patterns in these lists of quotients. Describe each pattern. Explain why you think it occurs. Predict the next three expressions in each list.

a) $\dfrac{x^2 + 2x + 1}{x + 1}$

$\dfrac{x^2 + 2x + 1}{x + 2}$

$\dfrac{x^2 + 2x + 1}{x + 3}$

b) $\dfrac{x^2 + 2x + 1}{x + 1}$

$\dfrac{x^2 + 2x + 1}{2x + 1}$

$\dfrac{x^2 + 2x + 1}{3x + 1}$

15. Divide. In each case, state whether the divisor is a factor of the dividend.

  a) $(x^2 + 5x + 6)$ by $(2x + 3)$

  b) $(6x^3 + 5x^2 - 3x + 1)$ by $(2x + 3)$

  c) $(-2x^3 + 4x^2 - 3x + 5)$ by $(4x^2 + 5)$

  d) $(2x^3 + 8x^4 - 3x^2 + 1 + 7x)$ by $(4x + 1)$

16. Choose one part of exercise 15. Write to explain how you completed the division, and how you determined whether the divisor is a factor.

17. Divide, then write the division statement in the form $P(x) = D(x)Q(x) + R(x)$.

  a) $(3x^3 + 4x^2 + 3x + 1)$ by $(5x + 2)$

  b) $(6x^3 - 2x^2 + 7x - 11)$ by $(3x^2 - 2)$

  c) $(7 + 8x + 10x^2 + 6x^3)$ by $(4x^2 + 3)$

  d) $(-11x^3 + 15x^2 + 12x^4 - 7x + 12)$ by $(3x^2 - 2)$

C 18. Divide. Investigate patterns in these lists of quotients. Describe each pattern. Explain why you think it occurs. Predict the next three expressions in each list.

  a) $\dfrac{x^2 - 1}{x - 1}$

  $\dfrac{x^3 - 1}{x - 1}$

  $\dfrac{x^4 - 1}{x - 1}$

  b) $\dfrac{x^2 + x + 1}{x + 1}$

  $\dfrac{x^3 + x^2 + x + 1}{x + 1}$

  $\dfrac{x^4 + x^3 + x^2 + x + 1}{x + 1}$

19. Determine a pattern, similar to the one in exercise 18a, in which the divisors are $x + 1$.

20. One factor of $4x^3 + 15x^2 - 31x - 30$ is $x - 2$. Determine the other factors.

21. Determine the value of k such that when $2x^3 + 9x^2 + kx - 15$ is divided by $x + 5$, the remainder is 0.

COMMUNICATING THE IDEAS

This section began with a comparison of division of whole numbers with division of polynomials. In your journal, write a description of some of the ways in which division of polynomials differs from division of whole numbers.

1. Calculate the surface area and volume of a sphere with diameter 3.5 m.

2. Calculate the radius, then the surface area of a sphere with volume 3.5 m^3.

3. Choose either exercise 1 or 2. Write to explain how you calculated the surface area.

4. Multiply.

   a) $(8xy)(-5y)$

   b) $(-3a^2b)(2ab^2)$

   c) $(3xy)(-2x^2y)(-4xy^2)$

5. Divide.

   a) $\dfrac{35m^4n^3}{-5mn^2}$

   b) $\dfrac{(8x^4y^5)(3x^2y)}{6x^2y^2}$

   c) $\dfrac{(4a^2b^2)^2(-3a^4b^2)}{(2ab)^3(6a^2b)}$

6. Simplify.

   a) $(3x^2 + 17xy) - (12x^2 - 3xy)$

   b) $(3m^2 - 5mn) - (3mn - 7n^2)$

   c) $2x(x + y) - 3x(2x - 3y)$

   d) $2a(3a - 5b) - a(2b + 3a)$

   e) $3xy(x - 2y) - 3x(2xy + 3y^2) - y(2x^2 + 5xy)$

   f) $5m(3mn - 2n^2) - 2(m^2n - 15mn^2) + 5n(-3mn - 5m^2)$

7. Simplify.

   a) $6(2x^2 - 5x) - 14(3x - x^2) + 3(x - x^2)$

   b) $-7(c - 3d + 5e) + 4(2c - 11d - 3e) - 5(3c - 7d + 2e)$

   c) $7m(2m - 5n + 3) + 2m(-3m + 9n - 4)$

   d) $4x^2(5x - 2y - 8) - 3x^2(4x - 8y - 2)$

8. Simplify.

   a) $(2x^2 - 7x - 4)(2x - 5)$

   b) $(8y^2 + 3y - 7)(3y + 4)$

9. Determine the perimeter and area of the shaded region of each rectangle.

   a)

   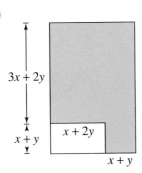

   b)

   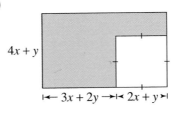

10. Choose one part of exercise 9. Write to explain how you determined the perimeter and area.

11. Factor.

a) $8m^3 - 4m^2$
b) $8y^2 - 12y^4 + 24$
c) $28a^2 - 7a^3$
d) $6a^2b^3c - 15a^2b^2c^2$
e) $30x^2y - 20x^2y^2 + 10x^3y^2$
f) $8mn^2 - 12mn - 16m^2n$

12. Factor.

a) $m^2 + 8m + 16$
b) $a^2 - 7a + 12$
c) $y^2 - 2y - 8$
d) $n^2 - 4n - 45$
e) $s^4 - 15s^2 + 54$
f) $k^4 - 9k^2 - 90$

13. Factor.

a) $a^2 + 14ab + 24b^2$
b) $m^2 + 9mn + 18n^2$
c) $s^2 + 20st + 36t^2$
d) $x^2 - xy - 20y^2$
e) $c^2 + 21cd - 100d^2$
f) $p^2 - 19pq - 120q^2$

14. Factor.

a) $4x^2 - 7x + 3$
b) $6a^2 - 13a - 5$
c) $21n^2 + 8n - 4$
d) $6r^4 - 31r^2 + 5$
e) $12t^2 - 15t - 18$
f) $56x^2 + 18x - 8$

15. Choose one part of exercise 14. Write to explain how you factored the trinomial.

16. Factor.

a) $b^2 - 36$
b) $81k^4 - 1$
c) $36x^2 - 49y^2$
d) $4a^2 - 9b^2$
e) $25m^2 - 81n^2$
f) $1 - 16s^4$
g) $196x^2 - 25z^2$
h) $256p^2 - 625q^2$
i) $289s^2 - 324t^2$

17. Factor.

a) $8a^2 - 72$
b) $150 - 6n^2$
c) $7x^4 - 7y^4$
d) $27m^3 - 12m$
e) $\frac{a^2}{36} - \frac{b^2}{49}$
f) $125p^2q^2 - 180q^2$

18. Factor.

a) $(2c - 5)^2 - 121$
b) $x^2 - (y + z)^2$
c) $a^2 - (b - c)^2$

19. Factor.

a) $-98m^3 + 32m$
b) $125x^2y^2 - 180y^2$
c) $128x^2y - 50y^3$

20. Factor.

a) $x^4 - 29x^2y^2 + 100y^4$
b) $m^4 - 38m^2n^2 + 72n^4$

21. Solve.

a) $x^2 - 4x - 21 = 0$
b) $x^2 + x - 56 = 0$
c) $4x^2 - 12x + 9 = 0$

22. Divide, then write the division statement.

a) $(2a^3 - 5a^2 - 9a + 18)$ by $(a + 2)$
b) $(2x^3 - 13x + 5x^2 - 28)$ by $(x + 3)$
c) $(6x^3 + 17x^2 - 26x + 8)$ by $(x + 4)$
d) $(32x^3 - 18x + 16x^2 + 9)$ by $(2x^2 + 1)$

6 Cumulative Review

1. Determine the sum of each arithmetic series.

   a) $3 + 8 + 13 + \ldots + 203$ b) $4 + 10 + 16 + \ldots + 328$

   c) $7 + 11 + 15 + \ldots + 459$ d) $15 + 20 + 25 + \ldots + 655$

2. a) Determine the indicated term of each arithmetic sequence.

   i) $5, 9, 13, \ldots\ t_6$ ii) $2, 11, 20, \ldots\ t_{13}$

   iii) $3, 6, 9, \ldots\ t_{41}$ iv) $2, 4, 6, \ldots\ t_{117}$

   v) $5, 10, 15, \ldots t_{69}$ vi) $11, 22, 33, \ldots t_{66}$

   b) Look at the results of the last four sequences in part a. Write to explain a different way to find the indicated term when the first term equals the common difference.

3. Determine the first five terms of each geometric sequence.

   a) The second term is –6 and the third term is 12.

   b) The third term is 6 and the fifth term is 3.

   c) The first term is 5 and the fifth term is 80.

   d) Choose one of parts a to c. Write to explain how you determined the terms.

4. From 1975 to 1995, the population of Alberta increased by approximately 11% every 5 years. Assume this trend continues. Copy and complete this table to predict the population of Alberta in 2000, 2005, and 2010.

Year	Population	5-year growth rate (%)	Increase in population	Population 5 years later
1995	2 747 000	11		
2000		11		
2005		11		
2010		11		

5. From the area of each square, determine the length of a side and the perimeter.

   a)
   Area 13.69 m²

   b)
   Area 576 mm²

6. Simplify without using a calculator.

   a) $\sqrt[3]{8}$ b) $\sqrt[3]{27}$ c) $\sqrt[3]{64}$ d) $\sqrt[3]{125}$

   e) $\sqrt[3]{216}$ f) $\sqrt[4]{16}$ g) $\sqrt[4]{81}$ h) $\sqrt[4]{1}$

CUMULATIVE REVIEW **395**

7. Some children decide to build the biggest snowball ever. To make your calculations easier, assume that the snowball is always spherical.

   a) After one hour, the snowball has a volume of 1.5 m^3. On average, how fast did the children add snow to the ball, in cubic metres per minute?

   b) Determine the radius of the snowball. How tall is the snowball?

   c) Determine the surface area of the snowball.

8. Determine the length of the line segment with each pair of endpoints.

   a) A(5, 8), B(−3, 2)   b) C(−5, 7), D(−5, −11)   c) E(4, 8), F(−8, −4)

9. Determine the perimeter of the triangle with vertices A(2, 2), B(5, 5), and C(8, 2). What kind of triangle is △ABC? Write to explain how you classified the triangle.

10. Determine the coordinates of the midpoint of the line segment with each pair of endpoints.

    a) M(5, −3), N(9, −5)   b) P(−6, 3), Q(10, −7)   c) R(−3, 2), S(1, −1)

11. The endpoints and slope of a line segment are given. Determine each value of k.

    a) A(3, k), D(−2, 4), slope $\frac{2}{5}$ b) B(2, 5), F(k, −1), slope $-\frac{3}{4}$

12. Each pair of numbers represents the slopes of parallel lines. Determine each value of k.

    a) $\frac{3}{4}, \frac{6}{k}$ b) $-\frac{5}{2}, \frac{k}{5}$ c) $\frac{4}{k}, \frac{2}{3}$ d) $\frac{k}{6}, \frac{3}{2}$ e) $\frac{k}{2}, \frac{5}{4}$

13. Suppose each pair of numbers in exercise 12 represents the slopes of perpendicular lines. Determine each value of k.

14. State the slope and y-intercept for each line.

    a) $y = -2x + 7$ b) $3x - 4y = 12$ c) $x + y = -2$
    d) $5x + 2y = -6$ e) $5 - 4x - 2y = 0$ f) $-2x - y + 3 = 0$

15. Write the equation of each line.

    a) with slope $-\frac{2}{3}$, y-intercept 4   b) with slope $\frac{6}{5}$, y-intercept −3

16. a) Determine the equation of the line that passes through each pair of points.

    i) P(4, 8) and Q(−3, 6)

    ii) R(−5, 10) and S(4, −6)

    iii) T(0, −2) and U(5, −3)

    b) Choose one line from part a. Write to explain how you determined its equation.

17. The rules defining three functions are given. For each function

   a) Make a table of values.   **b)** Graph the function.

   c) Decide if there are any input numbers that cannot be used.

      i) *Rule 1*: Output number is the same as the input number.

      ii) *Rule 2*: Add 3 to the input number.

      iii) *Rule 3*: Subtract 3 from the input number.

18. a) Write an equation to describe each function in exercise 17.

   b) Choose one function from part a. Write to explain how you determined its equation.

19. A person walks or runs along a straight track. Each graph illustrates the person's motion at some time. Write to describe the motion illustrated by each graph.

a) 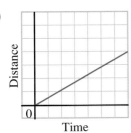   **b)** 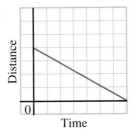   **c)**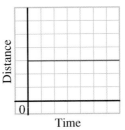

20. For each table of values, determine a rule that expresses y as a function of x.

a)

x	y
−2	−3
−1	−1
0	1
1	3
2	5

b)

x	y
−2	3
−1	0
0	−1
1	0
2	3

c)

x	y
−2	2
−1	2.5
0	3
1	3.5
2	4

21. Add.

   a) $(4x^3 - 2x^2 + 5x - 1) + (-2x^3 + 7x^2 - x + 4)$

   b) $(5x^4 + 3x^2 - 8x - 6) + (-3x^4 + x^3 + 4x + 2)$

   c) $\begin{array}{l} 4x^2 - 2x + 1 \\ 3x^2 + x - 6 \end{array}$ **d)** $\begin{array}{l} 5x^2y - 3xy + 2y^2 \\ -2x^2y + xy - 4y^2 \end{array}$

22. Subtract.

   a) $(5x^2 + 3x + 7) - (4x^2 + 6x + 5)$

   b) $(6y^3 - 3y^2 + 7y - 2) - (4y^3 - 4y^2 - 2y + 5)$

   c) $\begin{array}{l} 4x^2 - 2xy + 3y^2 \\ 3x^2 - 5xy - y^2 \end{array}$ **d)** $\begin{array}{l} a^2b^2 + 5a^3b - 3ab^3 \\ -2a^2b^2 - 3a^3b + ab^3 \end{array}$

RATIONAL EXPRESSIONS

7

Should Pop Cans Be Redesigned?

In 1996, Canadians drank 3 283 332 kL of soft drinks. It would take more than 9 billion aluminum cans to hold that much liquid. Consider producing these cans. How much might it cost? Would it be possible to reduce the cost?

 CONSIDER THIS SITUATION

The standard aluminum can holds 355 mL of pop. Soft drink companies have determined that each 1% reduction in the mass of a can would save approximately $20 million a year. Could pop cans be redesigned so that less aluminum is used?

Suppose you were designing a single-serving container for a soft drink.

- What advantage or disadvantage is there with each material?
  glass   aluminum   plastic
- What numerical information might you need to help you choose a material?
- What shape would you choose for the single-serving container? Why?

Should pop cans be redesigned? On pages 416 and 417, you will use rational expressions to develop a mathematical model to explore this question.

 FYI Visit www.awl.com/canada/school/connections

For information related to the above problem, click on <u>MATHLINKS</u> followed by <u>AWMath</u>. Then select a topic under It's All in the Packaging.

This replica of St. Peter's Basilica was constructed from approximately 10 million aluminum cans.

7.1 Evaluating Rational Expressions

Recall that a number that can be written as a fraction is a rational number. The numbers $\frac{2}{3}$, -5.25, $2\frac{1}{4}$, and 7 are examples of rational numbers. A rational number is defined as any number that can be written in the form $\frac{m}{n}$, where m and n are integers and $n \neq 0$.

In algebra, an expression involving variables represents a number. When the numerator or denominator (or both) of a fraction is a polynomial, the fraction is described as a rational expression.

> Any algebraic expression that can be written as the quotient of two polynomials is called a *rational expression*.

These are rational expressions.

$$\frac{3x + 5}{2x + 7} \qquad \frac{a^2 - 4a + 3}{2ab} \qquad n^3 + 27$$

> Observe that since $n^3 + 27$ can be written as $\frac{n^3 + 27}{1}$, it is a rational expression.

These expressions are not rational expressions.

$$\frac{2\sqrt{x} + 5}{4x} \qquad \frac{2m^2 + n}{3\sqrt{n}} \qquad 1 + 2^x$$

> Rational expressions cannot contain roots of variables, and they cannot contain variables in exponents.

To evaluate a rational expression, substitute a number for each variable in the expression.

Example 1

Evaluate each expression, if possible, for $x = -3$ and $y = 2$.

a) $\dfrac{5x + 2y}{x}$

b) $\dfrac{x + 4}{y - 2}$

Solution

Substitute for x and y in each expression.

a)
$$\begin{aligned}
\frac{5x + 2y}{x} &= \frac{5(-3) + 2(2)}{-3} \\
&= \frac{-15 + 4}{-3} \\
&= \frac{-11}{-3} \\
&= \frac{11}{3}
\end{aligned}$$

b)
$$\frac{x + 4}{y - 2} = \frac{-3 + 4}{2 - 2}$$
This reduces to $\frac{1}{0}$, which is not defined.

In *Example 1b*, we say that the expression $\frac{x+4}{y-2}$ is not defined when $y = 2$, since 2 is the value of y for which the denominator of the expression equals 0. Similarly, the expression in *Example 1a*, $\frac{5x+2y}{x}$, is not defined when $x = 0$. The values of the variable that would make the denominator 0 are called *nonpermissible* values.

A rational expression is not defined when its denominator is equal to 0.

Example 2

For what values of the variable is each rational expression not defined?

a) $\frac{3mn}{2m}$
b) $\frac{a^2 - 5a}{a - 3}$
c) $\frac{x}{x^2 + 1}$
d) $\frac{5x}{x^2 - 3x - 4}$

Solution

For each expression, let the denominator equal 0.

a) $\frac{3mn}{2m}$

If $2m = 0$, then $m = 0$

The expression $\frac{3mn}{2m}$ is not defined when $m = 0$.

b) $\frac{a^2 - 5a}{a - 3}$

If $a - 3 = 0$, then $a = 3$

The expression $\frac{a^2 - 5a}{a - 3}$ is not defined when $a = 3$.

c) $\frac{x}{x^2 + 1}$

Since x is a real number, $x^2 \geq 0$. Therefore, $x^2 + 1$ cannot equal 0. There are no real values of x for which the expression $\frac{x}{x^2 + 1}$ is not defined. We say that the expression is defined for all real values of x.

d) $\frac{5x}{x^2 - 3x - 4}$

Let $x^2 - 3x - 4 = 0$

Solve the quadratic equation.

$(x + 1)(x - 4) = 0$

Either $x + 1 = 0$ or $x - 4 = 0$

$\quad\quad\quad x = -1 \quad\quad\quad x = 4$

The expression $\frac{5x}{x^2 - 3x - 4}$ is not defined when $x = -1$ or when $x = 4$.

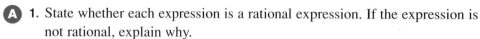

What does 12 divided by 3 mean? What does 12 divided by 0 mean? Why is division by 0 not defined?

7.1 EXERCISES

A 1. State whether each expression is a rational expression. If the expression is not rational, explain why.

a) $\frac{2}{5}$ b) $\frac{7}{-4}$ c) $\frac{a}{3}$ d) $\frac{\sqrt{a}}{3}$

e) $\frac{-8}{m}$ f) $\frac{1}{n}$ g) $\frac{x}{\sqrt{x}}$ h) $\frac{\sqrt{3x}}{2}$

2. State whether each expression is a rational expression. If the expression is not rational, explain why.

a) $\frac{x+4}{2x}$ b) $\frac{\sqrt{x}}{x-2}$ c) $x^2 - x - 6$ d) $\frac{3x-y}{x+3y}$

e) $\frac{x-\sqrt{x}}{3}$ f) $\frac{y}{\sqrt{2x}}$ g) $5x + 7y$ h) $-\frac{4}{x^3+1}$

i) $\frac{3}{5x-3}$ j) $\frac{1}{x^2-9}$ k) $\frac{3\sqrt{a}+1}{a+5b}$ l) $\frac{x+y}{\sqrt{3x}}$

3. Explain the difference between a rational number and a rational expression.

B 4. Evaluate each expression for $x = 4$.

a) $\frac{x}{2}$ b) $\frac{-3}{x}$ c) $\frac{x-4}{3x}$ d) $x^2 + 5x + 1$

e) $\frac{x^3+1}{5}$ f) $\frac{-6}{x^2-2}$ g) $\frac{1}{x^2+4x}$ h) $\frac{-3x-2}{3}$

5. Evaluate each expression in exercise 4 for $x = -1$.

6. Evaluate each rational expression for $x = 2$ and $y = -3$.

a) $\frac{x+2y}{2}$ ~ 2   b) $\frac{3x-4y}{6}$ -1   c) $\frac{x-y}{x}$ ~.5   d) $\frac{3x-2y}{xy}$ -2   e) $\frac{3x+2y}{x^2}$ 0

f) $\frac{x-2}{3y+1}$ 0   g) $\frac{x^3+1}{y-3}$ -1.5   h) $\frac{y^2+1}{x+3}$ +2.0   i) $\frac{x^2-9}{x^2-x-6}$ 1.25   j) $\frac{x^2-2x+1}{x^2-1}$ 8.33

7. Suppose you were to evaluate each expression in exercise 6 for $x = -2$ and $y = 3$. Which expressions could not be evaluated? Explain.

8. A cycling club organized a 72-km tour.

a) Let s represent the average speed of a cyclist in kilometres per hour, and t the time in hours. Write a formula for average speed. Identify the rational expression in the formula.

b) For each time, calculate the average speed to complete the tour.

  i) 4 h   ii) 4.5 h   iii) 5 h 15 min

9. The *Olympic Summer Games* database provides data about several swimming events, including the 100-m and 200-m butterfly and 400-m freestyle.

   a) In which of the above events would you expect the winner to have the fastest average speed? The slowest? Explain your thinking.

   b) Find the records for each event above in the database. Sort them to determine the fastest men's and women's time. For each fastest time, use the formula you wrote in exercise 8 to calculate the average speed in metres per second. Were your predictions in part a correct? If not, explain what might have affected the times.

   c) There are also 100-m and 200-m freestyle events at the Olympics. How do you think competitors' speeds in these events would compare with butterfly events of the same length? To check your prediction, find the 1996 Olympic winning times for the 100-m and 200-m freestyle events. Calculate the average speeds. Compare them with your answers from part b.

10. Consider the rational expression $\frac{x+4}{x-5}$.

   a) Evaluate this expression for $x = -4$.

   b) Evaluate this expression for $x = 5$.

   c) Explain why x can equal -4, but x cannot equal 5.

11. For which value(s) of x is each expression not defined?

   a) $\frac{3}{x}$
   b) $\frac{-2}{x+1}$
   c) $\frac{x+2}{2x}$
   d) $\frac{2x+1}{5}$
   e) $\frac{x^2}{2x-7}$

   f) $\frac{x^2-1}{x^3-8}$
   g) $\frac{3x-4}{x^2-9}$
   h) $\frac{5x-4}{x^3+1}$
   i) $\frac{6x-5}{2x(x+4)}$
   j) $\frac{-3}{x^3}$

12. Choose one part of exercise 11. Write to explain how you determined the value(s) of x for which the expression is not defined.

13. Explain why the expression $\frac{x-1}{3}$ is defined for all values of x including $x = 1$.

14. Some of these rational expressions do not have any restrictions on the variable. Which rational expressions are they? Explain.

   a) $\frac{x^2-1}{4}$
   b) $\frac{x^3+1}{2x}$
   c) $\frac{2x+5}{x^2-25}$
   d) $\frac{x^2+5x-24}{x^2+9}$

   e) $\frac{x^2+3x+2}{x^2+5}$
   f) $\frac{x^2-81}{x^3+8}$
   g) $\frac{x^2+25}{x^2-16}$
   h) $\frac{x^2+16}{x^3+1}$

15. Choose one part of exercise 14 that does not have a restriction on the variable. Write to explain how you know there is no restriction.

16. Which value(s) of the variable are not permitted for each rational expression below?

a) $\dfrac{x+3}{2x}$

b) $\dfrac{x+1}{3}$

c) $\dfrac{3x}{x^2+3x-4}$

d) $\dfrac{2a-1}{a-2}$

e) $\dfrac{m^2+4m}{3m-4}$

f) $\dfrac{(2x-3)(x+4)}{2x(x+5)}$

g) $\dfrac{3x+2}{x^2-9}$

h) $\dfrac{r^2+7r+6}{r^2-4r-12}$

i) $\dfrac{x^2+3x-18}{x^2+1}$

j) $\dfrac{x^2+5x+4}{x^2-36}$

k) $\dfrac{4}{x^3}$

l) $\dfrac{6}{x^3-1}$

m) $\dfrac{c^2+11c+24}{c^2-3c-4}$

n) $\dfrac{x^2+5x+6}{2x^2+7x+6}$

o) $\dfrac{a^2+3a+4}{3a^2-10a+8}$

17. Consider the rational expression $\dfrac{5x+y}{x-2y}$.

a) Suppose $x = 2$. What value of y would make the expression undefined?

b) Suppose $y = 3$. What value of x would make the expression undefined?

c) Copy and complete this table of values. List five ordered pairs that would make the expression undefined.

d) Plot these ordered pairs on a graph.

e) Write an equation to describe your graph.

f) How is this equation related to the denominator of the rational expression?

x	y
2	
	3

18. For which value(s) of the variables is each rational expression not defined?

a) $\dfrac{2x-y}{x-3y}$

b) $\dfrac{3x+5y}{2x-y}$

c) $\dfrac{6xy(4x-y)}{3x(2x-7y)}$

d) $\dfrac{(x+2y)(x-5y)}{(x-4y)(2x-y)}$

e) $\dfrac{x}{y^2+1}$

f) $\dfrac{4x-y}{x^2-49y^2}$

g) $\dfrac{9x^2-y^2}{x^2+y^2}$

h) $\dfrac{x^2+7xy+10y^2}{4x^2-9y^2}$

i) $\dfrac{x^2-3xy-40y^2}{x^2-7xy+10y^2}$

19. Create a rational expression for which each value of x is not permitted.

a) $x = 5$　　　　**b)** $x = -3$　　　　**c)** $x = 0$ or 4　　　　**d)** $x = 2$ or -1

20. Choose one part of exercise 19. Write to explain how you created the expression.

COMMUNICATING THE IDEAS

In your journal, write a paragraph to explain to your English teacher why there may be restrictions on the value(s) of the variable(s) in a rational expression. It may be helpful to include a brief definition of a rational expression.

INVESTIGATE

1. a) What are the prime factors of 10?

   b) What are the prime factors of 14?

   c) Which factor do 10 and 14 have in common?

2. Write each fraction in lowest terms.

   a) $\frac{10}{14}$
   b) $\frac{12}{15}$
   c) $\frac{24}{60}$
   d) $\frac{-15}{20}$

3. Use the results of exercise 2. How are factors related to writing fractions in lowest terms? Explain how you write a fraction in lowest terms.

4. Consider the rational expression $\frac{24xy^2}{60x^2y}$.

   a) What are the prime factors of the numerator?

   b) What are the prime factors of the denominator?

   c) Write the rational expression in lowest terms.

5. Consider the rational expression $\frac{x^2 - 4}{x^2 + 6x + 8}$.

   a) What are the prime factors of the numerator?

   b) What are the prime factors of the denominator?

   c) Write the rational expression in lowest terms.

Example 1

Simplify $\frac{21x^2}{35x}$. State the value(s) of the variable for which the result is true.

Solution

Use mental math to write the prime factors of the numerator and the denominator.

$$\frac{21x^2}{35x} = \frac{(3)(7)(x)(x)}{(5)(7)(x)}$$

The numerator and denominator have common factors of 7 and x.

Divide the numerator and denominator by their common factors.

Recall that we can only divide by x if $x \neq 0$.

$$\frac{21x^2}{35x} = \frac{(3)(x)}{5} \text{ if } x \neq 0$$

That is, $\frac{21x^2}{35x} = \frac{3x}{5}$ for all real values of x except $x = 0$.

In *Example 1*, an inspection of the denominator indicates that if $x = 0$ then the rational expression is undefined. That is why we say 0 is a nonpermissible value of x.

When we simplify monomials, we may not need to rewrite them in factored form if their factors are easily identified. However, it is more difficult to determine the factors of binomials and trinomials mentally. That is why it is important to write the factors of each polynomial in a rational expression.

Example 2

Simplify each rational expression. Write the nonpermissible value(s) for each variable.

a) $\dfrac{2x^2 + 6x}{3x}$ b) $\dfrac{x^2 - 5x - 6}{x^2 - 36}$ c) $\dfrac{25 - x^2}{2x^2 - 9x - 5}$

Solution

For each expression:
Factor the numerator and denominator.
Determine the nonpermissible values of x.
Divide the numerator and denominator by their common factors.

a) $\dfrac{2x^2 + 6x}{3x} = \dfrac{(2)(x)(x + 3)}{(3)(x)}$ ⟵ **Divide by the common factor x.**

Since $3x = 0$ when $x = 0$, then $x = 0$ is a nonpermissible value.

$\dfrac{2x^2 + 6x}{3x} = \dfrac{2(x + 3)}{3}$ if $x \neq 0$

b) $\dfrac{x^2 - 5x - 6}{x^2 - 36} = \dfrac{(x - 6)(x + 1)}{(x - 6)(x + 6)}$ ⟵ **Divide by the common factor $x - 6$.**

Since $x - 6 = 0$ when $x = 6$, and $x + 6 = 0$ when $x = -6$, then $x = 6$ and -6 are nonpermissible values.

$\dfrac{x^2 - 5x - 6}{x^2 - 36} = \dfrac{x + 1}{x + 6}$ if $x \neq 6$ and $x \neq -6$

c) $\dfrac{25 - x^2}{2x^2 - 9x - 5} = \dfrac{(5 - x)(5 + x)}{(x - 5)(2x + 1)}$

Since $x - 5 = 0$ when $x = 5$, and $2x + 1 = 0$ when $x = -\dfrac{1}{2}$, then $x = 5$ and $-\dfrac{1}{2}$ are nonpermissible values.

Rewrite the factor $5 - x$ as $-(x - 5)$.

$\dfrac{25 - x^2}{2x^2 - 9x - 5} = \dfrac{-(x - 5)(5 + x)}{(x - 5)(2x + 1)}$ ⟵ **Divide by the common factor $x - 5$.**

$\dfrac{25 - x^2}{2x^2 - 9x - 5} = \dfrac{-(5 + x)}{2x + 1}$, or $\dfrac{-5 - x}{2x + 1}$ if $x \neq 5$ and $x \neq -\dfrac{1}{2}$

When you work with rational expressions, you may find it inconvenient to write the restrictions on the variable for every expression. When we write an equation to state that two rational expressions are equal, we mean they are equal for all the values of the variable for which both expressions are defined.

DISCUSSING THE IDEAS

1. In *Example 1*, we found that $\frac{21x^2}{35x} = \frac{3x}{5}$ if $x \neq 0$.

   a) Is this result true for all values of x except 0?

   b) Substitute $x = 0$ in both expressions. Explain the results.

   c) Is $\frac{21x^2}{35x}$ the same expression as $\frac{3x}{5}$?

2. Why do you think the process discussed in this section is called "simplifying"?

3. In *Example 2b*, we divided the numerator and denominator by $x - 6$. Then, $x - 6$ did not appear in the denominator of the simplified expression. Why do we still need to say that $x = 6$ is a nonpermissible value?

4. In *Example 2b*, why can the rational expression $\frac{x+1}{x+6}$ not be simplified to $\frac{1}{6}$?

5. In *Example 2c*, explain why the factor $-(x - 5)$ is equal to $5 - x$.

7.2 EXERCISES

A 1. Reduce to lowest terms.

   a) $\frac{24x}{3}$

   b) $\frac{36x}{-4y}$

   c) $\frac{132a^4}{12a^2}$

   d) $\frac{-28m^3}{7m}$

   e) $\frac{5x^2}{15x}$

   f) $\frac{-20s^3}{-35s^3}$

   g) $\frac{102a^3}{17ab}$

   h) $\frac{216m^2}{-18mn}$

2. Reduce to lowest terms.

   a) $\frac{4ab}{8ac}$

   b) $\frac{6a^2c}{8ab}$

   c) $\frac{-18x^2y}{3xy^2}$

   d) $\frac{45a^2bc^3}{60ab^2c}$

   e) $\frac{-8x^2yz}{-24xyz^2}$

   f) $\frac{-15a^2bc^4}{27a^2b^3c^2}$

   g) $\frac{-9m^2n}{-m^5}$

   h) $\frac{-25a^3b^2c}{40ab^7}$

3. Reduce to lowest terms.

   a) $\frac{4x+8}{2x+4}$

   b) $\frac{a-9b}{3a-27b}$

   c) $\frac{2x-10}{3x-15}$

   d) $\frac{x-5}{10-2x}$

   e) $\frac{3m-12n}{20n-5m}$

   f) $\frac{16-4a}{32-8a}$

   g) $\frac{3a+12}{6a+24}$

   h) $\frac{7x+14}{5x+10}$

B 4. Simplify each expression. Identify the nonpermissible value(s) of x.

   a) $\frac{2x^2+6x}{5x}$

   b) $\frac{2x^2-10x}{4x-20}$

   c) $\frac{5x^2+7x}{3x}$

d) $\dfrac{4x^2 - 12x}{x - 3}$ **e)** $\dfrac{3x^2 - 6x}{14 - 7x}$ **f)** $\dfrac{9 - 3x}{x - 3}$

g) $\dfrac{x^2 + 7x + 12}{x + 4}$ **h)** $\dfrac{x^2 + x - 6}{2 - x}$ **i)** $\dfrac{x^2 - 10x + 25}{5 - x}$

5. Choose one part of exercise 4. Write to explain how you simplified the expression and identified the nonpermissible value(s).

6. The expression $\dfrac{2x^2 + 6x}{x^2 + 2x - 3}$ is given.

     a) Evaluate this expression for $x = 2$.

     b) Simplify this expression. Identify any nonpermissible value(s) of x.

     c) Evaluate your expression in part b for $x = 2$.

     d) Compare your answers to parts a and c.

     e) How can you check if you simplified an expression correctly?

7. Simplify.

     a) $\dfrac{5n - 5m}{3m - 3n}$ **b)** $\dfrac{c - d}{d - c}$ **c)** $\dfrac{2a - 2b}{3b - 3a}$

     d) $\dfrac{3xy - 18y^2}{12y^2 - 2xy}$ **e)** $\dfrac{10xy - 15x^2y}{6x^2 - 4x}$ **f)** $\dfrac{60a^2b^2 - 24ab}{16ab - 40a^2b^2}$

8. Simplify each expression, if possible. Identify any nonpermissible value(s) for each variable.

     a) $\dfrac{a^2 + 5a - 14}{a^2 - 6a + 8}$ **b)** $\dfrac{x - 3}{x^2 + 3x - 18}$ **c)** $\dfrac{m^2 - 7m + 10}{m - 2}$

     d) $\dfrac{r^2 - 9}{r^2 + 6r + 9}$ **e)** $\dfrac{x + 4}{x^2 - 16}$ **f)** $\dfrac{x - 3}{x^2 + 9}$

9. A graphing calculator was used to display the tables of values for two expressions in exercise 8. Identify the expression from exercise 8 that was used for each table. Y1 represents the given expression and Y2 represents its simplified form.

a)   **b)**

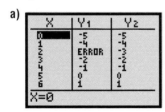

10. Simplify.

     a) $\dfrac{x^2 + 9xy + 18y^2}{2x^2 + 12xy}$ **b)** $\dfrac{m^2 - 9mn + 20n^2}{3m^2 - 15mn}$ **c)** $\dfrac{9a^2 - 16b^2}{6a^2 - 8ab}$

     d) $\dfrac{a^2 - ab - 6b^2}{a^2 + 2ab}$ **e)** $\dfrac{9x^2 - 4y^2}{3x^2 - 2xy}$ **f)** $\dfrac{m^2 + 2mn - 3n^2}{3m^2 + 9mn}$

     g) $\dfrac{x^2 - 25}{x^2 - 5x}$ **h)** $\dfrac{3m^2 - 15m}{3m^2 - 16m + 5}$ **i)** $\dfrac{8t^2 - 32}{2t^2 + 12t + 16}$

11. Simplify.

a) $\dfrac{x - 4}{2x^2 - 11x + 12}$

b) $\dfrac{2x - 7}{2x^2 - x - 21}$

c) $\dfrac{2x - 9}{2x^2 - 3x - 27}$

d) $\dfrac{c^2 - 5cd - 24d^2}{c^2 + 7cd + 12d^2}$

e) $\dfrac{x^2 + xy - 30y^2}{x^2 + 11xy + 30y^2}$

f) $\dfrac{a^2 + 10ab + 24b^2}{a^2 - 36b^2}$

g) $\dfrac{(x^2 - 1)(x^2 - 4)}{x^2 - 3x - 4}$

h) $\dfrac{(x^2 - 36)(x^2 - 25)}{x^2 + x - 30}$

i) $\dfrac{(e^2 - 81)(e^2 - 64)}{e^2 - e - 72}$

12. Simplify.

a) $\dfrac{7x^2 - 21x}{7x^2 - 28x + 21}$

b) $\dfrac{5x^2 - 20}{x^2 + 14x + 24}$

c) $\dfrac{x^3 - 9x^2 + 20x}{x^3 - 25x}$

d) $\dfrac{32 - 2a^2}{4a^2 - 44a + 112}$

e) $\dfrac{3x^2 - 75}{6x^2 + 30x}$

f) $\dfrac{2x^3 - 28x^2 - 102x}{18x - 2x^3}$

g) $\dfrac{2m^2 + 4m - 16}{3m^2 - 48}$

h) $\dfrac{3a^2 + 33a + 90}{6a^2 + 6a - 120}$

i) $\dfrac{3b^2 + 3b - 60}{2b^2 + 4b - 48}$

13. Choose one part of exercise 12. Write to explain how you simplified the expression.

14. Simplify each expression, if possible. If an expression cannot be simplified, explain why.

a) $\dfrac{4x + 2}{2}$

b) $\dfrac{2x + 6}{x + 3}$

c) $\dfrac{2x + 6x}{2}$

d) $\dfrac{4x^2 - 12x}{x - 3}$

e) $\dfrac{2x - 1}{2x - 1}$

f) $\dfrac{x - 7}{7 - x}$

g) $\dfrac{x + 5}{x - 5}$

h) $\dfrac{x^2 - 1}{x^2 + 1}$

15. Create a rational expression that simplifies to $\dfrac{x + 2}{2x}$, for which $x \neq 0$ and $x \neq 4$. Explain how you found the expression.

16. Three students were discussing how to simplify the expression $\dfrac{2x + 4}{x + 2}$.

George: $2x$ divided by x is 2. Four divided by 2 is 2. Then, $2 + 2$ equals 4.

Saleha: No, you must factor the numerator to $2(x + 2)$. Since the factor $x + 2$ is in the numerator and denominator, divide by this factor. They both become 1 and we get $\frac{2}{1}$, which is 2.

Roberta: No, divide 2 into the 2 in the denominator and the 2 in the numerator to get $\dfrac{x + 4}{x + 1}$.

Which student is correct? Explain how you know.

17. Simplify.

a) $\dfrac{2x^2 + 3xy + y^2}{3x^2 + 2xy - y^2}$

b) $\dfrac{9a^4 - 6a^3 + 15a^2}{3a}$

c) $\dfrac{16n - 8n^2 + 4n^3 - 4n^4}{2n^2}$

d) $\dfrac{x^2 - 4xy + 4y^2}{x^4 - 16y^4}$

e) $\dfrac{x^4 - y^4}{(x^2 + y^2)(x^2 - 5xy + 4y^2)}$

f) $\dfrac{16x^4 - y^4}{(4x^2 + y^2)^2(2x^2 + 3xy - 2y^2)}$

COMMUNICATING THE IDEAS

Explain how to simplify the rational expression $\dfrac{x^2 - 16}{x^2 + 2x - 24}$. Be sure to explain how factors are related to simplifying a rational expression.

Verifying Rational Simplifications

1. Work with a partner, using two graphing calculators. On the TI-83 graphing calculator, set the window to display $-9.4 \le X \le 9.4$ and $-6.3 \le Y \le 6.3$.

   a) Display the graph of $y = \dfrac{x^2 - 6x + 8}{x - 4}$. Check that $\dfrac{x^2 - 6x + 8}{x - 4}$ simplifies to $x - 2$. Have your partner display the graph of $y = x - 2$.

   b) Compare your displays *very carefully*. Are they identical, or is there a difference? Do they confirm that $\dfrac{x^2 - 6x + 8}{x - 4} = x - 2$? Explain.

   c) Predict what will happen on each graph if you use the trace key to move the cursor to the point where the two graphs differ. Use the two calculators to verify your prediction.

   d) When we write $\dfrac{x^2 - 6x + 8}{x - 4} = x - 2$, what value of x is not permitted? How do your results confirm that this value is not permitted?

   e) Use the tables feature to confirm the same results.

2. In exercise 1, there was a trinomial in the numerator of the rational expression and a binomial in the denominator. Suppose this situation is reversed. Use the same window settings as before.

   a) Display the graph of $y = \dfrac{x - 4}{x^2 - 6x + 8}$. Check that $\dfrac{x - 4}{x^2 - 6x + 8}$ simplifies to $\dfrac{1}{x - 2}$. Have your partner display the graph of $y = \dfrac{1}{x - 2}$.

   b) Compare the displays *very carefully*. Are they identical, or is there a difference? Do they confirm that $\dfrac{x - 4}{x^2 - 6x + 8} = \dfrac{1}{x - 2}$? Explain.

   c) Predict what will happen on each graph if you use the trace key to move through positive values of x, starting at 0. Use the two calculators to verify your prediction.

   d) When we write $\dfrac{x - 4}{x^2 - 6x + 8} = \dfrac{1}{x - 2}$, what two values of x are not permitted? How do your results confirm that both of these values are not permitted?

   e) Use the tables feature to confirm the same results.

3. In exercises 1 and 2, the simplified expressions were $x - 2$ and $\dfrac{1}{x - 2}$, respectively. Observe that each is the reciprocal of the other. How do the graphs of these two expressions compare?

7.3 Multiplying and Dividing Rational Expressions

1. Calculate each product. Write it in simplest form. Try to use the same method for each product.

   a) $\frac{4}{10} \times \frac{7}{2}$ b) $\frac{9}{10} \times \frac{7}{3}$ c) $\frac{16}{10} \times \frac{7}{4}$ d) $\frac{25}{10} \times \frac{7}{5}$

2. Consider the product $\frac{x^2}{10} \times \frac{7}{x}$.

   a) How are the products in exercise 1 related to this product?

   b) How do you think you would simplify $\frac{x^2}{10} \times \frac{7}{x}$?

   c) Simplify the product.

3. Calculate each quotient. Try to use the same method for each one.

   a) $\frac{2}{3} \div \frac{5}{9}$ b) $\frac{2}{4} \div \frac{5}{16}$ c) $\frac{2}{5} \div \frac{5}{25}$ d) $\frac{2}{6} \div \frac{5}{36}$

4. Consider the quotient $\frac{2}{x} \div \frac{5}{x^2}$.

   a) How are the quotients in exercise 3 related to this quotient?

   b) How do you think you would simplify $\frac{2}{x} \div \frac{5}{x^2}$?

   c) Simplify the quotient.

Rational expressions are multiplied and divided the same way as rational numbers are multiplied and divided.

Example 1

Simplify each expression.

a) $\frac{x^2}{6} \times \frac{2y}{3x}$

b) $\frac{10t^2(r+3)}{5(r-3)} \times \frac{2(r-3)}{rt}$

Solution

Use mental math to multiply, then divide.

a)
$$\frac{x^2}{6} \times \frac{2y}{3x}$$
$$= \frac{2x^2y}{18x}$$
$$= \frac{xy}{9}$$

Divide the numerator and denominator by their common factor $2x$.

b)
$$\frac{10t^2(r+3)}{5(r-3)} \times \frac{2(r-3)}{rt}$$
$$= \frac{20t^2(r+3)(r-3)}{5rt(r-3)}$$
$$= \frac{4t(r+3)}{r}$$

Divide the numerator and denominator by their common factor $5t(r-3)$.

To multiply rational expressions:

$$\frac{a}{b} \times \frac{c}{d} = \frac{ac}{bd}$$

Multiply the numerators.
Multiply the denominators.
Reduce to lowest terms.

Example 2

Simplify each expression.

a) $\frac{3a^3}{-5} \div \frac{(3a)^2}{10}$

b) $\frac{4(x-2)}{x^2} \div \frac{8(x-2)}{x(x+1)}$

Solution

Use mental math to multiply, then divide.

a)
$$\frac{3a^3}{-5} \div \frac{(3a)^2}{10}$$

Multiply by the reciprocal.

$$= \frac{3a^3}{-5} \times \frac{10}{9a^2}$$

$$= \frac{30a^3}{-45a^2}$$

Divide by the common factor $15a^2$.

$$= -\frac{2a}{3}$$

b)
$$\frac{4(x-2)}{x^2} \div \frac{8(x-2)}{x(x+1)}$$

$$= \frac{4(x-2)}{x^2} \times \frac{x(x+1)}{8(x-2)}$$

Divide by the common factor $4x(x-2)$.

$$= \frac{x+1}{2x}$$

To divide rational expressions:

$$\frac{a}{b} \div \frac{c}{d} = \frac{a}{b} \times \frac{d}{c}$$
$$= \frac{ad}{bc}$$

Multiply by the reciprocal.
Multiply the numerators.
Multiply the denominators.
Reduce to lowest terms.

When the polynomial in the numerator or denominator is a binomial or trinomial, you may need to factor it before you can simplify the expression.

Example 3

Simplify. Identify any nonpermissible values of the variables.

a) $\frac{x^2 + 7x + 10}{x^2 + x - 6} \times \frac{x+3}{x+5}$

b) $\frac{3m - 2n}{m^2 - 2mn + n^2} \div \frac{9m^2 - 4n^2}{3m^2n - 3mn^2}$

Solution

a) $\frac{x^2 + 7x + 10}{x^2 + x - 6} \times \frac{x+3}{x+5} = \frac{(x+2)(x+5)}{(x+3)(x-2)} \times \frac{(x+3)}{(x+5)}$

Note that $x + 3 = 0$ when $x = -3$, and $x - 2 = 0$ when $x = 2$, and $x + 5 = 0$ when $x = -5$

$$\frac{x^2 + 7x + 10}{x^2 + x - 6} \times \frac{x + 3}{x + 5} = \frac{x + 2}{x - 2}$$

This equation is not true if $x = -3, 2,$ or -5.

b) $\dfrac{3m - 2n}{m^2 - 2mn + n^2} \div \dfrac{9m^2 - 4n^2}{3m^2n - 3mn^2} = \dfrac{3m - 2n}{(m - n)(m - n)} \div \dfrac{(3m - 2n)(3m + 2n)}{3mn(m - n)}$ ①

Multiply by the reciprocal.

$\dfrac{3m - 2n}{m^2 - 2mn + n^2} \div \dfrac{9m^2 - 4n^2}{3m^2n - 3mn} = \dfrac{\cancel{(3m - 2n)}}{(m - n)\cancel{(m - n)}} \times \dfrac{3mn\cancel{(m - n)}}{\cancel{(3m - 2n)}(3m + 2n)}$ ②

$\dfrac{3m - 2n}{m^2 - 2mn + n^2} \div \dfrac{9m^2 - 4n^2}{3m^2n - 3mn} = \dfrac{3mn}{(m - n)(3m + 2n)}$

In ①, the denominators become 0 when $m = n$, $m = 0$, or $n = 0$.

In ②, the denominators become 0 when $m = \frac{2}{3}n$, or $m = -\frac{2}{3}n$.

This equation is not true if $m = n$, $m = 0$, $n = 0$, $m = \frac{2n}{3}$, or $m = \frac{-2n}{3}$.

DISCUSSING THE IDEAS

1. In exercise 1 of *Investigate*, can any of the products be simplified in more than one way? Explain your answer.

2. Are there other ways to simplify the products in *Examples 1* and *2* than those shown in the above solutions? If so, how do they compare with the methods above?

3. Explain why the quotient $\dfrac{\left(\dfrac{4(x - 2)}{x^2}\right)}{\left(\dfrac{8(x - 2)}{x(x + 1)}\right)}$ is equivalent to the expression in *Example 2b*.

7.3 EXERCISES

A 1. Simplify.

a) $\dfrac{5}{8} \times \dfrac{2a}{3}$ **b)** $\dfrac{m^2}{4} \times \dfrac{2}{m}$ **c)** $\dfrac{-3x}{10} \times \dfrac{5x}{9}$ **d)** $\dfrac{2c^2}{15} \times \dfrac{5}{3c}$

e) $\dfrac{-6t}{35} \times \dfrac{14t}{3}$ **f)** $\dfrac{9r^2}{4} \times \dfrac{8}{3r}$ **g)** $\dfrac{8t^3}{4} \times \dfrac{16}{2t}$ **h)** $\dfrac{25b^2}{7} \times \dfrac{14}{5b}$

2. Simplify.

a) $2 \div \dfrac{3}{x}$ **b)** $\dfrac{\left(\dfrac{2}{3}\right)}{x}$ **c)** $x \div \dfrac{2}{3}$ **d)** $\dfrac{3}{\left(\dfrac{2}{x}\right)}$

e) $-10 \div \dfrac{5}{x}$ **f)** $\dfrac{-\left(\dfrac{5}{x}\right)}{10}$ **g)** $-\dfrac{10}{x} \div 5$ **h)** $\dfrac{x}{\left(\dfrac{5}{10}\right)}$

3. Simplify. Identify any nonpermissible values of the variables.

a) $\dfrac{5}{8} \div \dfrac{3b}{4a}$ **b)** $\dfrac{x^2}{14} \div \dfrac{x}{2}$ **c)** $\dfrac{-6xy}{15} \div \dfrac{2x^2}{5}$ **d)** $\dfrac{9a}{4} \div \dfrac{3b}{2}$

e) $\dfrac{7m}{-3} \div \dfrac{5m}{-6}$ **f)** $\dfrac{15a}{2} \div \dfrac{25c^2}{3a^2}$ **g)** $\dfrac{3ab}{13c} \div \dfrac{9b^2}{52}$ **h)** $\dfrac{4st^2}{35} \div \dfrac{16s}{7t^2}$

4. Simplify.

a) $\dfrac{8t}{21s^2} \times \dfrac{3s}{4}$ **b)** $\dfrac{15x^2}{4x} \times \dfrac{6x^3}{5x^2}$ **c)** $\dfrac{14e}{8} \times \dfrac{12f^2}{49e^3}$ **d)** $\dfrac{-10x^3}{18x} \div \dfrac{-15x}{-27}$

e) $\dfrac{8n}{-21} \div \dfrac{-4n}{7n}$ **f)** $\dfrac{3x^2}{8y} \div \dfrac{9x}{28}$ **g)** $\dfrac{5a^2}{12b} \div \dfrac{25a}{6}$ **h)** $\dfrac{14st^3}{5s^2} \div \dfrac{4t}{15st^2}$

5. Simplify. Identify any nonpermissible values of the variables.

a) $\dfrac{3x^2}{2y} \times \dfrac{4y}{9x}$ **b)** $\dfrac{4a}{3b} \times \dfrac{9b^2}{6a}$ **c)** $\dfrac{2m}{9n} \div \dfrac{-4m}{3n^2}$

d) $\dfrac{3x^3y^2}{6xy} \times \dfrac{4xy}{5x^2y^2}$ **e)** $\dfrac{-8m^2n^5}{15mn^2} \div \dfrac{2m^4}{-25n^2}$ **f)** $\dfrac{4c^2d}{8cd} \div \dfrac{3c^2d^3}{6cd^3}$

6. Choose one part of exercise 5. Write to explain how you simplified the expression and identified the nonpermissible value(s) of the variable.

B **7.** Simplify.

a) $\dfrac{3a^2b}{12ab} \times \dfrac{8a^5b^4}{6ab^2}$ **b)** $\dfrac{-5x^2y}{(2xy)^3} \times \dfrac{-12x^2y^2}{-6x^2y}$ **c)** $\dfrac{2x^2y}{3xy} \times \dfrac{(6xy)^2}{4xy}$

d) $\dfrac{12mn^2}{9mn} \div \dfrac{(3mn)^2}{6mn^2}$ **e)** $\dfrac{2x}{3y} \times \dfrac{3y}{4z} \times \dfrac{4z}{5x}$ **f)** $\dfrac{(2m)^2}{5n} \times \dfrac{10m}{8n} \div \dfrac{15m}{(4n)^2}$

8. Each rational expression below is the product of two other rational expressions. For each, write what the two rational expressions might have been. Explain how you obtained your answers.

a) $\dfrac{1}{x}$ **b)** $\dfrac{x}{y}$ **c)** $\dfrac{3m}{-2n}$ **d)** $\dfrac{a}{2}$

e) $3b$ **f)** $\dfrac{-5}{xy}$ **g)** $\dfrac{-1}{abc}$ **h)** $\dfrac{4x}{3}$

9. Assume each rational expression in exercise 8 is the quotient of two other rational expressions. For each, write what the other two rational expressions might have been. Explain how you obtained your answers.

10. Simplify. Identify any nonpermissible values of the variables.

a) $\dfrac{2a}{a-3} \times \dfrac{7(a-3)}{4a}$ **b)** $\dfrac{5(x-2)}{8x} \times \dfrac{2x}{15(x-2)}$ **c)** $\dfrac{4(x-3)}{x+1} \div \dfrac{4}{x+1}$

d) $\dfrac{12m^2}{5(m+4)} \times \dfrac{10(m+4)}{3m}$ **e)** $\dfrac{3(s-2)}{4(s+5)} \div \dfrac{9(s-2)}{s+5}$ **f)** $\dfrac{3(5-a)}{2a} \div \dfrac{3(a-5)}{4(a+1)}$

11. Simplify.

a) $\dfrac{a+b}{3b} \times \dfrac{6b^2}{5(a+b)}$ **b)** $\dfrac{3x^2y}{12x} \times \dfrac{4xy^3}{2xy}$ **c)** $\dfrac{3y^3}{x^2-9} \times \dfrac{2x-6}{2y^2}$

d) $\dfrac{\left(\dfrac{3xy}{9x^2-12x}\right)}{\left(\dfrac{12y}{9x^2-16}\right)}$ **e)** $\dfrac{\left(\dfrac{10m^2n}{6m-9}\right)}{\left(\dfrac{25mn^2}{2m-3}\right)}$ **f)** $\dfrac{\left(\dfrac{4a^2-10}{a-3b}\right)}{\left(\dfrac{6a^2-15}{2a^2-18b^2}\right)}$

12. Simplify. Identify any nonpermissible values of the variables.

a) $\dfrac{15x}{2x+6} \div \dfrac{10x}{3x+9}$

b) $\dfrac{x^2-121}{x^2-4} \times \dfrac{x+2}{x-11}$

c) $\dfrac{5x-10}{6x+6} \div \dfrac{2x-4}{x+1}$

d) $\dfrac{y+2}{ay-by} \div \dfrac{y^2+2y}{ay^2-by^2}$

e) $\dfrac{3xy}{x^2-4} \times \dfrac{(x-2)^2}{4y^2}$

f) $\dfrac{(x+1)^2}{x^2-1} \times \dfrac{x^2-4}{(x+2)(x+1)}$

13. Each rational expression below is the product of two other rational expressions. For each, write what the other two rational expressions might have been. Explain how you obtained your answers.

a) $\dfrac{1}{x+1}$

b) $\dfrac{x-2}{x+2}$

c) $\dfrac{a}{a+b}$

d) $\dfrac{(x+y)^2}{x}$

14. Assume each rational expression in exercise 13 is the quotient of two other rational expressions. For each, write what the other two rational expressions might have been. Explain how you obtained your answers.

15. Simplify.

a) $\dfrac{a^2-3a-10}{25-a^2} \div \dfrac{a+2}{a+5}$

b) $\dfrac{x^2-2x-15}{x^2-9} \times \dfrac{x-3}{x-5}$

c) $\dfrac{x^2+x-2}{x^2-x} \times \dfrac{x^2+x}{x^2-1}$

d) $\dfrac{x^2-2x-15}{x^2-9} \times \dfrac{3-x}{x-5}$

e) $\dfrac{a^2+2a-15}{a^2-8a+7} \times \dfrac{a^2-5a-14}{a^2+7a+10}$

f) $\dfrac{x^2+5x+6}{x^2-5x+6} \div \dfrac{x^2-x-6}{x^2+x-6}$

C **16.** Simplify.

a) $\dfrac{x^2-16y^2}{6x^2y} \div \dfrac{x^2+xy-20y^2}{4x^3y^2}$

b) $\dfrac{a^2+11ab+30b^2}{a^2-25b^2} \times \dfrac{3a^2-15ab}{6a^2+36ab}$

c) $\dfrac{x^2+5xy+6y^2}{x^2+4xy-5y^2} \times \dfrac{x^2+3xy-10y^2}{x^2+xy-6y^2}$

d) $\dfrac{m^2-9mn+14n^2}{m^2+7mn+12n^2} \div \dfrac{3m^2-21mn}{4m^3+16m^2n}$

17. Simplify.

a) $\dfrac{3x^2+3x-6}{x^2y-7xy} \times \dfrac{x^2y-13xy+42y}{6x^2+12x}$

b) $\dfrac{x^2+5xy+6y^2}{x^2+7xy+10y^2} \times \dfrac{x^2+6xy+5y^2}{x^2+2xy-3y^2}$

c) $\dfrac{x+2y}{x-3y} \times \dfrac{x^2-9y^2}{x^2-4y^2} \div \dfrac{x+3y}{x-2y}$

d) $\dfrac{(3a+7b)^2}{2a-5b} \times \dfrac{4a^2-25b^2}{9a^2-49b^2} \div \dfrac{2a+5b}{3a-7b}$

18. Suppose $x = a+b$ and $y = a-b$. Write each expression in terms of a and b, then simplify.

a) $\dfrac{x^2-xy-12y^2}{x^2-2xy-3y^2} \times \dfrac{x^2+5xy+4y^2}{x^2-16y^2}$

b) $\left(\dfrac{3x-21y}{6x+12y}\right)^2 \div \dfrac{x^2-49y^2}{2x^2+8xy+8y^2}$

COMMUNICATING THE IDEAS

In your journal, write a brief note to compare multiplication and division of rational expressions with multiplication and division of rational numbers. Include examples to support your explanations.

Should Pop Cans Be Redesigned?

See page 398 for information about the aluminum pop can.

Is it possible to redesign the can so that less aluminum is used?
If so, how much money could the manufacturer save?

 **DEVELOP A MODEL**

Consider the can as a cylinder. Visualize changing the shape of the cylinder, but
keeping its volume as 355 mL.

A short, wide
design uses more
aluminum for the
top and bottom.

A tall, narrow
design uses more
aluminum for the
side.

For any height and radius, these two formulas apply:
    Volume: $355 = \pi r^2 h$ ①
    Surface area: $A = 2\pi r^2 + 2\pi rh$ ②

You will investigate what values of r and h will give the least possible value of
A, with volume 355 mL.

1. To express the surface area in terms of r

   a) Solve formula ① for h.

   b) Substitute the expression for h in formula ②, then simplify.

2. Copy and complete this table.
   To do this

   a) Use the expression from exercise
   1a to calculate the height for each
   radius.

   b) Use the expression from exercise
   1b to calculate the surface area
   for each radius.

Radius (cm)	Height (cm)	Surface area (cm²)
1.0		
2.0		
3.0		
4.0		
5.0		
6.0		
7.0		
8.0		

3. Examine your results for exercise 2.

   a) What values of r and h produce the least surface area?

   b) What is the least (minimum) surface area?

Complete exercises 4 and 5 if you have a graphing calculator or a computer with graphing software.

4. a) Use the equation from exercise 1b. Graph the surface area as a function of the radius, using an appropriate graphing window. You may need to change your window to get a suitable graph.

   b) Trace along the curve or otherwise to determine the radius for which the surface area is a minimum.

   c) Calculate the height of the can that has the minimum surface area.

5. Compare the radius and height you calculated in exercise 4. Describe how they appear to be related. What conclusion can you make about the diameter and the height of the can with the minimum surface area?

 LOOK AT THE IMPLICATIONS

6. An aluminum pop can has a radius of about 3.2 cm and a height of 11.0 cm.

   a) Calculate to confirm that the volume is approximately 355 mL.

   b) Calculate the surface area of the can.

   c) Calculate the difference between this area and the minimum area.

   d) Write the difference from part c as a fraction of the surface area of the can. Then write the fraction as a percent.

   e) How much could be saved by redesigning the cans this way?

7. Soft drink manufacturers know the savings that would result from using less aluminum. Yet they have not changed the standard aluminum can. Why? List possible reasons for keeping the aluminum can as it is.

 REVISIT THE SITUATION

8. The metal used to make the top and bottom of the can is thicker than the metal used to make the side. Modify the model to take this into account. How does this affect the results?

7.4 Adding and Subtracting Rational Expressions: Part I

1. Add. Try to use the same method each time.

   a) $\frac{5}{6} + \frac{1}{4}$ b) $\frac{5}{9} + \frac{1}{6}$ c) $\frac{5}{12} + \frac{1}{8}$ d) $\frac{5}{15} + \frac{1}{10}$

2. Consider the expression $\frac{5}{3a} + \frac{1}{2a}$.

   a) How are the expressions in exercise 1 related to this expression?

   b) How do you think you would calculate $\frac{5}{3a} + \frac{1}{2a}$?

   c) Simplify the expression in part b.

3. Subtract. Try to use the same method each time.

   a) $\frac{2}{3} - \frac{1}{9}$ b) $\frac{2}{4} - \frac{1}{16}$ c) $\frac{2}{5} - \frac{1}{25}$ d) $\frac{2}{6} - \frac{1}{36}$

4. Consider the expression $\frac{2}{a} - \frac{1}{a^2}$.

   a) How are the expressions in exercise 3 related to this expression?

   b) How do you think you would calculate $\frac{2}{a} - \frac{1}{a^2}$?

   c) Simplify the expression in part b.

In Section 7.2, rational expressions were reduced to lowest terms. To add and subtract rational expressions, it is sometimes necessary to raise them to higher terms to obtain the lowest common denominator.

Example 1

Write an expression equivalent to $\frac{x+5}{x}$ with each denominator.

a) $3x$ b) x^2

c) x^2y d) $x(x-2)$

Solution

For each expression, multiply the denominator by the factor necessary to get the given denominator. Then multiply the numerator by the same factor.

a) $\frac{x+5}{x} = \frac{x+5}{x} \times \frac{3}{3}$
$$= \frac{3(x+5)}{3x}$$
$$= \frac{3x+15}{3x}$$

b) $\frac{x+5}{x} = \frac{x+5}{x} \times \frac{x}{x}$
$$= \frac{x(x+5)}{x^2}$$
$$= \frac{x^2+5x}{x^2}$$

c) $\dfrac{x+5}{x} = \dfrac{x+5}{x} \times \dfrac{xy}{xy}$

$\qquad = \dfrac{xy(x+5)}{x^2y}$

$\qquad = \dfrac{x^2y + 5xy}{x^2y}$

d) $\dfrac{x+5}{x} = \dfrac{x+5}{x} \times \dfrac{x-2}{x-2}$

$\qquad = \dfrac{(x+5)(x-2)}{x(x-2)}$

$\qquad = \dfrac{x^2 + 3x - 10}{x(x-2)}$

Rational expressions are added and subtracted the same way that rational numbers are added and subtracted.

Example 2

Add or subtract.

a) $\dfrac{4}{3x} - \dfrac{7x}{6}$

b) $\dfrac{3}{8a} + \dfrac{5}{12a^2}$

c) $\dfrac{3y}{x} + \dfrac{7x}{y^2} - \dfrac{2x+1}{4y}$

Solution

a) The lowest common denominator is $6x$.

$\dfrac{4}{3x} - \dfrac{7x}{6} = \dfrac{4}{3x} \times \dfrac{2}{2} - \dfrac{7x}{6} \times \dfrac{x}{x}$

$\qquad = \dfrac{8}{6x} - \dfrac{7x^2}{6x}$

$\qquad = \dfrac{8 - 7x^2}{6x}$

b) The lowest common denominator is $24a^2$.

$\dfrac{3}{8a} + \dfrac{5}{12a^2} = \dfrac{3}{8a} \times \dfrac{3a}{3a} + \dfrac{5}{12a^2} \times \dfrac{2}{2}$

$\qquad = \dfrac{9a}{24a^2} + \dfrac{10}{24a^2}$

$\qquad = \dfrac{9a + 10}{24a^2}$

c) The lowest common denominator is $4xy^2$.

$\dfrac{3y}{x} + \dfrac{7x}{y^2} - \dfrac{2x+1}{4y} = \dfrac{3y}{x} \times \dfrac{4y^2}{4y^2} + \dfrac{7x}{y^2} \times \dfrac{4x}{4x} - \dfrac{2x+1}{4y} \times \dfrac{xy}{xy}$

$\qquad = \dfrac{12y^3}{4xy^2} + \dfrac{28x^2}{4xy^2} - \dfrac{xy(2x+1)}{4xy^2}$

$\qquad = \dfrac{12y^3 + 28x^2 - 2x^2y - xy}{4xy^2}$

In arithmetic, a fraction such as $\dfrac{2 + \frac{3}{4}}{5 - \frac{1}{3}}$, which has other fractions in the numerator and/or the denominator, is called a *complex fraction*. To simplify a complex fraction, we multiply the numerator and the denominator by the common denominator of the individual fractions.

$\dfrac{2 + \frac{3}{4}}{5 - \frac{1}{3}} = \dfrac{\left(2 + \frac{3}{4}\right)}{\left(5 - \frac{1}{3}\right)} \times \dfrac{12}{12}$

$\qquad = \dfrac{24 + 9}{60 - 4}$

$\qquad = \dfrac{33}{56}$

Sometimes, the numerator and denominator of a rational expression contain rational expressions.

Example 3

Simplify.

a) $\dfrac{\frac{1}{2} + x}{\frac{1}{3} - x}$
b) $\dfrac{\frac{1}{x} + 2}{\frac{1}{x} - 3}$

Solution

a) $\dfrac{\frac{1}{2} + x}{\frac{1}{3} - x}$

The denominator in the numerator is 2.
The denominator in the denominator is 3.

$$\dfrac{\frac{1}{2} + x}{\frac{1}{3} - x} = \dfrac{\left(\frac{1}{2} + x\right)}{\left(\frac{1}{3} - x\right)} \times \dfrac{6}{6}$$

> Multiply the numerator and denominator by their common denominator, 6.

$$= \dfrac{3 + 6x}{2 - 6x}$$

b) $\dfrac{\frac{1}{x} + 2}{\frac{1}{x} - 3}$

The denominator in both the numerator and the denominator is x.

$$\dfrac{\frac{1}{x} + 2}{\frac{1}{x} - 3} = \dfrac{\left(\frac{1}{x} + 2\right)}{\left(\frac{1}{x} - 3\right)} \times \dfrac{x}{x}$$

> Multiply the numerator and denominator by x.

$$= \dfrac{1 + 2x}{1 - 3x}$$

DISCUSSING THE IDEAS

1. When we raise a rational expression to higher terms, why must we multiply both the numerator and the denominator by the same factor?

2. In *Example 2a*, could you use other common denominators to simplify $\frac{4}{3x} - \frac{7x}{6}$? Would you get the same answer? Try it. Discuss your ideas with a partner.

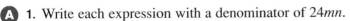

A **1.** Write each expression with a denominator of 24*mn*.

a) $\frac{-2}{3mn}$ b) $\frac{5m}{8mn}$ c) $\frac{-5}{6m}$ d) $-\frac{13}{24}$ e) $\frac{-1}{4}$

f) $\frac{3}{-8}$ g) $\frac{7m}{12}$ h) $\frac{-7m}{8n}$ i) $\frac{11n}{-12m}$ j) $\frac{5mn}{-6n}$

2. Write an expression equivalent to:

a) $\frac{2x}{3}$ with each denominator i) 6 ii) 12*x* iii) $3x^2$

b) $\frac{3m+1}{4}$ with each denominator i) −12 ii) 8*x* iii) $16x^2$

c) $\frac{a-7}{a}$ with each denominator i) 3*a* ii) a^2 iii) $5a^4$

d) $\frac{5y-3}{2y}$ with each denominator i) 6*y* ii) $4y^3$ iii) $-2y^2$

e) $\frac{4x+10}{2x^2}$ with each denominator i) $-4x^2$ ii) $20x^3$ iii) x^2

3. Write three rational expressions equivalent to $\frac{3}{x+4}$.

4. For each pair of expressions, write an equivalent pair with a common denominator.

a) $\frac{x}{2}, \frac{x}{3}$ b) $\frac{2}{x}, \frac{x}{5}$ c) $\frac{3}{2a}, \frac{2}{a}$ d) $\frac{5}{2a}, \frac{4}{3a}$

e) $\frac{5}{n}, \frac{2}{n^2}$ f) $\frac{1}{6x}, \frac{5}{8x}$ g) $\frac{x+1}{5x^2}, \frac{x-1}{4x}$ h) $\frac{x+2}{3x^2}, \frac{x-3}{2x}$

B **5.** Simplify. Identify any nonpermissible values of the variables.

a) $\frac{5}{x} - \frac{2}{x}$ b) $\frac{4}{3x} + \frac{2}{3x}$ c) $\frac{7}{4x} - \frac{5x}{4x}$ d) $\frac{2}{3m^2} - \frac{9m}{3m^2}$

e) $\frac{5b}{2a} + \frac{7b}{2a}$ f) $\frac{16x}{5y^2} - \frac{11x}{5y^2}$ g) $\frac{9a}{7b^2} + \frac{5a}{7b^2}$ h) $\frac{33s^2}{9t^3} - \frac{6s^2}{9t^3}$

6. Simplify.

a) $\frac{2a}{3} - \frac{4a}{5}$ b) $\frac{2}{3a} - \frac{4}{5a}$ c) $\frac{2a}{3} - \frac{4}{5a}$ d) $\frac{2}{3a} - \frac{4a}{5}$

e) $\frac{2}{3a} - \frac{4a}{5a^2}$ f) $\frac{2a}{3a^2} - \frac{4}{5}$ g) $\frac{3a}{2} - \frac{7}{6a^2}$ h) $\frac{3a}{2a} - \frac{7}{6a}$

7. Simplify. Identify any nonpermissible values of the variables.

a) $\frac{2}{x} + \frac{5}{2x}$ b) $\frac{4}{x} + \frac{27}{5x}$ c) $\frac{7}{10x} + \frac{4}{15x}$ d) $\frac{5}{2a} + \frac{3}{4a}$

e) $\frac{7}{8m} + \frac{5}{6m}$ f) $\frac{2}{9k} - \frac{5}{6k}$ g) $\frac{5}{9t} + \frac{2}{5t}$ h) $\frac{6}{7b} - \frac{5}{8b}$

8. Choose one part of exercise 7. Write to explain how you simplified the expression and identified the nonpermissible value(s) of the variable.

9. Write each rational expression as the sum of two rational expressions. Explain how you obtained each sum.

a) $\frac{2}{a}$

b) $\frac{x}{2}$

c) $\frac{1}{3x}$

d) $\frac{3}{x}$

e) $\frac{2}{3n}$

f) $\frac{3a}{4b}$

g) $\frac{1}{9ab}$

h) $\frac{7ab}{10c}$

10. Write each rational expression in exercise 9 as the difference of two rational expressions. Explain how you obtained each difference.

11. Simplify.

a) $\frac{7a}{10} - \frac{2a}{5} + \frac{3a}{10}$

b) $\frac{5m}{6} - \frac{3m}{4} + \frac{m}{8}$

c) $\frac{4x}{9} - \frac{2x}{3} + \frac{5x}{6}$

d) $\frac{7c}{12} - \frac{5c}{9} - \frac{5c}{6}$

e) $\frac{2e}{3} - \frac{5e}{6} + \frac{3e}{4}$

f) $\frac{3m}{8} - \frac{2m}{3} + \frac{5m}{6}$

12. Simplify. Identify any nonpermissible values of the variables.

a) $\frac{1}{2a} + \frac{1}{3a} + \frac{1}{4a}$

b) $\frac{2}{3x} - \frac{3}{4x} - \frac{1}{2x}$

c) $\frac{3}{8m} - \frac{2}{3m} + \frac{5}{6m}$

d) $\frac{7}{6x} - \frac{5}{2x} - \frac{1}{3x}$

e) $\frac{3}{4y} + \frac{2}{3y} - \frac{5}{6y}$

f) $\frac{3}{8y} - \frac{5}{6y} + \frac{1}{4y}$

13. Simplify.

a) $\frac{x+3}{x} + \frac{x-5}{x}$

b) $\frac{5m-2}{m} - \frac{4m+7}{m}$

c) $\frac{4a-2}{3a} + \frac{7a+11}{3a}$

d) $\frac{6x+7}{5x^2} - \frac{2x-19}{5x^2}$

e) $\frac{7m+4}{2m} - \frac{12m+11}{2m}$

f) $\frac{2x-8}{4x} - \frac{10x-7}{4x}$

14. Simplify.

a) $\frac{k-7}{4} - \frac{k+2}{5}$

b) $\frac{c+5}{3} + \frac{c-8}{2}$

c) $\frac{x+4}{3} - \frac{x+2}{4}$

d) $\frac{m-5}{4} + \frac{m+3}{6}$

e) $\frac{2a+3}{8} - \frac{5a-4}{6}$

f) $\frac{4x-7}{6} + \frac{2x-7}{9}$

15. Choose one part of exercise 14. Write to explain how you simplified the expression.

16. Simplify.

a) $\frac{x+2}{3x} + \frac{2x-5}{2x}$

b) $\frac{2n-7}{8n} - \frac{3n-4}{6n}$

c) $\frac{5a-9}{6a} - \frac{3a+1}{9a}$

17. Add or subtract.

a) $\frac{5a}{a} - \frac{5}{2a}$

b) $\frac{2}{3m} - \frac{1}{2n}$

c) $\frac{4}{x} + \frac{3}{xy}$

d) $1 + \frac{a}{2b}$

18. Simplify.

a) $\dfrac{\frac{1}{x}+4}{\frac{1}{x}-4}$

b) $\dfrac{4-\frac{1}{x}}{4+\frac{1}{x}}$

c) $\dfrac{x-\frac{1}{4}}{x+\frac{1}{4}}$

d) $\dfrac{\frac{1}{4}+x}{\frac{1}{4}-x}$

e) $\dfrac{\frac{3}{2x}+2}{\frac{4}{3x}-1}$

f) $\dfrac{5+\frac{2}{5x}}{3-\frac{3}{2x}}$

19. Add or subtract.

a) $\frac{3}{x} - \frac{2}{y} + 1$

b) $\frac{2}{a} - \frac{3}{b} + \frac{4}{c}$

c) $\frac{1}{2x} + \frac{3}{4y} - \frac{5}{6z}$

20. Add or subtract.

a) $\frac{7x}{6y} + \frac{5x}{3x}$

b) $\frac{5m}{3n} - \frac{4n}{3m}$

c) $\frac{3a}{5b} - \frac{4b}{3a}$

d) $\frac{3x}{2a} - \frac{4a}{5x}$

21. Write as a sum or a difference.

a) $\frac{12 - a}{3}$

b) $\frac{2x - 5}{10x}$

c) $\frac{x^2 + xy}{xy}$

d) $\frac{7x^2 + x + 1}{x}$

C **22.** Simplify.

a) $\left(a - \frac{1}{a}\right)\left(a - \frac{2}{a}\right)$

b) $\left(k + \frac{3}{k}\right)\left(k - \frac{5}{k}\right)$

c) $\left(2a - \frac{3}{a}\right)^2$

23. a) Simplify $\left(x + \frac{1}{x}\right)^2 - \left(x^2 + \frac{1}{x^2}\right)$.

   b) Suppose $x + \frac{1}{x} = 3$. Determine the value of $x^2 + \frac{1}{x^2}$.

24. Suppose $\frac{x}{y} = \frac{3}{2}$. Evaluate each rational expression.

a) $\frac{y}{x}$

b) $\frac{x + y}{y}$

c) $\frac{x + y}{x}$

d) $\frac{x + y}{x - y}$

25. Consider the product $\left(x + \frac{1}{x}\right)\left(x + \frac{2}{x}\right)$.

   a) Multiply the factors, then simplify.

   b) Simplify each factor, then multiply.

   c) Are the answers to parts a and b the same? Explain.

26. In *Example 3b*, we showed that the rational expression

$\dfrac{\frac{1}{x} + 2}{\frac{1}{x} - 3}$ simplifies to $\frac{1 + 2x}{1 - 3x}$. Compare the values of x for which these two

expressions are not defined. Is the following statement true or false? Give a
reason for your answer.

A rational expression is defined for all real values of x except those values
for which the denominator is 0.

COMMUNICATING THE IDEAS

In your journal, write a brief note to compare the addition and subtraction of rational
expressions with the addition and subtraction of rational numbers. Include examples to
illustrate your explanations.

Exploring the Lens Formula

The focal length of a lens is the distance from the centre of the lens to the point where incoming parallel light rays converge on the other side of the lens.

When an object is placed on one side of the lens, its image appears on the other side. The focal length, f, object distance, p, and image distance, q, are related by the formula called the *lens equation*:

$$\frac{1}{f} = \frac{1}{p} + \frac{1}{q}$$

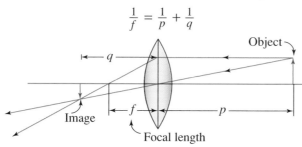

When any two of f, p, and q are given, you can use the lens equation to calculate the third. If this is constantly repeated, you can use the nomogram below instead of calculating. This nomogram is used with a straightedge. For example, assume that a lens has a focal length of 40 mm, and an object is placed 120 mm in front of it. Place a straightedge so that it passes through 40 on the f scale and 120 on the p scale. You will find that the straightedge passes through 60 on the q scale, indicating that the image is 60 mm from the lens.

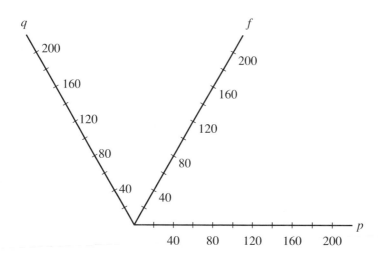

Mathematics & Science

In exercises 1 to 5, assume that the lens has a focal length of 40 mm.

1. a) Verify that $f = 40$, $p = 120$, and $q = 60$ satisfy the lens equation.

   b) Use the nomogram. Determine other values of p and q for the same value of f. Summarize your results in a table.

   c) Use your table. Draw a graph to show how the values of q depend on the values of p.

2. Suppose an object is placed far from the lens and is moved closer and closer to it. Use the nomogram to explain your answers to these questions.

   a) What do you know about the image distance when the object distance is very great?

   b) What happens to the image distance when the object distance is close to 40 mm?

   c) What happens to the image when the object is 40 mm from the lens? What happens when the object is less than 40 mm from the lens?

3. Use the graph in exercise 1. Explain your answers to the questions in exercise 2.

For more accurate results, you can use the lens equation instead of the nomogram. Substitute 40 for f in the lens equation to obtain an equation relating p and q. Use this equation in exercises 4 and 5.

4. Suppose an object is placed 60 mm in front of the lens. Substitute 60 for p, and calculate the image distance, q, to the nearest millimetre.

5. To determine values of q for other values of p, you can continue substituting for p and solving for q. Here is a more efficient method: solve your equation for q to obtain a formula to determine q for any value of p. Do this, then substitute some values of p into your formula. Calculate the corresponding values of q.

6. A camera lens has a focal length of 55 mm. The camera takes a picture of a flower that is 60 cm in front of the lens. Calculate the distance between the lens and the film.

7. A zoom lens in a camera is used to take a picture of the flower in exercise 6. The flower is 60.0 cm in front of the lens. The lens is adjusted so the distance between the lens and the film is 10.4 cm. Calculate the focal length of the lens at this setting.

7.5 Adding and Subtracting Rational Expressions: Part II

We can use the methods of Section 7.4 to add and subtract rational expressions with denominators that are not monomials.

Example 1

Simplify.

a) $\dfrac{m-4}{m-2} - \dfrac{m-10}{m-2}$ **b)** $\dfrac{5}{x} + \dfrac{6x}{x+4}$ **c)** $a - \dfrac{3}{a+b}$ **d)** $\dfrac{x+3}{x-5} - \dfrac{x-7}{x-2}$

Solution

a) Since the terms have a common denominator, the numerators can be combined.

$$\frac{m-4}{m-2} - \frac{m-10}{m-2} = \frac{(m-4) - (m-10)}{m-2}$$

$$= \frac{m-4-m+10}{m-2}$$

$$= \frac{6}{m-2}$$

b) The lowest common denominator is $x(x+4)$.

$$\frac{5}{x} + \frac{6x}{x+4} = \frac{5}{x} \times \frac{x+4}{x+4} + \frac{6x}{x+4} \times \frac{x}{x}$$

$$= \frac{5(x+4)}{x(x+4)} + \frac{6x(x)}{x(x+4)}$$

$$= \frac{5x+20}{x(x+4)} + \frac{6x^2}{x(x+4)}$$

$$= \frac{5x+20+6x^2}{x(x+4)}, \text{ or } \frac{6x^2+5x+20}{x(x+4)}$$

c) Since a can be written as $\frac{a}{1}$, the lowest common denominator is $a+b$.

$$\frac{a}{1} - \frac{3}{a+b} = \frac{a}{1} \times \frac{a+b}{a+b} - \frac{3}{a+b} \times \frac{1}{1}$$

$$= \frac{a(a+b)}{a+b} - \frac{(3)(1)}{a+b}$$

$$= \frac{a^2+ab}{a+b} - \frac{3}{a+b}$$

$$= \frac{a^2+ab-3}{a+b}$$

d) The lowest common denominator is $(x-5)(x-2)$.

$$\frac{x+3}{x-5} - \frac{x-7}{x-2} = \frac{x+3}{x-5} \times \frac{x-2}{x-2} - \frac{x-7}{x-2} \times \frac{x-5}{x-5}$$

$$= \frac{(x+3)(x-2)}{(x-5)(x-2)} - \frac{(x-7)(x-5)}{(x-5)(x-2)}$$

$$= \frac{x^2+x-6}{(x-5)(x-2)} - \frac{x^2-12x+35}{(x-5)(x-2)}$$

$$= \frac{x^2+x-6-x^2+12x-35}{(x-5)(x-2)}$$

$$= \frac{13x-41}{(x-5)(x-2)}$$

Frequently, we must factor the denominators of rational expressions so the lowest common denominator can be more readily determined.

Example 2

Simplify.

a) $\dfrac{7}{2x+4} - \dfrac{5}{3x+6}$ 　　　**b)** $\dfrac{b}{3-a} - \dfrac{b}{a+3} + \dfrac{1}{a^2-9}$

Solution

a) $\dfrac{7}{2x+4} - \dfrac{5}{3x+6} = \dfrac{7}{2(x+2)} - \dfrac{5}{3(x+2)}$

> The lowest common denominator is $6(x+2)$.

$$= \dfrac{7}{2(x+2)} \times \dfrac{3}{3} - \dfrac{5}{3(x+2)} \times \dfrac{2}{2}$$

$$= \dfrac{7 \times 3}{6(x+2)} - \dfrac{5 \times 2}{6(x+2)}$$

$$= \dfrac{21}{6(x+2)} - \dfrac{10}{6(x+2)}$$

$$= \dfrac{21-10}{6(x+2)}$$

$$= \dfrac{11}{6(x+2)}$$

b) $\dfrac{b}{3-a} - \dfrac{b}{a+3} + \dfrac{1}{a^2-9} = \dfrac{b}{3-a} - \dfrac{b}{a+3} + \dfrac{1}{(a-3)(a+3)}$

> Write $\dfrac{b}{3-a}$ as $\dfrac{-b}{a-3}$. Then the lowest common denominator is $(a-3)(a+3)$.

$$= \dfrac{-b}{a-3} \times \dfrac{a+3}{a+3} - \dfrac{b}{a+3} \times \dfrac{a-3}{a-3} + \dfrac{1}{(a-3)(a+3)}$$

$$= \dfrac{-b(a+3)}{(a-3)(a+3)} - \dfrac{b(a-3)}{(a-3)(a+3)} + \dfrac{1}{(a-3)(a+3)}$$

$$= \dfrac{-ba-3b}{(a-3)(a+3)} - \dfrac{ba-3b}{(a-3)(a+3)} + \dfrac{1}{(a-3)(a+3)}$$

$$= \dfrac{-ba-3b-ba+3b+1}{(a-3)(a+3)}$$

$$= \dfrac{-2ab+1}{(a-3)(a+3)}$$

When possible, the sum or difference of rational expressions should be reduced to lowest terms.

Example 3

Simplify.

$$\frac{x}{x^2 - 9x + 18} - \frac{x - 2}{x^2 - 10x + 24}$$

Solution

$$\frac{x}{x^2 - 9x + 18} - \frac{x - 2}{x^2 - 10x + 24} = \frac{x}{(x - 6)(x - 3)} - \frac{x - 2}{(x - 6)(x - 4)}$$

> The lowest common denominator is $(x - 6)(x - 3)(x - 4)$.

$$= \frac{x}{(x - 6)(x - 3)} \times \frac{x - 4}{x - 4} - \frac{x - 2}{(x - 6)(x - 4)} \times \frac{x - 3}{x - 3}$$

$$= \frac{x(x - 4)}{(x - 6)(x - 3)(x - 4)} - \frac{(x - 2)(x - 3)}{(x - 6)(x - 3)(x - 4)}$$

$$= \frac{x^2 - 4x}{(x - 6)(x - 3)(x - 4)} - \frac{x^2 - 5x + 6}{(x - 6)(x - 3)(x - 4)}$$

$$= \frac{x^2 - 4x - (x^2 - 5x + 6)}{(x - 6)(x - 3)(x - 4)}$$

$$= \frac{x^2 - 4x - x^2 + 5x - 6}{(x - 6)(x - 3)(x - 4)}$$

$$= \frac{x - 6}{(x - 6)(x - 3)(x - 4)}$$

> Divide numerator and denominator by $x - 6$.

$$= \frac{1}{(x - 3)(x - 4)}$$

DISCUSSING THE IDEAS

1. In *Example 2a*, what would have happened if we had not factored the denominators? Would the answer be the same? Try it. Discuss your ideas with a partner.

2. In *Example 2b*, explain why we can write $\frac{b}{3 - a}$ as $\frac{-b}{a - 3}$. How did we know that we should write $\frac{b}{3 - a}$ this way?

7.5 EXERCISES

A 1. Simplify.

a) $\frac{3m - 5}{m + 3} + \frac{m + 4}{m + 3}$

b) $\frac{2s + 7}{s - 5} - \frac{6s - 4}{s - 5}$

c) $\frac{5k - 9}{k - 4} - \frac{2k + 3}{k - 4}$

d) $\frac{2x + 7}{x + 6} + \frac{2x + 11}{x + 6}$

e) $\frac{4m - 9}{2m + 1} - \frac{m - 4}{2m + 1}$

f) $\frac{3a - 8}{a^2 + 4} - \frac{7a + 3}{a^2 + 4}$

2. Simplify.

a) $\frac{4}{a - 3} - \frac{1}{a}$

b) $\frac{2}{y - 5} - \frac{6}{y}$

c) $\frac{7}{m} - \frac{3}{m - 4}$

d) $\frac{2c}{c - 1} - \frac{5}{c}$

e) $\frac{3x}{x + 2} - \frac{6}{x}$

f) $\frac{3}{x} + \frac{5}{x + 2}$

3. Simplify. Identify any nonpermissible values of the variables.

a) $\dfrac{3}{2a} - 4$

b) $\dfrac{7}{y+1} - 2$

c) $4 - \dfrac{9}{n-5}$

d) $x - \dfrac{2}{x+4}$

e) $\dfrac{3}{s-8} - 2s$

f) $\dfrac{2w}{w+3} - 4w$

g) $\dfrac{4}{x-1} - (x-2)$

h) $\dfrac{4}{x-1} - x - 2$

i) $x - 5 + \dfrac{2}{x-3}$

j) $x + 3 + \dfrac{5}{x-2}$

k) $\dfrac{2}{x-4} - x - 8$

l) $4 - \dfrac{3}{x+2} - x$

B 4. Simplify.

a) $\dfrac{2}{x+5} + \dfrac{3}{x+2}$

b) $\dfrac{4}{x-3} - \dfrac{2}{x+1}$

c) $\dfrac{x}{x+1} - \dfrac{2}{x-1}$

d) $\dfrac{3x}{x+4} + \dfrac{2}{x+4}$

e) $\dfrac{5x}{x-1} - \dfrac{2x}{x+3}$

f) $\dfrac{2x}{x+5} + \dfrac{3x}{x-3}$

5. Choose one part of exercise 4. Write to explain how you simplified the expression.

6. Simplify.

a) $\dfrac{1}{x+1} - \dfrac{1}{x-1}$

b) $\dfrac{x-3}{x-2} + \dfrac{1}{x-3}$

c) $\dfrac{3x-1}{x+7} - \dfrac{2x+1}{x-3}$

d) $\dfrac{x-2}{x+2} + \dfrac{x+1}{x-4}$

e) $\dfrac{x+6}{x-3} + \dfrac{x-4}{x-5}$

f) $\dfrac{x+1}{x-2} - \dfrac{x-1}{x+2}$

7. Simplify.

a) $\dfrac{2}{x} - \dfrac{3x}{x-2}$

b) $\dfrac{7}{2(x+3)} - \dfrac{4}{5(x+3)}$

c) $\dfrac{3y}{2(y+9)} + \dfrac{5y}{3(y+9)}$

d) $\dfrac{5}{3(a-7)} - \dfrac{2}{3(a+1)}$

e) $\dfrac{3}{a+1} + \dfrac{1}{a-1}$

f) $\dfrac{3x}{x-2} - \dfrac{4x}{x-3}$

8. Write each rational expression as the sum of two rational expressions.

a) $\dfrac{2}{a+1}$

b) $\dfrac{x+2}{x-6}$

c) $\dfrac{2b}{a+b}$

d) $\dfrac{3xy}{2x-y}$

9. Choose one part of exercise 8. Write to explain how you determined the sum.

10. Write each rational expression in exercise 8 as the difference of two rational expressions.

11. Add the rational expressions in each list. Extend each pattern for two more examples.

a) $\dfrac{1}{x} + \dfrac{1}{x+1}$

$\dfrac{1}{x+1} + \dfrac{1}{x+2}$

$\dfrac{1}{x+2} + \dfrac{1}{x+3}$

$\vdots$

b) $\dfrac{1}{x+1} + \dfrac{2}{x+2}$

$\dfrac{2}{x+2} + \dfrac{3}{x+3}$

$\dfrac{3}{x+3} + \dfrac{4}{x+4}$

$\vdots$

c) $\dfrac{1}{x(x-1)} + \dfrac{1}{x(x+1)}$

$\dfrac{1}{x(x-2)} + \dfrac{1}{x(x+2)}$

$\dfrac{1}{x(x-3)} + \dfrac{1}{x(x+3)}$

$\vdots$

12. Make up a pattern similar to each of those in exercise 11.

13. Simplify. Identify any nonpermissible values of the variables.

a) $\dfrac{6}{2x + 4} + \dfrac{9}{3x + 6}$

b) $\dfrac{3}{5x - 10} + \dfrac{7}{2x - 4}$

c) $\dfrac{5x}{3x + 9} - \dfrac{9x}{2x + 6}$

d) $\dfrac{5x}{10x - 15} - \dfrac{4x}{16x - 24}$

e) $\dfrac{2x + 5}{3x - 12} - \dfrac{2x}{4 - x}$

f) $\dfrac{3x}{2x + 8} - \dfrac{2x - 3}{3x + 12}$

14. Simplify.

a) $\dfrac{4x}{x^2 - 9x + 18} + \dfrac{2x - 1}{x - 6}$

b) $\dfrac{x - 7}{x^2 - 2x - 15} - \dfrac{3x}{x - 5}$

c) $\dfrac{2x}{x - 2} - \dfrac{3}{x^2 - 4}$

d) $\dfrac{4x + 1}{x + 3} + \dfrac{x - 6}{x^2 - 9}$

e) $\dfrac{3x}{x - 1} - \dfrac{2x}{x^2 + x - 2}$

f) $\dfrac{8x - 3}{x^2 - 7x + 12} - \dfrac{2x + 1}{x - 4}$

15. Choose one part of exercise 14. Write the steps a student could follow to simplify the expression.

16. Simplify.

a) $\dfrac{3m}{2(m - 1)} - \dfrac{5m}{2(m + 1)}$

b) $\dfrac{x - 1}{x(x + 5)} + \dfrac{2x - 3}{x(x - 1)}$

c) $\dfrac{8a}{5(a - 2)} + \dfrac{5a - 1}{3(a + 3)}$

d) $\dfrac{5k}{4(k - 3)} - \dfrac{4k}{k - 4}$

e) $\dfrac{3x - 1}{2(x - 2)} + \dfrac{5x + 1}{x + 7}$

f) $\dfrac{4m + 3}{3(2m - 1)} - \dfrac{m - 5}{2(3m + 7)}$

17. Simplify.

a) $\dfrac{x + 3}{x^2 + 11x + 24} - \dfrac{2x + 10}{x^2 + 11x + 30}$

b) $\dfrac{m - 4}{m^2 - 8m + 16} + \dfrac{3m + 21}{m^2 + 12m + 35}$

c) $\dfrac{3x + 9}{x^2 + 5x + 6} - \dfrac{2x - 2}{x^2 + x - 2}$

d) $\dfrac{5m + 25}{2m^2 + 13m + 15} - \dfrac{10m - 20}{m^2 - 4}$

e) $\dfrac{4x^2 - 20x}{x^2 + 2x - 35} + \dfrac{3x - 6}{3x^2 - 10x + 8}$

f) $\dfrac{2x}{3x^2 - 11x + 6} - \dfrac{3x - 12}{3x^2 - 14x + 8}$

C 18. Write each rational expression as the sum of two rational expressions.

a) $\dfrac{2x - 4}{x(x - 4)}$

b) $\dfrac{2x}{x^2 - y^2}$

c) $\dfrac{3x + 11}{(x + 3)(x + 4)}$

19. Simplify.

a) $\dfrac{3x^2 + 6xy}{3x} - \dfrac{4y^2 - 2xy}{2y}$

b) $\dfrac{x^2 - 5xy + 6y^2}{x - 3y} - \dfrac{x^2 - xy - 12y^2}{x - 4y}$

c) $\dfrac{x^2 - 4xy - 21y^2}{3x - 21y} + \dfrac{x^2 + 2xy - 24y^2}{2x + 2y}$

d) $\dfrac{a - b}{a^2 + 2ab - 3b^2} + \dfrac{a + b}{a^2 - 2ab - 3b^2}$

20. Suppose you extended the lists in exercise 11 upward. Determine the previous three expressions in each list. Predict each sum. Add the expressions to confirm your prediction.

COMMUNICATING THE IDEAS

Write an expression that is the difference of two rational expressions. In your journal, write to explain how you wrote the difference as a single rational expression.

When an equation contains a rational expression, first identify the value(s) of the variable for which the expression is not defined. Then multiply both sides of the equation by a common denominator.

Example 1

The equation $\frac{x}{5} = \frac{2}{x} + \frac{x+3}{5}$ is given.

a) For what value of x is the equation not defined?   **b)** Solve the equation.

Solution

a) The equation is not defined when $x = 0$.

b)
$$\frac{x}{5} = \frac{2}{x} + \frac{x+3}{5}$$
$$5x\left(\frac{x}{5}\right) = 5x\left(\frac{2}{x}\right) + 5x\left(\frac{x+3}{5}\right)$$

> Multiply both sides by $5x$, $(x \neq 0)$.

$$x^2 = 10 + x^2 + 3x$$
$$-3x = 10$$
$$x = -\frac{10}{3}$$

The solution is $x = -\frac{10}{3}$.

Example 2

The average cost, A dollars, of producing n videos is given by the formula $A = \frac{12\ 000 + 3n}{n}$. How many videos would have to be produced for the average cost to be $3.25?

Solution

$$A = \frac{12\ 000 + 3n}{n}$$
$$3.25 = \frac{12\ 000 + 3n}{n}$$

> Substitute 3.25 for A in the formula.

$$3.25(n) = \left(\frac{12\ 000 + 3n}{n}\right)(n)$$
$$3.25n = 12\ 000 + 3n$$

> Multiply both sides by n, $(n \neq 0)$.

$$0.25n = 12\ 000$$
$$n = \frac{12\ 000}{0.25}$$
$$= 48\ 000$$

For the average cost to be $3.25, 48 000 videos must be produced.

If an equation has a single rational expression on each side, it can be solved by using the shortcut explained on page 182.

Example 3

Solve. $\dfrac{5}{x-2} = \dfrac{9}{3x-6}$

Solution

The equation is not defined when

$$x - 2 = 0 \quad \text{or} \quad 3x - 6 = 0$$
$$x = 2 \qquad\qquad x = 2$$

Assume that $x \neq 2$

$$5(3x - 6) = 9(x - 2)$$
$$15x - 30 = 9x - 18$$
$$6x = 12$$
$$x = 2$$

Since this is the value of x for which the equation is not defined, we

conclude that the equation $\dfrac{5}{x-2} = \dfrac{9}{3x-6}$ has no solution.

Example 3 underlines the importance of identifying the value(s) of the variable for which an equation is not defined.

Sometimes, a rational equation will simplify to a quadratic equation.

Example 4

Solve.

a) $2a + \dfrac{5}{2} = \dfrac{3}{a}$
b) $x + \dfrac{18}{x+7} = 4$
c) $\dfrac{5}{x-1} - \dfrac{12}{x^2-1} = 1$

Solution

a) $2a + \dfrac{5}{2} = \dfrac{3}{a}$

The equation is not defined when $a = 0$. Assume that $a \neq 0$.

$$(2a)(2a) + (2a)\left(\tfrac{5}{2}\right) = (2a)\left(\tfrac{3}{a}\right)$$
$$4a^2 + 5a = 6$$
$$4a^2 + 5a - 6 = 0$$

> Multiply each term by the common denominator $2a$.

Solve the quadratic equation.

$$(4a - 3)(a + 2) = 0$$

Either $4a - 3 = 0$ or $a + 2 = 0$

$$a = \tfrac{3}{4} \qquad a = -2$$

The solution is $a = \tfrac{3}{4}$ and -2.

b) $x + \dfrac{18}{x + 7} = 4$

The equation is not defined when $x + 7 = 0$, or $x = -7$.

Assume that $x \neq -7$.

$(x + 7)(x) + (x + 7)(\dfrac{18}{x + 7}) = (x + 7)(4)$ ← Multiply each term by the common denominator x + 7.

$\qquad\qquad x^2 + 7x + 18 = 4x + 28$

$\qquad\qquad x^2 + 3x - 10 = 0$

$\qquad\qquad (x + 5)(x - 2) = 0$

Either $x + 5 = 0 \qquad$ or $\quad x - 2 = 0$

$\qquad\qquad x = -5 \qquad\qquad\quad x = 2$

The solution is $x = -5$ and 2.

c) $\qquad \dfrac{5}{x - 1} - \dfrac{12}{x^2 - 1} = 1$

$\dfrac{5}{x - 1} - \dfrac{12}{(x - 1)(x + 1)} = 1$

The equation is not defined when

$x - 1 = 0$, $x = 1$; or when $x + 1 = 0$, $x = -1$.

Assume that $x \neq 1$ and $x \neq -1$.

$(x - 1)(x + 1)(\dfrac{5}{x - 1}) - (x - 1)(x + 1)\left(\dfrac{12}{(x - 1)(x + 1)}\right) = (x - 1)(x + 1)(1)$ ← Multiply each term by the common denominator (x − 1)(x + 1).

$\qquad\qquad\qquad 5(x + 1) - 12 = (x - 1)(x + 1)$

$\qquad\qquad\qquad 5x + 5 - 12 = x^2 - 1$

$\qquad\qquad\qquad -x^2 + 5x - 6 = 0$

$\qquad\qquad\qquad x^2 - 5x + 6 = 0$ ← Multiply each side of the equation by −1.

$\qquad\qquad\qquad (x - 3)(x - 2) = 0$

Either $x - 3 = 0$ or $x - 2 = 0$

$\qquad\qquad x = 3 \qquad\qquad x = 2$

The solution is $x = 3$ and 2.

DISCUSSING THE IDEAS

1. What other ways could you solve the equation in *Example 1*? Could you use the shortcut method? Explain.

2. In *Example 3*, observe that no mechanical errors were made, but the solution obtained was not correct. Explain how this is possible.

3. Suppose you reduce the right side of the equation in *Example 3* to lowest terms. Can you tell from the result that there is no solution?

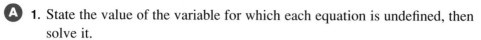

A 1. State the value of the variable for which each equation is undefined, then solve it.

a) $\frac{2}{a} = 4$ b) $\frac{a}{3} = 7$ c) $-\frac{5}{n} = 15$ d) $\frac{5c}{6} = -2$

e) $8 = \frac{24}{-x}$ f) $-15 = \frac{-3m}{2}$ g) $\frac{1}{2a} = -11$ h) $\frac{3}{4x} = 6$

i) $\frac{4}{3} = \frac{12}{x}$ j) $\frac{3}{5} = \frac{18}{x}$ k) $\frac{15}{x} = \frac{3}{4}$ l) $\frac{9}{x} = \frac{9}{8}$

m) $\frac{x}{3} = \frac{12}{x}$ n) $\frac{x}{8} = \frac{18}{x}$ o) $\frac{30}{x} = \frac{3x}{10}$ p) $\frac{90}{x} = \frac{2x}{5}$

2. State the value of the variable for which each equation is undefined, then solve it.

a) $\frac{1}{x} + \frac{2}{x} = \frac{3}{4}$ b) $\frac{5}{n} - \frac{2}{n} = \frac{1}{2}$ c) $5 - \frac{8}{2a} = \frac{11}{a}$

d) $\frac{2}{x} - \frac{7}{2x} = 6$ e) $\frac{3}{2a} = \frac{4}{3a} - \frac{1}{2}$ f) $\frac{4}{a} = \frac{1}{2} + \frac{3}{5a}$

g) $\frac{x}{3} = \frac{2}{x} + \frac{1}{3}$ h) $\frac{x}{5} = \frac{-2}{5} + \frac{3}{x}$ i) $\frac{x}{4} - \frac{7}{4} = \frac{2}{x}$

j) $\frac{x}{3} - \frac{2}{x} = \frac{1}{3}$ k) $\frac{2x}{7} - \frac{5}{7} = \frac{1}{x}$ l) $\frac{3x}{5} = \frac{2}{x} + \frac{x}{5}$

3. Choose one part of exercise 2. Write to explain how you determined the nonpermissible value of the variable, and how you solved the equation.

4. Solve.

a) $\frac{x}{5} + 3 = \frac{2}{x} + \frac{x+1}{5}$ b) $\frac{x}{3} + \frac{4}{x} = \frac{x+1}{3}$

c) $\frac{3x-2}{2} + 4 = \frac{13}{x} - \frac{1-6x}{4}$ d) $\frac{2(1-x)}{3} - 8 = \frac{1}{6x} - \frac{2x-3}{3}$

B 5. The average cost, A dollars, of producing n tape decks is given by the formula $A = \frac{40\,000 + 75n}{n}$. How many tape decks would have to be produced for the average cost to be $100?

6. In a regular polygon with n sides, the measure of each angle, a degrees, is given by the formula $a = 180 - \frac{360}{n}$.

   a) What is the measure of each angle in a regular decagon (10 sides)?

   b) Suppose each angle in a regular polygon measures 156°. How many sides does it have?

7. A 1-kg mass is suspended from n elastic bands, all the same size. The length to which they stretch, L centimetres, is given by the formula $L = 15.5 + \frac{45.5}{n}$.

   a) What is the length when there are 3 elastic bands?

   b) How many elastic bands are used when the length is 21.2 cm?

8. Determine the values of x for which each equation is undefined, then solve it.

a) $\dfrac{3}{x-2} = \dfrac{5}{2x+1}$

b) $\dfrac{5}{x+2} = \dfrac{4}{x-1}$

c) $\dfrac{7}{3x+5} = \dfrac{4}{2x+3}$

d) $\dfrac{3}{2x-4} = \dfrac{3}{x-2}$

e) $\dfrac{3}{5x-2} = \dfrac{2}{4x-5}$

f) $\dfrac{5}{2x+7} = \dfrac{-9}{4x+14}$

9. Choose one part of exercise 8. Write to explain how you determined the nonpermissible values of the variable, and how you solved the equation.

10. Solve.

a) $\dfrac{2x}{x-3} = 5$

b) $\dfrac{11x}{2x+3} = 7 + \dfrac{11x}{2x+3}$

c) $\dfrac{3x+3}{x+1} = 2$

d) $\dfrac{-2+x}{2x+1} = \dfrac{2}{3}$

e) $\dfrac{3x-1}{5x+4} = 4$

f) $\dfrac{2x+1}{3-2x} = \dfrac{3}{5}$

g) $\dfrac{2x}{x+3} = \dfrac{6x+5}{3x-1}$

h) $\dfrac{x-2}{x+3} = \dfrac{x+4}{x-1}$

i) $\dfrac{2x+1}{3x-2} = \dfrac{4x+3}{6x-5}$

j) $\dfrac{3x-5}{5x-3} = \dfrac{3x-1}{5x-1}$

k) $\dfrac{x-5}{1-x} = \dfrac{3-x}{x-1}$

l) $\dfrac{4x-3}{2x+1} = \dfrac{2x+1}{x-4}$

11. Solve.

a) $\dfrac{2x-3}{3x-4} = \dfrac{2x+7}{3x+4}$

b) $\dfrac{2x^2-2x+8}{2x-1} = x+1$

c) $\dfrac{3-x}{x-2} = 1 - \dfrac{2x-5}{x+2}$

d) $2 - \dfrac{x-1}{x+3} = -\dfrac{x-1}{x+3}$

e) $\dfrac{2(x-1)}{x-3} = \dfrac{x-4}{x-5} + 1$ long

f) $1 - \dfrac{x-5}{5x-1} = \dfrac{4(x-3)}{5x-2}$ long

g) $\dfrac{-8}{2-x} - \dfrac{20}{x^2-4} = 1$

h) $\dfrac{x}{x-3} + \dfrac{2}{x+3} = 0$

i) $x + \dfrac{6}{x+5} = 2$

j) $\dfrac{1}{x+2} + \dfrac{4}{2x-1} = 1$

k) $\dfrac{x+3}{3x-5} + 1 = \dfrac{3(x-2)}{x-1}$ long

l) $4 + \dfrac{1}{x-4} = 4$

m) $\dfrac{x^2}{x^2-4} = \dfrac{2x}{x+2}$

n) $\dfrac{3x^2}{x^2-1} = \dfrac{x}{x+1} + \dfrac{x}{1-x}$

o) $\dfrac{-9x^2}{x^2-25} = \dfrac{4x}{x-5} + \dfrac{x}{x+5}$

12. Solve.

a) $x + \dfrac{8}{x+5} = 4$

b) $x + \dfrac{30}{x+8} = 3$

c) $x - \dfrac{1}{x-2} = 2$

d) $x + \dfrac{6}{x+4} = 3$

e) $x + \dfrac{8}{x+6} = \dfrac{8}{x+6}$

f) $x - \dfrac{1}{x+4} = -4$

13. Solve.

a) $\dfrac{5}{x+1} + \dfrac{4}{3} = \dfrac{x+1}{x-1}$

b) $\dfrac{2m+3}{m+3} + \dfrac{1}{2} = \dfrac{m+1}{m-1}$

c) $\dfrac{a}{a+1} = \dfrac{1}{3} + \dfrac{a-1}{a+3}$

d) $\dfrac{3x+2}{2x+1} = \dfrac{3x+1}{x-1} - \dfrac{1}{3}$

e) $\dfrac{2x-1}{2x+1} + \dfrac{x+1}{x+3} = \dfrac{3x-1}{2x+1} + \dfrac{1}{6}$

f) $\dfrac{2x-3}{x-1} - \dfrac{x-1}{x+2} = \dfrac{2x-5}{x+2} + \dfrac{2-x}{1-x}$ unusual

COMMUNICATING THE IDEAS

In your journal, explain how solving an equation involving rational expressions is similar to solving an equation with rational coefficients. How is it different?

7.7 Applications of Rational Expressions

Many problems in engineering and science involve rational expressions. These expressions are almost always simpler than those with which you worked earlier in this chapter.

Example 1

A candy company sells candies in boxes. Each box has a square base with sides 17.5 cm and a volume of approximately 1070 cm^3. The company plans to redesign the box to have a smaller base. It must have a square base and contain the same volume.

a) Calculate the height of the box.

b) Suppose you know the decrease in the length of the base. Determine a formula to calculate the increase in height.

c) What is the increase in height for each decrease in the length of the base?
i) 1.5 cm   ii) 3.0 cm

Solution

a) Let h centimetres represent the height of the box.
Then the volume of the box is:
Base area × height = 1070
$$17.5^2 \times h = 1070$$
$$h = \frac{1070}{17.5^2}$$
$$\doteq 3.5$$
The height of the box is approximately 3.5 cm.

b) Let x centimetres represent the decrease in the length of the base.
Let y centimetres represent the increase in height.
The length of the base is now $(17.5 - x)$ centimetres.
Since the volume remains the same, the new height is
$$\frac{\text{volume}}{\text{base area}} = \frac{1070}{(17.5 - x)^2}$$
The increase in height is:
$y =$ new height $-$ original height
$$y = \frac{1070}{(17.5 - x)^2} - 3.5$$

c) **i)** Substitute 1.5 for x to obtain $y = \dfrac{1070}{16.0^2} - 3.5$

$$\doteq 0.7$$

ii) Substitute 3.0 for x to obtain $y = \dfrac{1070}{14.5^2} - 3.5$

$$\doteq 1.6$$

The height is increased by approximately 0.7 cm when the length of the base is decreased by 1.5 cm. The height is increased by approximately 1.6 cm when the length of the base is decreased by 3.0 cm.

Example 2

Each week, Angela flies her Cessna-150 500 km from Lethbridge to Moose Jaw. After a brief stopover, she returns to Lethbridge. On both trips, the air speed is 165 km/h. On the flight out there is a constant tail wind, and on the return trip a constant head wind of the same speed.

a) Suppose you know the wind speed. Determine a formula to calculate the total time for the round trip (not counting the stopover).

b) Calculate the time for the round trip for each wind speed.

    i) 30 km/h   **ii)** 40 km/h

Solution

a) Recall that for objects travelling at constant speed:

$$\text{Distance} = \text{speed} \times \text{time}$$

$$\text{Hence, time} = \frac{\text{distance}}{\text{speed}}$$

Let x km/h represent the wind speed. Then, Angela's speed is $(165 + x)$ km/h with a tail wind and $(165 - x)$ km/h with a head wind. Let t hours represent the total time.

Flight out time (hours): $\dfrac{500}{165 + x}$

Return trip time (hours): $\dfrac{500}{165 - x}$

Total time (hours): $t = \dfrac{500}{165 + x} + \dfrac{500}{165 - x}$

b) **i)** Substitute 30 for x to obtain $t = \dfrac{500}{195} + \dfrac{500}{135}$

$$\doteq 6.27$$

ii) Substitute 40 for x to obtain $t = \dfrac{500}{205} + \dfrac{500}{125}$

$$\doteq 6.44$$

The total time for the round trip, not counting the stopover, is approximately 6.27 h when the wind speed is 30 km/h, and approximately 6.44 h when the wind speed is 40 km/h.

Example 3

The average speed of an airplane is eight times as fast as the average speed of a train. To travel 1200 km, the train requires 14 h more than the airplane. Determine the average speeds of the train and the airplane.

Solution

Let x represent the average speed of the train in kilometres per hour. Hence, $8x$ represents the average speed of the airplane in kilometres per hour. Summarize the information in a table. Observe that four of the six spaces (distance and average speed for each) in the table can be completed at this stage. Use the distance, speed, time formula to complete the remaining two spaces (time for each).

	Distance (km)	Average speed (km/h)	Time (h)
Train	1200	x	$\dfrac{1200}{x}$
Airplane	1200	$8x$	$\dfrac{1200}{8x}$

Use the time formulas to form an equation. Since the train requires 14 h more than the airplane, we obtain:

$$\frac{1200}{x} - \frac{1200}{8x} = 14$$

$$8x\left(\frac{1200}{x}\right) - 8x\left(\frac{1200}{8x}\right) = 8x(14)$$

Multiply each side by the common denominator $8x$.

$$9600 - 1200 = 112x$$
$$112x = 8400$$
$$x = 75$$

The average speed of the train is 75 km/h, and the average speed of the airplane is 8×75 km/h, or 600 km/h.

1. In *Example 1b*, why was the expression written as $\dfrac{1070}{(17.5 - x)^2} - 3.5$ instead of $3.5 - \dfrac{1070}{(17.5 - x)^2}$?

2. In *Example 2*, observe that, in the solution, we did not add the expressions $\dfrac{500}{165 + x}$ and $\dfrac{500}{165 - x}$. If we had done so, the result would have been $\dfrac{165\,000}{27\,225 - x^2}$.
   a) Why was it not necessary to add the expressions?
   b) Check that the results are the same for the combined expression.

3. In *Example 2*, why is the total time longer when the wind speed is greater?

4. In *Example 3*, what are the advantages of using a table? Are there any disadvantages?

5. How could you check the solution for *Example 3*?

6. a) In *Example 3*, suppose x represented the average speed of the airplane. How would the solution change? Try solving the problem this way; then compare the result with the above solution.
   b) Suppose x represented the time taken by the airplane. How would the solution change? Try solving the problem this way, then compare the result with the above solution.

7.7 EXERCISES

A 1. In *Example 1*, the new style of box is to have a larger base.
   a) Suppose you know the increase in the length of the base. Change the formula in the solution to *Example 1b* to calculate the decrease in height.
   b) What is the decrease in height for each increase in the length of the base?
      i) 2.5 cm   ii) 5.0 cm

2. The company in *Example 1* also sells candies in a smaller box. The box has a square base with sides 12.4 cm, and volume approximately 535 cm^3. The company plans to redesign these boxes with a smaller square base, containing the same volume.
   a) Suppose you know the decrease in the length of the base. Change the formula in the solution to *Example 1b* to calculate the increase in height.
   b) What is the increase in height for each decrease in the length of the base?
      i) 1.5 cm   ii) 3.0 cm

3. In *Example 2*, suppose there was a head wind on the trip to Moose Jaw and a tail wind on the trip to Lethbridge. How would this affect the formula for the total time for the round trip? Explain.

4. In *Example 2*, suppose Angela sells her Cessna-150 and buys the faster Cessna-172, which has an air speed of 200 km/h.

a) Change the formula in the solution to *Example 2* so it applies to the Cessna-172.

b) Calculate the time for the round trip in the Cessna-172 for each wind speed.
i) 30 km/h   **ii)** 40 km/h

B 5. In *Example 2*, carry out the addition to confirm that
$$\frac{500}{165 + x} + \frac{500}{165 - x} = \frac{165\ 000}{27\ 225 - x^2}.$$

6. A square poster has an area of one square metre.

a) Suppose one dimension is increased by x metres. However, the area remains one square metre. Determine a formula to calculate the decrease in width.

b) What is the decrease in width for each increase in length?
i) 10 cm   **ii)** 25 cm

7. A poster has 400 cm² of printed matter. Its margins are 10 cm at the top and bottom, and 5 cm at each side.

a) Suppose the width of the poster is w centimetres. Determine formulas for its length and its area.

b) Determine the length and the area for each width.
i) 20 cm   **ii)** 25 cm

8. In *Example 1*, we determined the formula $y = \dfrac{1070}{(17.5 - x)^2} - 3.5$ for the increase in height.

a) Graph this equation, using appropriate window settings. Sketch the graph.

b) Use the trace or the tables feature. Determine the increase in height if the base length decreases by 2.5 cm, and by 5.0 cm. Show this information in your sketch.

c) Predict how the graph will change for the box in exercise 2. Use your calculator to confirm your prediction. Include the graph in your sketch.

9. A tuna can has a volume of 210 cm³ and a base radius of 4.2 cm. The company plans to redesign the can with a smaller base. The new can will have the same volume as the original can.

a) Suppose you know the decrease in the base radius. Determine a formula to calculate the increase in height.

b) What is the increase in height for each decrease in the base radius?
i) 0.5 cm   **ii)** 1.0 cm

10. A pasta sauce can has a capacity of 725 mL and a base radius of 4.2 cm. The company plans to redesign the can with a larger base. The new can will have the same volume as the original can.

a) Suppose you know the increase in the base radius. Determine a formula to calculate the decrease in height.

b) What is the decrease in height for each increase in the base radius?
   i) 1.0 cm   ii) 2.0 cm

11. Use the *Helicopters* database.

a) Find the sea level cruise speed of the Bell 407T helicopter.

b) The Bell 407T is used for a search and rescue mission. It must travel 145 nm to the site, pick up survivors, and return to base. On the flight out, there is a constant head wind, and on the return trip a constant tail wind of the same speed. Suppose you know the wind speed. Determine a formula to calculate the total time for the trip.

c) Calculate the time for the trip when the wind speed is 50 knots.

12. In *Example 2*, we determined the formula $t = \dfrac{500}{165 + x} + \dfrac{500}{165 - x}$ for the total time for the round trip.

a) Graph this equation using appropriate window settings. Sketch the graph.

b) Use the trace or the tables feature. Determine the total time for a wind speed of 20 km/h and a wind speed of 65 km/h. Show this information in your sketch.

c) Predict how the graph will change for the Cessna-172 in exercise 4. Use your calculator to confirm your prediction. Include the graph in your sketch.

13. The average speed of an airplane is five times as fast as the average speed of a passenger train. To travel 2000 km, the train requires 20 h more than the airplane. Determine the average speeds of the train and the airplane.

14. The average speed of an express train is 40 km/h faster than the average speed of a bus. To travel 1200 km, the bus requires 50% more time than the train. Determine the average speeds of the bus and the train.

COMMUNICATING THE IDEAS

Examples 1 to *3* all involved addition or subtraction of rational expressions. In *Examples 1* and *2* the addition or subtraction was not performed, although it was in *Example 3*. In applied problems like these, how can you tell if the operations in the rational expressions should be performed? Write your ideas in your journal.

Exploring Averages

Here is a puzzle that you might find perplexing. Try to solve the puzzle.

A road up one side of a hill is 12 km long, and it is 12 km down the other side of the hill. Suppose you can cycle up the hill at 6 km/h. How fast would you have to cycle down the other side to average 12 km/h for the entire trip?

If you are not sure how to solve the puzzle, here are some ideas you could try.

1

To determine the average speed, you need to know the total distance travelled and the total time taken.

$$\text{Average speed} = \frac{\text{total distance}}{\text{total time}}$$

What is the total distance?

For an average speed of 12 km/h, what is the total time? What was the time to go up the hill?

2

You could make a table, then solve the problem algebraically.

	Distance (km)	Speed (km/h)	Time (h)
Uphill			
Downhill		x	

Complete the cells in the table, and write an equation.

Does the equation have a solution? How can you tell?

1. Mark had seven mathematics tests. His average was 60%.

   a) On the next three tests, Mark averaged 80%. What was his average for all ten tests?

   b) Explain why the answer to part a is not 70%.

   c) Suppose Mark averages 60% on the first seven tests. Would it be possible for him to average 70% on all ten tests? If your answer is no, explain. If your answer is yes, what average mark would he need on the last three tests? Support your answer with some calculations.

2. An A320 flies 1800 km from Winnipeg to Vancouver against the wind at an average speed of 737 km/h. On the return trip, the average speed is 937 km/h.

   a) Calculate the average speed for the round trip. Do not take into account the stopover time in Vancouver.

   b) Explain why the answer to part a is not 837 km/h.

   c) Suppose the stopover time is 2 h. What would the average speed be for the round trip?

You should have found that there is no solution to the puzzle on the preceding page because it takes too much time to go up the hill. By completing the following exercises, you will discover how the average speed for the trip depends on the speed going down the hill. Use the information on the preceding page.

3. Suppose you cycle down the hill at 15 km/h.

   a) How long does it take to go down the hill?

   b) What is the total time for the trip?

   c) What is the average speed for the trip?

4. Suppose you cycle down the hill at x km/h. Write an expression for each time and speed.

   a) the time it takes to go down the hill

   b) the total time for the trip

   c) the average speed for the trip

5. a) Use your results from exercise 4. Determine some average speeds for a few speeds cycling down the hill. Summarize your results in a table.

   b) Use your table. Draw a graph to show how the average speed for the trip depends on the speed going down the hill.

6. Let X denote the speed down the hill. Define the equation $Y1 = 24/(2+12/X)$. Graph Y1 using appropriate window settings. Compare the result with your graph in exercise 5b.

Speed-Time Graphs

Refer to *Example 2* on page 437 as you complete these exercises.

1. **a)** In the solution to *Example 2*, we found expressions for Angela's times with a tail wind and with a head wind when the wind speed is x km/h. Graph the equations $y = \dfrac{500}{165 + x}$ and $y = \dfrac{500}{165 - x}$. Your graph should look similar to the one below left.

   b) Why do both curves start at the same point at the left side of the graph?

   c) Why does one curve go down to the right, while the other one goes up to the right?

   d) Why does the curve that goes up appear steeper than the one that goes down?

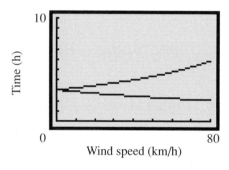
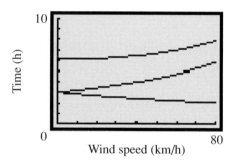

2. **a)** The total time for the return trip is represented by $y = \dfrac{500}{165 + x} + \dfrac{500}{165 - x}$. Graph this equation along with those in exercise 1. Your graph should look similar to the one above right.

   b) Why does this curve start at a different point at the left? How is this point related to the point where the other two curves start?

   c) Why is this curve above the other two?

3. Use the trace and zoom features.

   a) Suppose the wind speed is 35 km/h. Determine the time for the flight out, the time for the return trip, and the total time for the trip.

   b) What is the wind speed for each return trip?
   i) 1 h longer than the flight out   **ii)** double the time for the flight out

4. Exercise 4 on page 440 is similar to *Example 2* on page 437. How would the graphs above change if they were plotted using the corresponding expressions for the air speed in exercise 4?

1. Reduce to lowest terms.

a) $\dfrac{78a^3b}{6a^2}$

b) $\dfrac{9m+6}{3}$

c) $\dfrac{20y^2-8y}{4y}$

d) $\dfrac{36s^2-48s}{24s}$

e) $\dfrac{9ab-18a}{-3a}$

f) $\dfrac{6xy}{3x^2+15xy}$

g) $\dfrac{6m^2-15mn}{4m-10n}$

h) $\dfrac{7s^2-35st}{5s-25t}$

2. Simplify.

a) $\dfrac{36-12x}{4x-12}$

b) $\dfrac{a^2-25}{3a^2-15a}$

c) $\dfrac{n^2-10n+24}{n^2-6n+8}$

d) $\dfrac{a^2-9}{a^2+6a+9}$

e) $\dfrac{b^2-3b-28}{b^2-9b+14}$

f) $\dfrac{m^2-8m+15}{m^2-3m-10}$

3. Simplify.

a) $\dfrac{9a^3}{4}\times\dfrac{8}{3a}$

b) $\dfrac{-6mn}{15}\div\dfrac{2m^2}{5}$

c) $\dfrac{4x^2y}{8xy}\div\dfrac{3x^2y^3}{6xy^3}$

d) $\dfrac{15x^4y^2}{24xy^2}\times\dfrac{8x^2y}{5xy^2}$

e) $\dfrac{2m^2n}{3mn}\times\dfrac{(6mn)^2}{4mn}$

f) $\dfrac{2x}{3y}\times\dfrac{3z}{4y}\div\dfrac{4z}{5x}$

4. Simplify.

a) $\dfrac{12a^2}{5(a+4)}\times\dfrac{10(a+4)}{3a}$

b) $\dfrac{3(5-a)}{2a}\div\dfrac{3(a-5)}{8(a+1)}$

c) $\dfrac{6mn^2}{2(2m-5)}\times\dfrac{3(2m-5)}{9m^2n}$

d) $\dfrac{3x^2y(2x-y)}{15xy}\div\dfrac{2xy^2(2x-y)}{5x^2y^3}$

5. Choose one part of exercise 4. Write to explain how you simplified the expression.

6. Simplify.

a) $\dfrac{3b^3}{a^2-9}\times\dfrac{2a-b}{2b^2}$

b) $\dfrac{m^2-2m+1}{m^2-1}\div\dfrac{m^2-4}{m^2+3m+2}$

c) $\dfrac{x^2-3x-10}{25-x^2}\div\dfrac{x+2}{x+5}$

d) $\dfrac{y^2+2y-15}{y^2-8y+7}\times\dfrac{y^2-5y-14}{y^2+7y+10}$

7. Simplify.

a) $\dfrac{4}{3a}+\dfrac{5}{3a}$

b) $\dfrac{3}{2x}-\dfrac{5}{3x}$

c) $\dfrac{5}{m}+\dfrac{3}{4m}$

d) $\dfrac{5x}{6}-\dfrac{3x}{4}+\dfrac{x}{8}$

e) $\dfrac{2}{3a}-\dfrac{3}{4a}+\dfrac{1}{2a}$

f) $\dfrac{7}{6n}-\dfrac{5}{2n}+\dfrac{1}{5n}$

8. Simplify.

a) $\dfrac{a+3}{a}+\dfrac{a-5}{a}$

b) $\dfrac{3x-2}{x}-\dfrac{2x+7}{x}$

c) $\dfrac{y-5}{4}-\dfrac{y+2}{5}$

d) $\dfrac{a+2}{3a}+\dfrac{2a-5}{2a}$

e) $\dfrac{3x-7}{8x}-\dfrac{2x-4}{6x}$

f) $\dfrac{3}{4x}-\dfrac{2}{3x}+1$

g) $\dfrac{7x+3}{4x}-\dfrac{5x+2}{6x}$

h) $\dfrac{2m-9}{4m}+\dfrac{7m+5}{8m}$

i) $\dfrac{2a+3}{10a^2}-\dfrac{7a-4}{15a^2}$

9. Choose one part of exercise 8. Write to explain how you simplified the expression.

10. Add or subtract.

a) $\frac{2}{3a} + \frac{5}{b}$ b) $\frac{3}{4m} - \frac{5}{6n}$ c) $1 + \frac{s}{t}$ d) $2 + \frac{b}{2c}$

e) $\frac{9p}{7q} + \frac{7q}{6p}$ f) $\frac{3x}{5y} - \frac{2y}{3x}$ g) $\frac{3c}{7d} + \frac{5d}{3c}$ h) $\frac{8x}{9y} - \frac{3x}{5x}$

11. Simplify.

a) $\frac{5x - 9}{x - 4} - \frac{2x + 3}{x - 4}$ b) $\frac{2a}{a - 1} - \frac{5}{a}$ c) $\frac{7}{x + 1} - 2$

d) $\frac{m}{m + 1} - \frac{2}{m - 1}$ e) $\frac{y - 3}{y - 2} + \frac{1}{y - 3}$ f) $\frac{3a - 1}{a + 7} - \frac{2a + 1}{a - 3}$

12. Simplify.

a) $\frac{5a}{10a - 15} - \frac{4a}{16a - 24}$ b) $\frac{2m + 5}{3m - 12} + \frac{2m}{m - 4}$

c) $\frac{2k}{k - 2} - \frac{3}{k^2 - 4}$ d) $\frac{8b - 3}{b^2 - 7b + 12} + \frac{2b + 1}{4 - b}$

e) $\frac{8x}{5(x - 2)} + \frac{5x - 1}{3(x + 3)}$ f) $\frac{4x + 3}{3(2x - 1)} - \frac{x - 5}{2(3x + 7)}$

13. Simplify.

a) $\frac{x}{x + 2} - \frac{3}{x - 2}$ b) $\frac{3a}{8a - 12} - \frac{2a}{9 - 6a}$ c) $\frac{5a}{6a - 10} + \frac{4a}{15a - 25}$

d) $\frac{2x + 3}{2x - 8} + \frac{3x}{4 - x}$ e) $\frac{3m}{2(m - 2)} + \frac{2m - 1}{3(m + 2)}$ f) $\frac{3t}{3(t - 1)} + \frac{2t}{5(t + 1)}$

14. Solve.

a) $\frac{x}{5} + 3 = \frac{2}{x} + \frac{x + 1}{5}$ b) $\frac{y}{3} + \frac{4}{y} = \frac{y + 1}{3}$

c) $\frac{3x - 2}{2} + 4 = \frac{13}{x} - \frac{1 - 6x}{4}$ d) $\frac{2(1 - x)}{3} - 8 = \frac{1}{6x} - \frac{2x - 3}{3}$

e) $2 - \frac{x - 1}{x + 3} = \frac{x - 3}{x - 4}$ f) $\frac{3 - x}{x - 2} = 1 - \frac{2x - 5}{x + 2}$

15. Solve.

a) $x - \frac{6}{x - 2} = 7$ b) $\frac{x - 5}{3x - 1} = \frac{x + 5}{x - 4}$ c) $\frac{6}{x - 2} = \frac{21}{x^2 - 4} + 1$

16. Choose one part of exercise 15. Write to explain how you solved the equation.

17. Jay flew his airplane 500 km against the wind in the same time that it took him to fly it 600 km with the wind. The speed of the wind was 20 km/h. What was the average speed of the airplane?

18. Alex cycles 5 km to return a friend's bicycle, then walks home. The entire trip takes 1.5 h. Alex cycles four times as fast as he walks. How fast does he walk?

19. Nora jogs 3 km, then walks 2 km farther than she jogged. She jogs twice as fast as she walks. The total time is two hours. How fast does Nora walk?

1. **a)** Insert two numbers between 17 and 44, so the four numbers form an arithmetic sequence.

   b) Insert three numbers between 25 and 7, so the five numbers form an arithmetic sequence.

2. **a)** Write a formula for the general term of each sequence.

   i) 6, 9, 12, ... **ii)** 5, 0, −5, ... **iii)** 2, 1, 0, ...

   iv) 0, 1, 2, ... **v)** 1, 2, 3, ... **vi)** 2, 3, 4, ...

   b) What patterns do you see in the sequences and the general terms, for the last 3 sequences in part a? Write to explain what you notice.

3. **a)** For each arithmetic sequence, a later term in the sequence is given. Which term is it?

   i) 3, 7, 11, ... 71 **ii)** −5, −3, −1, ... 97 **iii)** 10, 14, 18, ... 234

   iv) 7, 6, 5, ... −104 **v)** 5, 10, 15, ... 2105 **vi)** 2, 9, 16, ... 184

   b) Choose one sequence from part a. Write to explain how you identified the term.

4. Determine the indicated term of each geometric sequence.

   a) 16, 32, 64, ... t_7 **b)** 2, 4, 8, ... t_{10} **c)** 4, 16, 64, ... t_5

   d) 0.328, 1.64, 8.2, ... t_6 **e)** $\frac{19}{144}, \frac{19}{24}, \frac{19}{4}, \ldots t_6$ **f)** $\frac{37}{27}, \frac{37}{9}, \frac{37}{3}, \ldots t_7$

5. Marcus earns $9.00/h working at the local arena. If he works more than 8 h a day, he is paid time and a half for any time over 8 h. If Marcus works more than 12 h a day, he is paid double time for any time over 12 h. He is paid time and a half for Saturday work and double time for Sunday work. Consider his schedule for this week:

Day	Mon.	Tues.	Wed.	Thurs.	Fri.	Sat.	Sun.
Hours worked	7	9	0	10	14	4	5

   a) Construct a table to show how much Marcus is paid for each day.

   b) Determine Marcus' total earnings and mean hourly wage for the week.

6. Simplify without using a calculator.

   a) $\sqrt{3^2}$ **b)** $\sqrt{15.6^2}$ **c)** $\sqrt[3]{21.9^3}$ **d)** $\sqrt{3^4}$ **e)** $\sqrt[3]{2^6}$ **f)** $\sqrt[15]{1.32^{15}}$

7. Determine each exact value.

   a) $[1^3]^{\frac{1}{2}}$ **b)** $[1^3 + 2^3]^{\frac{1}{2}}$ **c)** $[1^3 + 2^3 + 3^3]^{\frac{1}{2}}$

   d) $[1^3 + 2^3 + 3^3 + 4^3]^{\frac{1}{2}}$ **e)** $[1^3 + 2^3 + 3^3 + 4^3 + 5^3]^{\frac{1}{2}}$

8. a) Examine the results of exercise 7. Do you see a pattern? Explain.

  b) Use the pattern from part a. Write a formula for $[1^3 + 2^3 + \cdots + n^3]^{\frac{1}{2}}$.

  c) Use the formula in part b to determine $[1^3 + 2^3 + \cdots + 10^3]^{\frac{1}{2}}$.

9. Determine the coordinates of the midpoint of the line segment with each pair of endpoints.

  a) J(−5, −5), K(10, 4)   **b)** L(−3, −2), M(11, 5)   **c)** N(−9, 16), P(−3, −3)

10. A triangle has vertices A(3, 5), B(1, −6), and C(5, 4).

  a) Determine the slope of AB.

  b) Determine the coordinates of the midpoint of BC.

  c) Determine the length of the median from A to BC.

11. Quadrilateral ABCD has vertices A(1, 5), B(−3, −1), C(7, −3), and D(5, 3). Show that the midpoints of the sides of the quadrilateral are the vertices of a parallelogram.

12. Determine the equation of the line that passes through (−1, 2) and is:

  a) parallel to the line $2x + 5y = 10$ **b)** perpendicular to the line $3x − 5y = 15$

13. Determine the equation of each line.

  a) through A(−3, 2) with slope $-\frac{3}{5}$ **b)** through B(−2, 5) and parallel to the x-axis

  c) through C(−3, 5) and D(6, 2)   **d)** through E(5, 2) and perpendicular to the x-axis

14. The rules defining three functions are given. For each function:

  a) Make a table of values.

  b) Graph the functions on the same axes.

  c) Decide if there are any input numbers that cannot be used.

    i) *Rule 1*: Output number is the same as the input number.

    ii) *Rule 2*: Multiply the input number by 3.

    iii) *Rule 3*: Divide the input number by 3.

15. A garden centre charges a fixed amount for delivery and also an amount per cubic metre of top soil delivered. The total charges for different volumes of soil are listed in the table.

Volume of soil (m³)	Total cost ($)
4.6	112.40
7.2	161.80
8.1	178.90
10.3	220.70
13.5	281.50

a) Graph the data with *Total cost* on the vertical axis. Draw a line of best fit.

b) Determine the delivery charge.

c) Determine the cost for 1 m³ of top soil.

d) Write a formula for the total cost, y dollars, in terms of the volume of soil, x cubic metres.

e) What would be the total cost for each volume of soil delivered?

    i) 20 m³   **ii)** 15.7 m³

16. Graph the function represented by each equation. Identify the intercepts, slope, and range. If no restrictions are stated, assume the domain is the set of all real numbers.

a) $y = 2x - 1$ **b)** $y = 2x + 3$ **c)** $y = 3$

d) $y = \frac{x}{2} + 1$ **e)** $y = \frac{2}{3}x + 2$ **f)** $y = -x + 2; x \geq 0$

17. For each function, determine $f(0), f(1), f(2),$ and $f(-1)$, where possible.

a) $f(x) = 2x - 3$ **b)** $f(x) = 3x^2 - 2x + 1$ **c)** $f(x) = \sqrt{x - 1}$

d) $f(x) = x + \frac{1}{x}$ **e)** $f(x) = x + \frac{1}{x + 1}$ **f)** $f(x) = x^2 - \frac{1}{x + 2}$

18. a) Determine the circumference and area of a circle with radius 3 cm.

b) Suppose the radius of the circle in part a is doubled. What happens to the circumference and area?

c) Determine the surface area and volume of a sphere with radius 2 cm.

d) Suppose the radius of the sphere in part c is doubled. What happens to the surface area and volume?

e) Write to explain the results of parts b and d. Do you think the same results would be valid for all circles and all spheres? Is there a relation among the scale factors for circumference, area, surface area, and volume? Explain.

19. Evaluate each rational expression for $x = 2$.

a) $\frac{x + 3}{3 - x}$ **b)** $\frac{x^2 - 3x + 2}{x + 1}$ **c)** $\frac{2x + 5}{2x - 5}$

d) $\frac{x^3 + x^2 + x + 1}{x^2 + 2x - 3}$ **e)** $\frac{(x - 1)(x + 1)(x^2 + 1)}{x^4 - 1}$ **f)** $\frac{x^3 - 1}{x^2 + x - 2}$

20. In exercise 19, state any values of x for which each expression is undefined.

21. Simplify.

a) $\frac{x - 3}{x^2 - x - 6}$ **b)** $\frac{x^2 + 7x + 12}{x + 4}$ **c)** $\frac{6x^2 + x - 1}{3x - 1}$

d) $\frac{4x^2y^3}{6xy^2} \times \frac{9x^4y}{12x^5y^2}$ **e)** $\frac{-3x^3y^2}{(2x^2y^2)^2} \times \frac{14x^4y^4}{6x^3y^2}$ **f)** $\frac{2x}{5y} \times \frac{3w}{4z} \times \frac{2y}{3x} \times \frac{5z}{7w}$

22. Add or subtract.

a) $1 - \frac{1}{x + 1}$ **b)** $\frac{x + 1}{x + 2} - \frac{x}{x + 1}$ **c)** $\frac{x + y}{x + 2y} - \frac{x}{x + y}$

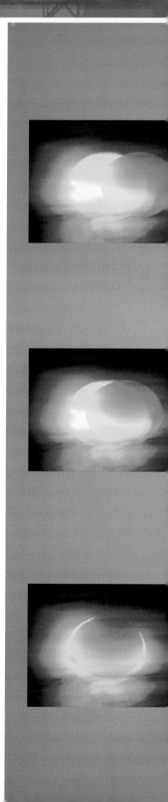

How Far Is the Sun?
How Large Is the Sun?

In this chapter, you will learn how to use trigonometry to measure inaccessible distances on Earth. We can also use trigonometry to measure distances in space.

 CONSIDER THIS SITUATION

The diagram on page 451 represents the sun. Suppose we took this diagram as a scale representation.

- Visualize a circle 1 mm in diameter that is 12 m away from this page. That little circle would represent Earth.

- Measure the diameter of the sun in the diagram. How many times as great as the diameter of Earth is the sun's diameter?

The information above is based on known astronomical measurements.

- How have astronomers been able to determine the distance to the sun, without leaving Earth?

- How did they determine the sun's diameter?

On pages 522 and 523, you will develop a mathematical model to determine these measurements. Your model will use trigonometry and algebraic equations.

 FYI Visit www.awl.com/canada/school/connections

For information related to the above problem, click on MATHLINKS followed by AWMath. Then select a topic under How Far to the Sun and Stars?

8.1 The Tangent Ratio

The word *trigonometry* means "triangle measurement." Using trigonometry, you can calculate the lengths of the sides and the measures of the angles in any triangle, provided you have enough information to construct the triangle. In practical situations, you can use trigonometry to calculate distances to, or between, remote objects.

Recall from a previous grade the three basic trigonometric ratios of tangent, sine, and cosine.

Trigonometry depends on the properties of similar triangles. When two triangles are similar:

• Corresponding angles are equal.

• The ratios of corresponding sides are equal.

The diagram contains 4 nested triangles.

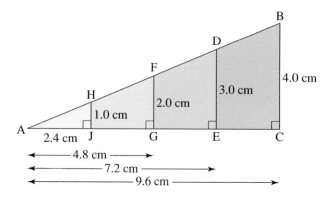

All these are right triangles with a common ∠A, so they must be similar.

To confirm this, examine the ratios of corresponding sides. In each triangle, compare the side opposite ∠A to the side adjacent to ∠A. Recall that these are the legs of the right triangle.

$$\frac{BC}{CA} = \frac{4.0}{9.6} \qquad \frac{DE}{EA} = \frac{3.0}{7.2} \qquad \frac{FG}{GA} = \frac{2.0}{4.8} \qquad \frac{HJ}{JA} = \frac{1.0}{2.4}$$

$$= 0.41\overline{6} \qquad\quad = 0.41\overline{6} \qquad\quad = 0.41\overline{6} \qquad\quad = 0.41\overline{6}$$

All the ratios are equivalent. They depend only on the measure of ∠A, and not on the size of the triangle. We call this ratio the *tangent* of the angle, and write it as tan A.

If ∠A is an acute angle in a right triangle, then

$$\tan A = \frac{\text{length of side opposite } \angle A}{\text{length of side adjacent to } \angle A}$$

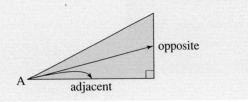

opposite

A   adjacent

To determine the tangent of an angle, you need a scientific calculator. Your calculator must be in degree mode. To check this, enter 45 and press [TAN].

- If your calculator displays 1, it is in degree mode.
- If your calculator does not display 1, you need to put it in degree mode.

On some calculators, you do this using [DRG]. If necessary, consult your calculator manual, or ask for help from another student or your teacher.

Finding tangents of angles

To determine tan 25°, press 25 [TAN] to display 0.466307658.

Rounded to 4 decimal places, tan 25° $\doteq$ 0.4663

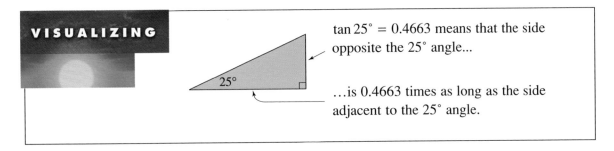

VISUALIZING

tan 25° = 0.4663 means that the side opposite the 25° angle...

...is 0.4663 times as long as the side adjacent to the 25° angle.

25°

Finding angles with a given tangent

You can also use a calculator to determine an angle when you know its tangent. For example, suppose tan A = 0.9528.

To get [TAN⁻¹], press [2nd] [TAN] or [INV] [TAN].

Press .9528 [TAN⁻¹] to display 43.61540642.
To the nearest degree, ∠A = 44°

The keying sequences are for the *TEXAS INSTRUMENTS TI-34* calculator. Other calculators may have different keying sequences. Refer to your calculator's manual to find out.

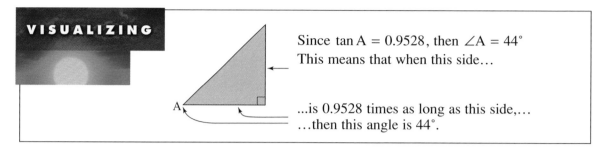

Since tan A = 0.9528, then ∠A = 44°
This means that when this side...

...is 0.9528 times as long as this side,...
...then this angle is 44°.

In the above examples, we found the tangent to 4 decimal places and the angle to the nearest degree. This is sufficient accuracy for the applications in this chapter.

Example 1

In right △ABC

a) Calculate tan A and ∠A.

b) Calculate tan C and ∠C.

Check your answers.

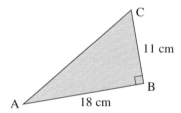

Solution

a) The side opposite ∠A is BC. The adjacent side is AB.

$$\tan A = \frac{11}{18}$$ Press: 11 ÷ 18 =
$$\doteq 0.6111$$ Then press: TAN⁻¹
$$\angle A \doteq 31°$$

b) The side opposite ∠C is AB. The adjacent side is BC.

$$\tan C = \frac{18}{11}$$
$$\doteq 1.6363$$
$$\angle C \doteq 59°$$

To check, confirm that the angle measures of the triangle add to 180°.

$$\angle A + \angle B + \angle C = 31° + 90° + 59°$$
$$= 180°$$

Example 2

In right $\triangle PQR$, $\angle R = 90°$, $\angle Q = 25°$, and $PR = 7.0$ cm. Determine the length of QR to the nearest millimetre.

Solution

Sketch the triangle using the measures given.

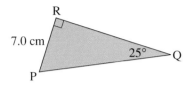

$$\tan 25° = \frac{7}{QR}$$ Multiply each side by QR.

$QR \times \tan 25° = 7$ Divide each side by $\tan 25°$.

$$QR = \frac{7}{\tan 25°}$$ Press: 7 $\boxed{\div}$ 25 $\boxed{\text{TAN}}$ $\boxed{=}$

$QR \doteq 15.0115$

QR is 15.0 cm to the nearest millimetre.

Trigonometry is a powerful tool for measuring distances that cannot be measured directly. *Example 3* involves measuring an inaccessible height. Many other examples will arise in this chapter, culminating in measuring the distance to the sun on page 522.

Example 3

A forest technician is collecting data about the heights of trees. She paces a distance of 15 m from the base of a tree and uses a clinometer to measure the angle of elevation to the top of the tree. The angle is 25°. The technician's eye is about 1.5 m above the ground. Assume the ground is level. How tall is the tree?

Solution

Sketch a diagram to show the given information.

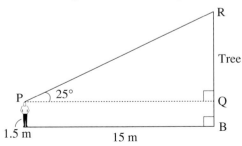

Determine the height QR, then add the height of the technician's eyes above the ground.

In $\triangle$PQR

$$\tan P = \frac{QR}{QP}$$

$$\tan 25° = \frac{QR}{15}$$

$$15 \times \tan 25° = QR$$

$$QR \doteq 6.9946$$

QR is approximately 7.0 m.

The approximate height of the tree is 7.0 m + 1.5 m = 8.5 m.

 MODELLING for **Measuring Inaccessible Heights**

The method in the solution of *Example 3* is a model for measuring an inaccessible height, such as the height of a tree.

- What assumptions does the model make about the tree and the surrounding terrain?

- What are some of the limitations of this model?

- Explain how the model could be applied to measure other distances that cannot be measured directly, such as the width of a river.

DISCUSSING THE IDEAS

1. Refer to *Example 1*.

   a) Explain how $\angle$A and $\angle$C are related.

   b) How are tan A and tan C related?

   c) Do you think relationships like those in parts a and b are true for all right triangles? How might you demonstrate this?

2. **a)** *Example 2* demonstrates one way to solve the equation $\tan 25° = \frac{7}{QR}$. Describe another way to solve this equation.

   b) Describe another way to complete *Example 2*, using a different equation that is easier to solve.

3. In *Example 3*, describe how you could determine the distance between the technician's eyes and the top of the tree.

A **1.** Use your calculator to determine each tangent to 3 decimal places.

    a) tan 15°   **b)** tan 33°   **c)** tan 60°   **d)** tan 85°

2. Use your calculator to determine ∠A to the nearest degree.

    a) tan A = 0.2   **b)** tan A = 1.2495   **c)** tan A = $\frac{3}{4}$ **d)** tan A = $\frac{29}{12}$

3. In each triangle

    a) Calculate tan A.   **b)** Determine ∠A.

i) 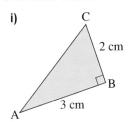 **ii)** 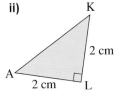 **iii)** 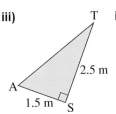 **iv)**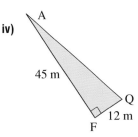

4. Select one triangle from exercise 3. Write to explain how you calculated tan A, and how you determined ∠A.

5. In each triangle, calculate tan P and tan Q. Then determine ∠P and ∠Q.

a) 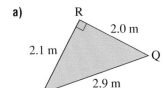 **b)** 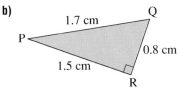 **c)**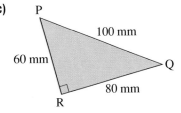

B **6.** In ΔDEF

    a) Calculate DF for each value of ∠E.

        i) 20°   **ii)** 30°   **iii)** 40°

    b) Determine ∠E for each value of DF.

        i) 8.0 m   **ii)** 12.0 m   **iii)** 16.0 m

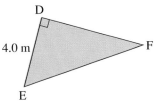

7. Draw a right triangle that satisfies each description. On your triangle, show the measure of each side and each angle of the triangle.

    a) ΔJKL, where tan J = $\frac{1}{2}$ and ∠K = 90°

    b) ΔPYT, where tan T = $\frac{17}{3}$ and ∠Y = 90°

8. Select one part of exercise 7. Write to explain your solution.

9. A tightrope walker attaches a cable to the roofs of two adjacent buildings (below left). The cable is 21.5 m long. The angle of inclination of the cable is 12°. The buildings are 21.0 m apart. The shorter building is 10.0 m high. What is the height of the taller building?

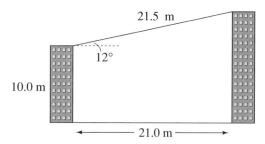

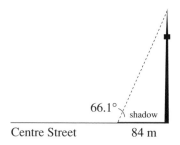

10. Around noon each day, the shadow of the Calgary Tower is aligned along Centre Street, which runs directly north from the tower (above right). One day, at noon, Syd paced the shadow and estimated its length to be approximately 84 m. Using a clinometer, he determined that the angle of elevation from the tip of the shadow to the top of the tower was 66.1°. Assume the clinometer is 1.5 m above the ground. Calculate the height of the tower to the nearest metre.

11. Suppose you are 133 m from the base of the Manitoba Legislative Building. The Golden Boy statue, on top of the building, has its highest point at the top of the hand holding the torch. The angle of elevation to this highest point is 30°. Calculate how high the top of the statue is above the base of the building. Identify your assumptions.

12. An airplane is flying at an altitude of 6000 m over the ocean directly toward a coastline. At a certain time, the angle of depression to the coastline from the airplane is 14°. How much farther does the airplane have to fly before it reaches the coastline?

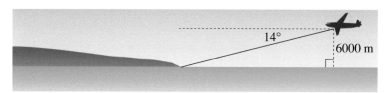

13. From a horizontal distance of 80.0 m, the angle of elevation to the top of a flagpole is 18°. Calculate the height of the flagpole.

14. The longest slide in the world is in Vermont, U.S.A. It drops 213 m in a horizontal distance of 1200 m.

a) How long is the slide?   **b)** What is its angle of inclination?

15. A PZL-Swidnik Mi-2 helicopter is taking off. Its angle of elevation from the eyes of an observer 150 ft from the launch pad is 38°. The observer's eyes are 5.5 ft above the ground.

a) Calculate the height of the helicopter to the nearest foot.

b) Use the *Helicopters* database. Find the record for the PZL-Swidnik Mi-2 and its rate of climb. Use this information and your answer to part a. Calculate the time since takeoff to the nearest second.

c) Select a different model of helicopter from the database. Write an exercise like this one for a classmate to solve.

16. A rectangle measures 7.7 cm by 5.5 cm. Determine the measure of the acute angle formed by the intersection of the diagonals.

17. Isosceles △PQR has a base QR 10 cm long. The height of the triangle is 4 cm. Determine the measures of the three angles in the triangle.

C 18. The diagram shows a cone in a cylindrical can, with the vertex of the cone just touching the bottom of the can. Determine the measure of the vertex angle, x, of the cone.

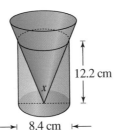

12.2 cm

x

8.4 cm

19. Determine the measure of the acute angle formed by each line and the x-axis.

a)
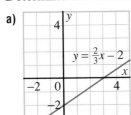
$y = \frac{2}{3}x - 2$

b)
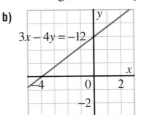
$3x - 4y = -12$

c)
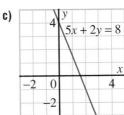
$5x + 2y = 8$

COMMUNICATING THE IDEAS

Create a problem that can be solved using the tangent ratio. Record the problem and your solution in your journal.

8.2 The Sine and Cosine Ratios

We revisit the similar triangles from Section 8.1 to investigate other ratios of sides. The length of each hypotenuse can be calculated using the Pythagorean Theorem.

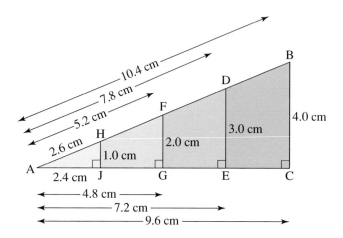

Write the ratio of each side opposite ∠A and the corresponding hypotenuse.

$$\frac{BC}{BA} = \frac{4.0}{10.4} \qquad \frac{DE}{DA} = \frac{3.0}{7.8} \qquad \frac{FG}{FA} = \frac{2.0}{5.2} \qquad \frac{HJ}{HA} = \frac{1.0}{2.6}$$

$$= \frac{1}{2.6} \qquad\qquad = \frac{1}{2.6} \qquad\qquad = \frac{1}{2.6} \qquad\qquad = \frac{1}{2.6}$$

All the ratios are equivalent. Each ratio is the *sine* of ∠A, and we write it as sin A.

Write the ratio of each side adjacent to ∠A and the corresponding hypotenuse.

$$\frac{AC}{AB} = \frac{9.6}{10.4} \qquad \frac{AE}{AD} = \frac{7.2}{7.8} \qquad \frac{AG}{AF} = \frac{4.8}{5.2} \qquad \frac{AJ}{AH} = \frac{2.4}{2.6}$$

$$= \frac{1.2}{1.3} \qquad\qquad = \frac{1.2}{1.3} \qquad\qquad = \frac{1.2}{1.3} \qquad\qquad = \frac{1.2}{1.3}$$

All the ratios are equivalent. Each ratio is the *cosine* of ∠A, and we write it as cos A.

If ∠A is an acute angle in a right triangle, then

$$\sin A = \frac{\text{length of side opposite } \angle A}{\text{length of hypotenuse}}$$

$$\cos A = \frac{\text{length of side adjacent to } \angle A}{\text{length of hypotenuse}}$$

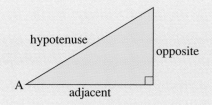

Finding sines and cosines of angles

To determine $\sin 43°$, press 43 $\boxed{\text{SIN}}$ to display 0.68199836

To determine $\cos 43°$, press 43 $\boxed{\text{COS}}$ to display 0.731353701

Rounded to 4 decimal places, $\sin 43° \doteq 0.6820$ and $\cos 43° \doteq 0.7314$

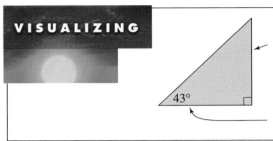

$\sin 43° = 0.6820$ means that the side opposite the 43° angle is 0.6820 times as long as the hypotenuse.

$\cos 43° = 0.7314$ means that the side adjacent to the 43° angle is 0.7314 times as long as the hypotenuse.

Finding angles with a given sine or cosine

Suppose $\sin A = \frac{2}{5}$, or 0.4. Press 0.4 $\boxed{\text{SIN}^{-1}}$ to display 23.57817848.

To the nearest degree, $\angle A = 24°$

Suppose $\cos A = \frac{2}{3}$. Press 2 $\boxed{÷}$ 3 $\boxed{=}$.

Then press $\boxed{\text{COS}^{-1}}$ to display 48.18968511.

To the nearest degree, $\angle A = 48°$

Example 1

A student measured the sides of right $\triangle ABC$, and obtained the measurements on the diagram.

a) Calculate $\sin A$, then $\angle A$ to the nearest degree.

b) Calculate $\cos A$, then $\angle A$ to the nearest degree.

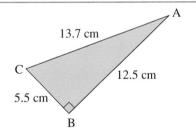

Solution

a) The side opposite $\angle A$ is BC. The hypotenuse is AC.

$$\sin A = \frac{BC}{AC}$$
$$= \frac{5.5}{13.7}$$ Press: 5.5 $\boxed{÷}$ 13.7 $\boxed{=}$
$$\angle A \doteq 23.6695$$ Then press: $\boxed{\text{SIN}^{-1}}$

$\angle A$ is 24° to the nearest degree.

b) The side adjacent to $\angle A$ is AB.

$\cos A = \dfrac{AB}{AC}$

$\qquad = \dfrac{12.5}{13.7}$ Press: 12.5 $\boxed{\div}$ 13.7 $\boxed{=}$

$\angle A \doteq 24.1596$ Then press: $\boxed{\cos^{-1}}$

$\angle A$ is 24° to the nearest degree.

Example 2

An architect plans a wheelchair ramp with a rise of 0.5 m. Safety standards indicate that the ramp should have an angle of inclination of 8°. How long will the ramp be?

Solution

Sketch the ramp from the side. The length of the ramp is the hypotenuse of a right triangle.

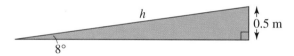

From the diagram, the side opposite the 8° angle is 0.5 m.
Let the hypotenuse be h.

$\sin 8° = \dfrac{0.5}{h}$

$h \times \sin 8° = 0.5$

$\qquad h = \dfrac{0.5}{\sin 8°}$ Press: 0.5 $\boxed{\div}$ 8 $\boxed{\text{SIN}}$ $\boxed{=}$

$\qquad h \doteq 3.5926$

The ramp will be about 3.6 m long.

Example 3

Use the information in $\triangle PQR$.

a) Determine the measures of the two acute angles to the nearest degree.

b) Determine the length of PQ to the nearest centimetre.

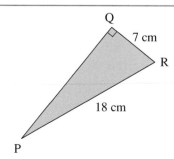

Solution

a) One given side is the hypotenuse. The other side is adjacent to $\angle R$.
Use the cosine ratio with $\angle R$.

$$\cos R = \frac{RQ}{RP}$$

$$= \frac{7}{18} \qquad \text{Press: } 7 \; \boxed{\div} \; 18 \; \boxed{=} \; \boxed{\text{cos}^{-1}}$$

$$\angle R \doteq 67.1146°$$

$\angle R$ is $67°$ to the nearest degree.

To determine $\angle P$, recall that the sum of the angles in a triangle is $180°$.

$$\angle P = 180° - \angle Q - \angle R$$

$$= 180° - 90° - 67°$$

$$= 23°$$

$\angle P$ is $23°$ to the nearest degree.

b) Use the Pythagorean Theorem.

$$PQ^2 = 18^2 - 7^2$$

$$= 275$$

$$PQ = \sqrt{275}$$

$$\doteq 16.5831$$

PQ is 17 cm to the nearest centimetre.

DISCUSSING THE IDEAS

1. In *Example 1*, $\angle A$ was shown to 4 decimal places in parts a and b. Why are these numbers not the same?

2. In *Example 2*, how did we know to use the sine ratio and not the cosine or tangent ratios?

3. In *Example 2*, suppose the architect needs to determine the run of the ramp, to ensure the available floor space will accommodate the ramp. Explain two different ways you could calculate the run.

4. What other way could we have solved *Example 3a*?

5. *Example 3b* can be solved using trigonometric ratios. Find as many different ways as you can to do this. Compare the results with the solution given.

A 1. In each triangle
  a) Calculate sin A.   **b)** Determine ∠A.

  i)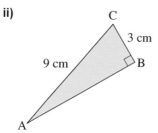

  ii)

  iii)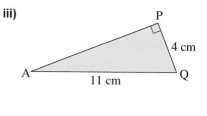

2. Select one triangle from exercise 1. Write to explain how you calculated sin A, and how you determined ∠A.

3. In each triangle
  a) Calculate cos A.   **b)** Determine ∠A.

  i)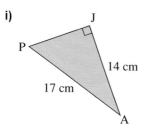

  ii)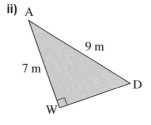

  iii)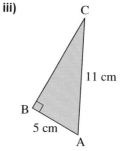

4. In each triangle, the measured lengths of the sides are labelled. Determine sin A, cos A, and ∠A. Determine sin C, cos C, and ∠C.

  a)   **b)**   **c)**

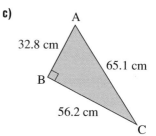

B 5. In △ABC
  a) Calculate AC for each value of ∠A.
    i) 34°   **ii)** 55°   **iii)** 79°
  b) Calculate ∠A for each value of AC.
    i) 6.0 m   **ii)** 11.0 m   **iii)** 16.0 m

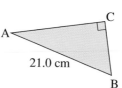

6. Draw a right triangle that satisfies each description. Label the measures of the sides and angles.

a) $\triangle$JKL where $\sin J = \frac{2}{3}$ and $\angle K = 90°$

b) $\triangle$PYT where $\cos T = \frac{1}{3}$ and $\angle Y = 90°$

7. Select one part of exercise 6. Write to explain your solution.

8. A 9.0-m ladder rests against the side of a wall (below left). The bottom of the ladder is 1.5 m from the base of the wall. Determine the measure of the angle between the ladder and the ground.

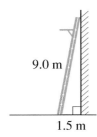

9.0 m

1.5 m

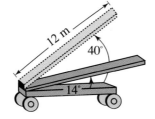

12 m   40°

14°

9. The inclination of a warehouse conveyor belt can be set between 14° and 40° (above right). The length of the conveyor is 12 m. What are the minimum and maximum heights the conveyor can reach?

10. The angle of inclination of a conveyor belt is 5°. The belt rises 0.75 m. What is the length of the conveyor?

11. A guy wire is fastened to the ground 40 m from the base of a TV tower. The wire makes an angle of 60° with the ground.

a) How high up the tower does the guy wire reach?

b) How long is the guy wire?

c) Suppose the same length of wire were attached, with a 45° angle of elevation. How high up the tower would the guy wire reach?

12. Rectangle ABCD measures 6 cm by 3 cm. Calculate the measures of the two acute angles formed at A by the diagonal AC.

Note: From now on, you may need to use the sine, cosine, or tangent ratios to solve each problem.

13. In isosceles $\triangle$ABC, AB = AC, $\angle C = 70°$, and BC = 4.0 cm. Calculate the height of the triangle to the nearest millimetre.

14. A wheelchair ramp is 8.2 m long. It rises 94 cm. Determine the angle of inclination of the ramp to 1 decimal place.

15. The angle of elevation of the sun is 68° when a tree casts a shadow 14.3 m long. How tall is the tree?

16. a) The guy wire for a jib sail ABC is 3.7 m long (below left). It is attached to the mast at point A. The foot of the guy wire is 1.1 m from the foot of the mast. What is the angle of inclination of the guy wire?

b) The mast AB is 3.5 m long. The boom BC is 3.2 m long. Determine ∠C to 1 decimal place.

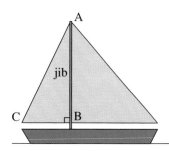

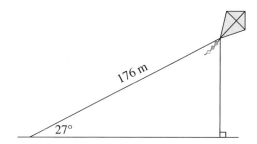

17. A person flying a kite has released 176 m of string (above right). The string makes an angle of 27° with the ground. How high is the kite?

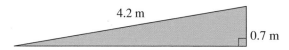

MODELLING the Height of a Kite

To complete exercise 17, you had to assume the kite string is straight. In reality, gravity drags the string down so that it curves up away from the ground at the kite.

* How does the curvature in the string affect your answer to exercise 17? Explain.
* How does the height of the hand holding the string affect the answer?

18. A wheelchair ramp is 4.2 m long. It rises 0.7 m. What is its angle of inclination to the nearest degree?

19. A surveyor, 31 m from a building, uses a transit to measure the angle of elevation to the top of the building. The angle of elevation is 37°. The transit is set at a height of 1.5 m.

a) Calculate the distance from the transit to the top of the building.

b) Calculate the height of the building.

20. At a point 28 m from a building, the angle of elevation to the top of the building is 65°. The observer's eyes are 1.5 m above the ground.

a) How tall is the building?

b) How far is the observer's eye from the top of the building?

21. Turn to Section 2.6, page 108, on relating the sides of special triangles.

   a) Describe the 30-60-90 Property in that section.

   b) Use the diagram on page 109. Determine an expression for each ratio.

       i) $\sin 30°$ ii) $\cos 30°$ iii) $\sin 60°$ iv) $\cos 60°$

   c) Use a calculator to verify the results of part b.

22. The altitudes of an equilateral triangle are 10.0 cm (below left). Calculate the lengths of the sides of the triangle.

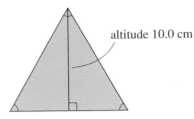

altitude 10.0 cm

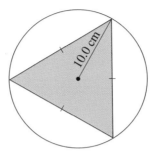

10.0 cm

C 23. An equilateral triangle is inscribed in a circle (above right). The radius of the circle is 10.0 cm. Calculate the side length of the triangle.

24. Calculate the perimeter of a regular pentagon inscribed in a circle with radius 5.4 cm.

25. Calculating Earth's circumference involves the use of trigonometry. Here is one possible approach: from the top of a mountain 5 km high, the angle between the horizon and the true vertical measures 87.73°. The true vertical is an imaginary line that connects the top of the mountain to the centre of Earth. Use the diagram to calculate the radius and the circumference of Earth. The diagram is not drawn to scale.

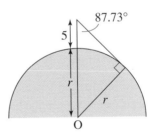

87.73°

5

r

r

O

COMMUNICATING THE IDEAS

In your journal, describe how the sine and cosine ratios are different from the tangent ratio. Explain when you would use the sine or cosine ratio to solve a problem rather than the tangent ratio.

The Spiral Tunnels

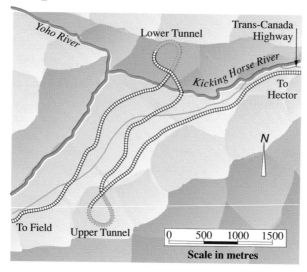

Yoho River

Lower Tunnel

Trans-Canada Highway

Kicking Horse River

To Hector

N

To Field

Upper Tunnel

0   500   1000   1500

Scale in metres

When British Columbia became a province on July 20, 1871, the federal government agreed to build a railway link to the rest of Canada. This railway was completed in 1885. The track through the mountains was kept to a maximum gradient of 2.2% (a rise of 2.2 m in 100 m of track) with one exception — the section of track between Field, B.C., and Hector, B.C. There, a rise of 297 m in only 6.6 km of track was necessary. This section became known as the Big Hill. Taking trains up and down this hill required additional locomotives, and on the downhill run there was always the danger of runaway trains.

The only way to reduce the gradient of Big Hill to 2.2% was to lengthen the track between the two towns. This was done in 1907–1909 when a pair of spiral tunnels was built into the mountains. These tunnels are the only ones of their kind in North America. As many as 15 trains a day pass through the tunnels in each direction. There is a lookout on the Trans-Canada Highway where you can see a long train passing under itself as it comes out of the tunnel before it has finished going in.

Visualize the track straightened out to form the hypotenuse of a right triangle. The track before and after the construction of the spiral tunnels would appear as shown in these diagrams.

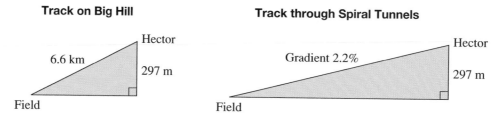

Track on Big Hill

Hector

6.6 km

297 m

Field

Track through Spiral Tunnels

Hector

Gradient 2.2%

297 m

Field

1. What was the gradient of the track on Big Hill?

2. What was the angle of elevation of the track on Big Hill?

3. What angle of elevation is needed for the gradient to be 2.2%?

4. To achieve a gradient of 2.2%, how long did the track have to be between Field and Hector?

5. By how much did the track have to be lengthened to reduce the gradient to 2.2%?

The train in the photograph is passing through Lower Tunnel, which is about 880 m long and curves through 288°. Each car in the train is about 12.2 m long.

6. Approximately how many cars are in the train?

7. What is the difference in height between the two sections of track in the photograph?

8. Suppose the speed of the train is 40 km/h. How long does it take to pass completely through the tunnel?

Mathematics & Geography

8.3 Solving Right Triangles

We can use the three trigonometric ratios of tangent, sine, and cosine to calculate unknown measures in any right triangle provided we know:

- the lengths of two sides, or
- the length of one side and the measure of one acute angle

When we determine all unknown measures in a right triangle, we *solve the triangle*.

Example 1

Solve $\triangle ABC$, given $AC = 5.0$ cm, $BC = 2.0$ cm, and $\angle B = 90°$.

Solution

Sketch the triangle.
Use the Pythagorean Theorem to determine AB.

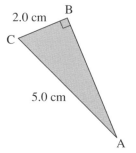

$AC^2 = AB^2 + BC^2$
$5.0^2 = AB^2 + 2.0^2$
$AB^2 = 25 - 4$
$ = 21$
$AB = \sqrt{21}$
$ \doteq 4.5826$

AB is approximately 4.6 cm.

Use the sine ratio to determine the measure of $\angle A$.

$\sin A = \dfrac{2.0}{5.0}$ $\qquad$ Press: 2 $\boxed{\div}$ 5 $\boxed{=}$ $\boxed{\text{SIN}^{-1}}$
$\quad \angle A \doteq 23.5782$

$\angle A$ is approximately 24°.

$\angle C = 90° - 24°$
$ = 66°$

$\angle C$ is approximately 66°.

Example 2

Solve $\triangle XYZ$, given $XY = 9.0$ m, $\angle Y = 90°$, and $\angle Z = 36°$.

Solution

$\angle X = 90° - 36°$
$\quad\ = 54°$

$\angle X$ is 54°.

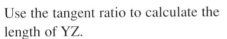

Use the tangent ratio to calculate the length of YZ.

$$\tan 36° = \frac{9.0}{YZ}$$
$$YZ \times \tan 36° = 9.0$$
$$YZ = \frac{9.0}{\tan 36°} \qquad \text{Press: } 9 \boxed{\div} 36 \boxed{\text{TAN}} \boxed{=}$$
$$\doteq 12.3874$$

YZ is approximately 12.4 m.

Use the sine ratio to determine the length of XZ.

$$\sin 36° = \frac{9.0}{XZ}$$
$$XZ = \frac{9.0}{\sin 36°}$$
$$\doteq 15.3117$$

XZ is approximately 15.3 m.

We can also use the tangent, sine, and cosine ratios to solve problems involving right triangles. Before solving a trigonometry problem, examine the information to select the appropriate ratio to use.

Example 3

Lighthouse Park is 7.0 km due north of Tower Beach, in Vancouver. A sailboat leaves Lighthouse Park on a bearing of 211°. When the boat is due west of Tower Beach, it tacks and sails straight to the beach. How far does the boat travel?

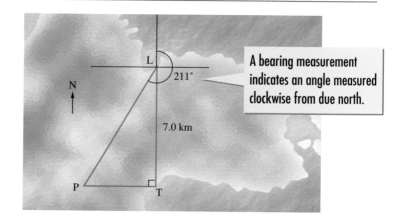

A bearing measurement indicates an angle measured clockwise from due north.

Solution

> **Think...**
>
> Triangle LPT is a right triangle. We know the length of LT. We can use the bearing measurement to determine $\angle L$. Then we will know a side and an acute angle in $\triangle LPT$, and can determine the other sides. The total distance travelled is LP + PT.

In $\triangle LPT$

$$\angle L = 211° - 180°$$
$$= 31°$$

LT is the side adjacent to the 31° angle. LP is the hypotenuse.
Use the cosine ratio.

$$\cos 31° = \frac{7.0}{LP}$$
$$LP \times \cos 31° = 7.0$$
$$LP = \frac{7.0}{\cos 31°}$$
$$\doteq 8.1664$$

Press: 7 ÷ 31 cos =
Do not clear the screen.

Use the Pythagorean Theorem.

$$PT^2 = LP^2 - LT^2$$
$$= 8.1664^2 - 7.0^2$$
$$\doteq 17.6906$$
$$PT \doteq 4.2060$$

Then press: x^2 − 7 x^2 =
Then press: $\sqrt{x}$

$$LP + PT \doteq 8.1664 + 4.2060$$
$$\doteq 12.3724$$

The total distance the boat travelled is approximately 12.4 km.

Example 4

The distance between two floors in a new home is 2.5 m. The horizontal length of the staircase is 2.8 m.

a) Determine the slope of the staircase to 4 decimal places.

b) Determine the angle of inclination of the staircase to 1 decimal place.

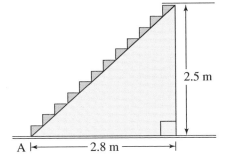

2.5 m

A |← ———— 2.8 m ————→|

Solution

a) Recall from Chapter 3 that slope is $\frac{\text{rise}}{\text{run}}$.
From the diagram, the rise is 2.5 m and the run is 2.8 m.
The slope of the stairs is $\frac{2.5}{2.8}$, or about 0.8929.

b) The rise and run are the measures of two legs of a right triangle.
Use the tangent ratio.

$$\tan A = \frac{2.5}{2.8}$$
$$\doteq 0.8929$$
$$\angle A \doteq 41.8°$$

The angle of inclination of the stairs is about 41.8°.

DISCUSSING THE IDEAS

1. In the last step of the solution to *Example 1*, explain why $\angle C$ can be subtracted from 90°.

2. In *Example 1*, describe a different way to determine each measure.

   a) $\angle A$ **b)** the length of AB

3. Describe a different way to solve the triangle in *Example 2*.

4. What other ways can we determine the lengths of the sides of the triangle in *Example 3*?

5. In *Example 4*, explain what it means for the slope of the stairs to be about 0.9.

6. Turn to *Linking Ideas* on pages 170 and 171 of Chapter 3. Does the staircase in *Example 4* meet Blondel's rule? Does it meet the CMHC standard? Discuss any implications.

A **1.** For each triangle, determine sin A, cos A, and tan A.

a)
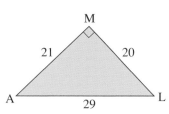

b)
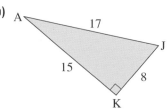

c)
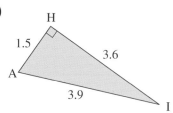

2. Solve each triangle.

a)
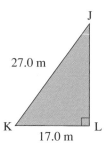

b)

c)
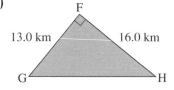

3. Solve each triangle.

a)
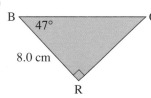

b)
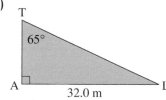

c)
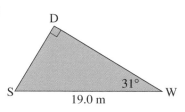

d)
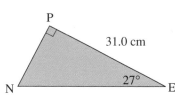

e)
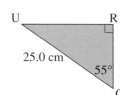

f)
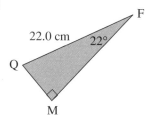

4. Solve each △XYZ, given ∠Y = 90°, and the following lengths and angle measures.

   a) XY = 24 m, XZ = 35 m   **b)** XY = 16 m, ∠X = 27°

   c) XZ = 51 cm, YZ = 13 cm   **d)** XZ = 72 cm, ∠Z = 52°

   e) YZ = 32 mm, ∠X = 64°   **f)** XY = 45 mm, YZ = 20 mm

5. Select one part of exercise 4. Write to explain how you solved the triangle.

B 6. A TV tower is supported by guy wires. For each wire, determine its slope and its angle of inclination.

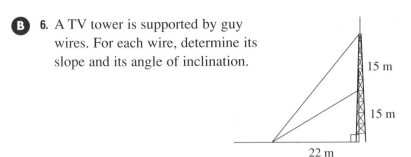

15 m

15 m

22 m

7. Shopping malls use ramps, steps, and escalators to move people from one level to another. For each conveyor, determine its slope and its angle of inclination.

a) ramp   b) steps   c) escalator

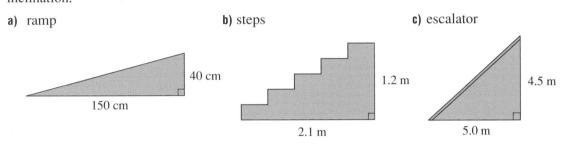

40 cm

150 cm

1.2 m

2.1 m

4.5 m

5.0 m

8. A mountain road rises 1 m for every 5 m along the road. Determine the angle of inclination of the road.

9. Surveyors with a highway planning crew measure a section of proposed roadway to be 4.4 km. They measure the angle of inclination over the 4.4 km to be 8.5°. Determine the change in altitude for drivers travelling from one end to the other.

10. As you drive along a road, you may see a sign that reads, "Hill grade 4%." Write to explain what this means.

MODELLING the Inclination of a Road

The inclination of a road up a hill or a mountain is often called its *gradient*. In exercises 8, 9, and 10, you modelled the gradients of roads. In *Linking Ideas* on page 468, you modelled the gradient of a railway track.

- Describe the model you have been using for the inclination of a road.
- Could the slope of a line be used as a model for the inclination of a road? Explain.

11. Prior to 1982, visitors to the Peace Tower in Ottawa had to ride two elevators. The Memorial Chamber at the base of the tower made a vertical ascent impossible. The elevator system now carries visitors up the first 24.2 m along a path inclined at 10° to the vertical. It then rises vertically for the rest of the trip.

a) How long is the elevator shaft that runs on the incline?

b) How far is the elevator displaced horizontally by the incline?

c) What is the slope of the incline? Give your answer to 2 decimal places.

12. Armstrong Point and Robert's Point are in Sidney on Vancouver Island. Fernie Island is 1750 m due north of Robert's Point. A sailboat left Robert's Point on a bearing of 016°. When it reached Coal Island, which is due east of Fernie Island, it turned and sailed straight to Fernie Island. The sailboat then returned to Robert's Point. How far did the boat travel?

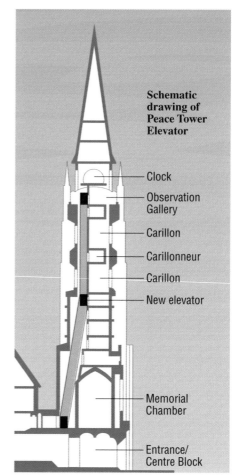

Schematic drawing of Peace Tower Elevator

- Clock
- Observation Gallery
- Carillon
- Carillonneur
- Carillon
- New elevator
- Memorial Chamber
- Entrance/ Centre Block

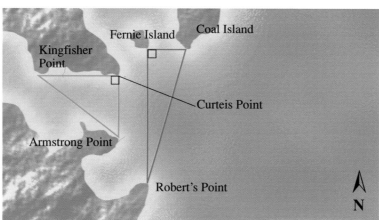

13. Curteis Point is 800 m due north of Armstrong Point. A sailboat left Armstrong Point on a bearing of 307°. When it reached Kingfisher Point, which is due west of Curteis Point, it turned and sailed straight to Curteis Point. The sailboat then returned to Armstrong Point. How far did the boat travel?

14. Grain is stored in a cone-shaped pile. The dimensions of the cone are shown. Calculate the angle of inclination of the side of the cone.

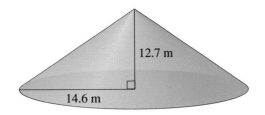

14.6 m

12.7 m

15. Construct these three figures. Make a cone from each figure by joining the straight edges and securing them with tape.

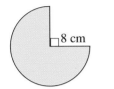 8 cm

 8 cm

8 cm

a) Calculate the base radius and the height of each cone.

b) Calculate the angle of inclination of the side of each cone.

c) Verify your results in parts a and b by measuring.

16. The Agusta A109C MAX helicopter is often used for emergency medical transport. An Agusta A109C MAX flies from its base B, due west to a small community C. It picks up two patients and a medical attendant. The helicopter then flies on a bearing of 056° to a hospital H. The hospital is 85 nm due north of the helicopter base.

a) Determine how far the helicopter travels on each leg of its journey.

    i) from B to C   **ii)** from C to H

b) Use the *Helicopters* database. Find the record for this model and its sea level cruise speed, fuel flow, fuel capacity, and hourly operating cost.

c) Based on the sea level cruise speed, how long will the helicopter take for this trip?

d) How much fuel will it use?

e) Calculate the operating cost for the time spent flying, to the nearest dollar.

f) Can the helicopter fly from the hospital to its base at B without refuelling? Explain.

C **17.** A cylindrical oil tank is 55.3 m high and 28.4 m in diameter. The top of the tank is reached by a spiral stairway that circles the tank once. Calculate the angle of inclination of the stairway to the nearest degree.

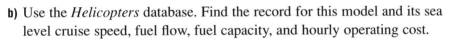
COMMUNICATING THE IDEAS

In your journal, write to explain what it means to solve a right triangle, and how this is similar to solving an equation. Illustrate your explanation with some examples.

Indirect Measurement

Some grade 10 students drew this scale diagram. It shows a pine tree, a flag, and a TV tower that are visible from their school. The students did all the measuring to draw the diagram, without leaving the school grounds.

In this chapter, you will learn how the students determined:

- the distance to each object
- the distances from any object to the other two objects
- the measures of the angles in $\triangle PFT$

Points X and Y are two points on the school grounds from where the three objects can be seen. The distance XY is 54.0 m.

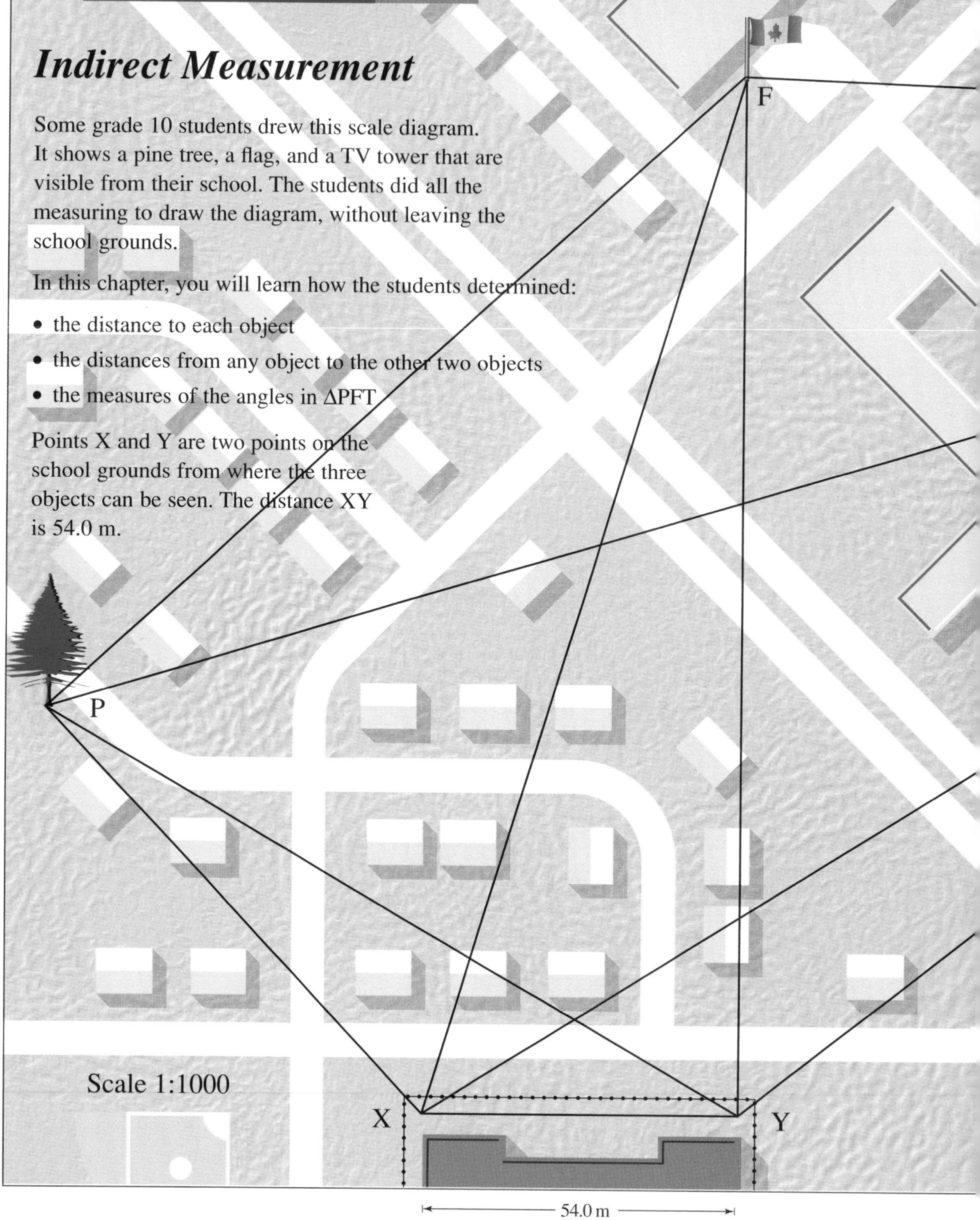

F

P

Scale 1:1000

X

Y

|← 54.0 m →|

Since the students did all the measuring from the school grounds, the only length they were able to measure was the distance XY. The line segment XY is called the *baseline*. The only angles the students were able to measure were the angles formed by the baseline and the lines of sight to particular objects.

T

You will need a protractor and a ruler to complete these exercises.

1. a) Measure ∠FYX and ∠FXY with your protractor.

   b) Use the result of part a. Calculate the distances from X and Y to the flag.

   c) Check your answers by measuring and using the scale.

2. Measure and record these angles.

   a) ∠PYX and ∠PXY   b) ∠TYX and ∠TXY

   Save your measurements for use later (page 511 exercise 14 and page 519 exercise 16).

Calculating the Speed of Earth's Rotation

Collected Wisdom is a weekly column in *The Globe and Mail*. Readers send in questions and other readers provide answers. Readers can contact *Collected Wisdom* by phone, fax, mail, e-mail, and the Internet. The following items appeared in this column in February, 1997.

February 6

The question: If Earth rotates at 1670 km/h at the equator, said Ottawa's **R.M.**, what is the speed of its rotation at Ottawa?

The answer: Strange to relate, many correspondents took this as an opportunity to make jokes about our nation's capital. For a serious reply, we go to **D.N.** of Gloucester, Ont., who also provided the handsome accompanying diagram.

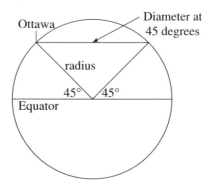

"The circumference of Earth at the equator is 40 076 km. This gives Earth a radius of 6378 km. Assuming that Earth is perfectly spherical (it isn't, but that would complicate the problem), we turn to the triangle in the diagram. Ottawa is at a latitude of about 45° north and, applying Pythagoras's theorem, we find that Earth's diameter at that point is 9020 km. Using *pi* times this diameter, we get a circumference of 28 338 km at Ottawa's latitude. Since a day has 24 hours, this means Ottawa is travelling at 1180 km/h."

February 13

Now, welcome to the second episode of *As the World Turns*, our investigation into the speed at which Ottawa is moving as Earth spins. **J.D.** of Wolfville, N.S., writes:

"Last week's response from **D.N.** of Gloucester, Ont., does not justify the answer of 1180 km/h, however correct that answer may be.

"Pythagoras's theorem … says nothing about the ratios between the lengths of the … sides. For this one must turn to trigonometry. The key to the problem is the cosine of the latitude."

Taking Ottawa's latitude as 45° and the cosine of 45° as 0.7071, "you can work out a solution by the route chosen by **D.N.**" You take Earth's radius at the equator (6378 km), divide it by 0.7071 and you get 9019 km, the diameter of the circle made by Ottawa's latitude. Multiply this by *pi* and you get 28 334 km, the circumference of the circle at Ottawa's latitude. Divide this by 24 (hours in the day) and you'll find that Ottawa is spinning around at 1180 km/h.

Use carefully drawn diagrams in your solutions to these problems.

1. Study the solution provided by **D.N.** in the February 6 column.

   a) Confirm that Earth's radius is 6378 km.

   b) What is the meaning of this statement?
      "… Earth's diameter at that point is 9020 km."

   c) Confirm that the 9020 km distance is correct.

   d) Confirm that the rest of the solution is correct.

2. In the February 13 column, **J.D.** claimed that **D.N.**'s response is not justified, although the answer is correct. Do you agree with this claim? Explain.

3. Study the solution provided by **J.D.** in the February 13 column.

   a) What does "… the diameter of the circle made by Ottawa's latitude" mean?

   b) Why does dividing Earth's radius by $\cos 45°$ give the diameter of this circle?

   c) Confirm that the rest of the solution is correct.

4. Determine the latitude of the location where you live. Could one or both of these methods be used to determine the speed of Earth's rotation at your location?

   a) If your answer is yes, calculate to determine this speed.

   b) If your answer is no, modify one method to determine the speed, and explain what you did.

5. Write a letter to the editor of the column explaining that both solutions are correct. In your letter, point out the advantages and disadvantages of the two solutions. Your letter should be brief so that it could be published in the following week's column.

8.4 Problems Involving More than One Right Triangle

Trigonometry has been used to solve a wide range of problems in mathematics, science, and industry. Some of these problems require two right triangles to model the situation.

Example 1

Two office towers are 50 m apart. From the 14th floor of the shorter tower, the angle of elevation to the top of the taller tower is 33°. The angle of depression to the base of the taller tower is 39°. Determine the height of the taller tower.

Solution

Let x metres represent the height of the taller tower above the 14th floor.
Let y metres represent the height of the taller tower below the 14th floor.

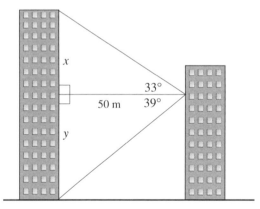

Use the tangent ratio in the upper triangle.

$$\tan 33° = \frac{x}{50}$$
$$x = 50 \times \tan 33°$$
$$\doteq 32.4704$$

Use the tangent ratio in the lower triangle.

$$\tan 39° = \frac{y}{50}$$
$$y = 50 \times \tan 39°$$
$$\doteq 40.4892$$

The height of the taller tower is $x + y = 32.4704 + 40.4892$, or 72.9596.
The taller tower is approximately 73 m tall.

Example 2

As part of a weekend expedition, an Adventurer's
Club proposes to climb a cliff overlooking a river.
To plan for the climb, a surveyor took some
measurements to calculate the height of the cliff.
From a point R on the shore directly across the
river, the angle of elevation to the top of the cliff is
∠TRB = 43°. From a point S, 30 m down the river,
∠BSR = 69°. Calculate the height of the cliff.

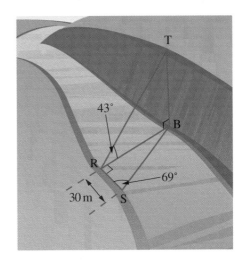

Solution

> **Think...**
>
> Triangle TRB is in a vertical plane and ∆RBS is in a horizontal plane. The height of
> the cliff is TB. To determine this height, we need to know the length of one side of
> ∆TRB. Side RB is also a side of ∆RBS. We can determine its length from that triangle.

In ∆RBS, use the tangent ratio
to determine RB.

$$\tan 69° = \frac{RB}{30}$$
$$RB = 30 \times \tan 69°$$
$$\doteq 78.15$$

In ∆TRB, use the tangent ratio
to determine the length of TB.

$$\tan 43° = \frac{TB}{78.15}$$
$$TB = 78.15 \times \tan 43°$$
$$\doteq 72.88$$

The height of the cliff is about 73 m.

 MODELLING for Measuring Inaccessible Heights

The method in the solution of *Example 2* is a model for measuring an inaccessible height in
a situation where the base of the object being measured cannot be approached.

- What assumptions about the cliff and the river shore are implied by this model?

- What are some limitations of this model?

- Explain how the model could be applied to measure other distances that cannot be
  measured directly, such as the width of a river.

1. The solutions to *Example 1* and *Example 2* involve the tangent ratio. What distances in these examples could you determine using the sine or cosine ratios? Why do you think these distances are not relevant to these two problems?

2. When you are solving a problem involving a right triangle, how do you know which trigonometric ratio to use?

8.4 EXERCISES

A 1. Calculate the length of BC.

a)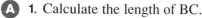

b)

c)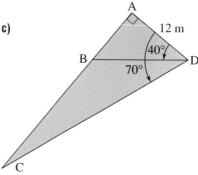

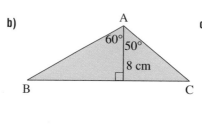

2. Calculate the measure of ∠ABC.

a) b) c)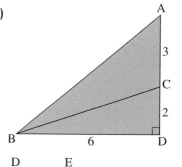

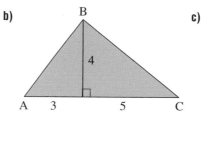

B 3. Three squares with sides 5 cm long are placed side by side.

a) Determine the measure of each angle to the nearest tenth of a degree.

   i) ∠CAB   ii) ∠DAB   iii) ∠EAB

b) Use your answers to part a. How are the three angles related?

c) Suppose the squares had a different side length. Would the relation in part b still hold? Explain.

4. Two office towers are 31.7 m apart (below left). From the shorter tower, the angle of elevation to the top of the taller tower is 27.5°. The angle of depression to the base of the taller tower is 48.2°. The diagram is not drawn to scale. Calculate the height of each tower.

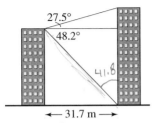

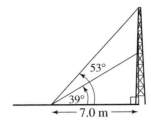

5. A tower is supported by a guy wire (above right). The angle of inclination of the guy wire is 39°. From the end of the guy wire, the angle of elevation to the top of the tower is 53°. The guy wire is fixed to the ground 7.0 m from the base of the tower. The diagram is not drawn to scale.

a) Calculate the length of the guy wire, and the height of the tower.

b) Calculate the distance from the top of the tower to where the guy wire is connected to the tower.

c) Choose part a or b. Write to explain how you calculated the measure.

6. Determine each value of x and y.

a)

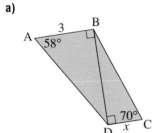

b)

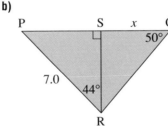

c)

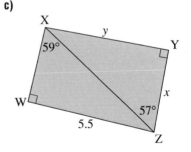

7. In △PQR (below left), calculate the length of PS. Try to find two different ways to do this.

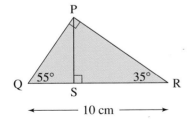

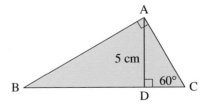

8. Calculate the area of △ABC (above right). Try to find two different ways to do this.

9. Two trees are 100 m apart. From a point midway between them, the angles of elevation to their tops are 12° and 16°. How much taller is one tree than the other? The diagram is not to scale.

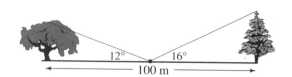

10. The world's longest suspension bridge is across the Humber Estuary in England. The towers of this bridge reach about 135 m above the level of the bridge. The angles of elevations of the towers measured from the centre of the bridge and either end are 10.80° and 18.65°, respectively. How long is the bridge?

11. All three triangles in the diagram have angles of 30°, 60°, and 90°, and AB = 4 cm.

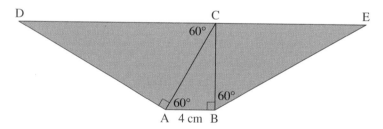

  a) Calculate the lengths of all the sides of ΔADC and ΔBEC.

  b) Determine the ratio of length DC to length CE.

  c) Determine the ratio of length AD to length BE.

12. From the top of a 120-m fire tower, a fire ranger observes smoke in two locations. One has an angle of depression of 6°, and the other has an angle of depression of 3°. Calculate the distance between the smoke sightings when they are as described below. The diagrams are not to scale.

  a) on the same side of the tower and in line with the tower

                    b) on opposite sides of the tower and in line with the tower

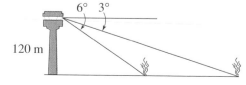
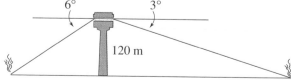

c) in perpendicular directions from the tower

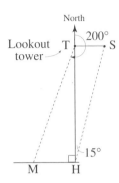

13. A person in a lookout tower T reports smoke at S, due east of the tower. She estimates the distance to the smoke is 22 km. A helicopter carrying firefighters lifts off from its base H, 82 km due south of the tower. From the helicopter base, the smoke is on a bearing of 015°. The person in the tower reports a second smoke sighting at M, on a bearing of 200°. From the helicopter base, the second smoke sighting is due west. The helicopter drops the firefighters at S, flies to T to pick up more firefighters, then flies to M. Calculate the total distance the helicopter travels.

For exercises 14 and 15, you will need an empty carton and either a metre stick or a yardstick.

14. **a)** Measure the length and height of the carton. Place the stick inside the carton as shown (below left). Use your measurements to determine the angle of inclination of the stick.

b) Place the carton on the floor near a wall, with the stick just touching the wall. Use the length of the stick and your answer to part a to calculate how high up the wall the stick reaches. Check your answer by measuring.

c) Give some reasons why your calculated result may differ from your measured result.

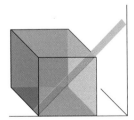 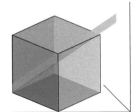

15. **a)** Place the metre stick along the main diagonal of the carton (above right). Then repeat exercise 14.

b) Explain why the angle of inclination in part a is less than the angle of inclination of the stick in exercise 14a.

16. Calculate the height of the cliff in the diagram (below left).

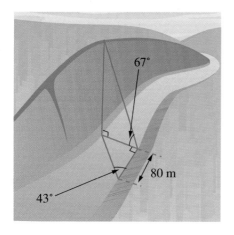

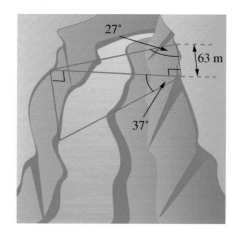

17. Calculate the width and the depth of the gorge in the diagram (above right).

Exercises 18 and 19 involve the Muttart Conservatory in Edmonton, which consists of two large and two small pyramids. As you can tell from the photograph, the outside walls of the pyramids are mostly glass.

18. In the small pyramids, each wall is a triangle with base 21.0 m and height 17.3 m. The person who cut the glass for the walls of the pyramid had to know the measures of the three angles in each triangular wall. Calculate these angles.

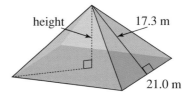

19. Calculate the height of the pyramid in exercise 18.

COMMUNICATING THE IDEAS

You have used trigonometry to solve many different problems that could be modelled by right triangles. Imagine that it is the end of the year, and you and your friend are reviewing everything you learned in this course. Prepare a summary of the steps you would use to solve a problem using trigonometry. Write the summary in your journal.

Sines and Cosines of Obtuse Angles

The sine and cosine of an acute angle were defined on page 460 as ratios of the side lengths of a right triangle. Since the largest angle in a right triangle is 90°, these definitions do not apply for obtuse angles. Hence, you might expect your calculator to give an error message if, for example, you try to determine sin 130° or cos 130°.

Press: 130 [SIN] to display: 0.766044443   Hence, sin 130° ≐ 0.7660

Press: 130 [COS] to display: −0.642787609   Hence, cos 130° ≐ −0.6428

Look at that! There *are* values for trigonometric ratios of angles greater than 90°. And, one of them is negative!

In the exercises below, you will use your calculator to determine sines and cosines of other obtuse angles. You will also look for patterns to discover how the results are related to the trigonometric ratios of acute angles. The reasons for the patterns are explained in the next section.

Sines of obtuse angles

1. a) Look at sin 130°. Use guess and check to find an acute angle that has the same sine. That is, find an acute ∠A such that sin A = 0.7660344443.

   b) How many values of A between 0° and 180° are there such that sin A = sin 130°? How do you know?

2. Find acute angles A such that:

   a) sin A = sin 100°   b) sin A = sin 120°   c) sin A = sin 150°

3. Summarize any patterns you found in the results of exercises 1 and 2.

Cosines of obtuse angles

4. Look at cos 130°.

   a) Explain why there is no acute angle that has the same cosine.

   b) Find an acute angle that has the same cosine, except for the sign. That is, find an acute ∠A such that cos A = 0.642787609.

5. Find acute angles A such that:

   a) cos A = −cos 100°   b) cos A = −cos 120°   c) cos A = −cos 150°

6. Summarize any patterns you found in the results of exercises 4 and 5.

8.5 Defining the Sine and the Cosine of an Obtuse Angle

Until now, we have used trigonometry to solve problems involving right triangles. In the next section, we will solve problems involving triangles that are not right-angled. Since some of these triangles might contain an obtuse angle, we will work with trigonometric ratios of obtuse angles.

The trigonometric ratios of acute angles were defined on pages 453 and 460 as ratios of the side lengths of a right triangle. Since right triangles do not contain obtuse angles, trigonometric ratios of obtuse angles have no meaning according to these definitions. Trigonometric ratios can be defined in another way that applies to both acute and obtuse angles. Working on a coordinate grid will help you to understand the new definitions.

Since the letter O looks like the number 0, we will use the letter A to represent the origin in this section.

INVESTIGATE

Consider a line segment AP 1 unit long. Its endpoints are the origin, A, and a point P(x, y) in the first quadrant. Triangle ANP is a right triangle with one vertex at the origin and hypotenuse 1 unit long. In this triangle, $\angle$PAN = 30°.

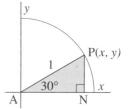

1. What is the length of AN, to 2 decimal places?

2. What is the length of NP, to 2 decimal places?

AN is the x-coordinate of P and NP is the y-coordinate of P.

3. Visualize the triangle changing, as $\angle$PAN changes from 0° to 90° and back. The measure of AP remains fixed at 1 unit. Describe how the coordinates of P change as it rotates from the x-axis to the y-axis and back.

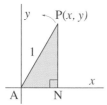

Rotating toward the y-axis

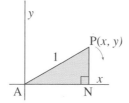

Rotating toward the x-axis

4. Start a diagram and a table similar to those below. For each ∠A, determine the sine and cosine, to 2 decimal places, then enter them in columns 2 and 3. Use the results to write the coordinates of point P(x, y) in column 4. Plot each point on your diagram.

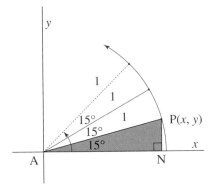

∠A	cos A	sin A	P(x, y)
15°			
30°			
45°			
60°			
75°			

5. Describe what happens to cos A and to sin A as AP rotates from the x-axis to the y-axis. Describe any other patterns you observe in your results of exercise 4.

6. Visualize line segment AP as it continues to rotate past the y-axis to the negative x-axis. What kind of angle is now formed between line segment AP and the positive x-axis? Extend the patterns in exercise 5 to predict the coordinates of P(x, y) in the table below. Copy and complete the diagram and column 4 of the table.

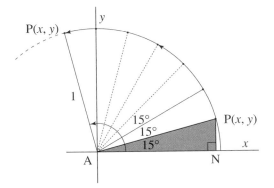

∠A	cos A	sin A	P(x, y)
105°			
120°			
135°			
150°			
165°			

7. In the first quadrant, P(x, y) corresponded to P(cos A, sin A). Suppose that relationship can be extended into the second quadrant. What are the cosine and sine of each angle in the table? Enter these in columns 2 and 3.

8. Use your calculator to confirm your results in exercise 7. Determine the cosine and sine of each angle.

  a) 105°   **b)** 120°   **c)** 135°   **d)** 150°   **e)** 165°

9. Your diagrams for exercises 4 and 6 also indicate that the points P(1, 0), P(0, 1), and P(−1, 0) fit the pattern you used to predict the sines and cosines of obtuse angles. What does this tell you about the following sines and cosines? Check with a calculator.

  a) $\cos 0°$, $\sin 0°$ **b)** $\cos 90°$, $\sin 90°$ **c)** $\cos 180°$, $\sin 180°$

10. Use your results from exercises 4 and 7.

  a) Look for pairs of angles that have the same sine. List the angle pairs, and their sines. How are the angles related?

  b) Use the angle pairs you identified in part a. How are the cosines of each angle pair related?

  c) What general observations can you make about the sines and cosines of acute and obtuse angles?

To give meaning to trigonometric ratios of obtuse angles, we redefine the trigonometric ratios using a coordinate grid. The diagrams below show a semicircle with centre the origin, A, and radius 1 unit. Point B is a fixed point on the semicircle with coordinates (1, 0). Point P(x, y) is any point on the semicircle. The angle formed by segments PA and AB is represented by ∠A.

We define:

- the *cosine* of ∠A is the x-coordinate of P: $\cos A = x$
- the *sine* of ∠A is the y-coordinate of P: $\sin A = y$

In these definitions, the radius AP must be 1 unit.

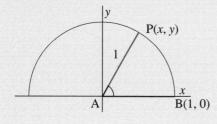 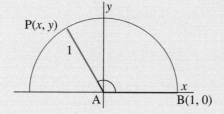

With these definitions, it does not matter if P is in the first or second quadrant. It does not matter if ∠A is acute or obtuse. In the first quadrant, ∠A is acute. In the second quadrant, ∠A is obtuse.

Finding sines and cosines of obtuse angles

To determine the sine and cosine of an obtuse angle, we use a calculator.
For example, to determine $\sin 144°$ and $\cos 144°$:
Press: 144 $\boxed{\text{SIN}}$ to display: 0.587785252
Press: 144 $\boxed{\text{COS}}$ to display: -0.809016994
Rounded to 4 decimal places: $\sin 144° = 0.5878$ and $\cos 144° = -0.8090$
$\sin 144°$ positive, but $\cos 144°$ is negative.

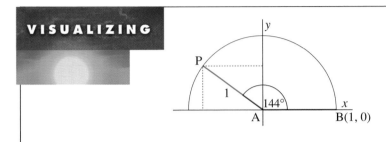

VISUALIZING

$\sin 144° = 0.5878$ means that when $\angle PAB = 144°$, the y-coordinate of P is 0.5878.

$\cos 144° = -0.8090$ means that when $\angle PAB = 144°$, the x-coordinate of P is -0.8089.

Finding angles with given sines and cosines

To determine an angle when its sine or cosine is known, we use a calculator.
Care must be taken in some situations, as the following example shows.

Example 1

For each equation below, determine the value(s) of $\angle A$ to the nearest degree. Assume that $0° \leq \angle A \leq 180°$. Draw a diagram to explain what each equation means.

a) $\cos A = 0.8$ **b)** $\cos A = -0.8$ **c)** $\sin A = 0.8$

Solution

In each case, draw a semicircle with centre $A(0, 0)$ and radius 1 unit.
Mark the point $B(1, 0)$ on the semicircle.

a) Press 0.8 $\boxed{\text{COS}^{-1}}$ to display: 36.86989765

Hence, $\angle A \doteq 37°$

Draw a vertical line at $x = 0.8$, to meet the semicircle at P. Join PA.

$\cos A = 0.8$ means that $\angle A$ is the angle where the x-coordinate of P is 0.8.

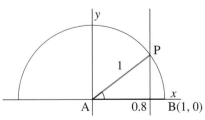

b) Press 0.8 [+/−] [cos⁻¹] to display:
143.1301024

Hence, ∠A ≐ 143°

Draw a vertical line at $x = -0.8$,
to meet the semicircle at P.

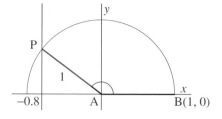

cos A = −0.8 means that ∠A is the
angle where the x-coordinate of P is −0.8.

c) Press 0.8 [sin⁻¹] to display: 53.13010235

Hence, ∠A ≐ 53°

Draw a horizontal line at $y = 0.8$, to meet the semicircle at P. Since this
line intersects the semicircle at two points, there are two possible
positions for P. Each position corresponds to an ∠A such that
sin A = 0.8. One ∠A is acute, and the other is obtuse.

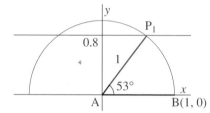 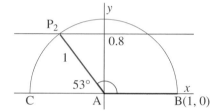

The calculator displays only one value of ∠A, for the acute angle. This
is approximately 53°. Since the calculator cannot display two different
results for the same keystrokes, we must use this angle to determine the
other angle.

Points P₁ and P₂ are symmetrically placed on either side of the y-axis.
This means that ∠P₂AC = 53°.
Hence, ∠P₂AB = 180° − 53°, or 127°.
Therefore, if sin A = 0.8, then ∠A ≐ 53° or 127°

sin A = 0.8 means that ∠A is the angle where the y-coordinate of P is
0.8. Since there are two positions for P, there are two values of ∠A.

In *Example 1c*, the equation sin A = 0.8 has two solutions. Since the second
solution was obtained by subtracting the first from 180°, the two solutions have
a sum of 180°. You will obtain similar results for any other equation that is
similar to sin A = 0.8.

Any equation of the form sin A = k, where $0° \leq \angle A \leq 180°$, has two solutions for $\angle A$. To determine these solutions:

- Use the $\boxed{\text{SIN}^{-1}}$ key to obtain one solution.
- Subtract the first solution from 180° to obtain the other solution.

Example 2

Given that $0° \leq \angle C \leq 180°$, determine the value(s) of $\angle C$, to 1 decimal place.

a) $\cos C = 0.5329$　　　　**b)** $\cos C = -0.5329$　　　　**c)** $\sin C = 0.5329$

Solution

a) The positive cosine indicates that the angle is acute.
Press: 0.5329 $\boxed{\text{COS}^{-1}}$ to display: 57.79839364
$\angle C \doteq 57.8°$

b) The negative cosine indicates that the angle is obtuse.
Press: 0.5329 $\boxed{+/-}$ $\boxed{\text{COS}^{-1}}$ to display: 122.2016064
$\angle C \doteq 122.2°$

c) The positive sine indicates that the angle could be acute or obtuse.
Press: 0.5329 $\boxed{\text{SIN}^{-1}}$ to display: 32.20160636
There are two solutions:
$\angle C \doteq 32.2°$　　　　　　$\angle C \doteq 180° - 32.2°$, or 147.8°

Relating sines and cosines

The diagrams on page 492 are repeated below, with perpendiculars drawn from $P(x, y)$ to the x-axis.

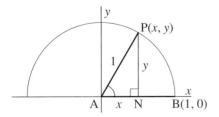 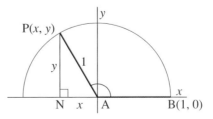

On the first diagram, $\triangle ANP$ is a right triangle with hypotenuse 1 unit and legs x units and y units. According to the Pythagorean Theorem, $x^2 + y^2 = 1$.

Since $x = \cos A$ and $y = \sin A$, this means that:

$$(\cos A)^2 + (\sin A)^2 = 1$$

For any $\angle A$, the sum of the squares of the sine and cosine is 1.

Example 3

Verify that $(\cos A)^2 + (\sin A)^2 = 1$ using:

a) $\angle A = 35°$ 　　　　　　 b) $\angle A = 160°$

Solution

a) Press: 35 $\boxed{\text{COS}}$ $\boxed{x^2}$ $\boxed{+}$ 35 $\boxed{\text{SIN}}$ $\boxed{x^2}$ $\boxed{=}$ to display: 1

b) Press: 160 $\boxed{\text{COS}}$ $\boxed{x^2}$ $\boxed{+}$ 160 $\boxed{\text{SIN}}$ $\boxed{x^2}$ $\boxed{=}$ to display: 1

DISCUSSING THE IDEAS

1. In the example on page 493, why is $\sin 144°$ positive and $\cos 144°$ negative?

2. How could you use your calculator to check the answers in *Examples 1* and *2*?

3. In *Example 1*, why wasn't the equation $\sin A = -0.8$ included?

4. a) In *Example 1c*, how do we know that P_1 and P_2 are symmetrical about the y-axis?

   b) In *Example 1c*, the calculator produced only one value of $\angle A$, when there are actually two. Give another example of a mathematical operation that has two answers, but the calculator gives only one answer.

5. In the second diagram on page 495, is x positive or negative? Does this affect the result that $(\cos A)^2 + (\sin A)^2 = 1$? Explain.

8.5 EXERCISES

A 1. Determine the sine and cosine of each angle.

   a) 120°　　 b) 165°　　 c) 75°　　 d) 96°　　 e) 158°　　 f) 43°　　 g) 0°　　 h) 180°

2. Predict whether each value will be positive or negative. Sketch each angle on a grid to show how you know.

   a) $\sin 98°$　　 b) $\sin 113°$　　 c) $\cos 62°$　　 d) $\cos 143°$　　 e) $\cos 92°$　　 f) $\sin 49°$

3. For exercise 2, write to explain how you can predict which values will be negative.

4. Determine the sine and cosine of each angle, to 3 decimal places. Then sketch each angle on a grid. Show the coordinates of the endpoint of the 1-unit line segment that forms that angle with the positive x-axis.

a) 110°   **b)** 95°   **c)** 138°   **d)** 73°   **e)** 142°   **f)** 35°   **g)** 172°   **h)** 54°

B 5. Each $\angle A$ is between 0° and 180°. Which equations result in two different values for $\angle A$?

a) $\sin A = 0.7071$ **b)** $\cos A = -0.5$ **c)** $\sin A = 0.9269$

d) $\cos A = -0.7071$ **e)** $\sin A = 0.8660$ **f)** $\cos A = -1$

g) $\sin A = \frac{3}{4}$ **h)** $\cos A = \frac{3}{4}$ **i)** $\cos A = -\frac{3}{4}$

6. For exercise 5, write to explain why there are two values of $\angle A$ for some equations, and only one value of $\angle A$ for others.

7. Given that $0° \le \angle C \le 180°$, determine the value(s) of $\angle C$.

a) $\sin C = 0.9063$ **b)** $\cos C = 0.5736$ **c)** $\cos C = -0.7321$

d) $\sin C = 0.4283$ **e)** $\sin C = 0.5726$ **f)** $\cos C = -0.3747$

g) $\sin C = \frac{1}{2}$ **h)** $\cos C = \frac{1}{2}$ **i)** $\cos C = -\frac{1}{2}$

j) $\sin C = \frac{2}{3}$ **k)** $\sin C = \frac{1}{4}$ **l)** $\cos C = -\frac{5}{6}$

8. Select two parts from exercise 7, one involving a sine and one involving a cosine. Write to explain how you decided the number of possible angles, and how you found them.

9. Verify that $(\cos A)^2 + (\sin A)^2 = 1$ for each $\angle A$.

a) 30°   **b)** 72°   **c)** 115°   **d)** 164°

C 10. The sine and cosine of an acute angle were defined on page 460 as ratios of the lengths of certain sides in a right triangle. They were also defined on page 492 as coordinates of a point $P(x, y)$ that is 1 unit from the origin in the first quadrant. Use a diagram similar to the first one on page 460. Show only two triangles. Explain why these two definitions are equivalent.

COMMUNICATING THE IDEAS

The trigonometric ratios were originally defined for acute angles in a right triangle. How would you define the sine and cosine ratios for obtuse angles? Write your definitions in your journal. Explain how the definitions apply to both acute and obtuse angles. Include diagrams with your explanation.

·WITH A GRAPHING CALCULATOR

Graphing Sines and Cosines

In *Investigate* on page 490, you graphed ΔANP in different positions and recorded the coordinates of P for angles increasing in increments of 15°. You can use a graphing calculator to do this for different angle increments. The calculator can also plot points from the tables on page 491. The results are graphs of the sines and cosines against their angle measures.

1. Ask your teacher for the program called CIRCLE1 from the Teacher's Resource Book. Enter the program in your calculator. When you run the program, the calculator asks for an angle increment: enter 15. When you press [ENTER], the calculator draws a quarter circle in the first quadrant. When it has finished, press [ENTER] five times. The diagram (below left) will gradually appear.

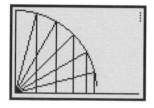 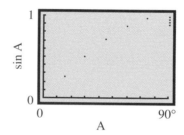

When you press [ENTER] again, a menu appears. Press 1 to select the sine graph. Press [ENTER] a few times. The graph (above right) will gradually appear. This graph shows the sines plotted against the angles. When the calculator has finished the graph, press [STAT] [ENTER] to see the table of values. Then press [2nd] [MODE] to get back to the usual screens.

2. a) Visualize what the screens above would look like if you use a smaller increment such as 10° or 5°. Run the program again to check.

   b) Predict what the cosine graph would look like. Use the program to check your prediction.

3. Ask your teacher for the program CIRCLE2 from the Teacher's Resource Book. This program is similar to CIRCLE1, except that it provides for angles up to 180°.

   a) Predict what the sine graph would look like for angles from 0° to 180°. Use the program to check your prediction.

   b) Repeat part a for the cosine graph.

In Section 8.3, you learned that you can solve any right triangle if you know:

- the lengths of two sides, or
- the length of one side and the measure of one acute angle.

In this section, you will determine unknown measures in triangles that are not right triangles. In the first example, two angles and one side are known.

Example 1

In △ABC, calculate the length of AC to the nearest millimetre.

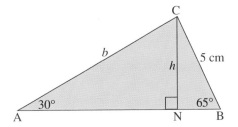

Solution

Represent the length of AC by b, to indicate it is the side opposite ∠B. △ABC is not a right triangle. Divide △ABC into two right triangles by constructing an altitude from one vertex to the opposite side. Construct the perpendicular from C to AB at N.

There are now two right triangles, △ANC and △BNC. They have a common side CN, whose length is represented by h.

Think...

Work from the known to the unknown. In △CNB we know one side and one acute angle. Hence, we can determine h.

Then, in △ANC we will know one acute angle and one side. Hence, we can determine b.

In △CNB, use the sine ratio to determine h.

$$\sin 65° = \frac{h}{5}$$
$$h = 5 \sin 65°$$
$$\doteq 4.5315$$

5 sin 65° means 5 × sin 65°

Now we know h, we know one side and one acute angle in △ANC.
Use the sine ratio to determine b.

$$\sin 30° \doteq \frac{4.5315}{b}$$
$$b \doteq \frac{4.5315}{\sin 30°}$$
$$\doteq 9.063$$

The length of AC is 9.1 cm to the nearest millimetre.

The next example is similar to *Example 1*, but one angle is obtuse.

Example 2

In △ADC, calculate the length of AC to the nearest millimetre.

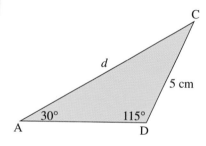

Solution

Construct the perpendicular from C to AD extended to N. Let h represent the length of CN.

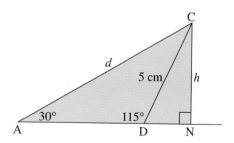

Think...

This creates two right triangles with a common side CN.

We can determine ∠CDN since we know ∠CDA.

The solution is similar to *Example 1*.

In △CDN, we know one side and one acute angle so we can determine *h*.

Then, in △ANC we will know one acute angle and one side so we can determine *d*.

Since ∠CDA = 115°, then ∠CDN = 180° − 115°, or 65°.

In △CDN, use the sine ratio to determine *h*.

$$\sin 65° = \frac{h}{5}$$
$$h = 5 \sin 65°$$
$$\doteq 4.5315$$

Now we know *h*, we know one side and one acute angle in △ANC.
Use the sine ratio to determine *d*.

$$\sin 30° \doteq \frac{4.5315}{d}$$
$$d \doteq \frac{4.5315}{\sin 30°}$$
$$\doteq 9.063$$

The length of AC is 9.1 cm to the nearest millimetre.

In *Example 1* and *Example 2*, two angles and one side were given. In the following example, two sides and the angle between them are given.

Example 3

In △ABC, calculate the length of CB to the nearest millimetre.

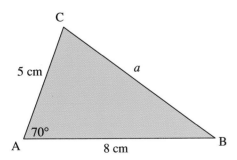

Solution

Construct the perpendicular from C to AB at N.
Let h represent the length of CN.

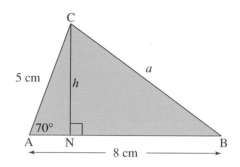

Think...

Work from the known to the unknown.
In $\triangle$ANC we know one side and one acute angle.
Hence, we can determine h.
This is not enough information to solve $\triangle$BNC, but we haven't used AB = 8 cm.
In $\triangle$ANC we can calculate the length of AN, so we can find the length of NB.
Then, in $\triangle$BNC we will know the lengths of CN and NB.
We can calculate the length of CB using the Pythagorean Theorem.

In $\triangle$ANC, use the sine ratio to determine h.

$$\sin 70° = \frac{h}{5}$$
$$h = 5 \sin 70°$$
$$\doteq 4.6984$$

In $\triangle$ANC, use the cosine ratio to determine the length of AN.

$$\cos 70° = \frac{AN}{5}$$
$$AN = 5 \cos 70°$$
$$\doteq 1.7101$$

AN is approximately 1.7101 cm.
Hence, NB is approximately $8 - 1.7101$, or 6.2899 cm.

In $\triangle$BNC, use the Pythagorean Theorem to determine a.

$$a^2 \doteq 4.6984^2 + 6.2899^2$$
$$\doteq 61.6378$$
$$a = \sqrt{61.6378}$$
$$\doteq 7.8510$$

The length of CB is 7.9 cm to the nearest millimetre.

1. In *Example 1*, describe how you would determine the length of AB and the measure of ∠C.

2. Compare the solution of *Example 2* with the solution of *Example 1*. After ∠CDN was determined, the calculations in *Example 2* are identical to those in *Example 1*. Explain why.

3. Compare the solutions of *Example 1* and *Example 3*. The first step in both solutions is the same. In *Example 3*, we calculated the length of AN next. Why did we not do this in *Example 1*?

4. In *Example 3*, we determined the length of AN using the cosine ratio. What other way is there to determine the length of AN?

5. In *Example 3*, describe how you would determine the measures of ∠B and ∠C.

8.6 EXERCISES

B 1. Determine the length of each side AB.

a)

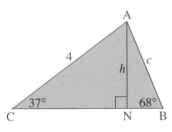

b)

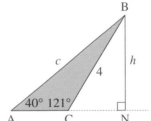

2. Calculate the length of each side BC.

a)

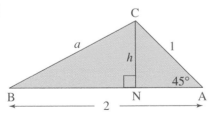

b)

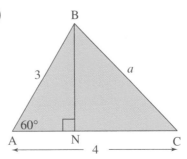

3. This triangle has some measures the same as the triangle in *Example 3*, but one angle is obtuse.

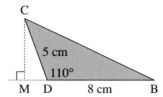

a) Calculate the length of CB to the nearest millimetre.

b) Compare your solution with the solution of *Example 3*. Many calculations are identical. There should be only one place where the solutions are slightly different. Explain why this is the only place where the solutions are different.

4. The diagram (below left) shows a roof truss. It is needed to span 10.0 m. One piece of the truss is 7.0 m long, and set at an angle of 35°. How long is the other piece of the truss?

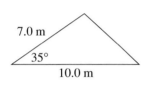

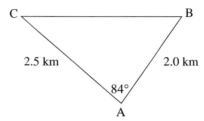

5. The diagram (above right) shows a radar station at A. It is tracking ships at B and C. How far apart are the two ships?

6. Determine the measures of all the angles in each isosceles triangle.

a)

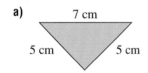

b)

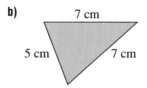

7. Solve each △PQR.

a) ∠P = 105.0°, *p* = 12.0 cm, *q* = 9.0 cm

b) ∠Q = 46.4°, *r* = 21.0 m, *p* = 29.0 m

COMMUNICATING THE IDEAS

In your journal, write to describe how it is possible to use trigonometry to determine unknown measures in a triangle that is not right-angled. Refer to one or more examples from your work in this section.

8.7 The Sine Law

In Section 8.6, you calculated unknown sides and angles in triangles that are not right-angled. Each case involved a careful sequencing of steps. You probably found that you were repeating similar steps, but working with new numbers each time.

We shall examine one example from Section 8.6, then formulate a general rule to produce a shortcut for the process. Consider *Example 1* from that section. The measures of two angles and one side of $\triangle ABC$ were given. The problem was to calculate the length of another side.

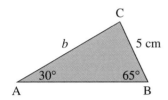

The length of AC is approximately 9.063 cm. Let us examine how the given numbers 5, 65°, and 30° were combined to get this answer. Turn to the solution on pages 499 and 500. Some calculations from that solution are repeated below.

From the first part of the solution:

$h = 5 \sin 65°$
$\doteq 4.5315$

From the second part of the solution:

$b \doteq \dfrac{4.5315}{\sin 30°}$
$\doteq 9.063$

You can see that 5 was multiplied by $\sin 65°$, and the product was divided by $\sin 30°$. Hence, we can write the length of AC in this form:

$b = \dfrac{5 \sin 65°}{\sin 30°}$

Observe where the numbers 5, 65°, and 30° appear in this expression. A similar result will be obtained for any other given numbers in a diagram like that above. Hence, for the triangle below, we can use this expression to determine a formula for b. The length of each side is represented by the lowercase letter corresponding to the uppercase letter that represents the opposite vertex.

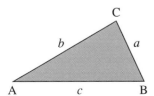

We replace 5 with a, 65° with B, and 30° with A. The result is:

$$b = \frac{a \times \sin B}{\sin A}$$

$$b \times \sin A = a \times \sin B$$

Multiplying both sides by sin A

$$\frac{b \times \sin A}{ab} = \frac{a \times \sin B}{ab}$$

$$\frac{\sin A}{a} = \frac{\sin B}{b}$$

Dividing both sides by *ab*

Both sides of this equation have the form $\frac{\text{sine of an angle in } \triangle ABC}{\text{length of side opposite that angle}}$.

On the left side of the equation, the angle and side length are the dimensions given in the first diagram on page 505. On the right side, the angle is given in the diagram. Since ∠C could have been given instead of ∠B, the right side of this equation could also be $\frac{\sin C}{c}$. These expressions of equal ratios yield the *Sine Law*.

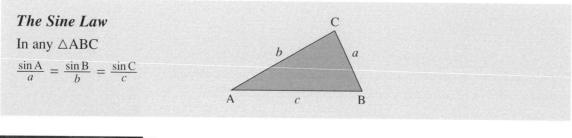

The Sine Law

In any △ABC

$$\frac{\sin A}{a} = \frac{\sin B}{b} = \frac{\sin C}{c}$$

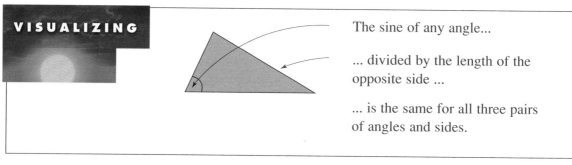

VISUALIZING

The sine of any angle...

... divided by the length of the opposite side ...

... is the same for all three pairs of angles and sides.

To use the Sine Law, you need to determine the sine of one angle divided by the length of the opposite side. Hence, you must know the measure of an angle and the length of the opposite side.

Example 1

In △PQR, ∠Q = 115°, PQ = 4.5 cm, and PR = 10.8 cm. Calculate ∠R to the nearest tenth of a degree.

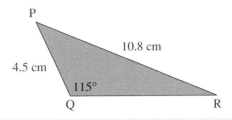

Solution

> **Think…**
>
> We know ∠Q and the length of the side opposite this angle. We also know the length of PQ. Hence, we can use the Sine Law to calculate the measure of the angle opposite PQ.

Use the Sine Law.

$$\frac{\sin R}{4.5} = \frac{\sin 115°}{10.8}$$

Multiply each side by 4.5.

$$\sin R = \frac{4.5 \sin 115°}{10.8}$$

Press: 4.5 $\boxed{\times}$ 115 $\boxed{\text{SIN}}$ $\boxed{\div}$ 10.8 $\boxed{=}$

$\sin R \doteq 0.377\ 628\ 24$ Then press: $\boxed{\text{SIN}^{-1}}$

$\quad ∠R \doteq 22.186\ 848$

∠R is approximately 22.2°.

In *Example 1*, two sides and one angle (opposite one of the given sides) were given. We used the Sine Law to calculate another angle. We can also use the Sine Law when two angles and the side between them are given. In this case we use the Sine Law to calculate another side. *Example 2* illustrates this.

Example 2

The Calgary Tower is located on the south side of 9 Ave. S. in downtown Calgary. A study group went to the intersection of 9 Ave. with 11 St. W. and measured the angle of elevation of the top of the tower; the result was 5.9°. They then drove 2.9 km to the intersection of 9 Ave. with 6 St. E. (which is east of the tower) and measured the angle of elevation from the other side; the result was 10.3°. Calculate the height of the tower.

Diagram not drawn to scale

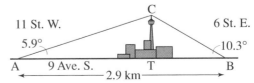

Solution

> ### Think...
>
> In $\triangle ABC$, we know two angles. We can calculate the third angle, which is opposite the known side. Hence, we can use the Sine Law to calculate the length of AC. Then we will know the hypotenuse and an acute angle in $\triangle ATC$, so we can calculate the length of CT.

In $\triangle ABC$, calculate $\angle C$.

$$\angle C = 180° - (5.9° + 10.3°)$$
$$= 163.8°$$

Use the Sine Law.

$$\frac{\sin 10.3°}{AC} = \frac{\sin 163.8°}{2.9}$$

$$AC \sin 163.8° = 2.9 \sin 10.3° \longleftarrow \boxed{\text{Divide each side by } \sin 163°.}$$

$$AC = \frac{2.9 \sin 10.3°}{\sin 163.8°}$$

Press: 2.9 $\boxed{\times}$ 10.3 $\boxed{\text{SIN}}$ $\boxed{\div}$ 163.8 $\boxed{\text{SIN}}$ $\boxed{=}$

$$\doteq 1.8586$$

Do not clear the screen.

AC is approximately 1.8586 km.

In right $\triangle ATC$, use the sine ratio.

$$\sin A = \frac{CT}{AC}$$

$$\sin 5.9° \doteq \frac{CT}{1.8586}$$

$$CT \doteq 1.8586 \sin 5.9°$$ Then press: $\boxed{\times}$ 5.9 $\boxed{\text{SIN}}$ $\boxed{=}$

$$\doteq 0.1910$$

The tower is approximately 0.191 km, or 191 m high.

Example 2 illustrates a common use of the Sine Law. Two angles and one side of the triangle were known. We used the Sine Law to determine another side.

MODELLING for Measuring Inaccessible Heights

The method in *Example 2* is another model for measuring a height where the base of the object being measured is not accessible.

- Could $\triangle BTC$ have been used instead of $\triangle ATC$? Explain.
- The points A and B, where the angles were measured, are on opposite sides of the tower. Would the model still apply if the angles were measured from two points on the same side of the tower? What are some limitations of this model?

1. In *Example 1*, how could you calculate the length of QR?

2. *Example 2* and exercise 10 on page 458 describe two different models for measuring the height of a tall object such as the Calgary Tower. Which model do you think is more useful? Explain.

3. Turn to *Example 3* in Section 8.6, page 501. Could the Sine Law be used to determine the length of CB? Explain.

4. What information must you know about a triangle to be able to use the Sine Law?

8.7 EXERCISES

A 1. Calculate the length of AB in each triangle.

a)
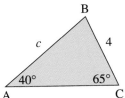

b)
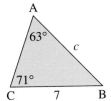

c)
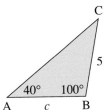

2. Choose one part of exercise 2. Write to explain how you calculated the length of AB.

3. Calculate the measure of ∠C in each triangle.

a)
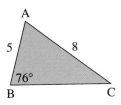

b)
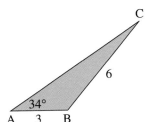

c)
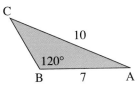

B 4. Determine the measure of ∠B in each triangle.

a)

b)
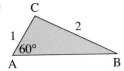

c)
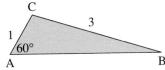

5. Choose one part of exercise 4. Write to explain how you calculated the measure of ∠B.

6. The three triangles in exercise 4 form a pattern. Draw the next triangle in the pattern. Determine the measure of ∠B in this triangle.

7. Repeat exercises 4 and 6 using triangles in which ∠A = 45°.

8. A bridge AB is to be built across a river (below left). Point C is located 62.0 m from B. Angle ABC is 74.0° and ∠ACB is 48.0°. How long will the bridge be? The diagram is not drawn to scale.

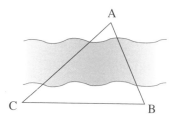

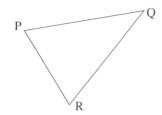

9. The diagram (above right) represents two ships at P and Q that are 32.0 km apart. Angle P is 68.0° and ∠Q is 42.0°. How far is each ship from a lighthouse at R?

10. Two cabins, A and C, are located 450 m apart on the bank of a river (below left). Across the river from the two cabins is a boathouse B. Calculate the width of the river.

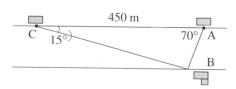

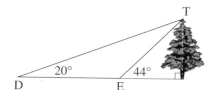

11. The diagram (above right) represents points D and E from which the lines of sight to the top of a tree at T make angles of 20° and 44°, respectively, with DE. The length of DE is 62.0 m. Calculate the height of the tree.

12. The diagram (below left) represents a triangular park measuring 251.0 m along one side. The other two sides form angles of 32.0° and 56.0° with the first side.

a) Determine the lengths of the other two sides.

b) Calculate the area of the park.

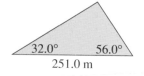

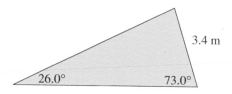

13. The diagram (bottom right on page 510) represents the roof lines of a ski chalet. They make angles of 26.0° and 73.0° with the horizontal. The shorter roof line is 3.4 m long.

   a) Calculate the length of the other roof line.

   b) This triangular side of the roof is to be finished with a special type of shingle. Calculate the area of the triangle.

14. Turn to *Mathematics File*, pages 478 and 479.

   a) Use your measurements of ∠YXP and ∠XYP. Calculate the distances from X and Y to the pine tree.

   b) Use your measurements of ∠TYX and ∠TXY. Calculate the distances from X and Y to the TV tower.

   c) Check your answers by measuring on the diagram and using the scale.

   Save your results for use in the next section (page 519, exercise 16).

15. Cattle Point is part of Victoria, B.C. Maynard Cove is 2.4 km from Cattle Point, on a bearing of 053°. A sailboat leaves Cattle Point on a bearing of 079°. After sailing for 1.8 km, the boat turns and heads directly to Maynard Cove, on a bearing of 008°.

   a) What is the bearing of Cattle Point from Maynard Cove?

   b) Calculate the total distance the boat travels to get to Maynard Cove.

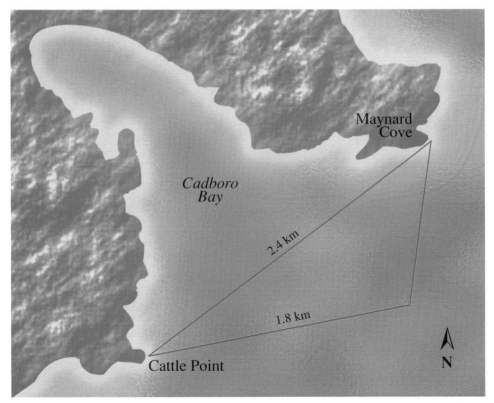

16. In △ABC, AD is the altitude from A to BC. Angle B = 48°, ∠C = 32°, and BC = 12.8 m; determine the length of AD.

17. In the diagram (below left), ABCD is a square with sides of length 1 unit. Point P is such that ∠PAD = ∠PCD = 30°.

a) Calculate the length of PD to 1 decimal place.

b) Calculate the length of PC to 1 decimal place.

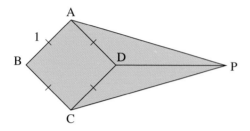
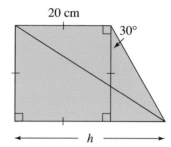

18. In the diagram (above right), calculate the length of *h*.

Ⓒ 19. Turn to *Example 1* and *Example 2* in Section 8.3, pages 470 and 471.

a) Try to use the Sine Law to solve the triangles in these two examples.

b) What happens when you try to use the Sine Law to solve a right triangle?

20. In △ABC, AB = 8 cm, AC = 6 cm, and ∠B = 40°

a) Solve the triangle.

b) In one step of your solution, you should realize that two different triangles satisfy these conditions. What step is this?

c) Explain why it is possible that two different triangles can have two equal angles and two pairs of equal sides.

21. Turn to the solution of *Example 1*, pages 499 and 500. Use this solution as a model. Derive the Sine Law for the triangle at the bottom of page 505.

COMMUNICATING THE IDEAS

Suppose your friend telephones you to discuss today's mathematics lesson. How would you explain the Sine Law to your friend? Write your ideas in your journal. Your explanation should include a description of the Sine Law and how it is used.

8.8 The Cosine Law

In the previous section we developed the Sine Law based on *Example 1* of Section 8.6, which was a case where the Sine Law can be applied. The Sine Law cannot be used in *Example 3* from Section 8.6, page 501, because the given angle is contained by the two given sides.

If we examine the steps in the solution of that example, we might be able to formulate a general rule that can be used in similar examples. In ΔABC, the measures of two sides and the angle between them were given. The problem was to calculate the length of the third side.

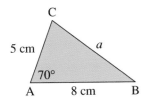

The length of CB is approximately 7.8510 cm. Let us examine how the given numbers 5, 8, and 70° were combined to get this answer. Turn to the solution on page 502. This is what happened to the given numbers.

5 was multiplied by $\sin 70°$: $5 \sin 70°$ ①

5 was multiplied by $\cos 70°$: $5 \cos 70°$ ②

② was subtracted from 8: $8 - 5 \cos 70°$ ③

① and ③ were squared
and added to obtain a^2: $a^2 = (5 \sin 70°)^2 + (8 - 5 \cos 70°)^2$

Observe where the numbers 5, 8, and 70° appear in this expression. Hence, for the triangle below, we can use this expression to obtain a formula for *a*.

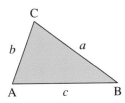

We replace 5 with *b*, 8 with *c*, and 70° with A. The result is:

$$a^2 = (b \sin A)^2 + (c - b \cos A)^2$$
$$= b^2(\sin A)^2 + c^2 - 2bc \cos A + b^2(\cos A)^2$$
$$= b^2[(\sin A)^2 + (\cos A)^2] + c^2 - 2bc \cos A$$

Expanding

Factoring

Recall from page 496 that $(\cos A)^2 + (\sin A)^2 = 1$ for any angle A. Hence, the above equation becomes:

$$a^2 = b^2(1) + c^2 - 2bc \cos A$$
$$a^2 = b^2 + c^2 - 2bc \cos A$$

The first part of this equation, $a^2 = b^2 + c^2$, looks like the Pythagorean Theorem, which applies to right triangles only. The expression $-2bc \cos A$ appears on the right side to compensate for the fact that $\triangle ABC$ is not a right triangle.

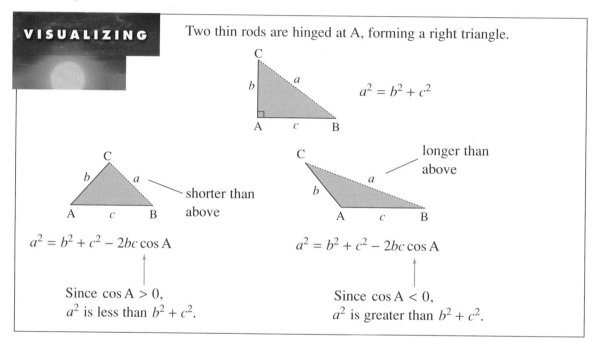

VISUALIZING Two thin rods are hinged at A, forming a right triangle.

$$a^2 = b^2 + c^2$$

shorter than above

longer than above

$$a^2 = b^2 + c^2 - 2bc \cos A$$

$$a^2 = b^2 + c^2 - 2bc \cos A$$

Since $\cos A > 0$,
a^2 is less than $b^2 + c^2$.

Since $\cos A < 0$,
a^2 is greater than $b^2 + c^2$.

Since $\triangle ABC$ does not have to be a right triangle, the equation $a^2 = b^2 + c^2 - 2bc \cos A$ applies to any of the three sides. Hence, it can be replaced by $b^2 = a^2 + c^2 - 2ac \cos B$ and $c^2 = a^2 + b^2 - 2ab \cos C$. These equations form the *Cosine Law.*

The Cosine Law

In any $\triangle ABC$

$$a^2 = b^2 + c^2 - 2bc \cos A$$
$$b^2 = a^2 + c^2 - 2ac \cos B$$
$$c^2 = a^2 + b^2 - 2ab \cos C$$

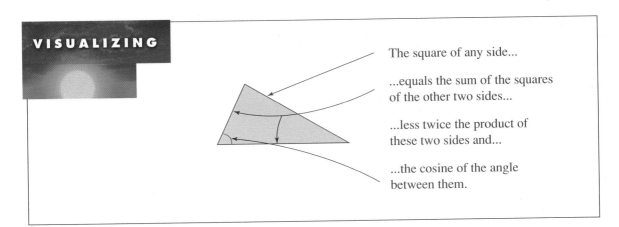

The square of any side...

...equals the sum of the squares of the other two sides...

...less twice the product of these two sides and...

...the cosine of the angle between them.

You can use the Cosine Law when you know the lengths of two sides and the angle between them.

Example 1

A tunnel is to be built through a hill to connect points A and B in a straight line. Point C is chosen to be visible from both A and B. Measurements are taken, and show ∠C = 67.5°, CA = 3.65 km, and CB = 5.18 km. Calculate the length of AB.

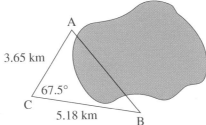

Solution

> **Think...**
> We know two sides and the angle between them. We can use the Cosine Law to calculate the length of the side opposite this angle.

$c^2 = a^2 + b^2 - 2ab \cos C$

$\quad = 5.18^2 + 3.65^2 - 2(5.18)(3.65) \cos 67.5°$

$\quad \doteq 25.6841$

$c \doteq \sqrt{25.6841}$

$\quad \doteq 5.0679$

AB is approximately 5.07 km.

Press: 67.5 cos × 2 × 5.18 ×

3.65 = +/− + 5.18 x² +

3.65 x² =

Then press: √x

The method in *Example 1* is a model for measuring an inaccessible distance, such as the length of a tunnel through a mountain.

- What assumptions about the tunnel and the terrain are implied by the model?
- What are some limitations of this model?
- Would the model apply if the measured angle were obtuse? Explain.
- Could the model be used to measure inaccessible heights? Explain.

The Cosine Law equation involves the lengths of all three sides of a triangle and the measure of one angle. This means that if you know the lengths of all three sides, you can use the Cosine Law to calculate the angles.

Example 2

In $\triangle PQR$, $RP = 7$, $RQ = 8$, and $PQ = 10$.
Calculate the measure of $\angle Q$ to
the nearest degree.

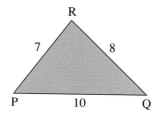

Solution

> **Think...**
>
> We know the lengths of all three sides. If we substitute these in the Cosine Law equation, we will get an equation for the cosine of one angle. We can solve the equation, then determine the angle.

To determine $\angle Q$, apply the Cosine Law to the side opposite this angle.

$$q^2 = p^2 + r^2 - 2pr \cos Q$$
$$7^2 = 8^2 + 10^2 - 2(8)(10) \cos Q$$
$$49 = 164 - 160 \cos Q$$
$$160 \cos Q = 115$$
$$\cos Q = \frac{115}{160} \qquad \text{Press: } 115 \boxed{\div}\ 160 \boxed{=}$$
$$\doteq 0.718\ 75 \qquad \text{Then press: } \boxed{\text{cos}^{-1}}$$
$$\angle Q \doteq 44.0486°$$

$\angle Q$ is approximately 44°.

1. In *Example 1*, we know the length of AB. What else could we calculate? Describe two different ways to do this.

2. In *Example 2*, how could you determine the measures of the other angles? Describe two different ways to do this.

3. What information must you know about a triangle to be able to use the Cosine Law?

4. When you are solving problems involving triangles, how can you tell whether to use the Sine Law or the Cosine Law?

8.8 EXERCISES

A 1. Calculate the length of AC in each triangle.

a)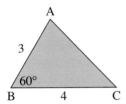

b)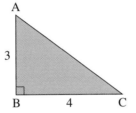

c)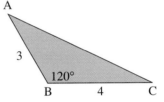

2. Choose one part of exercise 1. Write to explain how you calculated the length of AC.

3. Calculate the measure of ∠A in each triangle.

a)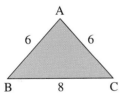

b)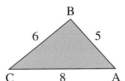

c)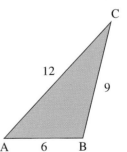

4. a) In *Example 1*, use the answer and the given information to calculate the measures of ∠A and ∠B to the nearest degree.

   b) Why would it be important for the people building the tunnel to calculate these angles?

B **5.** To determine the distance AB across a marsh, a surveyor locates a point C (below left). The measure of ∠C is 65°, and the lengths are as given in the diagram. How far is it across the marsh?

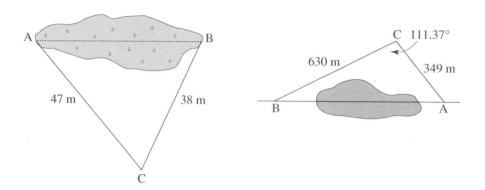

6. An electric transmission line is to go over a pond (above right). The power line will be supported by posts at points A and B. A surveyor measures the distance BC as 630 m, the distance AC as 349 m, and ∠BCA as 111.37°. What is the distance between the posts at A and B?

7. Determine the measure of ∠B in each triangle.

a)
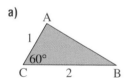

b)
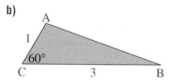

c)
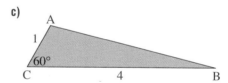

8. Choose one part of exercise 7. Write to explain how you calculated the measure of ∠B.

9. The three triangles in exercise 7 form a pattern. Draw the next triangle in the pattern. Determine the measure of ∠B in this triangle.

10. Repeat exercises 7 and 9 using triangles in which ∠C = 45°.

11. The lengths of the three sides of each triangle are consecutive numbers. Determine the measures of the angles in each triangle. What special property does each triangle have?

a)
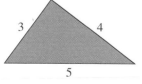

b)

12. The diagram (below left) represents a radar station at point A. It is tracking ships at B and C. For an observer on ship B, calculate the angle between the lines of sight to the ship at C and the radar station.

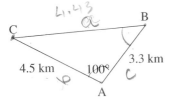
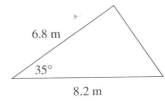

13. The diagram (above right) represents a roof truss spanning 8.2 m. One piece of the truss is 6.8 m long. It is set at an angle of 35°. Calculate the angles formed by the other two pieces.

14. Turn to exercise 12c, page 487. Suppose the smoke sightings are observed in directions that are 53° apart (below left). Calculate the distance between the smoke sightings. The diagram is not to scale.

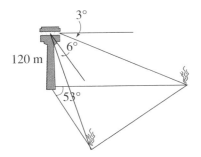
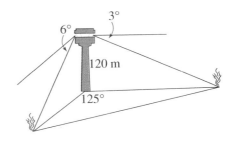

15. Repeat exercise 14, if the smoke sightings are observed in directions that are 125° apart (above right). The diagram is not to scale.

16. Turn to *Mathematics File*, pages 478 and 479, and exercise 14 on page 511.
   a) Calculate each distance to the nearest metre.
   i) from the pine tree to the flag
   ii) from the flag to the TV tower
   iii) from the pine tree to the TV tower
   b) Calculate, to the nearest degree, the angles in the triangle formed by the pine tree, the flag, and the TV tower.
   c) Calculate the area of the triangle in part a.

Note: From now on, you may need to use the sine, cosine, or tangent ratios, or the Sine Law, or the Cosine Law to solve each problem.

17. Turn to exercise 13 on page 487. Suppose the helicopter flew from the first smoke sighting to the second smoke sighting, not travelling to the look-out tower. Calculate the distance between the two smoke sightings.

18. From a certain point, the angle of elevation to the top of a church steeple is 10°. At a point 100 m closer to the steeple, the angle of elevation is 20°. Calculate the height of the steeple.

19. The highest waterfall in Canada is Della Falls on Vancouver Island. From a certain point, the angle of elevation to the top of the falls is 58°. At a point 41 m closer to the falls, the angle of elevation is 62°. Calculate the height of Della Falls.

20. A helicopter hovers directly above the landing pad on top of a 125-m high building. A person is standing 145 m from the base of the building. The angle of elevation to the helicopter from this person is 58°. How high is the helicopter above the landing pad?

21. A triangular park has sides of length 200 m, 155 m, and 172 m. Calculate the area of the park.

22. A farmer has a field in the shape of a triangle. From one vertex, it is 435 m to the second vertex and 656 m to the third vertex. The angle between the lines of sight to the second and third vertices is 49°. Calculate the perimeter and the area of the field.

23. Suppose the helicopter in exercise 13, page 487, is a Sikorsky S-76C, which is often used for passenger transport. Use information from the *Helicopters* database and your answer to exercise 13.

   a) How many firefighters can the Sikorsky S-76C carry?

   b) One nautical mile is 1852 m. What is the distance the helicopter travels in nautical miles?

   c) Assume the helicopter is flying at sea level cruise speed. How long will the trip take?

   d) How much fuel will be used? How much fuel remains in the fuel tank?

24. A radar tracking station locates a fishing trawler at a distance of 5.4 km, and a passenger ferry at a distance of 7.2 km. At the station, the angle between the lines of sight to the two ships is 118°. How far apart are the ships?

25. Two ships leave a port at the same time. One sails at 17 km/h on a bearing of 205°. The other sails at 21 km/h on a bearing of 243°. How far apart are the two ships after 2 h?

26. A box is 28 cm long, 28 cm wide, and 23 cm high. A metre stick is placed in the box, as described in exercise 14, page 487.

   a) Calculate how high up the wall the metre stick reaches.

   b) Calculate the distance from the carton to the wall.

27. Repeat exercise 26 with the stick placed as described in exercise 15, page 487.

28. ABCD is a square with side length 3 cm and △PBC is equilateral (below left). Determine the length of AP.

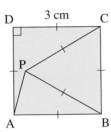

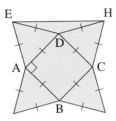

29. ABCD is a square with sides 1 unit long (above right). Equilateral triangles are constructed on the sides of the square.

a) Determine the length of EH.   **b)** Determine the area of △EDH.

C **30.** Turn to the solution of *Example 3*, page 502. Use this solution as a model. Derive the Cosine Law for the triangle at the bottom of page 513.

31. Angle parking allows more cars to park along a street than does parallel parking. However, the cars use more of the street width when angle parked.

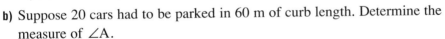

a) Each car requires a space 2.7 m wide. Determine the curb length required to park 20 cars when ∠A has each measure.

   i) 30°   **ii)** 50°   **iii)** 60°

b) Suppose 20 cars had to be parked in 60 m of curb length. Determine the measure of ∠A.

c) Each car requires a space 6.5 m long. Determine the length of roadway used for parking when ∠A has each measure.

   i) 30°   **ii)** 50°   **iii)** 60°

32. There is no "Tangent Law" in trigonometry. Why do you think this is?

COMMUNICATING THE IDEAS

Suppose your friend discusses today's mathematics lesson. How would you explain the Cosine Law to your friend? Write your ideas in your journal. Your explanation should include a description of the Cosine Law, how it is used, and how you can tell whether to use the Sine Law or the Cosine Law in a particular problem.

How Far Is the Sun?
How Large Is the Sun?

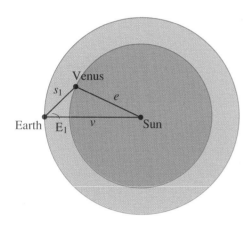

Have you heard the term "evening star"? People sometimes use this term to describe the first point of light in the western sky as the sun sets. Although this may look like a star, it is the planet Venus. The light you see is sunlight reflecting from Venus.

DEVELOP A MODEL

In our model we will assume that Earth and Venus move in circular orbits around the sun. They might appear as shown above. One month later, they would appear as shown at the right.

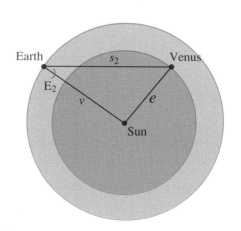

The distance v from Earth to the sun is the same in both cases. So is the distance e from Venus to the sun.

An observer on Earth can measure $\angle E_1$ and $\angle E_2$ on these two diagrams. Furthermore, by bouncing a radar signal off Venus, astronomers can calculate the distances s_1 and s_2. When these four measures are known, you can use trigonometry to calculate the distance v from Earth to the sun.

1. One evening, the planets appeared as in the first diagram. Astronomers measured $\angle E_1 = 31.8°$ and $s_1 = 53.1$ million km. Use this information and your knowledge of trigonometry. Write an equation relating v and e.

2. One month later, the planets appeared as in the second diagram. This time astronomers measured $\angle E_2 = 29.3°$ and $s_2 = 210.2$ million km. Use this information to write another equation relating v and e.

3. Combine the equations to obtain an equation involving v alone. Solve this equation to determine the distance from Earth to the sun.

 LOOK AT THE IMPLICATIONS

When we know the distance to the sun, we can use it to determine other distances.

4. Calculate the distance from Venus to the sun.

5. Astronomers can measure $\angle E_3$ in this diagram. The result is approximately 0.532°. Use this information and the distance to the sun from exercise 3. Calculate the diameter of the sun.

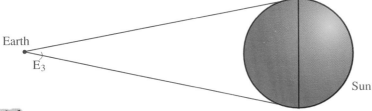

 REVISIT THE SITUATION

6. The radar signals astronomers use to calculate the distance to Venus travel at the speed of light, 3×10^5 km/s. In the diagrams on page 522, it took 5.90 min and 23.36 min, respectively, for the reflected signals to return to Earth. Use this information to confirm that the distances s_1 and s_2 are correct.

7. a) Calculate how far Earth travels in one year in its orbit around the sun.

   b) Calculate Earth's speed in kilometres per hour.

8. The orbits of Earth and Venus are so close to being circles that on the scale of the diagrams on page 522, you would not be able to tell they were not circles.

   a) Do you think the assumption that the orbits are circles has a significant effect on the results?

   b) Where in the calculations did we use the fact that the orbits are circles?

1. Use your calculator to determine each tangent.

   a) $\tan 27°$ **b)** $\tan 38°$ **c)** $\tan 65°$ **d)** $\tan 81°$

2. Use your calculator to determine $\angle A$ to the nearest degree.

   a) $\tan A = 1.4$ **b)** $\tan A = \frac{7}{4}$ **c)** $\tan A = 0.065$ **d)** $\tan A = \frac{20}{7}$

3. Determine each value of x.

   a) 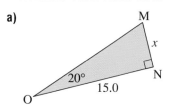   **b)** 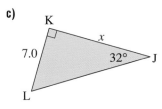   **c)**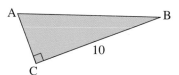

4. In $\triangle ABC$, determine $\angle B$ for each value of AC.

   a) 6   **b)** 12   **c)** 15

   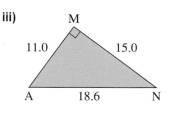

5. A rectangle measures 8 cm by 4 cm. Determine the measures of the two acute angles formed at a vertex by a diagonal.

6. A guy wire fastened 50 m from the base of a tower makes an angle of 55° with the ground. How high up the tower does the guy wire reach?

7. **a)** In each triangle, calculate $\sin A$ and $\cos A$.

   b) Determine $\angle A$.

   i) 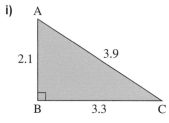   **ii)**   **iii)**

8. Determine the measures of the acute angles in each triangle.

   a) 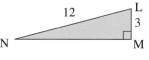   **b)** 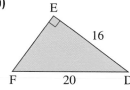   **c)**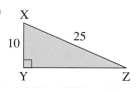

9. Determine the lengths of all sides not given.

a)

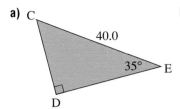

b)

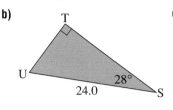

c)

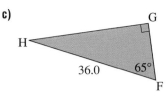

10. In △ABC, ∠A = 90° and AC = 8.0; determine the lengths of AB and BC for each angle measure.

   a) ∠C = 28°   b) ∠B = 54°   c) ∠B = 48°

11. From a distance of 60 m at ground level, the angle of elevation to the top of a flagpole is 32°. Determine the height of the flagpole to the nearest tenth of a metre.

12. When the foot of a ladder is 1.8 m from a wall, the angle formed by the ladder and the ground is 72°.

   a) How long is the ladder?

   b) How high up the wall does the ladder reach?

13. A tree casts a shadow 40 m long when the sun's rays are at an angle of 36° to the ground. How tall is the tree?

14. From the top of a building 70 m high, the angle of depression of a car on a road is 27°. How far is the car from the foot of the building?

15. Two office towers are 30 m apart. From the 15th floor (which is 40 m above the ground) of the shorter tower, the angle of elevation to the top of the other tower is 70°.

   a) Determine the angle of depression to the base of the taller tower from the 15th floor.

   b) Determine the height of the taller tower.

16. Given that 0° ≤ ∠B ≤ 180°, determine the value(s) of ∠B to the nearest degree.

   a) $\sin B = 0.3456$ b) $\cos B = 0.1234$ c) $\sin B = 0.5268$

   d) $\cos B = -\frac{2}{5}$ e) $\sin B = \frac{3}{8}$ f) $\cos B = -0.8652$

17. Look at the map on page 511. A sailboat leaves Cattle Point on a bearing of 025°. After sailing for 1.5 km, the boat turns and sails on a bearing of 165° for 1.7 km.

   a) How far is the sailboat from Cattle Point?

   b) In what direction would it have to sail to return directly to Cattle Point?

1. Andrea invests $1000 in a term deposit this year, and every year for the next 10 years. The interest rate is 8%, compounded annually.

   a) Copy and complete this table.

Year	Opening balance ($)	Interest rate (%)	Interest earned ($)	Annual investment ($)	Closing balance ($)
1	1000.00	8	80.00	1000.00	2080.00
2	2080.00	8			
⋮					
8					
9					
10					

   b) What is the total amount of money invested over the 10 years?

   c) What is the total amount of interest earned over the 10 years?

   d) Suppose Andrea had invested the entire amount in part b at the beginning and had made no annual investment. How much would she have after 10 years?

   e) Estimate the final closing balance if the interest rate had been 10%. Check your estimate by copying and completing the table for an interest rate of 10%.

2. a) Insert two numbers between 3 and 192, so the four numbers form a geometric sequence.

   b) Insert five numbers between 3 and 192, so the seven numbers form a geometric sequence.

3. Write to explain why you cannot use your calculator to evaluate each expression.

   a) 0^0 b) $\sqrt{-3}$ c) $\sqrt[4]{-2}$

4. The equations of the sides of a square are $y = 2x + 10$, $y = 2x - 10$, $y = -\frac{1}{2}x + 5$, and $y = -\frac{1}{2}x - 5$.

   a) Graph these lines on the same axes.

   b) Determine the coordinates of the centre of the square.

   c) Determine the equations of the lines through the centre of the square that are parallel to the sides.

5. Points P(−1, 3), Q(0, 6), R(3, 5), and S(3, 0) are the vertices of a quadrilateral.

   a) Determine the equations of the diagonals.

   b) Is PR the perpendicular bisector of QS?

   c) Is QS the perpendicular bisector of PR?

   d) Is PQRS a rhombus?

   e) Is PQRS a parallelogram?

6. Your boss offers you three different salary plans:
Plan A: $2000 per month
Plan B: $1000 per month + 10% of monthly sales
Plan C: 16% of monthly sales

   a) On the same axes, plot a graph of salary against sales for each plan.

   b) For each plan, calculate the monthly salary for each monthly sales.

   i) $5000   **ii)** $8000   **iii)** $12 000   **iv)** $15 000   **v)** $20 000

   c) Which plan represents direct variation?

   d) Write to explain the conditions under which each plan would be preferable. Which plan would you choose? Explain.

7. Expand, then simplify by factoring, if possible.

   a) $3(x - y) + 2(2x - y) - (x + y)$
   b) $4x^2(x + 2y) + 3x^2(3x + y) - 2x^2(3x - 5y)$
   c) $x(3x - 2y) - 2y(2x + y) - y(3x - 2y)$

8. Factor.

   a) $x(x + 2) + y(x + 2)$ **b)** $x^2(x + 4) + 2(x + 4)$

9. Solve.

   a) $\frac{x}{x + 1} - \frac{3}{3x - 1} = 2$ **b)** $\frac{1}{x + 2} - \frac{3}{x - 6} = 1$

   c) $\frac{x + 6}{3(x - 2)} = \frac{x + 3}{2x - 4}$ **d)** $\frac{1}{x} + \frac{1}{x + 1} = \frac{2x + 1}{5x}$

10. Solve each ΔPQR. Give the answers to 1 decimal place.

   a) $\angle Q = 75°, r = 8, p = 11$ **b)** $\angle R = 52°, r = 28, q = 25$
   c) $\angle P = 38°, \angle Q = 105°, p = 32$ **d)** $r = 17, p = 14, q = 26$

11. A market gardener has to fertilize a triangular field with sides of lengths 90 m, 45 m, and 65 m. The fertilizer is to be spread so that 1 kg covers 10 m². One bag of fertilizer has a mass of 9.1 kg. How many bags of fertilizer will be needed?

12. Solve each ΔXYZ.

   a) XY = 7, YZ = 5, and $\angle Y = 110°$ **b)** $x = 3.7, y = 4.1$, and $\angle X = 58°$

Should We Harvest Today or Wait?

Icewine is an expensive wine made from frozen grapes. To produce icewine, selected overripe grapes are picked before sunrise at below freezing temperatures and crushed instantly, while still frozen.

 CONSIDER THIS SITUATION

The timing of the harvest of grapes for icewine is important. The grapes should be harvested as late as possible, but before a killer frost. As each day passes, the quality of the grapes increases, but there is a greater risk of a heavy frost killing the grapes and reducing their quality. When should the grapes be picked?

Deciding when to harvest the grapes is something like a game. Suppose I offer you a chance to play this game with four coins.

> I'll toss four coins.
>
> On each toss, I'll pay you $1 unless all four coins are heads.
>
> If all four coins are heads, you pay me all your winnings, plus $5.00 more.

- Would you be willing to play this game?
- For how many turns would you be willing to play this game?
- Explain how deciding when to pick the grapes is like this game.

On pages 582 and 583, you will develop a mathematical model to explore situations like this. You will use your skills with sequences, graphing, functions, and probability.

 FYI Visit www.awl.com/canada/school/connections

For information related to the above problem, click on <u>MATHLINKS</u> followed by <u>AWMath</u>. Then select a topic under Should We Harvest or Wait?

INVESTIGATE

The Results of an Informal Survey

This newspaper article reports the results of an informal survey. Read the article, then complete the exercises.

The article includes data about the number of people who said thank you and the number who did not. Express these numbers as percents. Copy and complete these tables before you answer the questions that follow.

Hold that door – Survey finds older people are more polite

BILLY ARCHER and STEVEN CARALE
Kidsday Staff Reporters

Are people polite? Do they take the time to say thank you anymore, or are they just too busy? We wanted to find out so we conducted a test at a local shopping mall.

We spent an hour just holding the door for people going in and out of the mall. We wanted to see if people still had manners, and were kind enough to say thank you. Here is what we found out:

MALE, SAID THANK YOU, SAID NOTHING

Men, 69, 23.

KIDSDAY POLL

Teens, 4, 5.
Younger boys, 3, 10.
FEMALE, SAID THANK YOU, SAID NOTHING
Women, 94, 17.
Teens, 35,8.
Younger girls, 4, 6.

Most people were polite, but we did find that older people were more polite than younger people. We appreciate it when someone says thank you when we are being courteous.

Los Angeles Times-Washington Post

Percent who said thank you	
Women	
Teens	
Younger girls	

Percent who said thank you	
Men	
Teens	
Younger boys	

The following questions are ones you should ask about a survey like this. You may not be able to answer some of these questions. Answer as many as you can. For those you cannot answer, what additional information would you need?

1. The headline states a conclusion about people. Which people are they?

2. a) Which people were included in the survey?

   b) How were these people selected?

3. a) How many people were included in the survey?

   b) Do you think this is enough to justify the conclusion in the headline?

   c) Do you think there were enough people in each category?

4. How was the survey conducted?

5. Do you think the results are an accurate indication of people's politeness?

An advertisement contains this statement.

The advertisement tries to make you believe that a particular product is highly recommended by dentists. To obtain this information, the advertiser probably conducted a survey among dentists. This group of dentists is the *population*. The group of dentists who responded to the survey is a *sample*.

Here are some questions you should ask about this survey. Compare these with the questions in *Investigate*.

Questions about the population

To which dentists does the advertisement refer? The word "dentists" could refer to dentists in Edmonton, in Alberta, in Canada, in North America, or to some other group of dentists.

Questions about sample selection

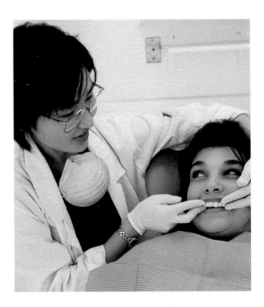

Which dentists were surveyed? How were they selected? Were they paid to field-test the gum produced by the company? If they were, their answers might have an unfair influence on the result. In this case, we say that the sample is *biased*.

Did every dentist in the population have an equal chance of being selected? If they did, the sample is a *random sample*. Random samples are used to reduce or eliminate the possibility of bias.

Questions about sample size

How many dentists were surveyed? The sample would not be large enough if only 4 dentists had been surveyed.

Questions about how the survey was conducted

What questions were asked? Were dentists asked to name their favourite gum? Were they asked to name the gum they considered least harmful for a person's teeth? Were they given a list of several gums from which they were asked to select one? Were they asked which type of gum they would recommend? Were they asked if they would recommend sugarless gum?

Questions about the accuracy of the results

Are the results an accurate indication of the gum preferences of all dentists in the population? If so, the results are *valid*.

Any conclusion or *inference* based on data, such as those discussed on the previous pages, should be made from the analysis of the results of a survey. A survey must be carefully planned and designed. A well-designed survey involves these steps:

- identifying the population
- selecting an unbiased sample
- preparing the survey question or item
- collecting data
- calculating the results
- making an inference about the population

Example

Suppose your school cafeteria wants to introduce a new dessert. Describe how to conduct a survey to decide which of three choices should be the new dessert.

Solution

Follow the steps above.

1. *Identify the population.* This is probably all students who buy lunch at the cafeteria.

2. *Select an unbiased sample.* You might decide that every 10th student who enters the cafeteria will be in your sample.

3. *Prepare the survey question or item.* Suppose the three choices are: fresh fruit salad, fruit pie, and frozen yogurt. A possible wording is:

> *You may choose only one dessert for the lunch menu in the cafeteria. Please circle <u>one</u> of the desserts below.*
>
> **Fresh fruit salad**   **Frozen yogurt**   **Fruit pie**

Since we are asking people to select from a list, and we are not asking a question, we describe the wording as an item.

4. *Collect the data.* Make copies of the item and ask students in the sample to circle their choice. Alternatively, you could rewrite the item as a question and record students' responses on a tally sheet.

5. *Calculate the results.* When you have all the responses, determine the total for each dessert.

6. *Make an inference about the population.* The most popular dessert is the one the cafeteria should introduce.

1. In *Investigate*, what other headline could you have written based on the data in the article?

2. In the *Example*, suppose the sample consists of all students sitting at a particular table or tables in the cafeteria. How might this affect the validity of the results?

3. In the *Example*, what would you do if two desserts tied for first choice? Explain your decision.

4. In the *Example*, how would you rewrite the item as a question?

9.1 EXERCISES

A 1. A survey result indicates that "…most Canadians feel that Quebec should remain a part of Canada." What are some of the questions you should ask about this survey?

2. A phone-in survey conducted in Calgary indicates that "…80% of Calgarians favour a Formula One race in the middle of the city." Explain why this result might not be valid.

3. Suppose most of the students in your grade were surveyed. The nearly unanimous conclusion was, "Canada is the mightiest hockey nation in the world."
   a) What was the intent of the survey?
   b) How do you think this could be determined?
   c) Do you think the result of the survey was valid? Explain.

4. Identify the population you would sample for an opinion on each topic.
   a) public transit fares
   b) minimum driving age
   c) the GST

5. Identify the segment of the school population you would sample for an opinion on each topic.
   a) student parking spaces
   b) fees for athletic teams
   c) cafeteria food

6. Identify the segment(s) of the population who would strongly bias a survey about each topic.

   a) gun control

   b) birth control

   c) gay rights

B 7. Read this newspaper excerpt. List the important elements of a well-planned survey that should have been reported along with the conclusion.

> A poll taken by Hallmark Cards found that 67 percent of women would let someone cut into the line at a supermarket checkout counter, compared with 47 percent of men.

8. A mail-in survey in a teen magazine concludes that high school students spend an average of $300 a month on clothing.

   a) How do you think the editors of the magazine obtained this information?

   b) Why should you question the conclusion?

   c) Describe how a survey might be conducted to determine the amount of money high school students spend on clothing.

9. At a fund-raising event for an upcoming election, a political party conducted a survey to estimate how many seats it might win. Explain why the results might not be valid.

10. A magazine poll indicates that its readers rate their social lives somewhere between "busy" and "hectic." Discuss any possible bias in the data.

11. Identify a problem associated with surveys based on a 5-point scale, such as "On a scale of 1 to 5, rate the taste of this new soft drink."

12. Professional pollsters take considerable care to avoid bias in the wording of survey questions.

   a) How would you word a biased survey question about the rating of a TV show?

   b) How would you reword the question to eliminate the bias?

13. The *Youth Health* database contains survey results for 11-, 13-, and 15-year-olds in 11 countries. Respondents were selected based on their grade level. In many countries, the survey was conducted later in the school year than originally planned. This meant that students were older than expected. This table lists the dates when the survey was administered and the mean ages of respondents.

Country	Administration date	Mean ages of respondents		
		Youngest age group	Middle age group	Oldest age group
Austria	May 1990	11.3	13.3	15.2
Belgium	March 1990	11.5	13.5	15.4
Canada	Feb-May 1990	12.1	14.1	16.0
Finland	March-May 1990	11.7	13.8	15.8
Hungary	May 1990	12.3	14.2	16.2
Norway	Nov-Dec 1989	11.4	13.4	15.4
Poland	Feb 1990	11.6	13.6	15.7
Scotland	Feb-March 1990	11.6	13.7	15.6
Spain	March 1990	11.8	13.8	15.8
Sweden	Nov-Dec 1989	11.4	13.4	15.4
Wales	March-April 1990	12.1	14.0	16.0

a) In which 3 countries were respondents the oldest? The youngest?

b) Which countries conducted the survey in the winter?

c) Open the database. Look through the questions students were asked. List 3 questions for which you think the differences in ages of respondents might have affected the data.

d) List 3 questions for which you think the differences in the time of year the survey was conducted might have affected the data.

C 14. Revise the survey question "Have you ever smoked?" to obtain a valid inference about the percent of smokers in the population. Explain how you wrote your question.

15. A local university bases its admission to the Faculty of Nursing on the applicants' marks in Mathematics 11. The university says that success in high school mathematics is a strong predictor of success in its first year program. Comment on the validity of this admission standard.

COMMUNICATING THE IDEAS

Use a dictionary to find the meanings of the words *sample*, *bias*, *random*, and *valid*. Use these meanings to write a short explanation of why it is important to have a random sample without bias, so the results will be valid.

Random Numbers

A list of digits, selected such that each digit has an equal chance of occurring is a list of random numbers. We can use random numbers to produce a random sample for a survey. Here are some ways to generate random numbers.

Using a Spinner

Divide a large cardboard circle into 10 equal sectors. Label the sectors from 0 to 9. To make a spinner, use a paper clip as shown. Spin it about a pencil. Write down the number on which it lands. In this way, you generate a list of random numbers.

Using a Telephone Directory

Turn to any white page of a telephone directory. Without looking, put a pencil point on a number. Write down the last four digits of this number. Continue down the page, listing the last four digits of each number to obtain as many random numbers as you need.

Using a Graphing Calculator

Ask your teacher for the program called RANDNUM from the Teacher's Resource Book. Enter the program in your calculator. When you run the program, the calculator will prompt you to enter the smallest and largest digits needed. The calculator will generate screens containing 42 random numbers between these two digits (inclusive). To display another screen, run the program again. The first screen below contains random numbers from 0 to 9 (simulating the spinner above). The second screen contains random numbers from 1 to 6 (simulating rolling a die).

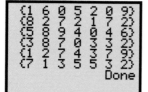

Using a Computer

Start a new spreadsheet document. Enter the text and numbers shown in the first and third rows.

	A	B	C	D	E
1	Random Number Generator				
2					
3	Lowest:	0	Highest:	9	
4					
5	9	6	8	5	5
6	0	5	2	0	4
7	5	0	4	7	2
8	3	3	4	8	7
9	3	5	3	1	1

In cell A5, enter this formula:

- for Microsoft Works and Microsoft Excel =INT(RAND()*($D$3–$B$3+1))+$B$3
- for ClarisWorks =INT(($D$3–$B$3+1)*RAND())+$B$3

Copy this formula down and to the right into a block of cells to generate a set of random numbers. These will be random numbers in which the lowest digit is the number in cell B3 and the highest is in cell D3. You can change these numbers to generate sets of random numbers between any two digits you want.

To get more random numbers, recalculate the spreadsheet. To do this, click on the menus at the top of the screen to find the one containing the Calculate Now command.

1. Why should you ignore the first three digits when you generate random numbers from a telephone directory?

2. What other ways can you think of to generate random numbers?

3. Why is it important that every digit has an equal chance of being selected?

4. What are the advantages and disadvantages of each method described above?

5. Use the RANDNUM program or the spreadsheet to simulate rolling a die and tossing a coin.

The most important part of survey design is the method of sampling. Careless sampling strategies can lead to biased samples that are not representative of the population. If you take care when designing the sampling process, you should obtain valid data.

Here are some common sampling methods.

Simple random sample

Every member of the population has an equal chance of being selected. For example, to select a random sample of 6 students from your class, each person is assigned a number. Then numbers are selected randomly using technology or some other method.

Stratified random sample

Every member of different segments of the population has an equal chance of being selected. For example, suppose the school population is divided into groups by grade (8s, 9s, 10s, 11s, 12s). The sample consists of 5 random samples, one from each grade. This ensures that members from each grade will appear in the sample.

Systematic sample

Every nth member of the population is selected. For example, to select a sample from your class, every 5th name is selected from the class list.

Cluster sample

Every member of a randomly drawn subdivision of the population is selected. For example, the school population is subdivided into groups by grade. One of the grades is then chosen randomly, and every student in that grade is in the sample.

Convenience sample

Every convenient member of the population is selected. For example, for a survey about shopping habits, people in a mall are asked questions. When one person has finished, another is approached.

Self-selected sample

Only interested members of the population will participate in the sample. For example, a radio station conducts a phone-in survey. Only those interested will phone the station to respond to the survey.

Example

You intend to survey your school population to determine whether the students would attend another dance this month. Identify each sampling technique below. Comment on the validity of the data you would get.

a) You ask every third person passing through the hall.

b) You distribute a questionnaire to each student and wait for responses.

c) You ask your friends.

d) You obtain all the class lists, and generate random numbers to identify who should be in your sample. Then you ask those students.

e) You use random numbers to select one class from each grade, then use random numbers again to pick names from each class. Then you ask those students.

Solution

a) This is a systematic sample. The data may not be valid if only certain students are likely to be in the hall at that particular time.

b) This is a self-selected sample. The data may not be valid because only those interested would respond.

c) This is a convenience sample. The data may not be valid because your friends may agree with you, or they may not want to offend you.

d) This is a simple random sample. The data should be valid because every student in the school has an equal chance of being selected.

e) This is a stratified random sample. Although every student does not have an equal chance of being selected, the data should be valid.

DISCUSSING THE IDEAS

1. Which two sampling methods do you think are least likely to be biased? Which two methods do you think are most likely to be biased? Explain.

2. Do you think a random sample could be biased? Explain.

3. In the *Example*, what difficulties might you encounter if you tried to use each sampling method?

4. In part e of the *Example*, why doesn't every student have an equal chance of being selected? Why should the data still be valid?

A 1. Many magazines conduct surveys and ask readers to mail in responses.

   a) Is the sample obtained a random sample? Explain.

   b) Do you think the data obtained would be valid? Explain.

2. The owners of a theatre would like to find out what percent of the patrons are returning customers. The customers enter the theatre one at a time and pass an attendant who takes their tickets. A suitable survey question can be posed at that time.

   a) What sampling method are you likely to use? Why?

   b) What question would you ask the customers?

3. A television program conducts a regular survey by asking viewers to call one of two 1-900 phone numbers to cast a vote. Is a sample obtained in this way a random sample? Explain.

4. People concerned about the environment think "green." You wish to determine if this affects people's preferences when they choose a colour for their car. How would you select a sample from the vehicles in a large and regularly full parking lot?

5. A school population comprises grades 8 through 12, distributed approximately as follows.

Grade	8	9	10	11	12
Students	300	400	450	450	400

   How would you choose a sample of the school population for a survey about the number of hours of extracurricular activities in which students are involved?

6. A shoe store conducts a survey of the most popular brands of athletic shoes before ordering for the new season. It asks the first 200 people in the mall who are wearing athletic shoes to respond.

   a) What sampling method is used?

   b) Do you think the data obtained would be valid? Explain.

7. The draft copy of the new course outlines for grades 10 to 12 is sent to all mathematics teachers in the province to solicit their comments. From the responses, the level of satisfaction with the outlines is determined and changes are made.

   a) Do you think teachers who do not agree with some aspect of the course outlines are more likely to respond than teachers who are in agreement?

   b) Do you think the data obtained would be valid?

B **8.** Which sampling methods are professional pollsters most likely to use? Why?

9. Do you think a telephone survey conducted by dialing telephone numbers at random would produce a random sample of the population? Explain.

10. Each summer, major league baseball holds an All-Star Game that pits the American League against the National League. The players in the starting lineups are chosen by baseball fans. For a few weeks prior to the All-Star Game, voting cards are distributed to people who attend all major league games. People punch holes in these cards to register their choices. Give some reasons why this method might not result in the best players playing in the All-Star Game.

11. The information in the *Youth Health* database was part of an international survey. Based on rates of return from previous surveys, the statisticians who administered the Canadian survey determined they needed data from 132 classes.

a) Because of cost limitations, only 1 class was chosen from the territories. The other classes were selected to represent the population proportionally. Use the information in this table to calculate how many classes were surveyed in each province.

Province	BC	AB	SK	MN	ON	PQ	NB	NS	PEI	NF
Percent of population	11.4	9.4	4.0	4.2	35.9	25.8	2.8	3.4	0.5	2.2

b) For each province, a master list of schools with enrolments of 30 or more students in grade 10 was created. Schools with enrolments over 2000 were listed twice so that classes from large schools would have similar chances of being selected as classes from smaller schools. After a school was selected, the classes in the designated grades were assigned numbers and a class was selected using a table of random numbers. If selected schools did not have grades 6 or 8 classes, classes from these grades were randomly selected from other schools in the same jurisdiction. Does this selection system seem fair? Explain.

12. To predict a winner in a provincial election, a company compiled a list of 2000 people from telephone directories, lists of car owners, and certain club membership lists. The company mailed questionnaires to these people. One hundred three people responded. Is this a random sample? Explain.

13. Choose one of these issues. Design a survey to determine the opinions of your school population on the issue. Your design should include:

   a) the question or questions you would ask

   b) a description of how you would select an unbiased sample

   c) a description of the steps you would take to conduct the survey

   Issue 1: Should a commercial franchise be allowed to run the school cafeteria?

   Issue 2: Would students like school to start and end half an hour earlier each day?

   Issue 3: Should students be required to participate in a school sport?

14. Choose one of the issues below. Design a survey to determine the opinions of people in your city or region on the issue. Your design should include:

   a) the question or questions you would ask

   b) a description of how you would select an unbiased sample

   c) a description of the steps you would take to conduct the survey

   Issue 1: Should all cyclists be required to wear helmets?

   Issue 2: Should people have to pay user fees for health care?

   Issue 3: Should people who own guns be required to have a license?

15. You wish to determine which of five movies people predict will be the winner of the Oscar for Best Picture. Design a survey question and a sampling strategy. Note any concerns you may have before the results are tabulated.

16. Suppose you want to choose a simple random sample of 6 students from your class. You assign a different number to each student. List as many different ways as you can that you could use to choose 6 of those numbers at random.

17. Work with a group of students. Find a newspaper or magazine article that reports provincial or national percents of people's opinions on a particular topic. Survey your school population on the same topic. Write a report of your findings, and include how they compare with the conclusions in the media report.

COMMUNICATING THE IDEAS

In your journal, write the most important condition for establishing a random sample, and explain why it is important.

Make a Sampling Box

For Section 9.3, you will need to generate samples. You can use the sampling box below, a graphing calculator, or the spreadsheet on page 544.

Materials required:

1 cardboard box with a clear lid, such as a greeting card box
This can be any standard size.

1 cardboard insert
This should have the same dimensions as the bottom of the box and be about 2 mm thick.

300 plastic 4-mm beads in two colours; for example,
90 red, 210 clear
Tape and glue
Single hole punch

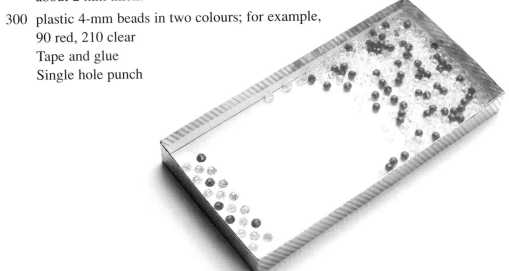

Punch 20 holes at one end of the cardboard insert, then glue it to the inside of the bottom of the box. Fill the box with the 300 beads, close the lid, and tape it shut.

The lid is taped so that the beads will not spill out. Do not glue the lid on permanently. If you wish to change the numbers of the two colours of beads later, the tape can be removed and the beads changed.

There are 90 red beads in 300 beads. We say that the percent of marked (red) beads is 30%. To obtain a sample of 20 beads, shake the box. Count the red beads that sit in the holes.

Since there are only two colours of beads in the box, we describe the beads as a *binomial population*.

Sampling Box Simulators

For Section 9.3, you will need to generate samples. You can use the sampling box on page 543 or you can use technology described below.

1. Ask your teacher for the program called SMPLSIM3 from the Teacher's Resource Book. Enter the program in your calculator. This program simulates the sampling box on page 543. It assumes that the percent of marked beads is 30% and that there are 20 beads in each sample. When you run this program, the calculator displays a screen similar to the one (below left). The 1s represent marked beads. Press ENTER to get results for more samples.

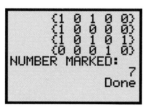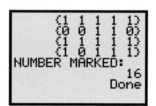

To obtain samples for other percents of marked beads, use the programs SMPLSIM1 (10% marked), SMPLSIM2 (20% marked), and so on. The screen (above right) shows a sample generated using SMPLSIM8 (80% marked).

2. a) Start a new spreadsheet document. This spreadsheet simulates the sampling box on page 543. It assumes 30% marked beads and 20 beads in each sample.

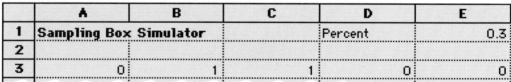

	A	B	C	D	E
1	Sampling Box Simulator			Percent	0.3
2					
3	0	1	1	0	0

   b) In cell A3, enter the formula below, then copy it into cells A3 to E6.
   - for Microsoft Works and Microsoft Excel =IF(RAND()*20<$E$1*20,1,0)
   - for ClarisWorks =IF(20*RAND()<$E$1*20,1,0)

   c) In cell E1, enter the percent of marked beads. Twenty digits, either 0 or 1, will be displayed in rows 3 to 6. The 1s represent marked beads.

   d) In cell B7, enter the formula below. It displays the total number of marked beads.
   - for Microsoft Works and Microsoft Excel =SUM(A3:E6)
   - for ClarisWorks =SUM(A3..E6)

   e) To generate more samples, recalculate the spreadsheet as explained on page 537.

LINKING IDEAS

Mathematics & Technology

Some experiments or survey questions have only two possible outcomes. For example:

- When you toss a coin, the outcomes are heads and tails.
- When you ask the question: "Did you pass your driving test?", the answer is yes or no.
- When a person writes, she uses either her left hand or her right hand.

Such outcomes are called *binomial outcomes*. You can use your sampling box, a graphing calculator, or a spreadsheet to simulate experiments and surveys involving binomial outcomes. You can use the results to make predictions about other experiments and surveys.

Suppose Churchill School has an enrollment of 2013 students. The school has extensive student profiles. The records show that 1410 students were born in Canada. This is approximately 70% of the school population. Suppose we take a random sample of 20 students. Since 70% of 20 is 14, we might expect 14 students to be Canadian-born, but the number could be more than 14 or less than 14. It could be any number from 0 to 20.

Suppose we take many random samples of 20 students. Which of the numbers from 0 to 20 of Canadian-born students are most likely? You will investigate this with the following experiment.

INVESTIGATE Samples Size 20 (70% Marked)

In the first part of this experiment, you will simulate the sampling of 20 students from a school in which 70% of the students are Canadian-born. You will take 100 samples and summarize the results.

1. List the numbers from 1 to 20 with a space beside each for tally marks.

Complete any of exercises 2, 3, or 4. Then go on to exercise 5.

2. Use your sampling box.

a) Adjust the number of marked beads to represent the percent of students born in Canada. Since this is 70%, use $0.70 \times 300 = 210$ marked beads and the rest (90) clear.

b) Shake the sampling box. Record the number of marked beads in your sample of 20. For example, if 9 beads are marked, put a tally mark beside 9.

c) Repeat this 99 times for 100 samples.

3. Use the SMPLSIM7 program in your graphing calculator (see page 544).

   a) When the calculator asks for the percent, enter 70.

   b) When the calculator asks for the number of samples, enter 100.

   c) Record the results. For example, if the first sample has 12 marked, put a tally mark beside 12.

   d) Repeat until you have the results for all 100 samples.

4. Use the spreadsheet on page 544.

   a) In cell E1, enter 70% as 0.7. The results of the first sample will be displayed.

   b) Record the number that is shown in cell B7. For example, if this number is 17, put a tally mark beside 17.

   c) Recalculate the spreadsheet and repeat part b. Keep doing this until you have the results for all 100 samples.

5. Summarize your results from exercise 2, 3, or 4 in a frequency table.

Number marked									
Frequency									

6. a) Combine the data from all students. Write each frequency as a percent.

   b) Which outcomes are most likely? Which outcomes are least likely?

In one set of 100 trials, these data were obtained.

Number marked	8	9	10	11	12	13	14	15	16	17	18
Frequency	1	0	2	5	12	19	21	18	11	8	3

These data are represented by the diagram below. This is called a *box-and-whisker plot*, or boxplot. In this diagram, the box represents the most likely 90% of the outcomes. The whiskers represent the least likely 10% of the outcomes.

90% boxplot for samples size 20, where the percent of marked beads in the population is 70%

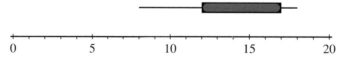

Number of marked beads in the sample

This boxplot was constructed using these steps.

Step 1 Identify the least likely 10% of the outcomes.
Since 10% of 100 is 10, we identify the 10 least likely outcomes. Look at the lowest numbers at both ends of the frequency row in the table. Starting at both ends, find the lowest numbers that add to 10, or close to 10. There is a 1, 2, and 5 under 8, 10, and 11 at the left, and a 3 under 18 at the right. Observe that $1 + 2 + 5 + 3 = 11$, and this is as close as we can get to 10.

Step 2 Identify the most likely 90% of the outcomes.
These are the outcomes from 12 to 17 marked beads.

Step 3 Draw a box to represent the most likely 90% of the outcomes.
Above a scale from 0 to 20, draw a box with its ends at 12 and 17.

Step 4 Draw the whiskers to represent the least likely 10% of the outcomes.
Draw a line at the left of the box to 8 and a line at the right to 18.

The result is called the 90% boxplot for samples of size 20 from a population in which 70% of the items are marked. This means that if we take many samples, there is a 90% probability that we would get 12 to 17 marked items. That is, we should expect 12 to 17 marked items 90% of the time.

This means that if we take a sample of 20 students from a school in which 70% of the students in a school were Canadian-born, we could predict that the sample would contain 12 to 17 Canadian-born students 90% of the time.

By adjusting the percent of marked beads in the sampling box (or using different probabilities in the graphing calculator or spreadsheet), you can create boxplots for any other percent.

INVESTIGATE **Samples Size 20 (Other Percents Marked)**

You will simulate the sampling of 20 students from a school in which other percents of the students are Canadian-born. Take samples as before, and draw the corresponding boxplots.

1. Work with 2 or 3 students. Each group generates 100 samples with a given percent of beads marked: 10%, 20%, 30%, 40%, 50%, …, 90%.

2. Draw a boxplot to represent the results for your group.

3. Combine your boxplot with those from other groups to produce a chart of boxplots similar to the one on page 552.

4. Compare the boxplots on page 552 with the chart you produced in exercise 3. Account for any differences you see among the boxplots.

You can use the boxplots to make estimates involving samples when information about a population is known.

Example 1

In Westview School, 40% of the students are younger than 15 years old. In a sample of 20 students, estimate how many students would be younger than 15.

Solution

Use your boxplots or the boxplots on page 552. Since we are dealing with a population in which 40% of the items are marked, locate 40% on the vertical axis. We want the boxplot next to this number. The boxplot on page 552 shows that between 4 and 12 items would likely be marked 90% of the time. Hence, there is a 90% probability that there would be from 4 to 12 students younger than 15 in a sample of 20 students.

90% boxplots for samples size 20

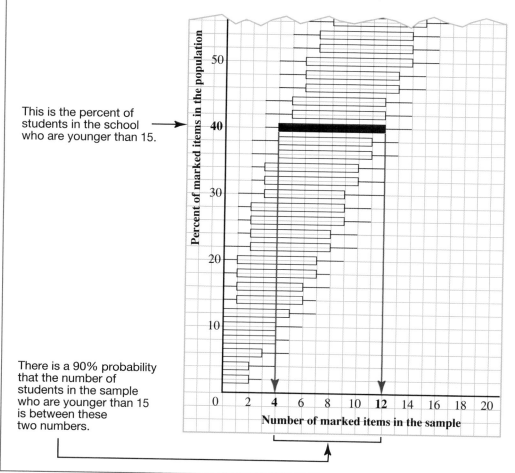

This is the percent of students in the school who are younger than 15.

There is a 90% probability that the number of students in the sample who are younger than 15 is between these two numbers.

Example 2

Repeat *Example 1*, for a sample of 100 students.

Solution

Use the boxplots on page 554. The boxplot for 40% marked items shows that between 32 and 48 items would likely be marked 90% of the time. There is a 90% probability that there would be between 32 and 48 students younger than 15 in a sample of 100 students.

The estimates in *Examples 1* and *2* are different from the kinds of estimates you have made in the past. They are expressed in terms of probability.

DISCUSSING THE IDEAS

1. Look at the boxplots on page 552. If you start at the lower left, the boxplots go up to the right. Explain why this occurs.

2. a) Suppose the boxplots on page 552 had been 80% boxplots instead of 90% boxplots. Explain how the boxes would differ from those on page 552.

   b) Suppose the boxplots had been 95% boxplots. What difference would that make? Why?

   c) What would happen if we tried to use 100% boxplots?

3. Compare the results of *Examples 1* and *2*. Since the sample size in *Example 2* is 5 times as large, why can't we solve *Example 2* by multiplying the estimates in *Example 1* by 5?

9.3 EXERCISES

Ⓐ 1. Thirty percent of the school population take theatre arts. In a random sample of 20 students, estimate how many students would take theatre arts.

2. In one school, 80% of the students take science. In a random sample of 20 students, estimate how many students would take science.

3. In another school, 18% of the students speak two languages. In a random sample of 100 students, estimate how many students would speak two languages.

4. A company makes computer disks. The quality control department finds 10% of the disks defective and must be rejected. How many defective disks would you expect to find in a random sample of each size?

a) 20 disks   **b)** 40 disks   **c)** 100 disks

5. An insurance company notes that 32% of the policyholders in a certain city have earthquake insurance coverage. In a random sample of 100, estimate the number of policyholders who would have earthquake insurance.

6. Approximately 28% of the Canadian population live in the western provinces. In a random sample of 100 Canadians, estimate the number of people who live in the western provinces.

B **7.** Suppose 46% of the school population are girls. You select random samples of 40 students many times. Could these samples contain 20 girls 90% of the time? Explain.

8. Suppose 10% of the school population are ESL students. Could a sample of 20 randomly selected students contain 5 ESL students 90% of the time? Explain.

9. Suppose 36% of the students at your school chew gum. Is it likely that a random sample of 40 students would contain 20 students who chew gum? Explain.

10. The probability of correctly answering a true/false question by guessing is 50%. Determine if, by guessing, you can correctly answer 25 out of 40 questions 90% of the time.

11. Suppose the probability of your team winning any game is 40%. Is it likely that your team would win 10 out of the remaining 20 games? Use a boxplot to support your answer.

12. a) In the *Youth Health* database, it is reported that 23% of 15-year-old Canadian girls and 20% of 15-year-old Canadian boys said they often had a bad temper in the preceding six-month period. In a sample of 20 girls and 20 boys of this age, estimate how many often have bad tempers.

b) Find the two records in the database that report students who said they often had bad tempers. Repeat part a for 15-year-olds from Finland, Scotland, and Poland. According to their responses, do boys or girls have worse tempers?

c) Choose a record about a different topic. Write an exercise like that in part a for a classmate to complete.

13. Suppose your basketball season record shows that you have made 60% of your foul shots. Assume that your success rate continues. How likely is it that you would make 30 out of your next 40 foul shots?

14. You have been late for class 30% of the time. How likely is it that you would be late no more than 3 times for the next 20 classes? Use a boxplot to support your answer.

15. Your dog awakens you late at night roughly 80% of the time. How likely, during the next 20 days, will your dog do this at least 14 times?

C 16. Look at the boxplots on page 552. Explain why there is no whisker

   a) at the left end of the boxplots near the bottom.

   b) at the right end of the boxplots near the top.

17. Find a baseball player's current batting average.

   a) Follow that player's performance during the next 20 at-bats. Express the performance during those 20 at-bats as a percent. Is the player reasonably consistent?

   b) Explain how a player's "slump" or a "hot streak" might be detected from batting average records.

18. The intervals of numbers represented by the boxes in the boxplots are called *90% confidence intervals*. We also say that 90% is the *confidence level* in stating the range of possible values. For example, in *Example 1* we can say that the 90% confidence interval for a sample of 20 is from 4 students to 12 students.

   a) Suppose 120 of the 300 beads in the sampling box are marked, and you use the box to take samples of size 20. Determine the 90% confidence interval, and explain what it means.

   b) Forty percent of the students in a school come to school by bus. Determine the 90% confidence interval for the number of students coming to school by bus that you would find in a sample of 20 students.

19. **a)** What would you have to do to obtain narrower confidence intervals?

   b) What would you have to do to obtain a higher confidence level?

COMMUNICATING THE IDEAS

Your friend telephones you to ask about tonight's mathematics homework. How would you explain to your friend what a boxplot is and why it is useful? Write your ideas in your journal, using an example to illustrate your explanation.

90% boxplots for samples size 20

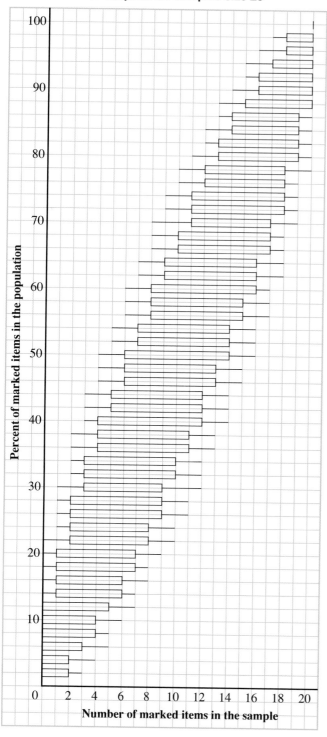

Percent of marked items in the population (vertical axis)

Number of marked items in the sample (horizontal axis)

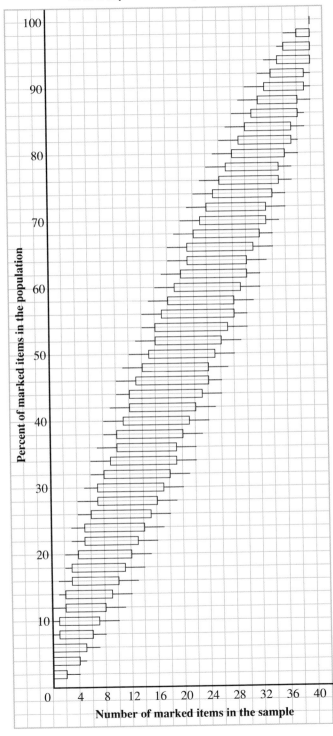

90% boxplots for samples size 40

Percent of marked items in the population

Number of marked items in the sample

90% boxplots for samples size 100

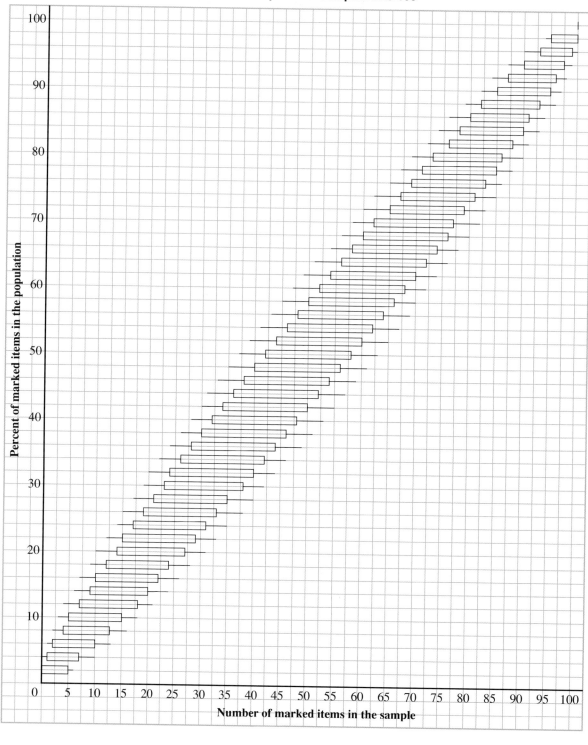

At the end of each year, Maclean's magazine reports the results of a year-end poll of general attitudes. Data are obtained from telephone interviews with 1500 Canadians aged 18 or over, selected randomly from all 10 provinces. The following information appeared in the magazine on December 30, 1996. Each result is expressed as a percent of the people polled.

GENERAL ATTITUDES

1. What is the most important problem facing Canada?

Unemployment 31	Morality/moral breakdown in society 1
Deficit/government spending 15	Native/aboriginal issues 1
National unity 9	Agricultural issues 1
Economy-general 6	Immigration/multiculturalism 1
Health care 5	Women's issues . . . . <1
Other social services 4	Free trade . . . . −
Taxes 3	Peace/defense . . . . −
Crime/violence/justice 2	International issues . . . . −
Education 2	Other 11
Environment 1	Don't know 5
Government-general 1	

National results ... are considered accurate to within three percentage points, 19 times out of 20.

Observe that 31% of the people polled feel that unemployment is the most important problem facing Canada. However, we cannot be certain that 31% of all Canadians feel the same way. We use the statement below the table to make a conclusion about how all Canadians feel. The statement means there is a probability of $\frac{19}{20}$ that the percent of all Canadians who feel that unemployment is the most important problem facing Canada is between 31% − 3% (or 28%) and 31% + 3% (or 34%).

Recall that a conclusion made about a population based on information from a sample is an inference. In the previous paragraph we made an inference about the percent of Canadians who feel that unemployment is the most important problem facing Canada. The inference is expressed in terms of probability. You can use boxplots to make similar inferences.

For example, suppose we visit a school, select a random sample of 20 students, and find that 6 students are left-handed. To make an inference about the number of left-handed students in the school, use the boxplots for samples size 20 on page 552. Since the sample contained 6 left-handed students, locate 6 on the horizontal axis. Look at the vertical line through this point. It passes through or touches several boxes, from the 14% box up to the 50% box.

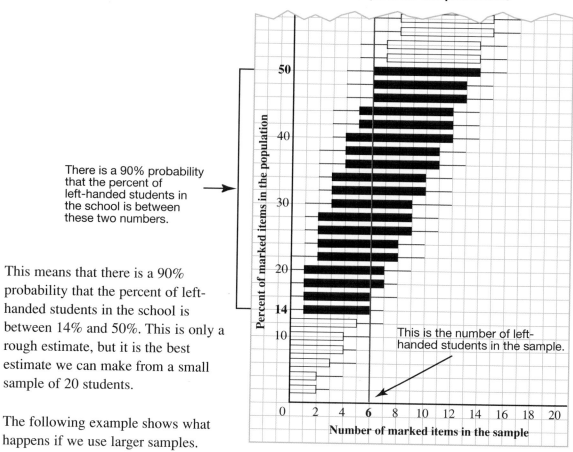

90% boxplots for samples size 20

Percent of marked items in the population

There is a 90% probability that the percent of left-handed students in the school is between these two numbers.

This is the number of left-handed students in the sample.

Number of marked items in the sample

This means that there is a 90% probability that the percent of left-handed students in the school is between 14% and 50%. This is only a rough estimate, but it is the best estimate we can make from a small sample of 20 students.

The following example shows what happens if we use larger samples.

Example

For each sample, make an inference about the percent of left-handed students in a school.

a) Twelve students in a sample of 40 students are left-handed.

b) Thirty students in a sample of 100 students are left-handed.

Solution

a) Use the boxplots for samples size 40 on page 553. A vertical line through 12 passes through or touches boxes from the 20% box to the 44% box. There is a 90% probability that the percent of left-handed students in the school is between 20% and 44%.

b) Use the boxplots for samples size 100 on page 554. A vertical line through 30 passes through or touches boxes from the 24% box to the 38% box. There is a 90% probability that the percent of left-handed students in the school is between 24% and 38%.

In each of the above samples, the percent of left-handed students is approximately 30%. The range for the estimates for the percent of left-handed students in the school is smaller for the larger sample sizes.

Sample size   **Result**

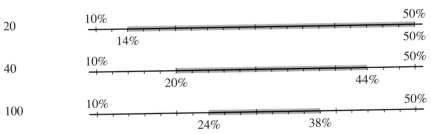

20

40

100

There is a 90% probability that the percent of left-handed students in the school lies in the coloured intervals.

DISCUSSING THE IDEAS

1. When we make an inference about a population based on information from a sample, why is it expressed in terms of probability?

2. Explain why you should get smaller ranges when you use larger sample sizes.

3. Suppose the boxplots on pages 552 to 554 had been 95% boxplots. What difference would this make to the coloured intervals in the diagram above? Explain.

9.4 EXERCISES

 1. Use the data in the table on page 555. Make an inference about the percent of Canadians who feel that the most important problem facing Canada is:
   a) deficit/government spending   b) national unity
   c) health care   d) taxes

2. In a randomly selected sample of 20 students, 5 said they like the food in the cafeteria. Use the boxplots to make an inference about the percent of students in the school who like cafeteria food.

3. Eight out of 20 randomly selected grade 10 students said they have a cat. Make an inference about the percent of grade 10 students who have a cat.

4. a) In Lakeview School, a survey showed that 8 out of 20 randomly selected grade 10 students spend less than one hour on homework. Use the boxplots to estimate the percent of the grade 10 students in Lakeview School who spend less than one hour on homework.

b) Repeat part a for a sample showing that 30 out of 40 grade 10 students spend less than one hour on homework.

c) Repeat part a for a sample showing that 75 out of 100 grade 10 students spend less than one hour on homework.

5. Thirty-two out of 40 randomly selected grade 10 students spent more than one hour preparing for their last math test. Make an inference about the percent of grade 10 students who spent more than one hour preparing for their last math test.

6. This article was published by Southam News in January 1997.

   a) Explain what the statement below the graph means.

   b) Identify a problem associated with this survey as it is reported here.

   c) The survey was conducted in the winter. How do you think the results might have been different if the survey had been conducted in the summer?

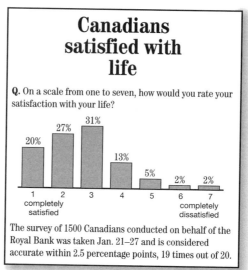

Canadians satisfied with life

Q. On a scale from one to seven, how would you rate your satisfaction with your life?

The survey of 1500 Canadians conducted on behalf of the Royal Bank was taken Jan. 21–27 and is considered accurate within 2.5 percentage points, 19 times out of 20.

B 7. Oak Ridges School has 820 students. A survey showed that 28 out of 40 randomly selected students go to school by bus.

   a) Make an inference about the percent of Oak Ridges students who go to school by bus.

   b) Use the result of part a. Estimate how many Oak Ridges students go to school by bus. Express your estimate in terms of probability.

8. The population of British Columbia is approximately 3.9 million. A random sample of 100 British Columbians includes 44 people who believe that a manlike creature called the Sasquatch exists.

   a) Make an inference about the percent of British Columbians who believe the Sasquatch exists.

   b) Use the result of part a. Estimate how many British Columbians believe the Sasquatch exists.

9. The population of Alberta is approximately 2.8 million. A random sample of 100 Albertans includes 84 people who believe that life exists elsewhere in the universe. Estimate the number of Albertans who believe that life exists elsewhere in the universe.

10. Suppose you estimate that 70% of students wear running shoes to school. A random sample of 20 students includes 11 who are wearing running shoes. Does this sample support your estimate? Explain.

11. According to a Reader's Digest/Roper Reports Canada poll in 1996, 90% of Canadians surveyed say they are concerned about our health-care system. The margin of error is plus or minus 2.8%, 19 times out of 20.

    a) Make an inference about the percent of Canadians who are concerned about our health-care system.

    b) Compare the result from this poll with the result about health care in the Maclean's poll on page 555. Explain why the results from the two polls are so different.

12. Suppose there are about 800 students in your school. Suppose you estimate that about 320 of the students at your school do not eat breakfast. In a random sample of 20 students, 12 eat breakfast. Does this sample support your estimate? Explain.

13. Collect newspaper and magazine articles that report survey results. Find examples of inferences that have been made about populations based on data from samples. Do you agree or disagree with the inferences? Explain.

14. Design a one-question survey about a topic in which you are interested. Word the survey question so the answer is yes or no. Conduct your survey with a random sample of people. Use the results to make an inference.

C 15. Explain how the size of a sample affects the size of the confidence interval obtained from the sample.

16. In an experiment to test for ability in extra sensory perception (ESP), a girl was shown 20 photographs of people, none of whom she knew. Below each photograph were 5 telephone numbers, one of which was the phone number of the person in the photograph. The girl had to identify the correct phone number. Suppose she correctly identified 9 phone numbers. Would you think that she had ESP ability? Use a boxplot to support your answer.

COMMUNICATING THE IDEAS

Compare the different ways boxplots were used in this section and in the preceding section. In any given problem, how can you tell which way to use a boxplot? Write your ideas in your journal, using examples to illustrate your explanations.

Estimating the Size of a Wildlife Population

To estimate the number of fish in a lake, wildlife biologists use a *capture-recapture* sampling technique. This method assumes that if enough members of a wildlife population are captured, tagged, and released, then any sufficiently large random sample of that population will contain about the same fraction of tagged animals as the entire population.

One year, biologists captured, tagged, and released 46 trout in a lake. The following year, they captured 20 trout and found 3 tagged trout. From these data, they estimated the number of trout in the lake as follows.

Let n represent the number of trout in the lake. The fraction of tagged trout in the lake is $\frac{46}{n}$. The fraction of tagged trout in the sample is $\frac{3}{20}$. We assume the sample is representative of the population, and write the equation:

$$\frac{3}{20} = \frac{46}{n}$$

> Solve the equation.

$$3n = 920$$
$$n = \frac{920}{3}$$
$$\doteq 307$$

This calculation indicates there might be about 300 trout in the lake. However, other random samples would contain other numbers of tagged trout, producing different results.

Since the trout in the lake are either tagged or untagged, they form a binomial population. The trout are like the coloured beads in the sampling box. Hence, we can use boxplots to obtain further information from our sample. Use the boxplots on page 552. Since the sample contained 3 tagged trout, locate 3 on the horizontal axis. A vertical line through 3 passes through boxes from the 6% box to the 34% box. There is a 90% probability that the percent of tagged trout in the lake is between 6% and 34%.

Assume that 6% of the trout were tagged. Since 46 trout were tagged, 46 is 6% of n.

Hence, $0.06n = 46$
$$n = \frac{46}{0.06}$$
$$\doteq 767$$

Assume that 34% of the trout were tagged. Since 46 trout were tagged, 46 is 34% of n.

Hence, $0.34n = 46$
$$n = \frac{46}{0.34}$$
$$\doteq 135$$

This indicates that there is a 90% probability that the number of trout in the lake is between 135 and 767. This is a rough estimate. In the following exercises you will investigate what happens to this estimate if you use a larger sample.

1. **a)** Suppose, the next year, the biologists captured 40 trout and found 6 tagged trout. Use the boxplots on page 553. Repeat the above calculations to estimate the number of trout in the lake.

   b) Suppose, the next year, the biologists captured 100 trout and found 15 tagged trout. Use the boxplots on page 554. Repeat the above calculations to estimate the number of trout in the lake.

   c) Compare the results of parts a and b with those above. Explain why a larger sample provides a more accurate estimate of the number of trout in the lake.

2. **a)** Suppose 80% boxplots had been used. Would the results of exercise 1 be the same or different? If you think they would be the same, explain your thinking. If you think they would be different, explain how they would be different, and why.

   b) Suppose the boxplots had been 95% boxplots. What difference would that make? Why?

3. A game warden nets and tags 250 trout in a lake in northern Alberta. Two months later, she nets 40 trout and finds 8 tagged trout. Use the boxplots to estimate the number of trout in the lake.

4. One of the world's largest surviving buffalo herds lives in Wood Buffalo National Park in the Northwest Territories and Alberta. One year, biologists captured, tagged, and released 300 buffaloes. The next year, they captured 100 buffaloes and found 4 tagged buffaloes. Estimate the number of buffaloes in the park.

5. Biologists study the migratory patterns of Canada geese. In October one year, they tagged 1425 geese as they headed south. In the same month the next year, they captured 100 geese and found 38 tagged geese. Estimate the number of Canada geese that fly through that region in October.

6. In exercises 1 to 5, what assumptions are you making about the animals and their movements? Answer this question by referring to one of the exercises.

9.5 Revisiting Probability

INVESTIGATE · Displaying Coin-Toss Results

Suppose 10 coins are tossed and they show 7 heads and 3 tails. This result is in the first row of the table below. Suppose the coins are tossed again; this time they show 4 heads and 6 tails. The result is in the second row. In this row, the numbers of heads and tails are combined to give results for 20 tosses.

Total number of tosses	Total number of heads	Total number of tails	Relative frequency of heads	Heads – tails
10	7	3	$\frac{7}{10} = 0.7$	$7 - 3 = 4$
20	$7 + 4 = 11$	$3 + 6 = 9$	$\frac{11}{20} = 0.55$	$11 - 9 = 2$

1. Suppose you repeat this experiment many times. As the total number of tosses gets larger, predict what will happen to each number.

   a) the relative frequency of heads

   b) the difference between the number of heads and the number of tails

In the following exercises you will try to confirm your predictions.

2. Toss 10 coins. Record the results in a table similar to the one above. Repeat the coin toss until you have results for 100 tosses.

3. Use the data in your table.

   a) Draw a graph of Relative frequency of heads against Total number of tosses.

   b) Draw a graph of Heads – tails against Total number of tosses.

4. Compare your graphs with those of your classmates. What conclusions can you make about the two graphs?

5. Suppose a coin is tossed thousands of times. Visualize what happens to the relative frequency of heads as the number of tosses gets larger. Explain.

6. Write down the number of heads obtained in 100 tosses from as many classmates as you can.

   a) What percent of the results were exactly 50 heads?

   b) What percent of the results were from 42 to 58 heads?

A graphing calculator was programmed to simulate tossing a coin. This graph shows one set of results for up to 189 tosses. There were 102 heads and 87 tails. The relative frequency of heads is $\frac{102}{189} \doteq 0.54$, which is slightly greater than 0.5. The graph shows this since the last point plotted is just above the middle line.

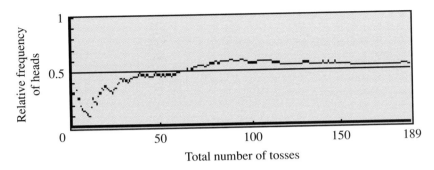

The graph shows that the relative frequency of heads is getting closer to $\frac{1}{2}$. This is predictable, because heads and tails are equally likely outcomes when you toss a coin. When we write P(heads) = $\frac{1}{2}$, we mean that the relative frequency of heads becomes closer to $\frac{1}{2}$ as the number of tosses gets larger.

Any outcome, or set of outcomes, of an experiment is an *event*. Getting heads when you toss a coin is an example of an event. Drawing a spade from a deck of cards is another example of an event. Since 13 of the 52 cards in a deck are spades, P(spade) = $\frac{13}{52}$, or $\frac{1}{4}$.

If an experiment has n equally likely outcomes of which r are favourable to event A, then the *probability* of event A is:

$$P(A) = \frac{r}{n}$$

Example 1

A card is drawn from a well-shuffled deck. Determine the probability that each card appears.

a) an ace   **b)** a red card   **c)** the 7 ♣   **d)** a face card

Solution

There are 52 cards and each has the same chance of being drawn.

a) There are 4 aces.

$P(\text{ace}) = \frac{4}{52}$

$\doteq 0.077$

b) There are 26 red cards.

$P(\text{red card}) = \frac{26}{52}$

$= 0.5$

c) There is one 7 ♣.

$$P(7 ♣) = \frac{1}{52}$$
$$\doteq 0.019$$

d) There are 12 face cards.

$$P(\text{face card}) = \frac{12}{52}$$
$$\doteq 0.231$$

In *Investigate*, you tossed a coin 100 times. When you compared your results with those of your classmates, you may have noticed that few students obtained 50 heads. However, most students probably obtained from 42 to 58 heads. The probability that this will happen is approximately 90%. You can tell this from the boxplots.

The outcomes for tossing a coin are like the beads in a sampling box with 50% marked beads. Use the boxplots for samples size 100 on page 554. Locate 50% on the vertical axis. The boxplot next to this number is the one we want. It is reproduced below with the labels on the axes changed to apply to this situation. This boxplot shows that between 42 and 58 heads should appear 90% of the time.

90% boxplots for samples size 100

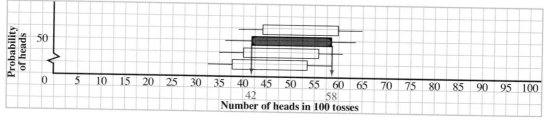

This means that there is a 90% probability that between 42 and 58 heads will appear when a coin is tossed 100 times. This is our estimate for the number of heads that should appear when a coin is tossed 100 times. Observe that it is expressed in terms of probability. Similar estimates can be made for other experiments, such as rolling a die.

Example 2

a) When a die is rolled, what is the probability of getting a 5?

b) Suppose a die is rolled 100 times. Use a boxplot to estimate the number of times a 5 should appear.

Solution

a) There are six equally likely outcomes. The probability that a 5 will appear is $P(5) = \frac{1}{6}$, or about 16.7%.

b) Use the boxplots for samples size 100. Locate 16% on the vertical axis. The boxplot next to this number is reproduced below with labels applying to this situation. It shows that between 10 and 22 fives would appear 90% of the time. Hence, there is a 90% probability that a 5 will appear from 10 to 22 times.

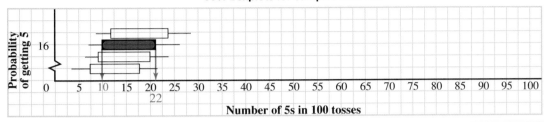

90% boxplots for samples size 100

Number of 5s in 100 tosses

DISCUSSING THE IDEAS

1. Does P(heads) = $\frac{1}{2}$ mean 5 heads will show when you toss 10 coins? Explain.

2. Are these events equally likely, or is one more likely than the other? Explain.

   Event 1: Five heads will show when 10 coins are tossed.

   Event 2: Fifty heads will show when 100 coins are tossed.

3. Suppose several heads show in a row when you toss a coin. If you were to continue to toss the coin, are several tails more likely to show in a row? Explain.

4. In *Example 2b*, the probability was rounded down to 16% because the 16% boxplot on page 554 is the closest one to 16.7%. Visualize how a 16.7% boxplot compares with the 16% boxplot. What changes, if any, do you think should be made to the answer to this example? Explain.

9.5 EXERCISES

 1. A person says that the probability of passing a test is 0.5 because there are two possible outcomes, pass or fail. Do you agree? Explain.

2. What three things do you need to know about an event to determine its probability?

3. A card is drawn from a well-shuffled deck of 52 cards. What is the probability that each card appears?

    a) a heart   **b)** black   **c)** a 5

    d) a red jack   **e)** an ace   **f)** a black 3, 6, or 9

4. One letter is selected at random from the word "mathematics." What is the probability that each letter is chosen?

   a) m

   b) e

   c) o

   d) a vowel

   e) a consonant

   f) t or h

5. A coin is tossed 40 times. Use a boxplot to estimate the number of times tails should appear.

6. A die is rolled 40 times. Use a boxplot to estimate the number of times a 6 should appear.

B 7. Bags A, B, and C contain green and red counters in the numbers shown. From which bag would you have the best chance of selecting a green counter in one draw?

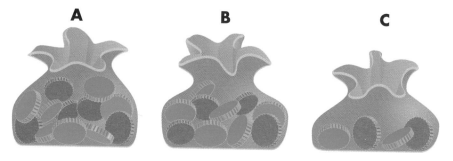

8. Slips of paper, numbered from 1 to 30, are placed in a bowl. One slip of paper is drawn at random. What is the probability that it has a number containing at least one 2?

9. **a)** When a die is rolled, what is the probability of getting a 1 or a 6?

   b) Suppose a die is rolled 100 times. Use a boxplot to estimate the number of times a 1 or a 6 should appear.

10. **a)** When two coins are tossed, what is the probability that both coins are heads?

    b) Suppose two coins are tossed 40 times. Use a boxplot to estimate the number of times two heads should appear.

11. This table shows the distribution of blood types among Canadians.

Blood type	O	A	B	AB
Percent of Canadians	45%	40%	11%	4%

   a) Determine the probability that a person selected at random will have type A blood.

   b) Use a boxplot to estimate the number of people in a sample of 100 Canadians who have each type of blood.

12. A card is drawn from a deck of cards, replaced, and the deck shuffled. This is done 100 times. Use a boxplot to estimate the number of times each card appears.

a) a black card   **b)** a heart   **c)** an ace   **d)** a face card

13. This graph shows the difference between the numbers of heads and tails for the same 189 coin tosses that were used to draw the graph on page 563.

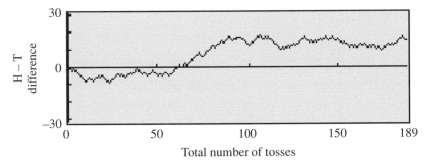

a) Compare the two graphs. In what ways are they similar? In what ways are they different? Account for the similarities and differences.

b) What happened from about the 60th to the 90th toss? How can you tell?

c) What happened from about the 90th toss to the last toss? How can you tell?

d) Suppose that after the 189th toss, there is a very long run of alternating heads and tails. How would this affect the two graphs?

MM MODELLING Coin Tosses

Three different models for coin tossing were presented in this section.

- Identify the three models.
- What are the major differences among the models?

C 14. A thumbtack is tossed 100 times. It lands point up 28 times. Use a boxplot to estimate the probability that the thumbtack lands point up.

COMMUNICATING THE IDEAS

Suppose your friend does not understand why boxplots can be used to solve probability problems. How would you explain why you can use a boxplot to estimate the number of times an event should occur when an experiment is repeated 100 times? Use an example to illustrate your explanation.

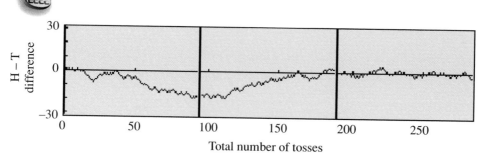

Will You Flip with Me?

The program called FLIPGAME is a game that simulates tossing a coin hundreds of times. To play the game, you choose heads or tails. Let's assume that you choose heads. If the coin lands heads up, you score a point. If it lands tails up, the calculator scores a point.

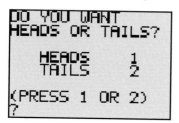

Ask your teacher for this program from the Teacher's Resource Book, and enter it in your calculator. After you respond to the initial prompts, the calculator graphs the results of the first 94 tosses. Press ENTER to continue. Each time you press ENTER, the calculator graphs the results of the next 95 tosses. After all tosses have been completed, a summary screen appears. The screens at the top of these two pages show an example of playing FLIPGAME with 650 tosses. This screen shows the result.

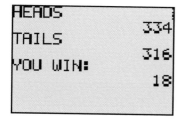

1. Play the game a few times, using different numbers of tosses.

2. Play the game several times, using the same number of tosses each time, and always choosing heads. Record the winning score for each game. Is there any pattern in the results?

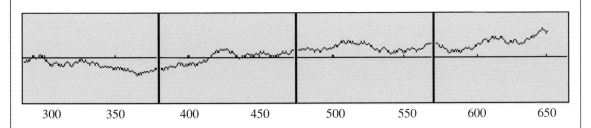

| 300 | 350 | 400 | 450 | 500 | 550 | 600 | 650 |

3. Repeat exercise 2 for a different number of tosses. Does the number of tosses appear to affect the results?

4. a) Is it possible for the graph produced by FLIPGAME to coincide with the horizontal line in the middle of the screen?

   b) When the graph coincides with the horizontal line in the middle, could it ever depart from it? Explain.

5. Closely examine the graph above. During the 650 tosses, how many times did the lead change? That is, how many times did the score change from being in your favour to being in the calculator's favour?

6. What is the maximum number of lead changes that could possibly occur in 650 tosses?

7. You probably found that the scores in FLIPGAME were relatively low. Ask your teacher for the program called FLIPMORE from the Teacher's Resource Book. This is a similar game that results in higher scores. Let's assume you choose heads. In this game, you score a point on each turn if the total number of heads that have appeared so far is greater than the total number of tails. Play this game a few times, using different numbers of tosses.

8. Repeat exercises 2 and 3 using FLIPMORE.

9. When you toss a coin many times, which of the following events is more likely, or are they equally likely? Explain your answer, based on your experience with the two games.

   Event 1: Either heads or tails is ahead most or all of the time.

   Event 2: Heads is ahead 50% of the time and tails is ahead 50% of the time.

Coincident Birthdays

Do you think it is likely that two classmates have the same birthday? How likely do you think this is?

To help you answer these questions, we will calculate the probability that everyone in the class has a different birthday.

Since there are 365 days in a year, the probability that another student has a birthday different from yours is $\frac{364}{365} \doteq 0.997\ 26$.

The probability that a third student has a birthday different from the first two is $0.997\ 26 \times \frac{363}{365} \doteq 0.991\ 80$.

The probability that a fourth student has a birthday different from the first three is $0.991\ 80 \times \frac{362}{365} \doteq 0.983\ 64$.

You should see a pattern developing. You can use this pattern to calculate more efficiently.

1. **a)** Record the calculations in a table similar to the one below.

   b) Explain why the numbers in each column are calculated the way they are.

   c) The numbers in the second column are decreasing and the numbers in the third column are increasing. Why do you think this is?

   d) Continue the calculations for a few more rows.

Number of students	Probability that all the birthdays are different	Probability that the birthdays are not all different
2	$\frac{364}{365} \doteq 0.997\ 26$	$1 - 0.997\ 26 = 0.002\ 74$
3	$0.997\ 26 \times \frac{363}{365} \doteq 0.991\ 80$	$1 - 0.991\ 80 = 0.008\ 20$
4	$0.991\ 80 \times \frac{362}{365} \doteq 0.983\ 64$	$1 - 0.983\ 64 = 0.016\ 36$
5	$0.983\ 64 \times \frac{361}{365} \doteq 0.972\ 86$	$1 - 0.972\ 86 = 0.027\ 14$

2. Are the numbers in the columns of the table in arithmetic sequence, in geometric sequence, or neither? Explain.

To determine the probability that two classmates have the same birthday, you will need to continue the calculations down to the row that corresponds to the number of students in your class. Answer any of exercises 3, 4, or 5 to do this. Then continue with exercise 6.

3. Use your calculator to continue the table down to the row that corresponds to the number of students in your class.

4. Ask your teacher for the program called BIRTHDAY from the Teacher's Resource Book. Enter the program in your calculator. When you run the program, the calculator will ask for the number of students in your class. Then it will complete a table similar to the one in exercise 1. You can view this table by pressing [STAT] [ENTER]. Scroll down the table to see the results for your class. To get back to the usual screens, press [2nd] [MODE].

5. a) Start a new spreadsheet document. Enter the text and formulas shown.

	A	B	C
1	Coincident Birthdays		
2			
3	Number	Probability that	Probability that
4	of	all the birthdays	they are not
5	students	are different	different
6	2	=364/365	=1-B6
7	=A6+1	=B6*(366-A7)/365	=1-B7

b) Copy the formulas in row 7 down far enough to get to the row corresponding to the number of students in your class.

6. Use your results from any of exercises 3, 4, or 5.

a) What is the probability that at least two classmates have the same birthday?

b) Suppose the probability that at least two students have the same birthday is approximately each number below. How many students would have to be in a class?

    i) 0.5   **ii)** 0.75   **iii)** 0.9

c) Draw a graph to show how the probability that at least two classmates have the same birthday depends on the number of students.

d) Explain why your graph represents a function. What are the domain and the range of this function?

7. Suppose all your classmates write a whole number between 1 and 100.

a) Estimate the probability that at least two students will pick the same number. Justify your estimate.

b) Calculate the probability in part a. Use any of these methods.

    i) using a calculator, as in exercise 3

    ii) by modifying the graphing calculator program in exercise 4

    iii) by modifying the spreadsheet in exercise 5.

9.6 Expectation

Suppose I charge you to play this game with me.

I'll toss two coins.

If they both show heads, I'll pay you $5.00.

If only one shows heads, I'll pay you $1.00.

If neither coin shows heads, I'll pay you nothing.

How much should you be willing to pay me to play this game?

For each event, we list the payoff and the probability.

Event	Payoff ($)	Probability
Both heads (HH)	5.00	$\frac{1}{4}$
Only one head (HT, TH)	1.00	$\frac{1}{2}$
No heads (TT)	0.00	$\frac{1}{4}$

Visualize what will probably happen if you play this game many times.

You should expect to win: $5.00 about $\frac{1}{4}$ of the time

$1.00 about $\frac{1}{2}$ of the time

nothing about $\frac{1}{4}$ of the time

We write: $(\$5.00 \times \frac{1}{4}) + (\$1.00 \times \frac{1}{2}) + (0 \times \frac{1}{4}) = \$1.25 + \$0.50$
$$= \$1.75$$

This means that if you play this game many times, you should expect to win an average of $1.75 per game. This is called your expectation.

Suppose you play many games.

If you pay more than $1.75 to play each game, you will probably lose some money.

If you pay less than $1.75, you will probably win some money.

If you pay $1.75, you will probably break even. The game would be fair to both of us.

You should not be willing to pay more than $1.75 to play this game.

Observe how the expectation is calculated in the above example.

> If there is a payoff for each event in an experiment, the *expectation* is the average amount you would expect to win if you perform the experiment many times. It is calculated as follows:
>
> Multiply the payoff for each event by the probability that it will occur.
> Add the results.

Example 1

In a game of chance, you pay to roll a fair die. You win the number of dollars on the top face of the die.

a) List the outcomes, their probabilities, and the payoff for each outcome.

b) What is the expectation for this game?

c) How much should you be willing to pay to play this game?

Solution

a)

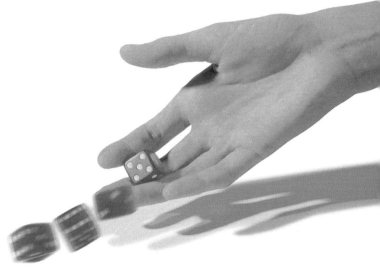

Outcome	Payoff ($)	Probability
1	1	$\frac{1}{6}$
2	2	$\frac{1}{6}$
3	3	$\frac{1}{6}$
4	4	$\frac{1}{6}$
5	5	$\frac{1}{6}$
6	6	$\frac{1}{6}$

b) The expectation for this game is:

$$\$1 \times \frac{1}{6} + \$2 \times \frac{1}{6} + \$3 \times \frac{1}{6} + \$4 \times \frac{1}{6} + \$5 \times \frac{1}{6} + \$6 \times \frac{1}{6} = \$3.50$$

c) You should not be willing to pay more than $3.50 to play this game. If you pay more than this amount, you will probably lose money in the long run.

Example 2

In a Super Scratch & Win contest, the prizes are as shown.

Prize	Value ($)	Probability of winning
Car	32 000.00	1 in 250 000
$5000 cash	5 000.00	1 in 80 000
VCR	350.00	1 in 4000
Muffin	1.20	1 in 10

What is the expectation for this contest?

Solution

The expectation for this contest is:

$$\$32\ 000 \times \frac{1}{250\ 000} + \$5000 \times \frac{1}{80\ 000} + \$350 \times \frac{1}{4000} + \$1.20 \times \frac{1}{10}$$
$$= \$0.128 + \$0.0625 + \$0.0875 + \$0.120$$
$$= \$0.398$$

The expectation is approximately 40¢.

Example 3

In a $500 investment, you have a 50% chance of gaining $300 and a 40% chance of losing your entire investment. There is a 10% chance of no gain or loss.

a) What is the expectation for this investment?

b) Explain the meaning of the answer in part a.

Solution

a) The expectation for this investment is:

$$\$300 \times 0.5 + (-\$500 \times 0.4) + \$0 \times 0.1 = \$150 - \$200$$
$$= -\$50$$

b) If many people participate in the investment, some people will gain $300, some will lose $500, and others will gain or lose nothing. The expectation −$50 means that an average loss of $50 per investor can be expected.

DISCUSSING THE IDEAS

1. In *Example 3a*, explain why −$500 was used in the calculation.

2. In *Example 3*, could the answer in part a mean that your expected loss is $50? Explain.

3. Expectation is sometimes called *expected value*. Which term do you think is more appropriate? Explain. Use one of the examples above to illustrate your answer.

A 1. A student writes an exam that has 25 true/false questions. The student knows 15 answers and guesses the answers to the remaining questions.

a) What is the expected number of correct answers?

b) Explain the meaning of the answer in part a.

2. A student writes an exam that has 100 multiple-choice questions. Each item has 5 possible answers. The student knows 57 answers and guesses the answers to the remaining questions.

a) What is the expected number of correct answers?

b) Explain the meaning of the answer in part a.

B 3. In a New Year's lottery, you get one free spin of a wheel to attempt to win a prize. The prizes and their probabilities are as follows:

Prize ($)	88	8	1
Probability	$\frac{1}{40}$	$\frac{1}{20}$	$\frac{1}{10}$

What is your expectation?

4. The local bakery distributes a Scratch & Win card with these prizes.

Prize	Value ($)	Probability of winning
Cake	10	1 in 100
Whole wheat loaf	3	1 in 25
Kaiser buns	2	1 in 20
25¢ off coupon	0.25	9 in 10

What is your expectation?

5. The Tandem Bike Rentals store relies on fine weather for good business. On sunny days in the summer, it makes an average of $1200. On cloudy days, the store makes an average of $800, and on rainy days it averages $400. The probabilities of sun, cloud, and rain are $\frac{1}{2}$, $\frac{1}{3}$, and $\frac{1}{6}$, respectively. What is the store's expectation?

6. Two students are playing "Double or Quits" with two coins. When the coins show two heads or two tails, student A wins 1 point. When the coins show one head and one tail, student B wins 1 point.

a) After 100 tosses, what is each player's expected score?

b) Is this a fair game? Explain.

7. In a game of chance, you pay to toss three coins. The number of heads that appear is the amount you win, in dollars. Calculate the expectation for this game.

8. Two cards are drawn from a deck of cards. If the cards are the same suit, you win $3. Otherwise, you lose $2. Calculate the expectation for this game.

9. Two dice are rolled. If a sum of 7 appears, you win $3. If a sum of 6 appears, you lose $4. Calculate the expectation for this game.

10. A radio station features a "weather prediction guarantee." People must register by mail to participate. Each day the station announces the predicted high temperature for the next day. If the actual high temperature differs from the prediction by more than 3°C, the station will pay $1000 to a randomly drawn participant.

   a) The station predicts that the probability it will have to pay the $1000 on any given day is 2%. What is its expectation for each day?

   b) Assume that 6400 people have registered on a particular day. What is the expectation that day for each person who is registered?

MODELLING a Radio Station's Weather Guarantee

In exercise 10, the radio station must have analyzed its chances of being wrong so it could predict the amount of money it might have to pay. This amount is modelled by the expectation calculation in part a.

- Is it possible that the radio station might have to pay more than it expected over a period of a few months? What steps could it take to prevent this happening?

- How do you think the station would estimate the probability that its prediction is wrong by 3°C?

11. Two games are described below. Decide if each game is fair to two players.

   a) I'll toss two coins.

      If they both show heads, I'll pay you $5.

      If only one shows heads, I'll pay you $1.

      If neither coin shows heads, you'll pay me $7.

   b) I'll toss three coins.

      If they all show heads, I'll pay you $3.

      If only two coins show heads, I'll pay you $2.

      If they all show tails, you'll pay me $6.

12. A motorcycle racer finds that there are two conflicting meets. She will only be able to participate in one of them.

If she enters the local event, it will cost her $100 for registration and travel costs. She estimates the probability of winning: 30% for 1st prize of $1000; 30% for 2nd prize of $500; and 40% for no prize money.

If she enters the provincial event, the registration and travel costs are $400. She estimates the probability of winning: 10% for 1st prize of $3000; 10% for 2nd prize of $1000; and 80% for no prize money.

In both events she has a 10% chance of serious injuries, which will cost her an additional $2000.

In which meet should she participate?

C **13.** At a carnival, one booth has a board 2 m by 1 m on which are mounted many circular tags:

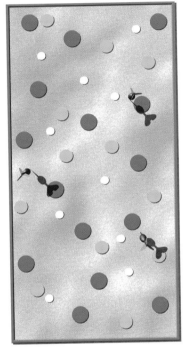

15 blue tags, each with a diameter of 10 cm

12 yellow tags, each with a diameter of 8 cm

10 white tags, each with a diameter of 5 cm

Contestants pay $1.00 to throw three darts at the board. These are the payoffs.

blue tag: $1.00

yellow tag: $2.00

white tag: $10.00

Make any assumptions that seem reasonable to help you answer these questions.

a) What is the probability that a dart hits a blue tag? A yellow tag? A white tag?

b) What is the contestant's expectation?

c) Would you be willing to play this game? Justify your answer.

COMMUNICATING THE IDEAS

Look up the word *expectation* in the dictionary. Examine the meanings carefully. Is the mathematical meaning included? If not, choose one meaning and describe what elements of the mathematical meaning are not present in the dictionary definition. In your journal, write a brief description of the mathematical meaning of expectation that should be included in the next edition of the dictionary.

9.7 Applications to Insurance

Insurance is an arrangement to pay money in case of a loss of property, an accident, sickness, or death. Insurance companies estimate the probabilities that situations such as these will occur. They use these probabilities to calculate the premiums for the insurance.

For example, you may be familiar with the insurance that pays the medical costs if a student is injured at school. Students who purchase this insurance pay a certain premium. The money the insurance company receives from the premiums is shared among the students who are injured.

To determine the premium for this insurance, the company analyzes the claims from previous years. For example, suppose it knows that in recent years the probability that a student makes a claim is 0.8%. The average amount paid out in claims is $1230.00. Based on this information, the expected loss per student is:

$$0.8\% \text{ of } \$1230.00 = 0.008 \times \$1230.00$$
$$= \$9.84$$

To take into account overhead costs and a reasonable profit, the company might charge a premium of $15 for this insurance.

Example

A car insurance company insures 250 000 drivers, and they all have liability coverage. The probability that a driver receives a claim in a particular year is 0.3%. The average liability claim payment is $100 000.

a) How many clients does the company expect will receive claim payments in one year?

b) How much does the company expect to pay out in claims?

c) What is the mean cost per driver?

d) What would be a fair premium for this insurance?

Solution

a) The number of claims is 0.3% of 250 000 = $0.003 \times 250\ 000$
$$= 750$$

The company expects that about 750 drivers will receive claim payments in one year.

b) The amount paid out in claims is $750 \times \$100\ 000 = \$75\ 000\ 000$

The company expects to pay approximately $75 000 000.

c) The mean cost per driver is $\frac{\$75\,000\,000}{250\,000} = \$300$

The mean cost per driver is $300.

d) The premium would have to be more than $300. A fair premium cannot be exactly determined because the company has to cover overhead costs and make a reasonable profit. There is insufficient information to estimate these.

DISCUSSING THE IDEAS

In part d of the *Example*, do you think a premium of $400 is fair? Explain.

9.7 EXERCISES

A 1. Rick was involved in a minor car accident that was his fault. It cost $850 to repair his car. His insurance policy has a $250 deductible clause. This means that Rick paid $250 and the insurance company paid $600 ($850 − $250).

   a) Why do you think insurance companies have deductible clauses?

   b) Before the accident, Rick was paying $199 a year for collision insurance. The next year, the insurance company increased this to $499. Do you think this is fair? Explain your answer.

   c) Suppose Rick chose a higher deductible, such as $500. How would that affect his premium? Explain.

2. When a teacher retires, she and her husband have to decide whether to continue their dental insurance. If they continue the insurance, the monthly premium will be $91.45, and their dental bills will be paid by the insurance company. If they do not continue the insurance, they will have to pay their own dental bills. What do you think they should do? Explain.

3. Maria bought a new VCR for $525. There was a one-year warranty on parts and labour. At the time of purchase, Maria had to decide whether to purchase an extended warranty. A one-year extended warranty costs $49.95. If she purchased the extended warranty she had to pay almost $575, but she would have a two-year warranty. If she did not purchase the warranty, she paid $525, and must pay her own repair bills after one year. What do you think she should do? Explain.

B **4.** Refer to exercise 3. The insurance company that offers the extended warranty estimates that the probability that a VCR will have to be repaired in the second year is about 5%, and that the average cost of a repair is $125. They also estimate that 8000 people will purchase the extended warranty.

   a) How many VCRs does the insurance company expect to repair in the second year?

   b) How much does the insurance company expect to pay to repair these VCRs?

   c) What is the mean cost per VCR?

   d) Do you think the $49.95 annual charge for the extended warranty is fair? Explain.

5. An insurance company insures 500 000 homes against loss from fire and theft. The maximum amount payable is $250 000. The probability that a homeowner makes a claim in a particular year is 0.6%. The average amount paid out in claims is approximately $15 000.

   a) How many claims does the company expect to pay out in one year?

   b) How much does the company expect to pay out in claims?

   c) What is the mean cost per home?

   d) What would be a fair premium for this insurance? Support your answer with some calculations and an explanation.

6. Home insurance companies offer extensions to their policies, called *riders*, for unusual or specific items such as expensive jewellery or computer equipment. For example, the insurance company in exercise 5 might charge an additional premium of $75 a year to insure a diamond ring for $5000. Compared to the cost of the insurance in exercise 5, do you think the cost of this rider is excessive? Explain.

7. Which of the following situations do you think are reasonable? Explain your answers.

   a) A homeowner pays $300 each year to insure a house for $200 000 against fire damage. The probability of making a claim is estimated to be 0.3%. Most claims are well below the policy limit.

   b) A professional hockey player's annual salary is $2 000 000. This must be paid even if he is injured and cannot play. The team estimates that the probability this will happen is 20%. The team pays an insurance company $250 000 per year to insure him against injury.

c) A person flies from Edmonton to Paris. She pays $5 for a $10 000 000 flight insurance to be paid to her dependents if she dies in a crash. The probability that this will occur is approximately 0.0002%.

d) A bank makes a loan of $100 000 at 8%. The bank estimates that the probability the loan will be repaid is 90%.

8. Mei-Lin has an older car worth about $2000, and has just received her insurance bill. The cost of the collision coverage, with a $250 deductible, is $99. She is considering dropping this coverage because it will only cover the cost of repairs up to $1750.

a) Mei-Lin feels that she is a good driver. She estimates the probability of a collision claim is about 0.5%. Do you think she should drop the collision coverage? Support your answer with some calculations.

b) What is another reason why Mei-Lin should drop the collision coverage?

9. Refer to the example on page 578. Suppose all the 250 000 drivers have $250 deductible collision insurance. The average amount the company pays out in collision claims is $4200. The probability that a driver has a claim in a particular year is 0.3%.

a) How many collision claims does the company expect to pay out in one year?

b) How much does the company expect to pay for these claims?

c) What is the mean cost per driver?

d) What would be a fair premium for this insurance?

C 10. In exercise 9, suppose the deductible for collision insurance is increased to $1000. What would be a fair premium for this insurance? Support your answer with some calculations.

COMMUNICATING THE IDEAS

Make a list of all the different types of insurance that are involved in the examples and exercises in this section. What do they have in common? How are they related to probability and expectation? Write your ideas in your journal.

Should We Harvest Today or Wait?

Page 528 presented a problem about growing grapes for icewine. As each day passes, the quality of the grapes increases, but there is a greater risk of a frost killing the grapes and reducing their quality. On which day does the risk of frost damage outweigh the gain from extra maturing time?

DEVELOP A MODEL

To solve this problem, we make assumptions about the value of the grape juice and the risk of frost. We will assume that in a particular year:

• The value of the grape juice is $2.00/L on October 1.

• The value of the grape juice increases by 15¢/L each day after October 1.

• The probability of a killer frost is 9% for any particular day.

• After a killer frost, the value of the juice is only $1.50/L.

1. a) Make a table similar to the one below. Complete the first two columns for several days.

Date	Value of juice (¢/L)	Expected gain that day (¢/L)	Expected loss that day (¢/L)	Net expected gain (¢/L)
Oct. 1	200			
Oct. 2	215			

b) Suppose there is no killer frost on October 1.

   i) What is the probability of no killer frost?

   ii) How much would you gain in value that day if there is no killer frost?

   iii) What is your expected gain in value? Enter your answer in column 3.

c) Suppose there is a killer frost on October 1.

   i) What is the probability of a killer frost?

   ii) How much would you lose in value that day if there is a killer frost?

   iii) What is your expected loss in value? Enter your answer in column 4.

   iv) Subtract the expected loss from the expected gain. Enter the result in column 5.

2. a) Repeat parts b and c of exercise 1 for several more days until the expected loss becomes greater than the expected gain.

   b) When does the risk of frost damage outweigh the gain from extra maturing time?

 LOOK AT THE IMPLICATIONS

3. a) Look at the numbers in the 4th column. What kind of sequence do these numbers form? Write an expression for the general term of this sequence.

   b) Let n represent the number of a day in October, that is, October n. Write expressions to represent the expected gain and the expected loss that day.

   c) Use your expressions to write an equation.

   d) Solve the equation for n. Does the result agree with your answer to exercise 2b?

4. a) Graph the net expected gain as a function of the number of days that have elapsed since October 1.

   b) Your graph should be a straight line. What does the slope of the line represent? What do the vertical and horizontal intercepts represent?

   c) Use the graph to determine an equation of the line.

 REVISIT THE SITUATION

5. For each situation, what changes would occur to your answer to exercise 2b and to the graph in exercise 4?

   a) The probability of a killer frost is 3% instead of 9%.

   b) The value of the juice increases by 25¢/L each day instead of 15¢/L.

   c) After a killer frost, the value of the juice is only 50¢/L instead of $1.50/L.

Our model assumes that the probability of frost is constant during the month. An improved model would provide for some variation in this probability from day to day.

1. Each country that participated in the Youth Health survey in exercise 13, page 534, designed its own questionnaire. To permit comparison, all countries included a set of core questions. However, for many questions about physical ailments and mental health, the Austrians provided different answer choices than the other countries. In the Austrian survey: "Often" was replaced by "Almost every day" and "Weekly"; "Sometimes" was "Nearly every week"; "Seldom" was "Once a month"; and "Never" was "Almost never."

   a) How do you think these different answer choices might have affected the results? Explain your ideas.

   b) Look through the records in the *Youth Health* database that relate to headaches, depression, bad temper, and difficulty getting to sleep. Do the results seem to reflect your ideas from part a?

2. Pinehurst School has 945 students. In a random sample of 20 students, 5 students have part-time jobs. In another random sample of 20 students, 3 students have part-time jobs. Use this information and the boxplots on page 552. Make an inference about the number of Pinehurst students who have part-time jobs.

3. The Maclean's magazine annual poll on page 555 includes results for individual provinces and other groups. The following statement occurred in an article about the poll.

   Accuracy ranges are wider for results from individual provinces and subgroups.

   a) Explain what the statement means.

   b) Why would accuracy ranges be wider for results from individual provinces and subgroups?

4. Two people conducted identical surveys at Centennial School to estimate the number of students who smoke. Although one person used a larger sample than the other, both samples showed the same percent of students who smoke. How will their inferences about the number of students who smoke differ?

5. A bag contains 30 marbles. Six are red, 9 are yellow, and 15 are blue. One marble is removed from the bag, at random. What is the probability of each outcome?

   a) red   b) yellow   c) blue

   d) not yellow   e) yellow or blue   f) green

6. There are two prizes in a raffle. The probability of winning the first prize of $10 000 is 0.001. The probability of winning the second prize of $1000 is 0.01. What should be a fair price for a ticket?

1. Write a formula for the general term of each arithmetic sequence.

   a) 2, 4, 6, …   **b)** 2, 6, 10, …   **c)** 10, 20, 30, …

   d) 3, 12, 21, …   **e)** 6, 5, 4, …   **f)** −23, −21, −19, …

2. Copy and complete each geometric sequence.

   a) 5, 7, ▬, ▬, ▬   **b)** 4, 6, ▬, ▬, ▬   **c)** 4, ▬, 6, ▬, ▬

   d) 7, ▬, ▬, ▬, 567   **e)** ▬, ▬, ▬, 384, 1536 **f)** ▬, ▬, ▬, 216, 648

3. **a)** One thousand dollars are invested at 12% interest, compounded annually. What is the value of the investment after 5 years?

   b) Repeat part a for an initial investment of $2000.00. How does the value after 5 years compare with that in part a?

   c) Repeat part a for an initial investment of $10 000.00. How does the value after 5 years compare with those in parts a and b? Write to explain what you notice.

   d) Suppose, for part a, the interest rate were doubled to 24%. How do you think the value of the investment after 5 years will compare with the value of the investment at 12%, after 5 years? Check your answer by calculating.

4. **a)** Start with any number. Add 2. Take the square root of the result. Add 2. Take the square root. Repeat this step until you know what the result will be if you carried on indefinitely.

   b) Repeat part a, but add 6 instead of 2.

   c) Repeat part a, but add 12 instead of 2.

   d) Repeat part a, but add 20 instead of 2.

   e) Look the results of parts a to d. Write to explain what pattern you see.

5. The equations of six lines are given. Find as many pairs of lines as you can that are parallel.

   a) $x - 2y - 10 = 0$ **b)** $4x - 3y - 12 = 0$ **c)** $6x + 8y - 24 = 0$

   d) $2x + y - 4 = 0$ **e)** $2x - 4y - 12 = 0$ **f)** $x - 2y - 5 = 0$

6. Refer to the equations in exercise 5. Find as many pairs of lines as you can that are perpendicular.

7. Wind chill factor was developed by scientists during World War II as a measure to indicate how cold it was. With a wind, the temperature was often not an accurate indicator. The wind chill considers the temperature as well as the wind speed. In Canada, the wind chill is often reported in watts per

square metre. To make it easier for people to understand, this measure is often translated into an equivalent temperature. Here is a chart of equivalent temperatures for several wind chill factors.

Wind chill factor (W/m²)	Equivalent temperature (°C)
1200	−11
1400	−18
1600	−25
2300	−50
2700	−66

a) Plot these data with *Wind chill factor* on the horizontal axis. Draw a line of best fit.

b) Write an equation to represent your line of best fit.

c) Use your equation. Determine the equivalent temperature for a wind chill of 2000 W/m².

d) The coldest wind chill on record occurred at Pelly Bay, NWT. The equivalent temperature was −92°C. Use your equation. Determine the wind chill factor.

8. An archaeologist can estimate a person's height from skeletal remains. The height is a linear function of the length of the humerus, the bone in the upper arm. The humerus of a 160-cm adult is about 30 cm long. The humerus of a 190-cm adult is about 40 cm long.

a) Write an equation that expresses the height h centimetres as a linear function of the humerus length l centimetres.

b) What is the approximate height of an adult whose humerus measures 38.2 cm?

9. Each function described below has an equation of the form $y = mx + b$. For each function, determine the values of m and b, then sketch the graph.

a) a line with slope $\frac{1}{2}$ and y-intercept 3

b) a line that passes through A(−1, 2) and has slope 2

c) a line that passes through B(2, 3) and has y-intercept 3

d) a line that passes through C(1, 4) and has x-intercept 2

e) a line that passes through D(2, −1) and E(4, 5)

10. A ball of putty, sphere A, has radius 4 cm. It is cut into two pieces of equal volume. These pieces are shaped into spheres B and C.

a) Without calculating, describe how the total surface area of B and C is related to the surface area of A.

b) Calculate the volumes of A, B, and C.

c) Calculate the radius of B and the radius of C.

d) Calculate the surface areas of A, B, and C.

e) Calculate the total surface area of B and C. Compare this area with the surface area of A. Do your calculations agree with your statement in part a?

11. Simplify.

a) $\dfrac{3x^2y^4}{72x^5y^3}$

b) $(15x^3y^2z)\left(\dfrac{1}{3}x^2y^3z^2\right)$

c) $(4x^2y)^2(3xy^3)^3$

d) $\dfrac{(3a^2b)^2(4abc^2)^3}{(18ab^5)(32a^6b^2c^4)}$

e) $\dfrac{(2ab)(3a^2b)^2}{(5a^3b)^2(4ab^2)^3}$

f) $\dfrac{(6a^2b^2)^3(3a^2bc^3)^2}{(ab^2)(4abc)^2}$

12. Factor.

a) $x^2 + 5x + 6$

b) $x^2 + x - 6$

c) $3x^2 + 15x + 12$

d) $5x^2 + 5x - 100$

e) $6x^2 - 9xy - 6y^2$

f) $24a^2 + 52ab + 24b^2$

g) $x^2 - 25$

h) $4x^2 - 100$

i) $49x^2 - 64y^2$

13. Simplify each expression. Identify the nonpermissible values of x.

a) $\dfrac{6x^2 + 9x}{3x}$

b) $\dfrac{3x^2 + 15x}{2x + 10}$

c) $\dfrac{4x^2 - 10x}{5 - 2x}$

d) $\dfrac{2x^2 + 10x + 12}{x + 3}$

e) $\dfrac{x^2 + x - 2}{4x^2 + 8x}$

f) $\dfrac{x^2 - 4}{x - 2}$

14. Add or subtract the rational expressions.

a) $\dfrac{1}{2} + \dfrac{1}{3} - \dfrac{1}{5}$

b) $\dfrac{2x}{3} - \dfrac{3x}{5}$

c) $\dfrac{2y}{3x} - \dfrac{y}{6x} + \dfrac{4y}{9x} - \dfrac{y}{9x}$

d) $\dfrac{x+1}{2x} + \dfrac{3x-2}{3x}$

e) $\dfrac{2x-3y}{6y} + \dfrac{x-y}{2x}$

f) $\dfrac{x}{2x+1} + \dfrac{1}{2(2x+1)}$

15. Two sides of a triangle have lengths 28.2 cm and 18.8 cm. The angle between them is 41.5°. Determine the lengths of the sides of an isosceles right triangle with the same area.

16. In a molecule of water, the two hydrogen atoms and one oxygen atom are bonded in the shape of a triangle. The nuclei of the atoms are separated by the distances shown. Calculate the bond angles. (Note that 1 Å = 10^{-8} cm)

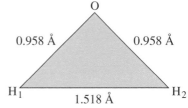

17. In one school, two different random samples of 20 students were asked if they believe some of the stories about flying saucers. These are the results:

Sample A: 4 students said yes.   Sample B: 12 students said yes.

a) Use only Sample A. Make an inference about the percent of students in the school who believe in flying saucers.

b) Repeat part a, using Sample B.

c) What problem is presented by the answers to parts a and b?

d) What could you do to solve this problem?

Answers

Chapter 1

Investigate, page 4

1. 2012, 2024, 2036   **2.** 2009, 2021, 2033

3. a) 2002, 2014, 2026   **b)** 2005, 2017, 2029
c) 2008, 2020, 2032

4. Add 12 repeatedly.

1.1 Exercises, page 7

1. a) 3; 21, 24, 27   **b)** −4; 13, 9, 5
c) 10; 57, 67, 77   **d)** −10; 1, −9, −19
e) 3; 1, 4, 7   **f)** 0.5; 3.3, 3.8, 4.3

2. a) 17, 20, 23, 26, 29   **b)** Row 1
c) Answers may vary.

3. $243

4. a) Answers may vary. 39 million years ago, 13 million years ago
b) In 13 million years time

5. a) i) 36   **ii)** −3   **iii)** 122
iv) −19   **v)** 60   **vi)** 44
b) Explanations may vary.

6. a) 112 000 ounces; 134 000 ounces
c) Answers may vary. The straight line would have a steeper slope.

7. $540 billion, $570 billion, $600 billion

8. a) 1896, 1900, 1904, 1908, 1912, 1916
b) 1916, because it coincided with World War I.
c) 1896, 1900, 1904, 1906, 1908, 1912, 1920

9. a) i) 2, 7, 12, 17, 22   **ii)** 37, 33, 29, 25, 21
iii) 5, 13, 21, 29, 37   **iv)** 50, 45, 40, 35, 30
v) −18, −8, 2, 12, 22, 32   **vi)** 43, 51.5, 60, 68.5, 77
b) Explanations may vary.

10. a) $\frac{46}{3}$, $\frac{68}{3}$ **b)** 21.25, 32.5, 43.75

11. a) 11, 14, 17, 20, 23   **b)** 25, 30, 35, 40, 45
c) 12, 10, 8, 6, 4

12. Answers may vary.

13. a) Explanations may vary.
b)

Sun	Mon	Tues	Wed	Thurs	Fri	Sat
1984		1985	1986	1987	1988	
1989	1990	1991	1992		1993	1994
1995	1996		1997	1998	1999	2000
	2001	2002	2003	2004		2005
2006	2007	2008		2009	2010	2011

c) i) Friday   **ii)** Wednesday
iii) Sunday   **iv)** Friday

d) Instructions may vary.

1.2 Exercises, page 13

1. a) 2, 5, 8, 11, 14   **b)** 7, 11, 15, 19, 23
c) −1, −4, −7, −10, −13   **d)** 12, 8, 4, 0, −4
e) −8, −3, 2, 7, 12   **f)** 5, −3, −11, −19, −27

2. a) $2n + 1$ **b)** 17   **c)** 51

3. a) $14 − 3n$ **b)** −4   **c)** −46

4. a) i) $3n + 2$ **ii)** $20 − 3n$ **iii)** $2n + 3$ **iv)** $12 − 2n$
b) Explanations may vary.

5. a) $t_n = 4n − 3$; 65   **b)** $t_n = 3n$; 63
c) $t_n = 5n − 9$; 56   **d)** $t_n = 47 − 6n$; −61
e) $t_n = 1 − 3n$; −29   **f)** $t_n = 17 − 8n$; −351

6. a) 9th   **b)** Assumptions may vary. 11 rows

7. 15th

8. a) 3300 ft, 4750 ft, 6200 ft, 7650 ft, 9100 ft
b) $t_n = 1850 + 1450n$ **c)** 7 min
d) Exercises may vary.

9. a) i) 13th   **ii)** 25th   **iii)** 21st
iv) 10th   **v)** 15th   **vi)** 25th
b) Explanations may vary.

10. a) 19, 22, 25, 28   **b)** $3n + 16$ **c)** 27

11. a) 1924, 1928, 1932, 1936, 1940, 1944
b) 1940, 1944; they coincided with World War II
c) 16   **d)** Answers may vary.

12. a)

2	4	6	8	10	12	14
7	9	11	13	15	17	19
12	14	16	18	20	22	24
17	19	21	23	25	27	29
22	24	26	28	30	32	34
27	29	31	33	35	37	39

b), c), d) Answers may vary.

13. a) $(2n − 1)(3n − 2)$; no   **b)** $6n^2$; no
c) $\frac{n}{2n + 1}$; no   **d)** $\frac{(2n − 1)(2n + 1)}{2n(2n + 2)}$; no

Linking Ideas: Mathematics and Science
Arithmetic Sequences in Astronomy, page 16

1. 1910, 1986   **2.** Answers may vary. 2061 or 2062
3. Yes, explanations may vary.   **4.** 29 times

1.3 Exercises, page 20

1. a) 153   **b)** 385   **c)** 244   **d)** 441

2. a) 44   **b)** 500

3. a) i) 210   **ii)** 342.5   **iii)** 290
iv) 180   **v)** 546   **vi)** −55.5
b) Explanations may vary.

4. $975

5. Job A

6. Job B pays more; explanations may vary.

7. 272

8. a) $1520 **b)** $29 885 **c)** 5th year
d) $209 195 **e)** Explanations may vary.

9. Assumptions may vary.
a) 44 **b)** 15 **c)** 345

10. a) 45 **b)** 4950
c) 499 500 **d)** 49 995 000

11. $S_n = \dfrac{n(n+1)}{2}$

12. a) i) 416 **ii)** 598 **iii)** 3604 **iv)** -110
b) Explanations may vary.

13. a) Answers may vary. **b)** 210

14. a) $9000 **b)** $18 000

15. $3 + 10 + 17 + 24$

16. a) 79; $4n - 1$ **b)** 820; $n(2n + 1)$
c) 125 **d)** 15

Investigate, page 23

2. a)

Number of folds	Number of layers	Area of rectangle
0	1	Areas
1	2	may
2	4	vary.
3	8	
4	16	
5	32	
6	64	

b) Answers may vary.

3. a) 1024
b) Answers may vary. $\dfrac{1}{1024}$ of the original area
c) Answers may vary. About 1000 times as thick as 1 piece of paper
d) Answers may vary. We assume the paper can be folded that number of times.

1.4 Exercises, page 27

1. a) 10; 2000, 20 000, 200 000 **b)** 0.5; 6, 3, 1.5
c) -2; -40, 80, -160 **d)** 3; 81, 243, 729

2. a) 400 **b)** 1600 **c)** 6400

3. a) i) 96 **ii)** -96 **iii)** 0.02 **iv)** 20 480
b) Explanations may vary.

4. a) About $3360 **c)** $20 000(0.7)^n$

5. a) 1.5 m, 1.13 m, 0.84 m, 0.63 m, 0.47 m **b)** 8

6. Explanations may vary.

7. a) i) 8 **ii)** 64 **iii)** 512

8. a) 4, 8, 16, 32, 64 **b)** 108, 36, 12, 4, $\frac{4}{3}$
c) 1, -3, 9, -27, 81 **d)** 1, 4, 16, 64, 256
e) 320, 160, 80, 40, 20, 10 **f)** 10, 50, 250, 1250, 6250

9. a) 6, 18 **b)** 20, 100, 500
c) Explanations may vary.

10. b) $8100, $7290, $6561, $5905; $5129, $4616, $4155, $3739
c) $0.9x$; $(0.9)^2x$; $(0.9)^3x$; $(0.9)^4x$

11. Answers may vary. Assume GST = PST = 7%; $1.026x$, $0.9234x$, $0.8311x$, $0.7480x$

12. a) About $56 100 **b)** About $78 700 **c)** $40 000(1.07)^n$

13. a) ar^{n-1} **b)** Answers may vary.

Mathematical Modelling: How Many Links Are Needed?, page 30

1. a) 800 000 **b)** 16 000 000
c) 320 000 000 **d)** 6 400 000 000, or 6.4 billion

2.

Person	Number of links	Number of people
A	0	100
B	1	2 000
C	2	40 000
D	3	800 000
E	4	16 000 000
F	5	320 000 000
G	6	6 400 000 000

3. Yes **4.** Answers may vary. **5.** Answers may vary.

6. a) 1 **b)** 4 **c)** 5

1.5 Exercises, page 35

1. a) 16 **b)** $\frac{1}{25}$ **c)** $\frac{1}{3}$ **d)** 4
e) 1.5 **f)** $\frac{16}{9}$ **g)** 2 **h)** 1

2. a) 1 **b)** $\frac{1}{9}$ **c)** $-\frac{1}{8}$ **d)** -1.5
e) $\frac{25}{9}$ **f)** 1 **g)** 10 000 **h)** 8

3. Explanations may vary.

4. a) x^7 **b)** a^7 **c)** b^9 **d)** m^9

5. a) x^2 **b)** y^4 **c)** n **d)** a^3

6. a) x^9 **b)** y^6 **c)** a^6b^6 **d)** x^2y^6

7. a) $\frac{1}{x^3}$, or x^{-3} **b)** $\frac{1}{c}$, or c^{-1} **c)** $\frac{1}{y^5}$, or y^{-5} **d)** a^{-4}, or $\frac{1}{a^4}$

8. Explanations may vary.

9. a) 2500 **b)** About 300

10. a) x **b)** d^{-5} **c)** a^4 **d)** 1
e) x^{-9} **f)** b^3 **g)** 1 **h)** 1

11. a) x^{-7}, or $\frac{1}{x^7}$ **b)** r^5 **c)** s^{10} **d)** 1
e) c **f)** x^2 **g)** b^{-5}, or $\frac{1}{b^5}$ **h)** t^{11}

12. a) x^{-6} **b)** y^2 **c)** m^{-6} **d)** c^{-9}
e) a^{-4} **f)** xy^{-2} **g)** x^4y^{-6} **h)** a^4b^{-4}

13. a) b^{-1} **b)** 1 **c)** x^2y^{-1}
d) xy^{-6} **e)** c^7d **f)** m^3n^{-4}

14. a) i) -32 **ii)** -4 **iii)** -64
iv) $-\frac{1}{32}$ **v)** -16 **vi)** 64
b) Explanations may vary.

15.

5^{-8}	0.000 002 56	5^1	5
5^{-7}	0.000 012 8	5^2	25
5^{-6}	0.000 064	5^3	125
5^{-5}	0.000 32	5^4	625
5^{-4}	0.001 6	5^5	3 125
5^{-3}	0.008	5^6	15 625
5^{-2}	0.04	5^7	78 125
5^{-1}	0.2	5^8	390 625
5^0	1.0	5^9	1 953 125
		5^{10}	9 765 625

a) 78 125   **b)** 390 625   **c)** 125
d) 390 625   **e)** 0.000 012 8   **f)** 25

16. a) d **b)** $\dfrac{4y^{10}}{25x^4}$ **c)** $a^{16}b^{-7}c^3$

Investigate, page 37

1. a) Explanations may vary. $7.20/h
  b) Multiply the number of hours by the hourly rate.
  c) $7.20x

2. $562.50

STATEMENT OF EARNINGS					
	Regular Hours	Regular Pay	Overtime Hours	Overtime Pay	Weekly Total
August 18–24	40	300.00	0	0.00	300.00
August 25–31	35	262.50	0	0.00	262.50
				Gross Pay	562.50

3. Explanations may vary. $583.20

1.6 Exercises, page 40

1. a) 7%; explanations may vary.
  b) 10.7%; explanations may vary.
  c) $14.33   **d)** $21.91
  e) Explanations may vary.

2.

STATEMENT OF EARNINGS					
	Regular Hours	Regular Pay	Overtime Hours	Overtime Pay	Weekly Total
September 1–7	15	108.00	2	21.60	129.60
September 8–14	8	57.60	0	0.00	57.60
				Gross Pay	187.20

3. a) 2 for a win, 1 for a tie, 0 for a loss

b) , c)

	Pts for b)	Pts for c)
Colorado	156	98
Dallas	145	96
Detroit	132	76
Anaheim	121	72
Phoenix	121	76
St. Louis	119	72
Edmonton	117	72
Chicago	115	68
Vancouver	112	70
Calgary	105	64
Toronto	98	60
Los Angeles	95	56
San Jose	89	54

4. a) 7%
  b)

Price ($)	GST ($)	PST ($)	Total ($)
50.00	3.50	4.25	57.75
243.29	17.03	20.68	281.00

  c) 12.5%

5. 8.5%

6. Answers may vary. With 7% GST and PST: $17.54; explanations may vary.

7. a)

Year	Canada car sales	U.S. car sales
1991	1 265 473	12 402 992
1992	1 206 510	12 918 633
1993	1 165 472	13 935 450
1994	1 233 920	15 109 523
1995	1 130 424	14 733 882
1996	1 173 624	15 088 976

  b) Explanations may vary.   **i)** No   **ii)** No

8. a) i) $73.14   **ii)** 124.16 marks   **iii)** 8865 yen
  b) i) $329.13   **ii)** £203.45   **iii)** 557 622 lira
  c) i) $112.80   **ii)** 472.76 francs   **iii)** 140.05 marks

9. a) i) $136.72   **ii)** 573.01 francs   **iii)** 12 121 yen
  b) i) $615.24   **ii)** 763.88 marks   **iii)** 762 381 lira
  c) i) $1105.85   **ii)** $808.85   **iii)** 98 035 yen
  d) Steps may vary.

10. a) Discount before and after taxes: $0.912x$
  b) Discount before taxes, store receives $0.80x$; discount after taxes, store receives $0.772x$

Problem Solving: How Many Kernels of Wheat?, page 44

1. , 2. a) , 6. a)

Square number	Kernels on square	Total kernels
1	1	1
2	2	3
3	4	7
4	8	15
5	16	31
6	32	63
7	64	127
8	128	255
9	256	511
10	512	1023

2. b) Answers may vary.
   c) The number is the power of 2 with an exponent that is 1 less than the number of the square.
   d) 9.2×10^{18}

3. a) 21st square
   b) 2 097 152 kernels; 4 194 304 kernels; 8 388 608 kernels; total of about 14.7 bushels

4. a) 51st square
   b) 2.3 years; 4.6 years; 9.2 years; total of 16 years production

5. a) 55th square   **b)** 529 years

6. b) Each number is 1 less than the power of 2 with an exponent that is equal to the number of the square.
   c) The number of kernels needed to fill a square is 1 more than the total so far.

7. a) 2.9×10^{10} years   **b)** 30th square

8. 5.8×10^{11} kernels per second

Investigate, page 46

1. a) , b) , c)

Year	Opening balance ($)	Interest rate (%)	Interest earned ($)	Closing balance ($)
1	1000.00	6	60.00	1060.00
2	1060.00	6	63.60	1123.60
3	1123.60	6	67.42	1191.02
4	1191.02	6	71.46	1262.48
5	1262.48	6	75.75	1338.23

   d) $1338.23   **e)** $338.23

2. $2676.45   **3.** $1276.28

Investigate, page 48

1. a) , b)

Year	Opening balance ($)	Interest rate (%)	Interest earned ($)	Annual investment ($)	Closing balance ($)
1	1000.00	6	60.00	1000.00	2060.00
2	2060.00	6	123.60	1000.00	3183.60
3	3183.60	6	191.02	1000.00	4374.62
4	4374.62	6	262.48	1000.00	5637.10
5	5637.10	6	338.23	1000.00	6975.33

   c) $6975.33   **d)** $975.33
   e) Answers may vary.

2. $12 612.41   **3.** $6801.92

1.7 Exercises, page 50

1. a) 3180.00; 6; 190.80; 3370.80
   b) 3387.80; 9; 304.90; 3692.70
   c) 4911.41; 7.5; 368.36; 5279.77

2. a) 4110.00; 5.5; 226.05; 2000.00; 6336.05
   b) 6788.54; 7; 475.20; 1500.00; 8763.74
   c) 6846.58; 6.75; 462.14; 3200.00; 10 508.72

3. a) $656.19   **b)** $0.06A; $1.06A   **c)** $3245.52
   d) $3337.37   **e)** Explanations may vary.

4. $1259.71

Year	Opening balance ($)	Interest rate (%)	Interest earned ($)	Closing balance ($)
1	1000.00	8	80.00	1080.00
2	1080.00	8	86.40	1166.40
3	1166.40	8	93.31	1259.71

5. a) $477.49   **b)** $5152.49

6. Explanations may vary. $2889.76

Year	Opening balance ($)	Interest rate (%)	Interest earned ($)	Annual investment ($)	Closing balance ($)
1	500.00	7.25	36.25	500.00	1036.25
2	1036.25	7.25	75.13	500.00	1611.38
3	1611.38	7.25	116.83	500.00	2228.21
4	2228.21	7.25	161.55	500.00	2889.76

7. a) $\$A\left(1 + \frac{i}{100}\right)$; $\$A\left(1 + \frac{i}{100}\right)^2$; $\$A\left(1 + \frac{i}{100}\right)^3$; $\$A\left(1 + \frac{i}{100}\right)^4$
   b) $\$A\left(1 + \frac{i}{100}\right)^n$

Linking Ideas: Mathematics and Technology
How Long Will It Take for an Investment to Double in Value?, page 53

1. b) Explanations may vary.   **e)** 10

2. a)

Interest rate (%)	Approximate doubling time (years)
4	18
5	14
6	12
7	10
8	9
9	8
10	7
11	7
12	6
13	6
14	5

b) Decreases

c) When interest rate is doubled, doubling time is halved. When interest rate is tripled, doubling time decreases to one-third of what it was.

d) All approximately 72.

e) Answers may vary. The interest rate as a percent multiplied by the doubling time in years is approximately 72.

Investigate, page 54

1. a) , b) , c)

Year	Opening balance ($)	Interest rate (%)	Interest charged ($)	Annual payment ($)	Closing balance ($)
1	25 000.00	9	2250.00	6427.31	20 822.69
2	20 822.69	9	1874.04	6427.31	16 269.42
3	16 269.42	9	1464.25	6427.31	11 306.36
4	11 306.36	9	1017.57	6427.31	5 896.62
5	5 896.62	9	530.70	6427.31	0.01

d) $7136.56

2. a) $12 854.62   **b)** $14 273.11

1.8 Exercises, page 58

1. a) 4620.66; 4218.56   **b)** 677.03; 7008.91
c) 2470.34; 5453.70

2. a) 40 066.79; 8.25; 3305.51; 6480.71; 36 891.59
b) 26 971.53; 11; 2966.87; 11 037.09; 18 901.31
c) 4248.96; 5.5; 233.69; 995.01; 3487.64

3. a) $2933.21; $7270.47; $721.63
b) $3175.20; $8070.22; $761.32

4. a) 8 years   **b)** $17 072.70   **c)** 7.5%

5. a) $136 581.60   **b)** $36 581.60

6. a) $9572.70   **b)** $10 290.65   **c)** $12 784.04

7. a) $277.97   **b)** $864.05

8. a) The loan is paid off in 7 years with a final payment of $8108.38.

Year	Opening balance ($)	Interest rate (%)	Interest charged ($)	Annual payment ($)	Extra payment ($)	Closing balance ($)
4	69 074.20	7.5	5180.57	17 072.70	20 000.00	37 182.07
5	37 182.07	7.5	2788.66	17 072.70		22 898.03
6	22 898.03	7.5	1717.35	17 072.70		7 542.78
7	7 542.68	7.5	565.70	8 108.38		0.00

b) $30 544.58   **c)** $6037.02

9. a) $11 892.13   **b)** $24 676.17   **c)** $38 419.02

10. a) $8322.30   **b)** $2322.30

11. a) $2782.70   **b)** $161 036.25

12. a) $6730.35
b)

Year	Opening balance ($)	Interest rate (%)	Interest charged ($)	Annual payment ($)	Closing balance ($)
4	28 044.08	8.5	2383.75	10 980.35	19 447.48
5	19 447.48	8.5	1653.04	10 980.35	10 120.17
6	10 120.17	8.5	860.21	10 980.35	0.03

c) $151.84   **d)** $23 822.17

13.

Year	Opening balance ($)	Interest rate (%)	Interest charged ($)	Annual payment ($)	Closing balance ($)
1	25 000.00	8	2000.00	7548.02	19 451.98
2	19 451.98	8	1556.16	7548.02	13 460.12
3	13 460.12	8	1076.81	7548.02	6 988.91
4	6 988.91	8	559.11	7548.02	0.00

a) Year 1: $19 451.98; Year 2: $13 460.12; Year 3: $6988.91; Year 4: $0.00
b) $882.57   **c)** $6471.21

14. a)

Month	Sales
June	15 000
July	30 000
August	60 000
September	120 000
October	240 000
November	480 000
December	960 000

b) 938   **c)** 2 224 688

15. a) $3650   **b)** $3300

Linking Ideas: Mathematics and Technology
Calculating Loan Payments, page 62

1. a) Less, $3559.44   **b)** Greater, $4516.86

2. a) $1203.17   **b)** $5288.45

3. a) $2545.16   **b)** $10 375.44

4. a) $38 628.68   **b)** $107 508.68

6. a) Option 1: $4203.36; option 2: $3320.11; option 3: can pay off in 5 or 6 more years, either making a payment of $5851.26 in year 10 or $2190.45 in year 11

b) Answers may vary; with option 1 you pay the least interest; with option 2 you pay the most interest.

7. a) Option 1: $16 502.26; option 2: $10 243.31; option 3: can pay off in 4 more years, the payment in year 8 being only $14 577.23

b) Answers may vary; with option 3 you pay the least interest, with option 2 you pay the most interest.

8. a) $222.45   **b)** $1477.37
c) i) $7151.12   **ii)** $4943.18   **iii)** $2563.83
d) Answers may vary; after 24 months, $1081.98 of what Karol has paid was interest.

9. 7.25%

10. a) $752.05   **b)** $125 614.23
c) $91 790.47   **d)** $37 665.52

11. a) He can pay off the loan after 276 months (23 years) with another extra payment of $442.48. The total interest paid would be $111 008.28, a savings of $14 605.95.
b) He can pay off the loan after 283 months. The total interest paid would be $115 414.68, a savings of $10 199.55.
c) He can pay off the loan in 218 months. The total interest paid would be $78 452.33, a savings of $47 161.90.

12. The balance at the beginning of the second loan is $92 113.48. His payments will be $1028.55.

1 Review, page 65

1. a) 18, 31   **b)** 116, 133, 150

2. a) i) 32   **ii)** 2048   **iii)** 4096   **iv)** 8192
b) Explanations may vary.

3. Explanations may vary.

4. a) x^3 **b)** x^{-3} **c)** x^2 **d)** x^{10}
e) x^{-3} **f)** 1   **g)** 1   **h)** x^{-4}

5. a) x^{-1} **b)** x^{-4} **c)** x^6 **d)** 1

6. a) $1163.06
b) i)

Year	Opening balance ($)	Interest rate (%)	Interest charged ($)	Annual payment ($)	Closing balance ($)
1	1000.00	8	80.00	500.00	1580.00
2	1580.00	8	126.40	500.00	2206.40
3	2206.40	8	176.51	500.00	2882.91
4	2882.91	8	230.63	500.00	3613.54
5	3613.54	8	289.08	500.00	4402.62

ii)

1	1000.00	12	120.00	500.00	1620.00
2	1620.00	12	194.40	500.00	2314.40
3	2314.40	12	277.73	500.00	3092.13
4	3092.13	12	371.06	500.00	3963.19
5	3963.19	12	457.58	500.00	4938.77

c) When interest rate is 15% in Year 5

i)

Year	Opening balance ($)	Interest rate (%)	Interest charged ($)	Annual payment ($)	Closing balance ($)
1	1000.00	15	150.00	500.00	1650.00
2	1650.00	10	165.00	500.00	2315.00
3	2315.00	10	231.50	500.00	3045.50
4	3046.50	10	304.65	500.00	3851.15
5	3851.15	10	385.12	500.00	4736.27

ii)

1	1000.00	10	100.00	500.00	1600.00
2	1600.00	10	160.00	500.00	2260.00
3	2260.00	10	226.00	500.00	2986.00
4	2986.00	10	298.60	500.00	3784.60
5	3784.60	15	567.69	500.00	4852.29

Chapter 2
2.1 Exercises, page 72

1. a) ±7 **b)** ±9 **c)** ±11
d) ±20 **e)** ±23 **f)** ±25

2. a) 8   **b)** 10   **c)** 12   **d)** 30   **e)** 40
f) 0.5   **g)** 0.2   **h)** 0.1   **i)** 0.04   **j)** 0.005

3. a) 1.414   **b)** 1.732   **c)** 7.232   **d)** 11.336   **e)** 21.703

4. a) 2   **b)** -3 **c)** 3
d) 2   **e)** 3   **f)** 0.1

5. a) 3.1 m, 12.4 m   **b)** 2.5 m, 10.1 m   **c)** 2.9 m, 11.7 m

6. a) 10 cm, 100 cm^2 **b)** 5 cm, 25 cm^2
c) 6.3 cm, 39.7 cm^2

7. a) 2.45   **b)** 3.32   **c)** 2.84
d) 11.14   **e)** 5.18   **f)** 6.33

8. a) $\sqrt{25}$ **b)** $\sqrt{9}$ **c)** $\sqrt{4}$
d) $\sqrt{16}$ **e)** $\sqrt{49}$ **f)** $\sqrt{1}$

9. a) Rounded: 7.7, 7.75, 7.746; truncated: 7.7, 7.74, 7.745
b) Answers may vary; the truncated answer will have the lower last number.
c) No, unless you use your calculator to square the number and investigate the result.

10. a) 3, 30, 300; $\sqrt{9\,000\,000}$
b) 2, 20, 200; $\sqrt[3]{8\,000\,000\,000}$
c) 2, 20, 200; $\sqrt[4]{16\,000\,000\,000\,000}$

11. a) 4   **b)** 5   **c)** 2   **d)** -1 **e)** 6
f) -10 **g)** 4   **h)** 10   **i)** 7   **j)** 10

12. Explanations may vary.

13. a) 3.5 m   **b)** 2.4 m

14. 9.3 cm

15. a) $\sqrt[3]{8}$ **b)** $\sqrt[3]{-27}$ **c)** $\sqrt[3]{1}$
d) $\sqrt[3]{-125}$ **e)** $\sqrt[3]{64}$ **f)** $\sqrt[3]{-1}$

16. Explanations may vary.

17. a) 79 500 cm^3, 43 cm
b) Assume a five-day work week. 306 cm^3, 6.7 cm
c) 395 700 cm^3, 73 cm

18. a) 44.7 cm   **b)** 447 cm

19. a) 5 b) 7 c) 6 d) 12 e) 2
f) 2 g) 6 h) 23 i) 0

20. Explanations may vary. 21. $\pm\sqrt{50}$

22. a) 3, $3\sqrt[3]{4}$, $3\sqrt[3]{16}$, 12; 3, 4.76, 7.56, 12
b) $12\sqrt[3]{4}$, $12\sqrt[3]{16}$, 48; 19.05, 30.24, 48

23. 4, $4\sqrt[4]{5}$, $4\sqrt[4]{25}$, $4\sqrt[4]{125}$; 4, 5.981, 8.944, 13.375

24. b) 4.7 miles, 7.5 km c) 56 320 000 m³, 383 m

Investigate, page 76

1. a) 1.732 050 808 b) Answers may vary.
2. a) 1.442 249 57 b) Answers may vary.
3.

x	$x^{\frac{1}{2}}$
1	1
2	1.414 213 562
3	1.732 050 808
4	2
9	3
16	4
25	5

x	$x^{\frac{1}{3}}$
1	1
2	1.259 921 05
3	1.442 249 57
8	2
27	3
64	4
125	5

4. The nth root of x

5. Predictions may vary.

x	$x^{-\frac{1}{2}}$
1	1
2	0.707 106 781
3	0.577 350 269
8	0.5
27	0.333 333 333
64	0.25
125	0.2

6.

x	$x^{\frac{2}{3}}$
1	1
2	1.587 401 052
3	2.080 083 823
8	4
27	9
64	16
125	25

The square root of the cube root of x

7. The reciprocal of the nth root of x; the mth power of the nth root of x

2.2 Exercises, page 80

1. a) 1 b) 2 c) 4 d) 8 e) 16
f) 0.5 g) 0.25 h) 0.125 i) 0.0625 j) 0.031 25

2. a) 4 b) 6 c) 10 d) 2 e) 4
f) 3 g) −4 h) 3 i) −3 j) −10

3. a) $\frac{1}{2}$ b) $\frac{1}{3}$ c) $\frac{1}{3}$ d) $\frac{1}{4}$ e) $-\frac{1}{4}$

4. a) $\sqrt[3]{4}$ b) $\sqrt[5]{16}$ c) $\sqrt[5]{64}$ d) $\sqrt[5]{256}$ e) $\sqrt[5]{1024}$
f) $\frac{1}{\sqrt[3]{4}}$ g) $\frac{1}{\sqrt[5]{16}}$ h) $\frac{1}{\sqrt[5]{64}}$ i) $\frac{1}{\sqrt[5]{256}}$ j) $\frac{1}{\sqrt[5]{1024}}$

5. Explanations may vary.

6. a) 27 b) 9 c) 8 d) 125 e) 4
f) 9 g) 216 h) 16 i) 1000 j) 400

7. a) $\frac{1}{9}$ b) $\frac{1}{8}$ c) $\frac{1}{27}$ d) $\frac{1}{8}$ e) $\frac{1}{1000}$

8. a) $9^{\frac{1}{2}}$ b) $4^{\frac{1}{2}}$ c) $16^{\frac{1}{2}}$
d) $1^{\frac{1}{2}}$ e) $100^{\frac{1}{2}}$ f) $64^{\frac{1}{2}}$

9. Explanations may vary.

10. a) $\frac{1}{81}$ b) $\frac{1}{64}$ c) 27 d) $\frac{1}{4}$ e) 343
f) $\frac{3}{4}$ g) $\frac{125}{343}$ h) $\frac{1}{16}$ i) $\frac{9}{4}$ j) $\frac{8}{27}$

11. Explanations may vary.

12. a) 1.778 b) 10.814 c) 3.659 d) 44.313 e) 1.102

13. a) 20 cm, 400 cm² b) $\sqrt[3]{2}$ m, $\sqrt[3]{4}$ m² c) $\sqrt[3]{V}$ m, $\sqrt[3]{V^2}$ m²

14. a) 5 cm, 125 cm³ b) $\sqrt{2}$ m, $\sqrt{8}$ m³ c) $\sqrt{A}$ m, $\sqrt{A^3}$ m³

15. a) $27^{\frac{1}{3}}$ b) $(-1)^{\frac{1}{3}}$ c) $(-8)^{\frac{1}{3}}$
d) $(-64)^{\frac{1}{3}}$ e) $1^{\frac{1}{3}}$ f) $(-27)^{\frac{1}{3}}$

16. a) $x^{\frac{3}{2}}$, $\sqrt{x^3}$ b) $m^{\frac{4}{3}}$, $\sqrt[3]{m^4}$ c) y^2, $\sqrt{y^4}$
d) b^4, $\sqrt{b^8}$ e) $x^{\frac{1}{2}}$, $\sqrt{x}$ f) $m^{\frac{2}{3}}$, $\sqrt[3]{m^2}$
g) d, $\sqrt{d^2}$ h) $p^{\frac{2}{5}}$, $\sqrt[5]{p^2}$

17. a) St. John's: −5.9°C; Sydney: −7.1°C; Charlottetown: −8.6°C b) Charlottetown, Sydney, St. John's
c) St. John's: −15.1°C; Sydney: −14.7°C; Charlottetown: −15.9°C; Charlottetown still feels the coldest, but when you consider the wind chill, St. John's feels colder than Sydney.

18. a) $\sqrt{x}$, $x^{\frac{1}{2}}$ b) $\sqrt{x^4}$, x^2 c) $\sqrt[4]{x}$, $x^{\frac{1}{4}}$
d) $\sqrt[6]{x}$, $x^{\frac{1}{6}}$ e) $\sqrt[3]{3x^5}$, $(3x^5)^{\frac{1}{6}}$ f) $\sqrt[6]{x^5}$, $x^{\frac{5}{6}}$

19. a) $\sqrt{x^5}$, $x^{\frac{5}{2}}$ b) $\sqrt[4]{x^5}$, $x^{\frac{5}{4}}$ c) $\sqrt[10]{x^{19}}$, $x^{\frac{19}{10}}$
d) $\sqrt[15]{x^{19}}$, $x^{\frac{19}{15}}$ e) $\sqrt[6]{x^{11}}$, $x^{\frac{11}{6}}$ f) $\sqrt{x^2}$, x^1

20. Explanations may vary.

21. a) Descriptions and explanations may vary.
b) Approximations: $\sqrt{2}$, $(\sqrt{2})^3$; exact: $(\sqrt{2})^0$, $(\sqrt{2})^2$, $(\sqrt{2})^4$
c) $(\sqrt{2})^2 = \sqrt{4}$; $(\sqrt{2})^3 = \sqrt{8}$; $(\sqrt{2})^4 = \sqrt{16}$
d) $2^{\frac{1}{2}} = 2^{0.5}$; $2^{\frac{2}{2}} = 2^1$; $2^{\frac{3}{2}} = 2^{1.5}$; $2^{\frac{4}{2}} = 2^2$

22.

Decimal form	1	0.707	0.5	0.354	0.25
Radical form	$(\sqrt{2})^0$	$\frac{1}{(\sqrt{2})^1}$	$\frac{1}{(\sqrt{2})^2}$	$\frac{1}{(\sqrt{2})^3}$	$\frac{1}{(\sqrt{2})^4}$
Power form	$2^{\frac{0}{2}}$	$2^{-\frac{1}{2}}$	$2^{-\frac{2}{2}}$	$2^{-\frac{3}{2}}$	$2^{-\frac{4}{2}}$

23. a) 19.95 b) 199.5 c) 1995 d) 19 950
e) 0.1995 f) 0.019 95 g) 0.001 995 h) 0.000 199 5

Mathematical Modelling: Population Growth and Deforestation, page 84

1. Answers may vary. A geometric sequence
a)

Year	1960	1970	1980	1990	2000	2010
Population (billions)	3.2	3.8	4.5	5.3	6.3	7.5

b) 7.5 billion

2. a) Answers and explanations may vary. An arithmetic sequence
b)

Year	1960	1970	1980	1990	2000	2010
Rain forest area (billion hectares)	2.10	1.98	1.87	1.75	1.63	1.52

3. Answers may vary.   4. Answers may vary.

5. a) 5.3×10^8 m³   b) Answers may vary. 1.125×10^9 m³
   c) Answers may vary.

6. Answers may vary. About 1.7% per year; about 0.65% per year

7. Between 1.35 billion hectares and 1.43 billion hectares

2.3 Exercises, page 90

1. a) 1.0 kg   b) Estimates may vary.

2. a) 1.17 kg; 0.86 g   b) 0.13%; 2.9%
   c) Answers may vary.

3. a) Elephant: 3.4 kg; cat: 0.034 kg; shrew: 0.000 34 kg
   b) Each animal is 1000 times as massive as the one below it. Each brain is 100 times as massive as the one below it.
   c) 0.054%, 0.54%, 5.4%
   d) Each percent is 10 times the percent of the animal below it. The mass of a smaller animal's brain is a greater percent of its body mass than that of a larger animal.
   e) $\dfrac{\text{Mass of brain}}{\text{Mass of body}} = m^{-\frac{1}{3}}\%$

4. b)

	A	B	C	D
1	Brain Masses of Mammals			
2	Mammal	Body	Brain	Percent
3		mass	mass	of body
4	Mouse	1	0.01	0.01
5	Gerbil	8	0.04	0.005
6	Marten	64	0.16	0.0025
7	Porcupine	512	0.64	0.00125
8	Seal	4096	2.56	0.000625
9	Cow	32768	10.24	0.0003125
10	Elephant	262144	40.96	0.00015625

5. a) Elephant: 27, 6; cat: 152, 34; shrew: 852, 189
   b) They increase.   c) 4.5   d) No

6. a) 0.24 m²   b) 3.1 m²   c) Answers may vary.

7. a) 0.25 m²   b) 2.6 m²   c) Answers may vary.

8. All measurements are in Earth days.
   a) Mercury: 88; Venus: 224; Earth: 364; Mars: 689; Jupiter: 4340; Saturn: 10 700; Uranus: 30 900; Neptune: 60 800; Pluto: 91 300

9. a)

Year	Population
1930	13
1931	45
1932	154
1933	528
1934	1 816
1935	6 245
1936	21 471
1937	73 819
1938	253 796
1939	872 576
1940	3 000 000

10.

Year	1980	1981	1982	1983	1984	1985	1986
Population (thousands)	39	59	91	138	210	320	487

11. a) $11.84m^{0.25}$
    b) Elephant: 106 years; cat: 19 years; shrew: 3 years

12. 1.1×10^{37}; unreasonable; explanations may vary

13. a) i) 88.4 min, 1.5 h   ii) 96.6 min, 1.6 h
    iii) 1440 min, 24 h
    b) Satellite has the same period as Earth.

Linking Ideas: Mathematics and Science
Bird Eggs, page 94

1. , 2. b) , 3. b)

Bird	Hummingbird	Hen	Ostrich
Mass of bird (g)	3.6	2000	113 000
Mass of egg (g)	0.74	96.4	2154.5
Mass of shell (g)	0.03	8.5	286
Incubation time (d)	11.3	32.4	63.6
Shell thickness (mm)	0.045	0.41	1.7
Egg mass as % of body mass	20.6	4.82	1.91
Shell mass as % of egg mass	4.6	8.8	13.3

2. a) $0.277m^{-0.23}$ b) See table above.

3. a) $0.0482e^{0.132}$ b) See table above.

4. $0.0113m^{0.872}$

5. d)

	A	B	C	D	E	F
1	Some Properties of Bird Eggs					
2		Hummingbird	Pigeon	Hen	Goose	Ostrich
3	Mass (g)	3.6	280	2000	4500	113 000
4	Egg mass (g)	0.74	21	96	180	2155
5	Fraction of bird	21%	8%	5%	4%	2%
6	Shell mass (g)	0.03	1.53	8	17	286
7	Fraction of egg	5%	7%	9%	10%	13%
8	Incubation time (d)	11	23	32	37	64
9	Shell thickness (mm)	0.04	0.21	0.41	0.55	1.70

6. 830 kg

2.4 Exercises, page 99

1. a) 8.9   b) 10.7   c) 12.0
   d) 1.7   e) 4.8   f) 9.2

2. a) $\sqrt{8}$ cm   b) $\sqrt{13}$ cm   c) $\sqrt{10}$ cm   d) $\sqrt{29}$ cm

3. 4.0 cm, 5.8 cm   4. 27.50 m

5. About 87 m   6. About 40 m

7. Answers may vary.
   a) $\sqrt{20}$ cm, $\sqrt{5}$ cm   b) 1 cm, $\sqrt{8}$ cm
   c) 8 cm, 6 cm   d) 10 cm, $\sqrt{44}$ cm
   e) 1 cm, $\sqrt{3}$ cm   f) 5 cm, $\sqrt{24}$ cm

8. Explanations may vary.

10. a) 6.2 cm   b) $d = \sqrt{a^2 + b^2 + c^2}$

11. a) No

12. a) $\sqrt{\pi} : 1$ b) $\sqrt{2\pi} : 1$

2.5 Exercises, page 104

1. a) Rational   b) Irrational   c) Irrational
   d) Irrational   e) Irrational   f) Irrational

2. a) Rational   **b)** Irrational

3. a) 999999   **b)** No

4. a) All natural numbers are also integers.
    b) All integers are also rational numbers.

5. $\sqrt{21}$, $\sqrt{200}$, $\sqrt{2.5}$

6. Explanations may vary.

7. $\sqrt{3}$, $\sqrt{24}$

8. a) 41 275, $\sqrt[3]{8}$, $\sqrt{121}$ **b)** All of part a and -6
    c) All of part b and $\frac{3}{5}$, $0.2\overline{17}$, $-2\frac{1}{4}$, 6.121 121...
    d) $3\sqrt{2}$, 6π

9. a) Yes   **b)** Yes   **c)** No   **d)** Yes   **e)** No

10. Explanations may vary.

11. Answers may vary.
    a) -1 **b)** 0   **c)** 5   **d)** $-\frac{5}{2}$

13. No

Mathematics File: Significant Digits, page 107

1. a) 61 275 cm^3, 61 000 cm^3 **b)** 58 830.625 cm^3
    c) 63 781.875 cm^3
    d) i) 61 000 cm^3; 59 000 cm^3; 64 000 cm^3
    ii) 60 000 cm^3; 60 000 cm^3; 60 000 cm^3
    e) 1

2.6 Exercises, page 110

1. About 43%   **2.** 1 : 3 : 4

3. a) 6, 5.20   **b)** 9, 7.79   **c)** 5, 8.66
    d) 2.5, 4.33   **e)** 1, 2   **f)** 2, 4

4. Explanations may vary.

5. Check to see if the lengths of the sides of each triangle
satisfy the Pythagorean Theorem.

6. a) 10   **b)** 6.93   **c)** 7.5
    d) 10.39   **e)** 19.05   **f)** 6.06

7. a) $\sqrt{3}$ cm, 2 cm   **b)** $5\sqrt{3}$ cm, 10 cm
    c) $3\sqrt{3}$ cm, 6 cm   **d)** $9\sqrt{3}$ cm, 18 cm
    e) $2\sqrt{3}$ cm, 4 cm   **f)** $10\sqrt{3}$ cm, 20 cm

8. a) 0.5 cm, $0.5\sqrt{3}$ cm   **b)** 2.5 cm, $2.5\sqrt{3}$ cm
    c) 1.5 cm, $1.5\sqrt{3}$ cm   **d)** 4.5 cm, $4.5\sqrt{3}$ cm
    e) 1 cm, $\sqrt{3}$ cm   **f)** 5 cm, $5\sqrt{3}$ cm

9. a) 1 cm, 2 cm   **b)** 2 cm, 4 cm
    c) 4 cm, 8 cm   **d)** 0.25 cm, 0.5 cm
    e) $0.1\overline{6}$ cm, $0.\overline{3}$ cm

10. Explanations may vary.

11. a) 6 cm   **b)** $6\sqrt{3}$ cm

12. $5\sqrt{3}$ cm, $10\sqrt{3}$ cm   **13.** $48\sqrt{3}$ cm^2, 83.1 cm^2

14. Explanations may vary.   **15.** $(20 + 10\sqrt{3})$ cm, 37.321 cm

16. 128 cm   **17.** $48\sqrt{3}$ cm^2; 83.138 cm^2

18. 10.529 cm, 9.118 cm

19. $6\sqrt{2}$ cm, 72 cm^2; 8.485 cm, 72 cm^2 **20.** 8258

Problem Solving: Squaring the Circle, page 115

1. a) 19.6 cm^2 **b)** 4.4 cm

2. a) πr^2 cm^2 **b)** $\sqrt{\pi}r$ cm

3. a) $9r^2$ cm^2 **b)** $4r^2$ cm^2

4. b) 1 cm^2 **5.** Greater

Investigate, page 116

1. Yes

2. a) 1.732 050 808, 2.236 067 978, 3.872 983 346
    b) Yes

3. Examples may vary.   **4.** $\sqrt{a} \times \sqrt{b} = \sqrt{a \times b}$

2.7 Exercises, page 119

1. a) $2\sqrt{14}$ **b)** $\sqrt{154}$ **c)** -12
    d) $2\sqrt{10}$ **e)** -12 **f)** $70\sqrt{6}$

2. Answers may vary.
    a) $\sqrt{4} \times \sqrt{6}$ **b)** $\sqrt{9} \times \sqrt{2}$ **c)** $\sqrt{5} \times \sqrt{9}$
    d) $\sqrt{4} \times \sqrt{7}$ **e)** $\sqrt{2} \times \sqrt{36}$ **f)** $\sqrt{4} \times \sqrt{15}$
    g) $\sqrt{3} \times \sqrt{13}$ **h)** $\sqrt{5} \times \sqrt{13}$ **i)** $\sqrt{16} \times \sqrt{6}$
    j) $\sqrt{4} \times \sqrt{30}$ **k)** $\sqrt{9} \times \sqrt{14}$ **l)** $\sqrt{5} \times \sqrt{21}$

3. a) $4\sqrt{2}$ **b)** $5\sqrt{2}$ **c)** $3\sqrt{3}$ **d)** $4\sqrt{6}$
    e) $2\sqrt{2}$ **f)** $5\sqrt{3}$ **g)** $3\sqrt{7}$ **h)** $3\sqrt{6}$
    i) $2\sqrt{19}$ **j)** $4\sqrt{5}$ **k)** $2\sqrt{22}$ **l)** $10\sqrt{2}$

4. Instructions may vary.

5. a) $\sqrt{18}$ **b)** $\sqrt{12}$ **c)** $\sqrt{20}$ **d)** $\sqrt{50}$
    e) $\sqrt{45}$ **f)** $\sqrt{75}$ **g)** $\sqrt{32}$ **h)** $\sqrt{16}$
    i) $\sqrt{36}$ **j)** $\sqrt{48}$ **k)** $\sqrt{80}$ **l)** $\sqrt{100}$

6. Explanations may vary.

7. a) $2\sqrt[3]{2}$ **b)** $3\sqrt[3]{2}$ **c)** $2\sqrt[3]{3}$ **d)** $5\sqrt[3]{3}$
    e) $3\sqrt[3]{4}$ **f)** $5\sqrt[3]{2}$ **g)** $4\sqrt[3]{3}$ **h)** $4\sqrt[3]{2}$
    i) $2\sqrt[3]{4}$ **j)** $3\sqrt[3]{3}$ **k)** $2\sqrt[3]{5}$ **l)** $3\sqrt[3]{5}$

8. Explanations may vary.

9. a) 2.4495, 7.7460, 24.4949, 77.4597; $\sqrt{60\ 000} \doteq 244.949$,
    $\sqrt{600\ 000} \doteq 774.597$
    b) 1.5811, 5.0000, 15.8114, 50.0000; $\sqrt{25\ 000} \doteq 158.114$,
    $\sqrt{250\ 000} = 500.0000$
    c) 3.1623, 4.4721, 6.3246, 8.9443; $\sqrt{160} \doteq 12.6492$,
    $\sqrt{320} \doteq 17.8886$

10. a) i) 4.899   **ii)** $2\sqrt{6}$ **iii)** 4.899
    b) i) 59.161   **ii)** $10\sqrt{35}$ **iii)** 59.161
    c) i) 101.823   **ii)** $72\sqrt{2}$ **iii)** 101.823

11. a) $12\sqrt{3}$ **b)** $105\sqrt{2}$ **c)** $-192\sqrt{3}$
    d) $40\sqrt{15}$ **e)** $84\sqrt{2}$ **f)** $165\sqrt{2}$

12. a) $-108\sqrt{6}$ **b)** $15\sqrt{15}$ **c)** $-14\sqrt{\frac{2}{21}}$
    d) -2 **e)** $-10\sqrt{0.21}$ **f)** $132\sqrt{0.4}$

13. Explanations may vary.

14. a) $12\sqrt{3}$ **b)** 120   **c)** 60
    d) $280\sqrt{6}$ **e)** $60\sqrt{21}$ **f)** 1008

15. Answers may vary.
  a) $2\sqrt{3} \times 3\sqrt{5}$ **b)** $2\sqrt{6} \times 4\sqrt{7}$ **c)** $3\sqrt{5} \times 5\sqrt{4}$
  d) $6\sqrt{2} \times 2\sqrt{5}$ **e)** $4\sqrt{4} \times 3\sqrt{3}$ **f)** $5\sqrt{3} \times 2\sqrt{7}$

16. Answers may vary.
  a) $2\sqrt{3} \times 3\sqrt{3}$ **b)** $3\sqrt{2} \times 4\sqrt{2}$ **c)** $2\sqrt{2} \times 3\sqrt{2}$
  d) $3\sqrt{2} \times 10\sqrt{2}$ **e)** $3\sqrt{5} \times 2\sqrt{5}$ **f)** $3\sqrt{7} \times \sqrt{7}$

17. Explanations may vary.

18. a) $\sqrt{192}$, $\sqrt{768}$ **b)** $\sqrt{48}$, $\sqrt{96}$
  c) $\sqrt{54}$, $\sqrt{162}$ **d)** $\sqrt{5000}$, $\sqrt{50\,000}$

19. a) $2\sqrt{15}$, $3\sqrt{7}$, $4\sqrt{6}$, $7\sqrt{2}$
  b) $2\sqrt{10}$, $\sqrt{43}$, $4\sqrt{3}$, $5\sqrt{2}$, $2\sqrt{13}$
  c) $3\sqrt{7}$, $2\sqrt{17}$, $6\sqrt{2}$, $4\sqrt{5}$, $2\sqrt{21}$

20. Explanations may vary.

21. a) $2\sqrt{6}$, 5, $3\sqrt{3}$, $4\sqrt{2}$ **b)** $2\sqrt{7}$, $\sqrt{31}$, $4\sqrt{2}$, $3\sqrt{5}$
  c) $6\sqrt{2}$, $5\sqrt{3}$, $2\sqrt{19}$, $4\sqrt{5}$, $3\sqrt{10}$

22. Instructions may vary.

23. $3\sqrt{2}$ cm $\doteq 4.243$ cm, 18 cm^2; 3 cm, 9 cm^2

24. a) , **c)**

Side length (cm)	Area (cm²)
3	9
$3\sqrt{2}$	18
6	36
$6\sqrt{2}$	72
12	144
$12\sqrt{2}$	288
24	576
$24\sqrt{2}$	1152

  b) Explanations of patterns may vary.

25. a) 17.321, 173.21, 1732.1   **b)** 0.173 21, 0.017 321
  c) 3.4642, 5.1963, 6.9284, 8.6605   **d)** 0.866 05, 0.577 $3\overline{6}$

26. 3 square units

Investigate, page 123

1. 3, 3, 3; yes

2. 2.646 751 311, 2.646 751 311, 2.646 751 311; yes

3. Examples may vary.   **4.** $\dfrac{\sqrt{a}}{\sqrt{b}} = \sqrt{\dfrac{a}{b}}$

2.8 Exercises, page 127

1. a) $2\sqrt{2}$ **b)** $\sqrt{7}$ **c)** $2\sqrt{3}$ **d)** $3\sqrt{5}$ **e)** $3\sqrt{3}$
  f) $\dfrac{5\sqrt{2}}{2}$ **g)** 4 **h)** 1 **i)** $\dfrac{\sqrt{2}}{7}$ **j)** $\dfrac{1}{9}$

2. Answers may vary.
  a) $\dfrac{3}{2}$ **b)** $\dfrac{\sqrt{2}}{3}$ **c)** $\dfrac{\sqrt{6}}{3}$ **d)** $\dfrac{\sqrt{7}}{5}$ **e)** $\dfrac{1}{2\sqrt{3}}$
  f) $3\sqrt{6}$ **g)** $\dfrac{1}{4\sqrt{5}}$ **h)** $\dfrac{12}{7}$ **i)** $\dfrac{\sqrt{3}}{3\sqrt{2}}$ **j)** 1

3. a) i) 2.236 **ii)** $\sqrt{5}$ **iii)** 2.236
  b) i) 7.071 **ii)** $5\sqrt{2}$ **iii)** 7.071
  c) i) 1.225 **ii)** $\dfrac{\sqrt{6}}{2}$ **iii)** 1.225
  d) i) 1.414 **ii)** $\sqrt{2}$ **iii)** 1.414
  e) i) 0.816 **ii)** $\dfrac{\sqrt{6}}{3}$ **iii)** 0.816

4. a) $\dfrac{1}{\sqrt{320}}$, $\dfrac{1}{\sqrt{1280}}$ **b)** $\dfrac{2}{\sqrt{1458}}$, $\dfrac{2}{\sqrt{13\,122}}$

---

5. Explanations may vary.

6. Answers may vary.
  a) $\dfrac{\sqrt{6}}{6}$ **b)** $\dfrac{3\sqrt{5}}{5}$ **c)** $2\sqrt{2}$ **d)** $2\sqrt{5}$
  e) $\sqrt{3}$ **f)** $\dfrac{3\sqrt{7}}{14}$ **g)** $\dfrac{\sqrt{15}}{4}$ **h)** $\dfrac{8\sqrt{6}}{15}$

7. Answers may vary.
  a) $\dfrac{15\sqrt{15}}{5\sqrt{3}}$ **b)** $\dfrac{8\sqrt{4}}{2\sqrt{2}}$ **c)** $\dfrac{36\sqrt{42}}{6\sqrt{6}}$
  d) $\dfrac{4\sqrt{20}}{2\sqrt{2}}$ **e)** $\dfrac{25\sqrt{30}}{5\sqrt{5}}$ **f)** $\dfrac{4\sqrt{6}}{2\sqrt{3}}$

8. Explanations may vary.

9. a) $2\sqrt{3}$, 3.46 **b)** $\dfrac{20}{\sqrt{3}}$, 11.55 **c)** $16\sqrt{3}$, 27.71

10. $\dfrac{2}{\sqrt{3}}$ cm, $\dfrac{4}{\sqrt{3}}$ cm

11. a) 3.5 cm   **b)** 6.0 cm

12. 8.32 cm, 7.21 cm

13. Answers may vary.
  a) 2 **b)** 2 **c)** $2\sqrt{2}$ **d)** $\dfrac{3}{\sqrt{2}}$ **e)** $\dfrac{5\sqrt{3}}{3}$
  f) $\dfrac{4\sqrt{14}}{15}$ **g)** $\sqrt{15}$ **h)** 2 **i)** $\dfrac{4\sqrt{5}}{3\sqrt{6}}$ **j)** $\sqrt{5}$

14. No; examples may vary.

15. Examples may vary.
  a) Yes; $a = 4$, $b = 9$ **b)** Yes; $a = 3$, $b = 5$ **c)** No

Investigate, page 129

1. No

2. a) 1.732 050 808, 2.236 067 978, 2.828 427 125   **b)** No

3. In general, $\sqrt{a} + \sqrt{b} \neq \sqrt{a+b}$

4. a) $\sqrt{2} + \sqrt{2} = 2\sqrt{2}$ **b)** $\sqrt{2} + 2\sqrt{2} = 3\sqrt{2}$
  c) $\sqrt{2} + 3\sqrt{2} = 4\sqrt{2}$

5. $\sqrt{2} + 4\sqrt{2} = 5\sqrt{2}$

6.

Expression	Radical	Expression	Radical
$\sqrt{2} + \sqrt{2}$	$2\sqrt{2}$		
$\sqrt{2} + 2\sqrt{2}$	$3\sqrt{2}$	$\sqrt{2} + \sqrt{8}$	$\sqrt{18}$
$\sqrt{2} + 3\sqrt{2}$	$4\sqrt{2}$	$\sqrt{2} + \sqrt{18}$	$\sqrt{32}$
$\sqrt{2} + 4\sqrt{2}$	$5\sqrt{2}$	$\sqrt{2} + \sqrt{32}$	$\sqrt{50}$
$\sqrt{2} + 5\sqrt{2}$	$6\sqrt{2}$	$\sqrt{2} + \sqrt{50}$	$\sqrt{72}$
$\sqrt{2} + 6\sqrt{2}$	$7\sqrt{2}$	$\sqrt{2} + \sqrt{72}$	$\sqrt{98}$

2.9 Exercises, page 132

1. a) $2\sqrt{7}$ **b)** $16\sqrt{6}$ **c)** $-6\sqrt{13}$
  d) $-25\sqrt{19}$ **e)** $33\sqrt{3}$ **f)** $5\sqrt{15}$

2. a) $-4\sqrt{5}$ **b)** $3\sqrt{10}$ **c)** $18\sqrt{2}$
  d) $13\sqrt{6} - 6\sqrt{2}$ **e)** $3\sqrt{5} - 6\sqrt{10}$ **f)** $8\sqrt{5} - 11\sqrt{2}$

3. a) $5\sqrt{10}$ **b)** $6\sqrt{2}$ **c)** $-3\sqrt{3}$ **d)** $-\sqrt{5}$
  e) $2\sqrt{2}$ **f)** $-2\sqrt{6}$ **g)** $-2\sqrt{5}$ **h)** $8\sqrt{6}$

4. Explanations may vary.   **5.** $\sqrt{10}$

6. a) i) 22.517 **ii)** $13\sqrt{3}$ **iii)** 22.517
  b) i) 22.627 **ii)** $16\sqrt{2}$ **iii)** 22.627
  c) i) -2.236 **ii)** $-\sqrt{5}$ **iii)** -2.236

7. Explanations may vary.

8. a) $7\sqrt{6}$ **b)** $3\sqrt{7}$ **c)** $2\sqrt{5}$
d) $9\sqrt{3}$ **e)** $11\sqrt{3}$ **f)** $-4\sqrt{2}$

9. Answers may vary.
a) $3\sqrt{8} + 4\sqrt{2}$ **b)** $\sqrt{3} + 2\sqrt{12}$ **c)** $2\sqrt{7} + 3\sqrt{28}$
d) $2\sqrt{5} + 2\sqrt{20}$ **e)** $6\sqrt{11} + 2\sqrt{44}$ **f)** $3\sqrt{6} + 2\sqrt{54}$

10. Answers may vary.
a) $6\sqrt{8} - 2\sqrt{2}$ **b)** $4\sqrt{12} - 3\sqrt{3}$ **c)** $12\sqrt{7} - 2\sqrt{28}$
d) $6\sqrt{20} - 6\sqrt{5}$ **e)** $9\sqrt{44} - 8\sqrt{11}$ **f)** $17\sqrt{6} - 4\sqrt{24}$

11. Explanations may vary.

12. a) $5\sqrt[3]{2}$ **b)** $9\sqrt[3]{2}$ **c)** $3\sqrt[3]{3}$
d) $2\sqrt[3]{3}$ **e)** $16\sqrt[3]{5}$ **f)** $29\sqrt[3]{4}$

13. a) $4\sqrt{2}$ cm **b)** $8\sqrt{2}$ cm^2 **c)** $(8\sqrt{2} + 4)$ cm

14. $12\sqrt{7}$ cm^2, $(12 + 4\sqrt{7})$ cm

15. a) $10\sqrt{3}$ **b)** $13\sqrt{3}$ **c)** $24\sqrt{2}$ **d)** $-8\sqrt{2}$
e) $13\sqrt{6}$ **f)** $6\sqrt{5}$ **g)** $15\sqrt{2}$ **h)** $9\sqrt{7}$
i) $10\sqrt{6} - 2\sqrt{7}$ **j)** $6\sqrt{7} - 24\sqrt{2}$

16. a) $-33\sqrt{3}$ **b)** $7\sqrt{7} + \sqrt{11}$ **c)** $\sqrt{3} - 5\sqrt{5}$
d) $47\sqrt{2} - \sqrt{7}$ **e)** $-6\sqrt{3} - 9\sqrt{2}$ **f)** $41\sqrt{6} + 36\sqrt{7}$
g) $-16\sqrt{3} - 2\sqrt{2}$ **h)** $-2\sqrt{7} - 16\sqrt{2}$

17. Explanations may vary.

18. About 5 km **19.** About 4 km

20. No **21.** $(8 + 8\sqrt{3})$ cm

22. a) No **b)** $4\sqrt{3}$ cm

23. Examples may vary. $a = 0$ or $b = 0$

2.10 Exercises, page 138

1. a) $\sqrt{10} - \sqrt{14}$ **b)** $\sqrt{33} + \sqrt{6}$ **c)** $\sqrt{78} - \sqrt{30}$
d) $6\sqrt{15} + 6\sqrt{21}$ **e)** $\sqrt{30} - 5\sqrt{2}$ **f)** $12\sqrt{22} + 20\sqrt{26}$

2. a) $\sqrt{39} + 13$ **b)** $9 - 18\sqrt{2}$ **c)** -48
d) $6\sqrt{10} + 8\sqrt{15}$ **e)** $36\sqrt{3} - 72\sqrt{2}$ **f)** $36 - 40\sqrt{3}$

3. a) $\dfrac{2\sqrt{5}}{5}$ **b)** $\dfrac{7\sqrt{11}}{11}$ **c)** $\dfrac{4\sqrt{3}}{3}$
d) $\dfrac{5\sqrt{14}}{14}$ **e)** $2\sqrt{30}$ **f)** $\dfrac{12\sqrt{35}}{35}$
g) $3\sqrt{10}$ **h)** $\dfrac{5\sqrt{21}}{3}$

4. a) 8.456 **b)** 3.543 **c)** -2.461
d) -2.349 **e)** 130.395 **f)** 77.367

5. a) $5 + 3\sqrt{3}$ **b)** $1 - 5\sqrt{7}$ **c)** $19 - 11\sqrt{2}$
d) $2 + 6\sqrt{3}$ **e)** 3 **f)** $9\sqrt{2} - 12$
g) $11 - 6\sqrt{2}$ **h)** $4 + 2\sqrt{3}$ **i)** $33 - 20\sqrt{2}$

6. Explanations may vary.

7. a) $5 + 3\sqrt{3}, 9 + 5\sqrt{3}, 15 + 7\sqrt{3}; 23 + 9\sqrt{3}, 33 + 11\sqrt{3},$
$45 + 13\sqrt{3}$
b) $3 + 2\sqrt{2}, 4 + 2\sqrt{3}, 5 + 2\sqrt{4}; 6 + 2\sqrt{5}, 7 + 2\sqrt{6},$
$8 + 2\sqrt{7}$

8. a) $14 + 7\sqrt{2}$ **b)** $52 - 38\sqrt{2}$ **c)** $2 - 7\sqrt{3}$
d) $25 - 11\sqrt{7}$ **e)** 4 **f)** 18
g) $65 - 6\sqrt{14}$ **h)** $18 + 12\sqrt{2}$ **i)** $98 - 24\sqrt{5}$

9. a) $5 + 2\sqrt{6}$ **b)** 1 **c)** 18 **d)** 6
e) $32 - 10\sqrt{7}$ **f)** -18 **g)** $57 + 12\sqrt{15}$
h) 33 **i)** 147 **j)** 63

10. Explanations may vary.

11. a) 7.323 **b)** 1.022 **c)** 11.904

12. Descriptions and explanations may vary.
a) $\dfrac{\sqrt{5} + \sqrt{3}}{2}, \dfrac{\sqrt{6} + \sqrt{3}}{3}, \dfrac{\sqrt{7} + \sqrt{3}}{4}; \dfrac{\sqrt{8} + \sqrt{3}}{5}, \dfrac{\sqrt{9} + \sqrt{3}}{6},$
$\dfrac{\sqrt{10} + \sqrt{3}}{7}$
b) $2 - \sqrt{2}, 3 - \sqrt{6}, 4 - \sqrt{12}; 5 - \sqrt{20}, 6 - \sqrt{30}, 7 - \sqrt{42}$

13. a) $\dfrac{6 + 4\sqrt{3}}{3}$ **b)** $\dfrac{35 - 3\sqrt{7}}{7}$ **c)** $\dfrac{20 - 2\sqrt{5}}{5}$
d) $2\sqrt{6} - 1$ **e)** $\dfrac{8\sqrt{30} + 5}{5}$ **f)** $3\sqrt{5} - 1$
g) $\dfrac{10\sqrt{3} + 3\sqrt{2}}{3}$ **h)** $\dfrac{\sqrt{6}}{2}$

14. a) $\sqrt{5} + \sqrt{2}$ **b)** $\dfrac{5\sqrt{7} - 5\sqrt{3}}{4}$ **c)** $\dfrac{7 + \sqrt{5}}{4}$
d) $\dfrac{\sqrt{30} - \sqrt{5}}{5}$ **e)** $\dfrac{6\sqrt{2} + \sqrt{30}}{7}$ **f)** $\dfrac{\sqrt{105} + \sqrt{70}}{5}$
g) $\sqrt{5} - \sqrt{2}$ **h)** $\dfrac{15 - 3\sqrt{5}}{10}$

15. a) $3 + 2\sqrt{2}$ **b)** $\dfrac{7 + 3\sqrt{5}}{2}$ **c)** $4 - \sqrt{15}$
d) $\dfrac{\sqrt{6}}{2}$ **e)** $\dfrac{32 + 7\sqrt{6}}{10}$ **f)** $\sqrt{15}$

16. Explanations may vary.

17. Form of expressions may vary.
a) $\dfrac{\sqrt{3} + \sqrt{5}}{\sqrt{15}}$ **b)** $\dfrac{\sqrt{3} - 1}{\sqrt{6}}$ **c)** $\dfrac{\sqrt{2} + 1}{\sqrt{6}}$
d) $\dfrac{2\sqrt{5} - 3\sqrt{7}}{\sqrt{35}}$ **e)** $\dfrac{3\sqrt{3} + 2\sqrt{2}}{6}$ **f)** $\dfrac{5\sqrt{3} - 4}{\sqrt{24}}$
g) $\dfrac{7\sqrt{3} - 4\sqrt{5}}{2\sqrt{15}}$ **h)** $-\dfrac{9}{2\sqrt{6}}$

18. Explanations may vary.

19. a) $(5\sqrt{3} + 5)$ cm; $(5\sqrt{3} - 5)$ cm
b) 25 cm^2; 50 cm^2 **c)** $(75 + 25\sqrt{3})$ cm^2

20. a) $(6 + 4\sqrt{3})$ cm; 12.928 cm
b) $(36 + 21\sqrt{3})$ cm^2; 72.373 cm^2

21. a) $\dfrac{8\sqrt{3}}{2 + \sqrt{3}}$ cm; 3.713 cm **b)** $\dfrac{192}{7 + 4\sqrt{3}}$ cm^2; 13.785 cm^2

2 Review, page 141

1. a) 5 **b)** 3.7 **c)** 3
d) 27 **e)** 16 **f)** 2

2. a) $\dfrac{3}{4}$ **b)** 8 **c)** $\dfrac{1}{100}$ **d)** $\dfrac{1}{20}$ **e)** $\dfrac{64}{27}$

3. Explanations may vary.

4.

Year	1956	1961	1966	1971	1976	1981	1986	1991	1996
Population	36	58	95	154	250	406	659	1069	1736

5. Explanations and assumptions may vary.

6. a) 4.12 **b)** 6.71 **c)** $6.06, 3.50$ **d)** $6.93, 8.00$

8. a) Irrational **b)** Proofs may vary.

9. a) $3\sqrt{2}$ **b)** $9\sqrt{3}$ **c)** $10\sqrt{5}$ **d)** $7\sqrt{7}$

10. a) $30\sqrt{2}$ **b)** -72 **c)** $-300\sqrt{5}$
d) $\dfrac{3\sqrt{3}}{2}$ **e)** $\dfrac{35}{6}$ **f)** $\dfrac{\sqrt{30}}{180}$

11. a) $\dfrac{\sqrt{2}}{\sqrt{3}}; \dfrac{4}{9\sqrt{3}}, \dfrac{4\sqrt{2}}{27}$ **b)** $\dfrac{\sqrt{3}}{\sqrt{5}}; \dfrac{3}{5\sqrt{5}}, \dfrac{3\sqrt{3}}{25}$
c) $\left(\dfrac{1}{2}\right)^{\frac{1}{7}}; \left(\dfrac{1}{2}\right)^{\frac{4}{7}}; \left(\dfrac{1}{2}\right)^{\frac{5}{7}}$ **d)** Not geometric; $\dfrac{\sqrt{5}}{6}, \dfrac{\sqrt{6}}{7}$

12. Explanations may vary.

13. a) 0.717 **b)** 3.414 **c)** -1.604
d) 0.581 **e)** 0.697 **f)** 5.464

14. a) $5\sqrt{5}$ **b)** $7\sqrt{3}$ **c)** $2\sqrt{7}$

MATHEMATICS 10

d) $4\sqrt{30}$ e) $4\sqrt{5}$ f) $23\sqrt{2}$

15. a) 0.318   b) 2.898   c) 1.678

16. a) $\dfrac{3\sqrt{6} - \sqrt{3}}{3}$ b) $\dfrac{5\sqrt{6} - \sqrt{10}}{2}$

   c) $\dfrac{6\sqrt{6} + 6\sqrt{2} - 3 - \sqrt{3}}{2}$ d) $\dfrac{3\sqrt{10} + 3\sqrt{6} - 5\sqrt{15} - 15}{2}$

17. Explanations may vary.

18. a) $4\sqrt{3} - 7$ b) $\dfrac{2\sqrt{6} - 9}{19}$ c) $3 - 2\sqrt{2}$ d) $4 + \sqrt{6}$

19. a) $\dfrac{\sqrt{2} + \sqrt{3}}{\sqrt{6}}$ b) $\dfrac{3\sqrt{6} - 2\sqrt{7}}{\sqrt{42}}$ c) $\dfrac{\sqrt{5} + 3\sqrt{6}}{3\sqrt{2}}$ d) $\dfrac{1}{2\sqrt{6}}$

2 Cumulative Review, page 143

1. a) 3, 3, 3, 3, 3   b) 3, 8, 13, 18, 23
   c) 3, 13, 23, 33, 43   d) 3, 28, 53, 78, 103
   e) 3, 78, 153, 228, 303   f) 3, 753, 1503, 2253, 3003

2. a) $2n - 1$ b) $4n - 3$ c) $8n - 7$
   d) $3n + 1$ e) $4n + 2$ f) $13 - 2n$

3. a) 55   b) 110   c) 165
   d) 220   e) 275   f) 330

4. a) 16   b) 13 122   c) 177 147   d) 0.007 812 5

5. a) $x^6 y^3$ b) y^8 c) $x^7 y^3$
   d) $x^3 y^{-1}$ e) x^2 f) $x^{-18} y^{-19}$

6.

Date (January each year)	1993	1994	1995	1996	1997	1998	1999
Number of hosts (millions)	1.3	2.4	4.6	8.6	16.2	30.4	57.2

7. a) $\dfrac{1}{\sqrt{3}}$ cm   b) $\dfrac{2}{\sqrt{3}}$ cm   c) $\dfrac{4}{\sqrt{3}}$ cm   d) $\dfrac{8}{\sqrt{3}}$ cm
   e) Geometric sequence with common ratio 2; $\dfrac{16}{\sqrt{3}}$ cm, $\dfrac{32}{\sqrt{3}}$ cm

Chapter 3

Investigate, page 146

1. a) 6   b) 7   c) 13

2. a) Answers may vary.
   b) Subtract 2 from 8.   c) $2 - (-5) = 7$
   d) Answers may vary. No, if you know the distance is always positive

3. a) 4   b) 3

5. Right triangle; 5

6. Pythagorean Theorem; $AC = \sqrt{AB^2 + BC^2}$

7. $(3, -7)$; 13   8. $(-2, 5)$ 9. $\sqrt{73}$

3.1 Exercises, page 149

1. a) 5 units   b) 7 units   c) 11 units
   d) 12 units   e) 8 units   f) 14 units

2. AB: 7 units, CD: $2\sqrt{10}$ units, EF: $3\sqrt{10}$ units, GH: $\sqrt{73}$ units, JK: $\sqrt{85}$ units

3. Estimates may vary.
   a) $\sqrt{20}$ units   b) $\sqrt{52}$ units   c) $\sqrt{50}$ units
   d) $\sqrt{137}$ units   e) $\sqrt{29}$ units   f) $\sqrt{28.25}$ units

4. a) HA: 5 units, HT: 12 units, AT: 13 units. DI: 5 units,

IP: 3 units, DP: $\sqrt{34}$ units. BE: 6 units, EG: 3 units, BG: $\sqrt{45}$ units. RK: $\sqrt{10}$ units, RM: $\sqrt{40}$ units, MK: $\sqrt{50}$ units. SJ: $\sqrt{20}$ units, JL: $\sqrt{20}$ units, SL: $\sqrt{40}$ units

   b) △HAT: 30 square units; △DIP: 7.5 square units; △BEG: 9 square units; △MRK: $\dfrac{\sqrt{629}}{2}$ square units; △SJL: 10 square units

5. a) $\sqrt{98}$ units   b) $\sqrt{80}$ units   c) $\sqrt{89}$ units   d) $\sqrt{85}$ units

6. a) AB: $\sqrt{53}$ units, BC: $\sqrt{106}$ units, AC: $\sqrt{53}$ units; △ABC is isosceles.
   b) PQ: $\sqrt{53}$ units, QR: 5 units, PR: $\sqrt{52}$ units; △PQR is scalene.
   c) JK: $\sqrt{72}$ units, KL: $\sqrt{68}$ units, JL: $\sqrt{68}$ units; △JKL is isosceles.

7. Explanations may vary.

8. a) AB, CD: 9 units; BC, DA: 8 units; perimeter: 34 units
   b) JK, LM: $3\sqrt{10}$ units; KL, MJ: $\sqrt{10}$ units; perimeter: $8\sqrt{10}$ units
   c) PQ, RS: $2\sqrt{13}$ units; QR, SP: $4\sqrt{13}$ units; perimeter: $12\sqrt{13}$ units

9. a) AC, BD: $\sqrt{145}$ units; area ABCD: 72 square units
   b) JL, KM: 10 units; area JKLM: 30 square units
   c) PR, QS: $2\sqrt{65}$ units; area PQRS: 104 square units

10. a) AB: 5 units, BC: 5 units, AC: 10 units
    b) AB: 10 units, BC: 5 units, AC: 15 units
    c) AB: $4\sqrt{5}$ units, BC: $2\sqrt{5}$ units, AC: $6\sqrt{5}$ units; in each case, AB + BC = AC; A, B, and C lie in a straight line.

11. A(6, 4), C(−3, 1)   12. Explanations may vary.

13. a) 3   b) 180 m   c) About 134 m

14. a) i) 4   ii) $\sqrt{17}$ iii) $\sqrt{20}$ iv) 5
    b) 0   c) 2, 8   d), e) Explanations may vary.

15. a) 10   b) About 180 m

16. a) Answers may vary.
    i) (1, 0), (1, 1)   ii) (1, 2), (4, 6)
    iii) (1, 2), (8.5, 2)   iv) (1, 2), (5, −11)
    v) (0, 0), (20, 0)   vi) (1, 2), (11.5, 2)

17. a) Answers may vary.
    i) (1, 2), (2, 3)   ii) (2, 2), (4, 3)
    iii) (5, 5), (3, 2)   iv) (4, 3), (8, 8)
    v) (1, 0), (3, −5)   vi) (−2, −1), (−6, −3)

18. a) Heather: 54 m; Kim: 76 m   b) 120 m   c) No

19. Coast guard cutter

20. a) AB: 134 nm; BC: 72 nm; CA: 160 nm
    b) 1 h 32 min, 1 h 12 min
    c) Answers may vary. $1320.20 will just cover operating costs.
    d) Answers may vary. Yes   e) Answers may vary.

21. 526 m, 410 m; 1365 m; 105 313 m^2

22. a) (7, 0), (7, 13), (0, 13)   b) 1.84 m

23. a) (7, 4)   b) 2.2 m

24. P(0, −3), R(5, 7), S(2, 1)   25. (2, 0), (8, 0)

26. a) (0, 3)   b) (0, 2)   c) (0, 1)

27. Explanations may vary.

28. a) Ottawa to Vancouver: 2651; Ottawa to Quebec City: 812; Ottawa to Washington: 1431
b) Calgary

Linking Ideas: Mathematics and Technology
Shortest Networks, page 154

2. The y-coordinate stays the same; the x-coordinate increases from 50 to 60.

4. 71.1

5. 120°, 60°; 60°, 30°, 90°

Investigate, page 156

1. At 5

2. a) Answers may vary. **b)** $\frac{9+1}{2} = 5$

3. (3, 5); (5, 1) **4.** Answers may vary.

5. (3, 1); each coordinate is the mean of the corresponding coordinates of A and C.

3.2 Exercises, page 158

1. AB: (1, 4); CD: (4, 2.5); EF: (−1, 0); GH: (2, −1); JK: (0, 5)

2. a) (3, 4) **b)** (−2, 2) **c)** (4, 5)
d) (1, 1.5) **e)** (−1.5, 1.5) **f)** (−2, 2.5)
g) (2, −0.5) **h)** (−2, −2) **i)** (−3.5, 1.5)

3. JK: (4, 2); KL: (7, 5); JL: (3, 6)

4. a) (2, 1) **b)** 5

5. a) AC: (6, 1); BD: (6, 1)
b) PR: (5, 2); QS: (5, 2), the diagonals of a rectangle intersect at their midpoints.

6. EG: (5, 0.5); FH: (5, 0.5), the diagonals of a parallelogram intersect at their midpoints.

7. a) M(1, 4) **b)** All lengths are 5 units.

8. a) M(6, 5), N(2, 2)
b) MN: 5 units; AB: 10 units; AB = 2MN

9. a) DF: (3.5, −1.5), EF: (−0.5, −1), DE: (1, 2.5)
b) Answers may vary. The triangles are similar; the area of the smaller triangle is one-quarter the area of the larger triangle.

10. b) Median from P is $\sqrt{45}$ units; median from Q is $\sqrt{72}$ units; median from R is 9 units.

11. (−2, 1), (0, 4) **12.** Explanations may vary.

13. a) i) (0, −4.5) **ii)** (0, −4.5) **iii)** (0, −4.5)
b) Answers may vary.
i) (−2, −4.5), (2, −4.5) **ii)** (−1, −5), (1, −4)
iii) (−2, −6), (2, −3)

14. Answers may vary. The x-coordinate of B is the opposite of the x-coordinate of A. The y-coordinate of B is double the y-coordinate of M.

15. a) B(6, −6) **b)** B(2, −9) **c)** B(−6, −2) **d)** O(0, 0)

16. Explanations may vary.

17. a) Answers may vary.

i) (4, 6), (2, 4) **ii)** (4, −3), (6, −5)
iii) (−6, −1), (0, −7) **iv)** (−7, 0), (−3, 4)

18. a) (−1, 0), (2, −4), (5, −8)
b) (1.5, −4), (−2, −1), (−5.5, 2)
c) (0.5, −3.5), (3, −1), (5.5, 1.5)
d) (−2.5, −1.5), (0, 0), (2.5, 1.5)

19. a) (14, −4) **b)** (10, 10) **c)** (−8, 9) **d)** (0, 0)

20. a) (2, 4) **b)** (7, 3) **c)** $\sqrt{26}$

21. a) (8.9, 4.4) **b)** $108\ 490 **c)** $216\ 980

22. a) $\left(\frac{x_2 + 2x_1}{3}, \frac{y_2 + 2y_1}{3}\right)$, $\left(\frac{2x_2 + x_1}{3}, \frac{2y_2 + y_1}{3}\right)$
b) $\left(\frac{x_2 + 3x_1}{4}, \frac{y_2 + 3y_1}{4}\right)$, $\left(\frac{x_2 + x_1}{2}, \frac{y_2 + y_1}{2}\right)$, $\left(\frac{3x_2 + x_1}{4}, \frac{3y_2 + y_1}{4}\right)$

23. a) 6, 8 **b)** C(6, 8) is the midpoint of the square.

24. (2, 2) **25.** (1, 6), (5, −4), (−3, −2)

Investigate, page 162

1. a) Roof 1 **b)** Roof 2 **c)** Least to greatest: 2, 3, 1

2. a) , b)

House	rise	run	rise/run
1	11	5	2.2
2	5	8	0.625
3	7	6	$1.1\overline{6}$

c) Answers may vary. The steepest slope has the greatest $\frac{\text{rise}}{\text{run}}$.

3.3 Exercises, page 166

1. $\frac{1}{3}$ **2.** $\frac{5}{3}$

3. a) AB: $\frac{5}{8}$; AC: $\frac{3}{8}$; AD: $\frac{1}{8}$; AE: 0; AF: $-\frac{1}{4}$; AG: $-\frac{5}{8}$
b) PQ: $-\frac{7}{5}$; PR: $-\frac{7}{2}$; PS: undefined; PT: $\frac{7}{2}$; PU: $\frac{7}{4}$; PV: 1

4. Explanations may vary.

5. Answers may vary; rise = 6, run = 2

6. 24

7. a) $-\frac{11}{8}$ **b)** 3 **c)** $-\frac{5}{2}$
d) Undefined **e)** $\frac{2}{3}$ **f)** 0

8. −7

9. a) AB: −1; BC: $\frac{9}{2}$; AC: $\frac{4}{7}$ **b)** RS: $\frac{1}{3}$; ST: $\frac{5}{2}$; RT: $-\frac{7}{5}$
c) LM: $-\frac{5}{4}$; MN: 0; LN: undefined
d) EF: −5; FG: 2; EG: 1

10. Answers may vary. (3, 0), (4, 3)

11. a) Answers may vary.
i) (4, 7), (2, 1) **ii)** (−3, 3), (−2, −2)
iii) (0, 0), (5, 3) **iv)** (−3, −3), (−6, 2)
v) (1, 3), (3, −3) **vi)** (0, 0), (1, 5)

12. a) 4 **b)** 3.5; no **c)** Explanations may vary.

13. a) A, B, C

b) A: 10 500 ft, 1312.5 ft/min; B: 20 000 ft, 1538 ft/min; C: 20 000 ft, 1818 ft/min
c) No **d)** Answers may vary. **e)** Answers may vary.

14. a) About 10.2 m/s; Bailey's average speed
  b) About 10.4 m/s; Johnson's average speed
  c) Answers may vary. Johnson had the greater average speed.

15. a) Answers may vary. The graph would consist of 2 line segments.
  b) Johnson's time is greater.
  c) Bailey's time would be greater.

16. All figures can be drawn.

17. a) 7   **b)** 5   **c)** 4   **d)** 0

18. a) 6   **b)** −2   **c)** −1   **d)** 4

19. Explanations may vary.

20. a) $4\sqrt{3}$; $\frac{1}{\sqrt{3}}$; 8   **b)** $8\sqrt{3}$; 8; $\frac{1}{\sqrt{3}}$ **c)** $\frac{12}{\sqrt{3}}$; $\frac{1}{\sqrt{3}}$; $\frac{24}{\sqrt{3}}$

21. a) (−2, 0)   **b)** (1, 0)   **c)** (2, 0)
  d) (−8, 0)   **e)** (7, 0)   **f)** (16, 0)

22. a) (0, 13)   **b)** (0, −2)   **c)** (0, 7)
  d) (0, 5.5)   **e)** (0, 3.25)

23. (−2, 0), (8, 0)   **24.** (0, 6), (8, 0) and (0, −6), (−8, 0)

Linking Ideas: Mathematics and Construction
Building the Best Staircase, page 170

1. a) There is 1 more riser than tread.
  b) Answers may vary.   **c)** C   **d)** A

2. a) $x + 2y = 620$ **b)** B, C   **c)** No   **d)** 50 mm

3. a) 220 mm; 0.91   **b)** 185 mm; 0.74   **c)** 14

4. Explanations may vary.

Linking Ideas: Mathematics and Technology
Length, Midpoint, and Slope of a Line Segment, page 172

1. b) B4: Subtract the y-coordinates. B5: Subtract the x-coordinates.
  c) B6: Applies the distance formula to calculate the length of the line segment. B7: Calculates the mean of the x-coordinates to determine the x-coordinate of the midpoint of the line segment. C7: Calculates the mean of the y-coordinates to determine the y-coordinate of the midpoint of the line segment. B8: Divides the difference in y-coordinates by the difference in x-coordinates to calculate slope.

5. a) Tables may vary.   **b)** Answers may vary.

6. a) $(4, 4\sqrt{3})$ **b)** 60°; it's a 30-60-90 triangle

7. a) 10π
  b) Perimeter of polygon is less than circumference.

8. a) 3   **b)** $\sqrt{24}$, $\sqrt{21}$, 4
  c) 1.005, 1.049, 1.157, 1.414   **d)** 1.157, 1.049, 1.005

9. 31.344

10. a) 3.1344; explanations may vary.
  b) Less than π; explanations may vary.
  c) About 0.2%

Investigate, page 175

1. a) Parallel   **b)** Equal   **c)** Equal   **d)** Equal

2. No   **3.** Parallel line segments have the same slope.

3.4 Exercises, page 177

1. AB ∥ CD; SR ∥ QP; explanations may vary.

2. a) AB: 2; CD: 2; AB ∥ CD
  b) EF: $\frac{3}{8}$; GH: $\frac{2}{5}$ **c)** RS: $-\frac{3}{4}$; TU: $-\frac{2}{3}$

3. a) Slopes AB: $\frac{3}{4}$; CD: $\frac{3}{4}$; AB ∥ DC; BC: −5; DA: −5; BC ∥ AD; ABCD is a parallelogram.
  b) Slopes PQ: $-\frac{7}{4}$; SR: $-\frac{7}{3}$; QR: $\frac{2}{3}$; SP: $\frac{8}{13}$; PQRS is not a parallelogram.
  c) Slopes JK: −3; LM: −3; JK ∥ ML; KL: $-\frac{3}{8}$; JM: $-\frac{3}{8}$; KL ∥ JM; JKLM is a parallelogram.

4. Explanations may vary.

5. a) i) $\frac{2}{3}$ **ii)** $\frac{2}{3}$ **iii)** $\frac{2}{3}$ **iv)** $\frac{2}{3}$
  b) The slopes are equal.   **c)** Answers may vary. (10.5, 7)

6. c) Slopes JK, LM: $\frac{4}{3}$

7. a) M(4, 1), N(1, −1)   **b)** MN: $\frac{2}{3}$; AB: $\frac{2}{3}$; MN ∥ AB
  c) P(−1, 3); slopes PM: $-\frac{2}{5}$; BC: $-\frac{2}{5}$; PN: −2; AC: −2; PM ∥ BC and PN ∥ AC

8. a) Midpoint AB: (2, 1); midpoint BC: (5, 5); midpoint CD: (0, 6); midpoint DA: (−3, 2)
  b) Parallelogram
  c) Opposite sides are parallel with slopes $\frac{4}{3}$ and $-\frac{1}{5}$.

9. D(0, 5.5)   **10.** D(0, 6)   **11.** Answers may vary.

12. a) Answers may vary.
  i) (7, 3), (−6, 0)   **ii)** (0, 6), (2, 0)
  iii) (5, 4), (3, −1)   **iv)** (5, 2), (10, −1)
  b) Explanations may vary.

13. −5   **14.** 9

15. a) (−12, −3), (6, −5), (4, 5)   **b)** (13, −11), (3, 1), (−11, 7)

16. a) (0, 0), (10, 0), (4, 6)   **b)** (5, 10), (5, 0), (−11, 6)

Mathematics File: Solving Equations of the Form $\frac{a}{b} = \frac{c}{d}$, page 182

1. a) 6   **b)** 9   **c)** 7.5   **d)** 3.5
  e) 20   **f)** 6.4   **g)** 10   **h)** 0.5

2. b) $9.\overline{3}$ **d)** 20   **f)** 6   **h)** 12.5

3. Answers may vary.

4. a) −21   **b)** −4   **c)** −3.6   **d)** −7.5

5. a) 6   **b)** −6   **c)** −10   **d)** $-\frac{8}{7}$
  e) 8   **f)** −7.5   **g)** 3   **h)** $\frac{6}{7}$

6. a) (−7.5, 0)   **b)** (7.5, 0)   **c)** (−2, 0)   **d)** (2, 0)

7. a) 3.5   **b)** Explanations may vary.

8. D(4, 0)

Investigate, page 184

1. a) Perpendicular   **b)** Opposites
c) Equal   **d)** Negative reciprocals

2. No

3. Perpendicular line segments have slopes that are negative reciprocals.

3.5 Exercises, page 187

1. a) CD $\perp$ AB   **c)** GF $\perp$ EF

2. a) $-\frac{3}{2}$ **b)** $-\frac{8}{5}$ **c)** $\frac{4}{3}$ **d)** 2 **e)** 3

3. $\frac{3}{4}$, $-\frac{4}{3}$; -4, $\frac{1}{4}$; 2, $-\frac{1}{2}$

4. a) Slopes OB: $\frac{2}{3}$; CD: $-\frac{3}{2}$; OB $\perp$ CD
b) Slopes HI: $\frac{1}{3}$; JK: -3; HI $\perp$ JK
c) Slopes LM: $-\frac{7}{4}$; NP: $\frac{3}{5}$

5. Explanations may vary.

6. c) Slope JK: $\frac{4}{3}$, slope LM: $-\frac{3}{4}$

7. a) $\triangle$DEF is a right triangle.
b) $\triangle$ABC is a right triangle.
c) $\triangle$PQR is a right triangle.
d) $\triangle$KLM is not a right triangle.

8. a) ABCD is a rectangle.   **b)** JKLM is not a rectangle.
c) PQRS is a rectangle.

9. a) -2 **b)** $\frac{1}{2}$ **c)** -1
d) $\frac{2}{5}$ **e)** 0 **f)** Undefined

10. a) $\frac{1}{2}$ **b)** -2 **c)** 1
d) $-\frac{5}{2}$ **e)** Undefined   **f)** 0

11. a) $-\frac{1}{3}$ **b)** 2 **c)** -1 **d)** 1 **e)** 1
f) -8 **g)** 4 **h)** -10 **i)** -6 **j)** 35

12. Answers may vary.
a) C(-3, 5)   **b)** C(2, 10)   **c)** C(2, 3)
d) C(-1, 6)   **e)** C(-2, 7)   **f)** C(4, 5)

13. Explanations may vary.

14. a) P(9, 0)   **b)** P(0, 6)

15. a) All sides $\sqrt{65}$ units   **b)** Rhombus
c) Both midpoints (1, 1)   **d)** Slopes AC: -2; BD: $\frac{1}{2}$
e) The diagonals of a rhombus bisect each other at right angles.

16. a) (4, 2)   **b)** (2, -4)   **c)** (5, 1)   **d)** (5, -2)

17. Explanations may vary.

18. a) Answers may vary.
    i) (0, 0), (-3, 13)   **ii)** (1, 1), (4, 2)
    iii) (2, 3), (4, -2)   **iv)** (-1, 2), (2, 7)
b) Explanations may vary.

19. a) C(4, 5), D(5, 1)   **b)** C(7, 10), D(10, 3)
c) C(1, 10), D(10, 9)   **d)** C(b, $a + b$), D($a + b$, a)

20. T(0, 0) or T(5, 0)

21. (-1, 0), (7, 0), (2, 0), (3, 0)

22. a) (7, 3), (5, -2); (3.5, 1.5), (-1.5, 3.5); (-3, 7), (-5, 2)
b) (4, 8), (7, 4); (3.5, 4.5), (-0.5, 1.5), (-4, 2), (-1, -2)

Problem Solving: Square Patterns on a Grid, page 190

1. a) 15   **b)** 12   **c)** Explanations may vary.

2. a) 5   **b)** 16, 8, 4, 2, 1: 4, $\sqrt{8}$, 2, $\sqrt{2}$, 1
c) Geometric with common ratio $\frac{1}{2}$; geometric with common ratio $\frac{1}{\sqrt{2}}$

3. Answers may vary.

4. a) Answers may vary. The squares have the same area.
b) 4   **c)** Answers may vary.

5. a) 8

6. a) 8   **b)** 15, 24   **c)** $\frac{(n + 3)(n - 1)}{4}$

7. Answers may vary.

8. a) Coordinates (1, 1) with lower left vertex as the origin.
b) $\sqrt{2}$, 2   **c)** 1, $\sqrt{2}$ **d)** Diagrams may vary.

3 Review, page 192

1. a) 15 units   **b)** 14 units   **c)** $10\sqrt{2}$ units
d) $5\sqrt{2}$ units   **e)** $4\sqrt{5}$ units   **f)** $8\sqrt{2}$ units

2. a) 8 units, 6 units   **b)** 28 units
c) 10 units   **d)** 48 square units

3. a) Right isosceles triangle   **b)** Right scalene triangle

4. Explanations may vary.

5. a) (3, 2)   **b)** (-3, -1)   **c)** (-4, -3)
d) (9, -5)   **e)** (1.5, 6)   **f)** (3.5, 2.5)

6. a) KL: (-1, 4); LM: (1, -2); KM: (5, 1) **b)** 10 units

7. a) $\frac{5}{2}$ **b)** 0   **c)** $-\frac{13}{7}$
d) $\frac{2}{3}$ **e)** $-\frac{5}{7}$ **f)** -3

8. a) ST: -2; PQ: -2 **b)** ST: $\frac{\sqrt{5}}{2}$ units; PQ: $\sqrt{5}$ units

10. a) AB: $\frac{3}{8}$; AC: 4   **b)** $-\frac{9}{5}$ **c)** $-\frac{6}{13}$

11. Slopes AB: 3; BC: $\frac{1}{6}$; CD: 6; AD: $\frac{1}{5}$; ABCD is not a parallelogram.

12. Explanations may vary.

13. a) $-\frac{3}{5}$ **b)** 4 **c)** $-\frac{7}{4}$ **d)** 0.3 **e)** -8

14. a) $\frac{5}{3}$ **b)** $-\frac{1}{4}$ **c)** $\frac{4}{7}$ **d)** $-\frac{10}{3}$ **e)** $\frac{1}{8}$

15. a) 15 **b)** -8 **c)** $-\frac{4}{3}$ **d)** -3

16. a) $-\frac{20}{3}$ **b)** 2 **c)** 12 **d)** $\frac{25}{3}$

17. a) $-\frac{6}{7}$ **b)** $-\frac{2}{3}$ **c)** Undefined

18. a) $\frac{7}{6}$ **b)** $\frac{3}{2}$ **c)** 0

19. Slopes AB: 1; BC: $\frac{5}{-6}$; CD: -5; DA: $\frac{1}{6}$; neither

20. a) S(5, 0)   **b)** S(0, -15)

21. a) S(13, 0)   **b)** S(0, $\frac{13}{3}$)

22. a) $6\sqrt{5}$ units   **b)** (2, 0)   **c)** $12\sqrt{17}$ units
d) $2\sqrt{85}$ units   **e)** (2, 0)

3 Cumulative Review, page 194

1. a) 25   **b)** 40   **c)** -11
d) 13   **e)** 70   **f)** 31

2. a) 7, 23, 39, 55   **b)** 7, 15, 23, 31, 39, 47, 55

3. a) 1.75, 7, 28, 112, 448 **b)** 5, 10, 20, 40, 80
c) 7, 21, 63, 189, 567 **d)** 6, −6, 6, −6, 6
e) 3, 6, 12, 24, 48 or 3, −6, 12, −24, 48
f) 4.5, 9, 18, 36, 72 or 4.5, −9, 18, −36, 72

4. a) i) $3.50 **ii)** $4.00 **iii)** $5.49
b) Sale price $39.99
i) $2.80 **ii)** $3.20 **iii)** $45.99
c) $45.99
d) Answers are the same. It doesn't matter to the customer when the discounts are applied.

5. a) Estimates may vary. Julianna's is worth more ($1126.83). Emile's is worth $1120.
b) Julianna's is worth more ($10 892.58). Emile's is worth $9646.29.

6. a) 2 **b)** 2 **c)** 2 **d)** 2 **e)** 2
f) 2 **g)** 2 **h)** 2 **i)** 2

7. a) 2 **b)** 3 **c)** About 7.1 m square

8. a) $5\sqrt{3}$ cm **b)** $25\sqrt{3}$ cm^2

9. b) From 1972 on, the slopes of the line segments joining the record times decrease slightly with each new record.
c) Answers may vary.
d) Answers may vary; when you join the first and last points with a line segment, there does not appear to be a limit to how quickly the race can be run. The answer in part c is more realistic since there is a limit to how well the human body can perform.
e) The slopes of line segments joining the record times decrease from 1960.

10. a) DE: $\sqrt{72}$, EF: $\sqrt{68}$, DF: $\sqrt{68}$
b) DE: (−1, 1), EF: (3, 2), DF: (0, 5)
c) DE: −1, EF: 4, DF: $\frac{1}{4}$

11. a) B(1, 6) **b)** B(7, 0) **c)** B(8, 7) **d)** B(6, 4)

12. (6, 4); 5 units

13. K(6, 0); (0, 4)

14. a) 2 **b)** 1

Chapter 4

4.1 Exercises, page 202

1. a) Answers may vary.

i)

x	y
0	3
−1.5	0
1	5
2	7

ii)

x	y
0	5
$\frac{5}{3}$	0
1	2
2	−1

iii)

x	y
0	12
3	0
1	8
2	4

iv)

x	y
0	−4
6	0
3	−2
−3	−6

v)

x	y
0	10
4	0
2	5
6	−5

vi)

x	y
0	−7.5
5	0
1	−6
3	−3

b) Answers may vary.

2. a) Tables of values may vary. **b)** Answers may vary.

3. a) Answers may vary.

n	C (¢)
0	70
10	270
20	470
30	670
40	870
50	1070
60	1270
70	1470
80	1670
90	1870
100	2070

c) $15.70
d) 46

4. a) Answers may vary.

n	C ($)
0	300
2	340
4	380
6	420
8	460
10	500
12	540
14	580

c) 12

5. a) Table of values may vary.

t (h)	d (km)
0	280
0.5	230
1	180
1.5	130
2	80
2.5	30
2.8	0

b) Explanations may vary; 280 km
c) Explanations may vary; 2.8 h
d) 80 km; 200 km
e) Explanations may vary.

6. b) Explanations may vary. **7. b)** Answers may vary.

8. a) Table of values may vary.

F	C
−40	−40
−22	−30
−4	−20
14	−10
32	0
50	10
68	20
86	30

b) i) 194°F **ii)** 248°F **iii)** 392°F
c) i) 32°C **ii)** 49°C **iii)** 93°C
d) i) −7°C **ii)** −18°C **iii)** −23°C **iv)** −29°C
e) −40°

9. a) Table of values may vary.

x	y
−3	18
−2	16
−1	14
0	12
1	10
2	8
3	6

c) Answers may vary. **d)** Answers may vary.

10. Explanations may vary. The points should not be joined, because n is a natural number.

11. Answers may vary.　　**12.** Answers may vary.

Linking Ideas: Mathematics and Technology
Investigating Hall Rental Costs, page 206

1. a) $241.45　　**b)** $378.45　　**c)** $652.45　　**d)** $789.45

2. a) 18　　**b)** 54　　**c)** 91　　**d)** 127

3. $610.25　　**4.** Answers may vary.

4.2 Exercises, page 210

1. a) 2　　**b)** $-\frac{3}{2}$　　**c)** $\frac{1}{2}$

2. a) Answers may vary.
　　i) (1, 7), (2, 10)　**ii)** (1, 6), (−1, 2)　**iii)** (1, 5), (−1, 3)
　　iv) (2, 5), (−2, 3)　**v)** (3, 5), (−3, 3)　**vi)** (1, 4), (−1, 4)
　　vii) (2, 3), (−2, 5)　**viii)** (1, 3), (−1, 5)　**ix)** (1, 2), (−1, 6)
　　x) (4, 3), (−4, 5)　**xi)** (1, 1), (−1, 7)　**xii)** (0, 0), (0, 3)
b) Answers may vary.

4. Answers may vary.
　　a) i) (1, 4), (2, 5)　　**ii)** (1, 2), (2, 1)
　　b) i) (5, 3), (8, 5)　　**ii)** (4, −2), (6, −5)
　　c) i) (3, −4), (−3, −1)　**ii)** (2, −1), (3, 1)
　　d) i) (3, 4), (0, 4)　　**ii)** (5, 1), (5, 2)

5. Answers may vary.
　　a) (0, 0), (1, 3)　　**b)** (0, 0), (3, 4)　　**c)** (0, 0), (1, −2)
　　d) (0, 0), (−5, 2)　**e)** (0, 0), (−4, 3)

6. Answers may vary. There is an infinite number of lines for each slope.

7. Answers may vary.
　　a) i) (6, 3), (0, 7)　　**ii)** (1, 2), (5, 8)
　　b) i) (1, 3), (−1, 1)　　**ii)** (1, 1), (−1, 3)
　　c) i) (−1, 2), (−3, −4)　**ii)** (1, −2), (−5, 0)
　　d) i) (3, 0), (3, 2)　　**ii)** (2, 1), (4, 1)

10. a) 4　　**b)** Explanations may vary.　　**11.** 1

12. a) All slopes are $\frac{2}{3}$.　　**b)** All slopes are $-\frac{1}{2}$.
　　c) Slopes are $\frac{1}{3}$, $\frac{3}{7}$, and $\frac{1}{5}$.　　**13.** −2

14. Answers may vary.
　　a) (3, −1), (0, 11)　　**b)** (6, −7), (11, −14)
　　c) (16, 18), (27, 28)　**d)** (−5, −5), (−2, −8)

15. a) 2, $\frac{7}{4}$, $\frac{13}{7}$　　**b)** A, B, and C are not collinear.
　　c) Answers may vary. Try to draw a straight line through the 3 points.

16. a) Yes　　**b)** No　　**c)** No

17. c) The lines are parallel.　　**18.** 0.24 or −0.24

19. a) (0, −3)　　**b)** (0, 17)　　**c)** (0, 7) or (0, 1)

Exploring with a Graphing Calculator:
Investigating *y* = *mx* + *b*, page 213

1. a) The second screen　　**b)** Answers may vary.

2. b) Explanations may vary. Each line passes through the point (0, 1).
　　c) A horizontal line through the point (0, 1)

3. b) Explanations may vary. The lines are parallel.
　　c) A line with slope 0.5, passing through the origin

4. *m* is the slope, *b* is the *y*-intercept

6. $y = x + 4$, $y = -x + 4$, $y = x - 4$, $y = -x - 4$

Investigate, page 214

1. b) All pass through (0, 0); all have different slopes

2. a) 1, 2, $\frac{1}{2}$, 0, −1, −2, $-\frac{1}{2}$
　　b) The slope is the coefficient of *x*.

3. *m* is the slope of the line.

4. b) i) All pass through (0, 5); all have different slopes
　　ii) All are parallel; all have different *y*-intercepts

5. a) i) 2, 1, $\frac{1}{2}$, 0, $-\frac{1}{2}$, −1, −2　　**ii)** 2
　　b) The slope is the coefficient of *x*.

6. *m* is the slope and *b* is the *y*-intercept.

4.3 Exercises, page 217

1. a) 3, 5　　**b)** −2, 3　　**c)** $\frac{2}{5}$, −4　　**d)** $-\frac{1}{2}$, 6
　　e) −4, −7　**f)** $\frac{3}{8}$, $-\frac{5}{2}$　**g)** $\frac{4}{3}$, −2　**h)** $\frac{9}{5}$, 1

2. a) $y = 2x + 3$　　**b)** $y = -x + 4$　　**c)** $y = \frac{2}{3}x - 1$
　　d) $y = -\frac{4}{5}x + 8$　**e)** $y = -3x + \frac{5}{2}$　**f)** $y = 3$

3. a) $\frac{1}{2}$, 1, $y = \frac{1}{2}x + 1$　　**b)** $\frac{3}{2}$, −2, $y = \frac{3}{2}x - 2$
　　c) −2, 1, $y = -2x + 1$

4. a) i) $y = -x + 2$　**ii)** $y = -\frac{3}{2}x - 3$　**iii)** $y = \frac{2}{3}x$
　　b) Explanations may vary.

5. b) Answers may vary.

6. b) (6, 0)　　**c)** $3\sqrt{5}$ units
　　d) 9 square units, $(9 + 3\sqrt{5})$ units

7. a) (1, 6)　　**b)** 27 square units

8. (−2, 7), (4, 4), (1, −2)

9. a) i) $b - m = 2$　　**ii)** $b \div m = -3$
　　b) i) All the lines intersect at the point (−1, 2).
　　ii) All the lines intersect at the point (3, 0).
　　c) Explanations may vary.

10. a) i) −5　**ii)** 7　**iii)** −11　**iv)** 8
　　b) Answers may vary.

11. a) $\frac{1}{4}$　　**b)** −5　　**c)** −2　　**d)** $\frac{1}{5}$

12. a) $mb = 1$　　**b)** Answers may vary.
　　c) Explanations may vary.

13. Answers may vary.　　**14.** $-\frac{1}{2}$

15. b) Yes　　**c)** Explanations may vary.

16. a) $\frac{1}{3}$
　　b) The slope of a line perpendicular to a line with slope *m* is $-\frac{1}{m}$.

17. a) $y = x + 8$, $y = x + 4$, $y = x$, $y = x - 4$, $y = x - 8$,
　　$y = -x - 8$, $y = -x - 4$, $y = -x$, $y = -x + 4$, $y = -x + 8$
　　b) $y = \frac{1}{3}x - 2$, $y = \frac{1}{3}x + 2$, $y = -\frac{1}{3}x - 2$, $y = -\frac{1}{3}x + 2$

18. a) The lines have the same *y*-intercept at (0, 3).
　　b) The lines have the same *x*-intercept at (−3, 0).

19. $y = -\frac{1}{3}x + \frac{4}{3}$ **20.** $y = x + 5$ **21.** $y = -\frac{1}{3}x + \frac{2}{3}$
22. a) $y = \frac{2}{3}x - 1$ **b)** $y = -\frac{4}{3}x + 5$

Linking Ideas: Mathematics and Technology
Patterns in Equations and Lines, page 221

1. $y = -4x + 16$

2. a) $y = -3x + 9$ **b)** $y = -2x + 4$ **c)** $y = -x + 1$

3. a) The y-intercept is b. The slope is $-\sqrt{b}$.
b) Answers may vary.

4. Answers may vary.   **5.** Answers may vary.

Linking Ideas: Mathematics and Science
How Slope Applies to Speed and Acceleration, page 222

1. 600 km/h   **2. a)** 600   **b)** Equal   **c)** $d = 600t$

3.

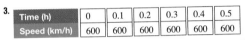

Time (h)	0	0.1	0.2	0.3	0.4	0.5
Speed (km/h)	600	600	600	600	600	600

4. 0   **5. a)** 0   **b)** Equal   **c)** $v = 600$

6. a) 9.8   **b)** 1   **c)** 9.8   **d)** 9.8 m/s^2

7. a) 9.8   **b)** Equal   **c)** $v = 9.8t$

8. Explanations may vary.

4.4 Exercises, page 226

1. a) i) $\frac{y-2}{x-1}$ **ii)** $\frac{y-1}{x-1}$ **iii)** $\frac{y-2}{x-3}$
b) i) $\frac{2}{3}$ **ii)** 2   **iii)** $-\frac{1}{2}$
c) i) $2x - 3y + 4 = 0$ **ii)** $2x - y - 1 = 0$ **iii)** $x + 2y - 7 = 0$

2. a) i) $3x - y - 1 = 0$ **ii)** $7x - y + 30 = 0$
   iii) $4x + y - 16 = 0$ **iv)** $2x - y + 3 = 0$
   v) $3x + 5y + 15 = 0$ **vi)** $2x - 3y - 16 = 0$
   vii) $7x + 2y + 7 = 0$ **viii)** $4y - 3 = 0$
b) Explanations may vary.

3. B, D, E, F, G

4. a) $x - y + 6 = 0$, $3x - y + 14 = 0$, $x + 2y = 0$,
   $2x + 5y - 2 = 0$ **b)** Explanations may vary.

5. a) $3x + 4y + 6 = 0$ **b)** $3x - 4y - 5 = 0$
c) $2x + 5y - 1 = 0$

6. a) $x - 2y + 2 = 0$ **b)** $2x - 3y + 6 = 0$
c) $2x + y - 3 = 0$ **d)** $x + 3y + 6 = 0$
e) $x + 4y - 8 = 0$ **f)** $2x + 7y - 41 = 0$

7. a) $2x - y - 3 = 0$ **b)** $3x + y + 7 = 0$
c) $7x - 2y - 39 = 0$ **d)** $2x - y - 1 = 0$
e) $2x - 3y - 11 = 0$ **f)** $8x - 3y + 20 = 0$

8. RS: $5x + 3y - 15 = 0$; ST: $3x - y - 9 = 0$;
   RT: $x - 5y + 25 = 0$

9. a) i) $3x - y - 5 = 0$ **ii)** $2x + 5y - 18 = 0$
   iii) $2x - y - 6 = 0$ **iv)** $4x + 3y - 12 = 0$
   v) $3x - 4y - 12 = 0$ **vi)** $2x + y - 5 = 0$
   vii) $15x - 18y + 10 = 0$ **viii)** $2x - 5y + 10 = 0$
b) Answers may vary.

10. a) AB: $2x - 3y + 12 = 0$; BC: $3x + 2y + 18 = 0$;
   CD: $2x - 3y - 14 = 0$; DA: $3x + 2y - 8 = 0$
b) AC: $5x - y + 4 = 0$; BD: $x + 5y + 6 = 0$

11. a) i) -4 **ii)** 6   **iii)** -19 **iv)** -8 **v)** -2
b) Answers may vary.

12. a) $x - 3y - 6 = 0$ **b)** $x - 3y + 24 = 0$
c) $x - 3y + 9 = 0$ **d)** $9x + 13y - 59 = 0$

13. a) -6 **b)** 22   **c)** 4

14. b) AB: $2x - y = 0$; CD: $x - 2y = 0$; EF: $3x + y = 0$;
   GH: $x + 3y = 0$; They pass through the origin.
c) AI: $x - 2y + 12 = 0$; CJ: $2x - y - 12 = 0$;
   KL: $x + 3y - 48 = 0$; MN: $3x + y - 48 = 0$; (12, 12)

15. c) $5x + 2y - 20 = 0$

16. b) i) $3x - 10y + 19 = 0$ **ii)** $4x - y - 27 = 0$
   iii) $3x + 2y - 19 = 0$

17. a) PQ: $x - 2y - 2 = 0$; QR: $3x + 4y - 26 = 0$;
   PR: $3x - y + 14 = 0$
b) Perpendicular bisector of PQ: $2x + y + 1 = 0$;
   QR: $4x - 3y + 7 = 0$; PR: $x + 3y - 2 = 0$
c) $(-1, 1)$
d) AP: $5\sqrt{2}$ units, AQ: $5\sqrt{2}$ units, AR: $5\sqrt{2}$ units

4.5 Exercises, page 233

1. a) 3, 4, $-\frac{4}{3}$ **b)** 4, -2, $\frac{1}{2}$ **c)** -2, -1, $-\frac{1}{2}$

2. $x + 2y - 3 = 0$, $2x + 5y - 6 = 0$

3. $x + y + 2 = 0$, $3x + 4y + 8 = 0$

4. a) i) 5, -5 **ii)** 3, 9   **iii)** -4, 2
   iv) -6, 4   **v)** 6, -8 **vi)** 3, $\frac{6}{5}$
b) Explanations may vary.

5. a) -2 **b)** 1   **c)** $\frac{3}{2}$ **d)** $\frac{1}{3}$ **e)** 2   **f)** $\frac{1}{4}$

6. a) i) $\frac{3}{4}$, -3 **ii)** $\frac{5}{2}$, -5 **iii)** -2, 3
   iv) $-\frac{2}{5}$, -4 **v)** $-\frac{1}{2}$, $\frac{5}{2}$ **vi)** $\frac{4}{7}$, $\frac{15}{7}$
b) Explanations may vary.   **7.** Answers may vary.

10. a) None   **b)** None
c) $2x - 3y + 12 = 0$, $4x - 6y - 9 = 0$

11. $x + 2y - 16 = 0$ **12.** $x - 2y + 3 = 0$

13. Explanations may vary.

14. a) 9   **b)** -6 **c)** -6 **d)** 14

15. a) $2x - 3y - 12 = 0$, $2x + 3y + 12 = 0$, $2x - 3y + 12 = 0$
b) 48 square units, $8\sqrt{13}$ units

16. a) $x + 2y = 0$, $2x - y - 4 = 0$
b) $2x - y = 0$, $2x - y - 8 = 0$, $2x - y + 12 = 0$,
   $2x - y - 20 = 0$

17. $x - 2y - 7 = 0$

18. a) $2x - y - 8 = 0$, $5x - 7y - 20 = 0$
b) $x - y + 2 = 0$, $4x - 3y + 6 = 0$
c) $2x + 4y - 9 = 0$, $3x + 6y + 5 = 0$

19. Explanations may vary.

20. a) $-\frac{A}{B}$ **b)** $-\frac{C}{B}$ **c)** $-\frac{C}{A}$

21. a) (6, 2), $(-3, 8)$, $(0, -7)$
b) $3\sqrt{13}$ units, $3\sqrt{26}$ units, $3\sqrt{13}$ units

c) $-\frac{2}{3}, \frac{3}{2}, -5$ **d)** Right isosceles triangle
e) 58.5 square units, $3\sqrt{13}(2 + \sqrt{2})$ units

Problem Solving: The Vanishing Square Puzzle, page 236

1. Answers may vary.   **2.** Answers may vary.

3. Answers may vary.

4. a) 1 square unit
c) Side lengths: $\sqrt{194} \doteq 13.9$, $\sqrt{29} \doteq 5.4$, $\sqrt{73} \doteq 8.5$;
    perimeter: $\sqrt{194} + \sqrt{29} + \sqrt{73} \doteq 27.9$
d) 0.072 units   **e)** 0.186 units

6. a) The sum of the angles in a triangle is 180°.
b) Answers may vary.

Mathematical Modelling: It's All in the Packaging, page 239

1. 1; $2x + 3y - 60 = 0$ **2. c)** Answers may vary.

3.

x	y
0	20
3	18
6	16
9	14
12	12
15	10
18	8
21	6
24	4
27	2
30	0

11 solutions

4. $y = -\frac{2}{3}x + 20$; explanations may vary.

5. a is the x-intercept and b is the y-intercept.

6. a) $3x + 5y - 90 = 0$ **b)** $19x + 30y - 570 = 0$

x	y
0	18
5	15
10	12
15	9
20	6
25	3
30	0

x	y
0	19
30	0

7. , 8. , 9. Answers may vary.

Exploring with a Graphing Calculator: Sequences and $Ax + By + C = 0$, page 240

1. $-\frac{A}{B} = -\frac{L_1}{L_2}$, $-\frac{C}{B} = -\frac{L_3}{L_2}$

2. Answers may vary.   **3.** Answers may vary.

4 Review, page 241

1. Tables may vary.

a)

x	y
-2	-15
-1	-12
0	-9
1	-6
2	-3

b)

x	y
-4	7
-2	6
0	5
2	4
4	3

c)

x	y
-6	-4
-3	0
0	4
3	8
6	12

2. a) Tables may vary.

n	C
0	1000
1000	1200
2000	1400
3000	1600
4000	1800
5000	2000
6000	2200
7000	2400
8000	2600
9000	2800
10 000	3000

c) 9250 km   **d)** 11 750 km

3. $l_1 : -\frac{3}{2}$; $l_2 : -3$; $l_3 : \frac{4}{3}$; $l_4 : \frac{1}{4}$

5. b) Explanations may vary.

6. The points are collinear.

7. a) 4, −3   **b)** $-\frac{5}{3}, 7$ **c)** $-\frac{9}{4}, -3$

8. a) $y = -\frac{1}{2}x - 4$ **b)** $y = \frac{4}{3}x - 6$ **c)** $y = -\frac{3}{2}x + \frac{3}{4}$

9. b) Explanations may vary.

10. a) 1   **b)** 5

11. a) 8   **b)** $\frac{3}{4}$

12. C, F, G   **13.** $3x - 2y + 17 = 0$, $5x + 2y + 7 = 0$

14. a) $x - y + 2 = 0$ **b)** $11x + 9y - 19 = 0$
c) $4x - y + 2 = 0$ **d)** $4x + 7y + 11 = 0$

15. a) $3x + 4y - 35 = 0$ **b)** $2x - 5y + 15 = 0$

16. a) PQ: $x - 3y = 0$; QR: $2x + 3y - 27 = 0$;
    PR: $4x - 3y + 9 = 0$
b) $x - 3y + 18 = 0$ **c)** $3x + y - 16 = 0$ **d)** $y = 3$

17. a) $-5, 2, \frac{2}{5}$ **b)** 3, 12, −4   **c)** $\frac{14}{3}, -2, \frac{3}{7}$

19. a) $-\frac{14}{3}$ **b)** $\frac{7}{6}$

4 Cumulative Review, page 243

1. a) 1, 7, 13, 19, 25   **b)** 3, 7, 11, 15, 19
c) −10, −2, 6, 14, 22   **d)** 2, 9, 16, 23, 30
e) −15, −5.5, 4, 13.5, 23   **f)** 5, 22, 39, 56, 73

2. a) 8   **b)** −16   **c)** $\frac{1}{64}$
d) $0.\overline{4}$ **e)** 4374   **f)** 14.332 723 2

3. a) $\frac{2}{3}$ **b)** 32   **c)** $\frac{1}{16}$ **d)** $\frac{27}{8}$ **e)** $\frac{4}{3}$
f) −2   **g)** $\frac{2}{3}$ **h)** $\frac{1}{1\ 000\ 000}$ **i)** 10   **j)** $\frac{9}{7}$

4. a) 4.47 cm   **b)** 7.48 mm
c) 5.00 cm, 8.66 cm   **d)** 3.00 m, 5.20 m

5. a) $\sqrt{2}$ **b)** $7\sqrt{3} + 19\sqrt{2}$ **c)** 10

6. The points are collinear.

7. a) -1 **b)** $\frac{1}{2}$ **c)** $-\frac{8}{9}$

8. D$\left(-\frac{2}{3}, 0\right)$ **9.** Right isosceles triangle

10. b) Answers may vary. **c)** Answers may vary.

Chapter 5

Investigate, page 247

Rule **1. a)** Table of values may vary.

x	y
−2	1
−1	2
0	3
1	4
2	5

c) All input numbers can be used. **d)** $y = x + 3$

Rule **2. a)** Table of values may vary.

x	y
−2	−6
−1	−3
0	0
1	3
2	6

c) All input numbers can be used. **d)** $y = 3x$

Rule **3. a)** Table of values may vary.

x	y
−2	−1
−1	1
0	3
1	5
2	7

c) All input numbers can be used. **d)** $y = 2x + 3$

Rule **4. a)** Table of values may vary.

x	y
−2	12
−1	11
0	10
1	9
2	8

c) All input numbers can be used. **d)** $y = 10 - x$

Rule **5. a)** Table of values may vary.

x	y
0	0
1	1
4	2
9	3
16	4

c) Input numbers cannot be negative. **d)** $y = \sqrt{x}$

5.1 Exercises, page 250

1. a) $y = x + 3$ **b)** $y = 3x$ **c)** $y = 2x + 1$

2. Explanations may vary. **a)** False **b)** False

3. a) Answers may vary.
 b) Yes. Negative values for $\sqrt{x}$ and 0 for $\frac{1}{x}$

4. Explanations may vary.

5. a) Tables of values may vary.

i)

x	y
−2	0.5
−1	1.5
0	2.5
1	3.5
2	4.5

ii)

x	y
−2	−1
−1	−0.5
0	0
1	0.5
2	1

iii)

x	y
0	2
1	3
4	4
9	5
16	6

c) Any input number can be used in parts i and ii. Negative input numbers cannot be used in part iii.
 d) Explanations may vary.

6. Answers may vary.
 i) (−4, −1.5), (−3, −0.5), (−2, 0.5), (−1, 1.5), (0, 2.5)
 ii) (−4, −2), (−2, −1), (0, 0), (2, 1), (4, 2)
 iii) (0, 2), (1, 3), (4, 4), (9, 5), (16, 6)

7. a)

A	D
6	1
7	2
8	3
9	4
10	5
11	6
12	7

b) $D = A - 5$

9. Answers may vary.
 a) (−4, −4), (−3, −3), (−2, −2), (−1, −1), (0, 0)
 b) (−4, 3), (−3, 3), (−2, 3), (−1, 3), (0, 3)
 c) (0, −1), (1, 0), (2, 3), (3, 8), (4, 15)

10. b) 25%; 2.9% **c)** 6.7 **d)** Answers may vary.
 e) Small **f)** Answers may vary.

11. a) Table of values may vary.

D	T ($)
0	25
100	25
200	40
300	55
400	70
500	85

c) $T = 25$ if $D \leq 100$; $T = 10 + 0.15D$ if $D > 100$
 d) The horizontal part of the graph would extend to 200 km.

12. b) Explanations may vary.

13. b) $1422 **c)** $1617 **d)** Answers may vary.

14. b) Explanations may vary.

15. b) 70% **c)** 52 m

16. b) Explanations may vary.

17. a) iii **b)** i **c)** ii **d)** iv

18. a) $\text{BMI} = \dfrac{70}{(\text{height (m)})^2}$

c) Answers may vary.   **d)** Answers may vary.

19. a) i) Graph will include points at $h = 0.5, 1.5, 2.5, \ldots$
    ii) Graph will include points at $h = 0.25, 0.50, 0.75, 1.25, 1.50, \ldots$

b) Yes. Explanations may vary.

5.2 Exercises, page 258

1. b

2. Answers may vary.
    a) $x + y = 7$, x and y are whole numbers
    b) $y = 2x$, x and y are real numbers

3. b   **4. a)** iii   **b)** Speed up or down the slide

5. a) Domain: $\{-4, -2, 0, 2, 4\}$, range: $\{-4, -2, 2, 6, 8\}$
    b) Domain: $-8 \leq x \leq 8$, range: $2 \leq y \leq 6$
    c) Domain: $-5 \leq x < 6$, range: $\{-2, 2, 6\}$

6. a) ii   **b)** Answers may vary.

7. a) 1 h 10 min   **b)** About 57% **c)** Answers may vary.
    d) Time; the time Habiba walked
    e) Distance; the distance from home Habiba walked

8. Answers may vary.

9. a) The fixed cost of $150 to produce the newsletter
    b) $\dfrac{3}{20}$; $\dfrac{1}{10}$ **c)** The cost per newsletter
    d) $180   **e)** Answers may vary.

10. a) 400 lb
    b) Decreases by 225 lb; flies for 1.5 h; stops to fill up tank
    c) On the ground for 0.5 h, flies for 0.75 h, on the ground for 0.25 h
    d) -150; the amount of fuel used per hour
    e) Schweizer 300 (250-C20W); 132 nautical miles; 66 nautical miles   **f)** Answers may vary.

11. a) Answers may vary.
    b) Domain: $t \geq 0.5$ min, range: $n \geq 0$
    c) Slope is 2.3; the number of pages printed per minute
    d) I-intercept: 0.5; warm-up time for printer

12. a) Mercury; Pluto   **b)** Mercury; Pluto
    c) Neptune; Jupiter's speed is about double.
    d) No other planets exist.
    e) As their distances from the sun increase, their orbital velocities decrease.

13. b) $y = 90 - x$ **c)** As x decreases, y increases.
    d) Domain: $45 < x < 90$, range: $0 < y < 45$

14. a) The car travelled for 3 h, stopped for 1 h, travelled for 3 h, filled the tank, stopped for 1 h, travelled for 2 h
    b) 11.7 L/100 km

15. a) iii   **b)** Velocity: part ii; acceleration: part i

16. a) $C = 2\pi r$ **e)** $A = s^2$

17. a) Its value drops from $3500 to $1500.   **b)** $1000
    c) The purchase price of the computer
    d) Yes; explanations may vary.   **e)** -2000; -500; -400

Linking Ideas: Mathematics and Technology
Some Examples of Functions, page 265

1. Graph 1: x-intercepts: 2, -2; y-intercept: 2; Graph 2: x-intercept: none; y-intercept: 2; Graph 3: x-intercept: 0; y-intercept: 0; Graph 4: x-intercepts: $-2, 0, 2$; y-intercept: 0

2. Graph 1: $y = \dfrac{4 - x^2}{2}$; Graph 2: $y = \dfrac{2}{x^2 + 1}$; Graph 3: $y = \dfrac{4x}{x^2 + 1}$; Graph 4: $y = \dfrac{x(x^2 - 4)^2}{4}$

3. Graph 1: domain: all real numbers, range: $y \leq 2$; Graph 2: domain: all real numbers, range: $0 < y \leq 2$; Graph 3: domain: all real numbers, range: $-2 \leq y \leq 2$; Graph 4: domain: all real numbers, range: all real numbers

Investigate, page 266

2. $f = s + 80$; $f = 2s + 90$

3. a) $92   **b)** $108   **c)** $168   **d)** $192

4. Answers may vary.

5.3 Exercises, page 269

1. Answers may vary. $y = x + 2$, $y = 2x$, $y = x^2$

2. a) $y = x + 5$ **b)** $y = 2x - 1$ **c)** $y = x^2 + 2$

3. a) $y = 2x$ **b)** $y = 10x + 1$

4. a) $y = 3x$ **b)** $y = -\dfrac{1}{2}x$

5. Explanations may vary.

6. a) i)

x	y
-3	7.5
-1.5	6
2.5	2
6	-1.5

ii)

x	y
-4	13
-2	1
1	-2
3	6

    b) i) $y = -x + 4.5$ **ii)** $y = x^2 - 3$
    c) Answers may vary.   **i)** $(1, 3.5), (0, 4.5), (-1, 5.5)$
    ii) $(0, -3), (-1, -2), (2, 1)$; The points lie on the graph.

7. b) 2.07   **c)** 2.07 g/cm^3 **d)** Equal
    e) Answers may vary.

8. a) $y = x + 2$ **b)** $y = 2x + 1$ **c)** $y = -x - 3$
    d) $y = x^2 + x$ **e)** $y = x^3$ **f)** $y = \dfrac{12}{x}$

9. Explanations may vary.

10. a) $y = -2x + 5$ **b)** $y = x^2 + 3$ **c)** $y = \dfrac{12}{x}$ **d)** $y = (x - 3)^2$

11. a) i) 14   **ii)** 20   **iii)** 26   **iv)** 32   **v)** 38
    b) $y = 6x + 2$
    c) i) 10   **ii)** 12   **iii)** 14   **iv)** 16   **v)** 18
    d) $y = 2x + 6$ **e)** $y = 4x + 4$

12. b) 84.2 g   **c)** 1.34 g/mL   **d)** Answers may vary.

13. b) Answers may vary.   **c)** 24-piece pizza

14. a) 1 048 576
    b) One gigabyte is 2^x, where x is the exponent of the least power of 2 that exceeds 1 000 000 000.
    c) 1 073 741 824

15. a)

x	y
1	2
2	4
3	6
4	8
5	10
6	24

b) Same, except it does not have $x = 6$ and its value for y.

c) No; explanations may vary.

16. a) $y = 5x$ **b)** $5 \times 7 = 35 \neq 755$

c) $y = 5x + (x - 1)(x - 2)(x - 3)(x - 4)(x - 5)(x - 6)$

d) $y = 5x + \frac{13}{144}(x - 1)(x - 2)(x - 3)(x - 4)(x - 5)(x - 6)$

e) Answers may vary.

Investigate, page 274

3. All are straight lines.

4. Answers may vary. All can be rewritten in the form $y = mx + b$.

5.4 Exercises, page 279

1. a) Linear; slope: 4, y-intercept: 0

b) Linear; slope: -1, y-intercept: 2

c) Not linear   **d)** Linear; slope: $\frac{1}{5}$, y-intercept: $\frac{1}{2}$ **e)** Not linear

f) Linear; slope: $\frac{1}{3}$, y-intercept: $-\frac{4}{3}$ **g)** Not linear

h) Not linear   **i)** Linear; slope: $\frac{1}{2}$, y-intercept: 0

2. a) x-intercept: $\frac{5}{2}$, y-intercept: -5, slope: 2, domain: all real numbers, range: all real numbers

b) x-intercept: 0, y-intercept: 0, slope: $\frac{-3}{2}$, domain: $\{-4, -2, 0, 2, 4, 6\}$, range: $\{-9, -6, -3, 0, 3, 6\}$

c) x-intercept: $\frac{8}{3}$, y-intercept: -2, slope: $\frac{3}{4}$, domain: all real numbers, range: all real numbers

d) x-intercept: none, y-intercept: 5, slope: 0, domain: all real numbers, range: $\{5\}$

e) x-intercept: 4, y-intercept: 4, slope: -1, domain: $x \geq 0$, range: $y \leq 4$

f) x-intercept: none, y-intercept: -3, slope: 0, domain: all real numbers, range: $\{-3\}$

3. b) 0.05; the rate of commission

c) I-intercept: 800. There is no s-intercept for the given domain. The I-intercept is the base salary.

d) $S \geq 0$, $I \geq 800$ **e)** Answers may vary.

5. a) i) Domain: $x \leq 0$, range: $y \leq 2$; $y = 2x + 2$

ii) Domain: all real numbers, range: all real numbers; $y = \frac{-3}{4}x + 3$

iii) Domain: $\{1, 2, 3, 4\}$, range: $\{-5, -3, -1, 1, 3\}$; $y = 2x - 7$

b) Explanations may vary.

6. a) $l = 12 - w$ **b)** $0 < w \leq 6$

7. a) $y = x + 90$, $0 < x < 90$ **b)** $y = 230 - x$, $0 < x < 180$

c) $y = 180 - 2x$, $0 < x < 90$

8. a) About 37 s   **b)** Interpolate

c) About 720 m   **d)** 15; the speed of the car in m/s

e) $d = 15t + 50$ **f)** About 77 s   **g)** 1550 m

9. a)

Time in cab (minutes)	0	1	2	3	4	5	6
Total fare ($)	2.50	3.00	3.50	4.00	4.50	5.00	5.50

b) Arithmetic   **d)** 2.50; the fixed fare in dollars

e) $f = 0.5t + 2.5$ **f)** $10.00

10. b) $1.50   **c)** $11.00

11. Explanations may vary.

12. b) $25; $12/day   **c)** Explanations may vary.

13. b) Domain: $0 \leq t \leq 30$, range: $20 \leq v \leq 35$

14. b) $T = -6.5A + 15$

c) i) $-30.5\,°C$ **ii)** About 2.3 km

15. b) About 115 min; travelling at the same speed

c) About -2.05

d) About 2.05 nm/min; speed and slope are opposites

e) $d = 235 - 2.05t$

f) 123 knots; the Enstrom TH-28 or 480 (250-C20W)

g) About $263 U.S.

h) The d-intercept would change.

16. b) No   **c)** Answers may vary. $C = 0.2d + 2.6$

d) $134.20   **e)** $220.40

f) Part e; it is an extrapolation.

g) Answers may vary.

17. b) Answers may vary. $P = 23\,500 - 2300A$

c) Does not describe the data well.

d) No, it is low.

e) Answers may vary. Condition, mileage, make

18. a) $m = \frac{2}{5}$, $b = 7$ **b)** $m = \frac{-4}{3}$, $b = 10$

c) $m = -2$, $b = 4$ **d)** $m = 3$, $b = -6$

19. a) Domain: all real numbers, range: all real numbers, x-intercept: -17.5, y-intercept: 7

b) Domain: all real numbers, range: all real numbers, x-intercept: 7.5, y-intercept: 10

c) Domain: all real numbers, range: all real numbers, x-intercept: 2, y-intercept: 4

d) Domain: all real numbers, range: all real numbers, x-intercept: 2, y-intercept: -6

20. b) $d = \frac{3}{2}s - 25$ **c)** 125 m   **d)** 113 km/h

21. b) $F = 300D + 200$ **c)** 16 days

d) First term: 500, common difference: 300

22. a) $x + y = 8$ **b)** $2 < x < 8$

Exploring with a Graphing Calculator: The Line of Best Fit, page 287

1. b) $V = 6.6I + 4.9$; 6.6 ohms   **d)** 43.8 V   **e)** 1.95 A

2. c) $C = 0.3d + 2.6$ **d)** Answers may vary.

Problem Solving: Use a Graph, page 289

1. $81\,363.64 at 6.1%, $48\,636.36 at 8.3%

2. Answers may vary.

3. a) No   **b)** Close to $9000   **c)** Answers may vary.

4. 0.45 h

5. 44.1 kg of lean beef, 5.9 kg of fat trim

6. 300 km

7. 320 mL of the 50% solution, 80 mL of the 100% solution

8. 3.3 min

Investigate, page 290

1. Table of values may vary.

Month	Cost ($)
0	0
1	400
2	800
3	1200
4	1600
5	2000
6	2400
7	2800
8	3200
9	3600
10	4000
11	4400
12	4800

3. a) Doubles; triples   **b)** Answers may vary.

4. a) Linear   **b)** 0   **c)** 400; monthly leasing cost in dollars

5. $C = 400n$

5.5 Exercises, page 295

1. a) y is doubled.   **b)** y is halved.

2. a) i) 5   **ii)** −3   **iii)** $\frac{1}{2}$ **iv)** $\frac{3}{2}$

b) i)

x	y
1	5
2	10
3	15
4	20
5	25

ii)

x	y
1	−3
2	−6
3	−9
4	−12
5	−15

iii)

x	y
2	1
4	2
6	3
8	4
10	5

iv)

x	y
2	3
4	6
6	9
8	12
10	15

c) i) $y = 5x$ **ii)** $y = -3x$ **iii)** $y = \frac{1}{2}x$ **iv)** $y = \frac{3}{2}x$

3. Answers may vary.

4. a) i) Yes; $C = \pi d$
   ii) Yes; for example, $T = 0.07P$, for tax of 7%
   iii) No   **iv)** No
   v) Yes; for example, $C = 0.55n$, where gas is 55¢/L
   b) Explanations may vary.

5. a) No; line does not pass through the origin
   b) $S = 250n + 275$

6. a) i)

x	y
−6	−4
−3	−2
0	0
3	2
9	6

ii)

x	y
−4	2
−2	1
0	0
4	−2
6	−3

iii)

x	y
0	0
3	4
6	8
9	12
12	16

iv)

x	y
0	0
2	5
4	10
8	20
12	30

b) i) $y = \frac{2}{3}x$ **ii)** $y = -\frac{1}{2}x$ **iii)** $y = \frac{4}{3}x$ **iv)** $y = \frac{5}{2}x$

7. a) $28   **b)** About 28.6 km   **c)** $A = 1.4d$

8. a) $y = \frac{2}{3}x$ **b)** 14   **c)** 22.5

9. a) $V = \frac{3}{40}m$ **b)** 4.5 L

10. a) $d = 0.275s$ **b)** 27.5 m
   c) Explanations may vary.

11. a)

Time in hours	0	1	2	3	4	5	6	7	8
Total area lost (acres)	0	240	480	720	960	1200	1440	1680	1920

b) Yes   **c)** Arithmetic   **d)** $A = 240t$ **e)** 174 days

12. a) $d = 0.2s$ **b)** 3 km   **d)** 2.4 km
   e) 17.5 km/h

13. a) 31.25 t   **b)** 312.5 t   **c)** 2000 t

14. b) The origin
   c) Answers may vary. $v = 1.53 \times 10^{-5}d + 150.5$
   d) The increase in recession velocity per light year
   e) i) 6.4×10^8 light years   **ii)** 1.5×10^{10} light years

15. a) No   **b)** No   **d)** No
   e) i) 22   **ii)** 41   **iii)** 60   **f)** 1

16. a) $y = \sqrt{2}x$ **b)** Yes

17. $\frac{\sqrt{5}}{2}$

18. a) $d = 4.9t^2$ **b)** 122.5 m   **c)** 1102.5 m

Mathematics File: Arithmetic Sequences and Linear Functions, page 301

1. a) i) $3n + 2$ **ii)** $20 - 3n$ **iii)** $2n + 3$ **iv)** $12 - 2n$
   b) i) $t_n = 4n - 3$; 65   **ii)** $t_n = 3n$; 63
   iii) $t_n = 5n - 9$; 56   **iv)** $t_n = 47 - 6n$; −61
   v) $t_n = 1 - 3n$; −29   **vi)** $t_n = 17 - 8n$; −351
   c) 9th; Assumptions may vary. 11 rows
   d) i) 13th   **ii)** 25th   **iii)** 21st
   iv) 10th   **v)** 15th   **vi)** 25th
   e) i) 19, 22, 25, 28   **ii)** $3n + 16$ **iii)** 27

2. a)

Number of hours worked	1	2	3	4	5
Pay ($)	6.50	13.00	19.50	26.00	32.50

b) $a = 0$, $d = 6.50$; $t_n = 6.50n$ **c)** Same

3. a)

Number of tickets sold	1	2	3	4	5
Profit ($)	−995	−990	−985	−980	−975

b) −1000, $d = 5$; $t_n = 5n - 1000$ **c)** Same

Mathematical Modelling: Pushing Your Physical Limits, page 303

1. b) Answers may vary.

2. a) 39.7 mL/kg/min   **b)** 42.4 mL/kg/min

3. a) 60.3 beats/min   **b)** 138.9 beats/min

4. As the heart rate increases, VO_2 max decreases; high VO_2 max, low recovery heart rate

5. , 6. , 7. Answers may vary.

5.6 Exercises, page 307

1. a) −3   **b)** −8   **c)** 0.75

2. a) 2   **b)** 14   **c)** $\frac{1}{2}$

3. a) i) 6, $7 - 2\sqrt{2}$, 0.75   **ii)** −4, $6 + 7\sqrt{2}$, −1
 b) Explanations may vary.

4. b) i) Domain: all real numbers, range: all real numbers
 ii) Domain: $x \geq 1$, range: $f(x) \geq 0$

5. a) 2, 4, 2   **b)** −1.5, 3, 1   **c)** 2, −1, 2

6. a) (−2, 2), (1, 4), (3, 2)   **b)** (−2, −1.5), (1, 3), (3, 1)
 c) (−2, 2), (1, −1), (3, 2)

7. a) $3m - 5$ **b)** $12x - 5$ **c)** $6x - 10$ **d)** $\frac{6}{x} - 5$ **e)** $6x - 2$

8. a) $5k + 1$ **b)** $5x - 4$ **c)** $10x + 6$ **d)** $21 - 15x$

9. a) 1, −27, $4\sqrt{3} - 7$, −5   **b)** 25, 186, $15 + \sqrt{3}$, −6.5
 c) $\sqrt{11}$, undefined, $\sqrt{6\sqrt{3} - 1}$, $\sqrt{2}$
 d) 4.5, 24.8, $3 + \frac{1}{\sqrt{3}}$, 2.25   **e)** 4, −150, $3\sqrt{3} - 3$, −0.125
 f) 1.6, $\frac{20}{9}$, $\frac{4\sqrt{3}}{2\sqrt{3}+1}$, 1

10. a) i) Domain: all real numbers, range: all real numbers
 ii) Domain: all real numbers, range: $f(x) \geq 1$
 iii) Domain: all real numbers, range: $f(x) \geq -2.25$
 b) Explanations may vary.

12. a) $2x^2 + 7x$ **b)** $2x^2 + 11x + 9$ **c)** $2x^2 + 15x + 22$
 d) $8x^2 + 6x - 5$ **e)** $18x^2 + 9x - 5$ **f)** $2x^2 - 3x - 5$

13. Answers may vary.
 a) $f(x) = x + 3$ **b)** $f(x) = x + 5$ **c)** $f(x) = x^2 - 2$
 d) Explanations may vary.

14. Answers may vary.

15. a) $m = \frac{2}{5}$, $b = -3$ **b)** $m = 3$, $b = 2$ **c)** $m = 0$, $b = -3$
 d) $m = \frac{8}{3}$, $b = \frac{-13}{3}$ **e)** $m = \frac{-3}{2}$, $b = 5$

16. a) 1   **b)** −6   **c)** $\frac{-3}{2}$

17.

x	y
−6	−9
−1.5	0
1	5

18. a) i) 1   **ii)** 1   **b)** 1   **c)** $n \neq 0$

20. b) $[g(x)]^n$

21. a) Table of values may vary.

x	7^x
1	7
2	49
3	343
4	2 401
5	16 807
6	117 649
7	823 543
8	5 764 801
9	40 353 607
10	282 475 249
11	1 977 326 743
12	13 841 287 201

b) 7^4, 7^8, 7^{12}, ...   **c)** $x = 4n$, where $n = 1, 2, 3, 4, ...$

Mathematics File: Absolute Value, page 310

1. a) 29   **b)** 12   **c)** 107
 d) 15   **e)** 6.7   **f)** 1.8

2. a) 4   **b)** 6   **c)** 8   **d)** 7
 e) 5   **f)** 27   **g)** 10   **h)** −4

3. a) 30   **b)** −12   **c)** −36
 d) 22   **e)** −9   **f)** 1

4. a) −8, 6   **b)** −4, 5   **c)** No solution
 d) 5   **e)** 0   **f)** No solution

5. b) For $x \geq 0$, $y = |x|$ same as $y = x$. For $x < 0$, $y = |x|$ same as $y = -x$.

7. a) Graph 1: $y = x - 2$
 b) Graph 2: $y = |x - 2|$, Graph 3: $y = |x| - 2$, Graph 4: $y = ||x| - 2|$

5.7 Exercises, page 316

1. Yes   **2.** No   **3.** No   **4.** It may be.

5. a) i) Domain: {1, 2, 3, 4}, range: {1, 4, 9, 16}   **ii)** Function
 b) i) Domain: {2, 4, 7, 10}, range: {3, 6, 9}   **ii)** Function
 c) i) Domain: {0, 3, 6}, range: {−6, −3, 0, 3, 6}
 ii) Not a function
 d) i) Domain: {1, 2, 3, 5, 8}, range: {2, 4, 6, 10} **ii)** Function

6. a) i) Domain: $-2 \leq x \leq 3$, range: $-4 \leq y \leq 3$
 ii) Not a function
 b) i) Domain: all real numbers, range: $y \geq 0$ **ii)** Function
 c) i) Domain: all real numbers, range: {−2, 2}   **ii)** Function
 d) i) Domain: $1 \leq x \leq 5$, range: $-5 \leq y \leq -1$
 ii) Not a function
 e) i) Domain: $x \geq -1$, range: all real numbers
 ii) Not a function
 f) i) Domain: {−4, −3, −2, −1, 0, 1, 2, 3},
  range: {−3, −2, 0, 2, 3, 4}
 ii) Function
 g) i) Domain: all real numbers, range: all real numbers
 ii) Function
 h) i) Domain: $-2 \leq x \leq 7$, range: $-1 \leq y \leq 5$ **ii)** Function

7. Explanations may vary.

8. a) i) Domain: all real numbers, range: $y \geq 0$ **ii)** Function
  b) i) Domain: all real numbers, range: $y \geq -1$ **ii)** Function
  c) i) Domain: {1, 3, 5, 7, 9}, range: {10, 16, 22, 28, 34}
   ii) Function
  d) i) Domain: {7}, range: all real numbers
   ii) Not a function

9. a) Domain: all real numbers, range; all real numbers;
   function
  b) Domain: $x \geq 1$, range: all real numbers; not a function
  c) Domain: $x \geq 0$, range: $y \leq 4$; function
  d) Domain: all real numbers, range; all real numbers;
   function
  e) Domain: all real numbers, range; $y \geq 0$; function
  f) Domain: $x \geq 3$, range: $y \geq 0$; function

10. a) No   **b)** Yes

11. True   **12. a)** Not a function

Mathematics File: Classifying Functions, page 319

1. Answers may vary.   **2.** Answers may vary.

5 Review, page 320

1. a) Tables of values may vary.

i)

x	y
−4	−9
−3	−7
−2	−5
−1	−3
0	−1
1	1
2	3
3	5
4	7

ii)

x	y
−4	−8
−3	−1
−2	4
−1	7
0	8
1	7
2	4
3	−1
4	−8

iii)

x	y
0	3
1	4
4	5
9	6
16	7

  c) i) All input numbers can be used.
   ii) All input numbers can be used.
   iii) Only positive numbers and 0 can be used.
  d) i) $y = 2x - 1$ **ii)** $y = 8 - x^2$ **iii)** $y = \sqrt{x} + 3$

2. b) Explanations may vary.

3. Answers may vary. **b)** $C = 1.1e + 4.0$ **c)** 6.8 L/100 km
  d) 10.6 L/100 km   **e)** Part d

4. a)

Time (h)	Earnings ($)
0	0.00
8	76.00
16	152.00
24	228.00
32	304.00
40	418.00
48	532.00
50	560.50

  b) Answers may vary.

5. a) $C = 5000 + 20n$ **b)** $13 000
  d) i) $45   **ii)** $40   **iii)** $38.52   **iv)** $35.63

6. a) $h_2 = \frac{2}{3}h_1$ **b)** 2 m   **c)** 120 cm

7. a) 20   **b)** 25   **c)** $11 - 4\sqrt{2}$

8. a) (1, 3)   **b)** (−10, −261)
  c) $(\sqrt{2}, -5 + 6\sqrt{2})$ **d)** (0.5, 1.5)

9. a) Domain: {−1, 0, 2, 3, 4, 6}, range: {−1, 1, 2, 3, 5}
  b) Domain: $x \geq -1$, range: $y \geq 1$
  c) Domain: $-5 \leq x \leq 5$, range: $-3 \leq y \leq 3$
  d) Domain: $-3 \leq x \leq 3$, range: $-3 \leq y \leq 3$
  e) Domain: $x \geq -2$, range: all real numbers
  f) Domain: all real numbers , range: all real numbers

10. a) Not a function   **b)** Function   **c)** Not a function
  d) Not a function   **e)** Not a function   **f)** Function

5 Cumulative Review, page 322

1. a) 55   **b)** 100   **c)** 145
  d) 190   **e)** 235   **f)** 280

2. Each sum is 45 greater than the previous sum.

3. a)

Year	Value ($)
0	81 000
1	64 800
2	51 840
3	41 472
4	33 178
5	26 542
6	21 234
7	16 987
8	13 590
9	10 872
10	8 697

  b) i) About 2.5 years **ii)** About 6 years   **iii)** About 10 years

4. a) 120.970 **b)** 936.363 **c)** 6.082   **d)** 0.273   **e)** 0.037
  f) 1.999   **g)** −1.236   **h)** 8.979   **i)** 2.201   **j)** 9.906

5. a) $\sqrt{2} : 1$ **b)** 2 : 1

6. a) $\frac{6}{5}, -\frac{5}{6}$ **b)** $\frac{-7}{6}, \frac{6}{7}$ **c)** $\frac{7}{5}, \frac{-5}{7}$

8. b) $y = x$ or $y = x + 8$

9. a) ii   **b)** iv   **c)** i
  d) iii   **e)** vi   **f)** v

10. a) Answers may vary.
  b) Explanations may vary.

11. a) $y = 10 - x$ **b)** $y = x + 3$

Chapter 6

6.1 Exercises, page 328

1. a) 1257 cm^2; 4189 cm^3 **b)** 314 cm^2; 524 cm^3
  c) 2903.3 cm^2; 14 710.2 cm^3

2. a) 172.0 cm^2, 212.2 cm^3 **b)** 58.1 cm^2, 41.6 cm^3
  c) 43.0 cm^2, 26.5 cm^3 **d)** 1372.2 cm^2, 4780.1 cm^3

3. Answers may vary.
  a) 20 cm   **b)** 1257 cm^2; 4189 cm^3

4. a) 8.1×10^9 km^2; 6.9×10^{13} km^3
b) 3.2×10^7 km^2; 1.7×10^{10} km^3

5. a) 2.2 m   **b)** 1.8 cm   **c)** 6.9 mm

6. a) 27.4 mm   **b)** 10.2 cm   **c)** 1.1 m

7. Explanations may vary.

8. a) 11.9 cm   **b)** 1790.49 cm^2 **c)** 7124 cm^3

9. a) 50.3 cm^2; 96.0 cm^2; Answers may vary. The edge of the box is equal to the diameter of the ball.
b) π : 6   **c)** 33.5 cm^3; 64 cm^3
d) π : 6; The ratios are equal.

10. a) 1.6×10^7 km^2; 6.2×10^9 km^3
b) 6.1×10^{12} km^2; 1.4×10^{18} km^3

11. a) 219.4 cm^3 **b)** 62.4 m^3 **c)** 2.37×10^5 mm^3

12. c) Answers may vary. As the radius increases, the surface area increases faster; for example, when the radius doubles, the surface area quadruples.

13. c) Answers may vary. As the radius increases, the volume increases faster; for example, when the radius doubles, the volume is multiplied by 8.

14. a)

Volume of air pumped in (m³)	10	30	40	50	60	70
New diameter (m)	4.036	4.702	4.974	5.219	5.442	5.650
New surface area (m²)	51.18	69.46	77.72	85.57	93.04	100.28
New volume (m³)	34.43	54.43	64.43	74.43	84.43	94.43

b) Descriptions may vary. The graph is a curve that gets less steep as the volume of air pumped in increases.
c) Descriptions may vary. The graph is a straight line (with a positive slope) that intersects the vertical axis at the original volume.

15. a)

Diameter (m)	1.0	2.0	3.0	4.0	5.0	6.0
New diameter (m)	3.397	3.588	4.025	4.675	5.464	6.335
New surface area (m²)	36.25	40.45	50.89	68.67	93.82	126.07
New volume (m³)	20.53	24.19	34.14	53.50	85.41	133.12

b) Descriptions may vary. The graph is a curve that gets steeper as the original diameter increases.
c) Descriptions may vary. The graph is a curve that gets steeper as the original diameter increases.

16. The sphere; explanations may vary.

17. a) 9   **b)** 27

18. a) Quadrupled; multiplied by 8
b) Multiplied by 9; multiplied by 27
c) Multiplied by n^2; multiplied by n^3

19. a) $A = \pi d^2$; $V = \dfrac{\pi d^3}{6}$ **b)** $A = \dfrac{C^2}{\pi}$; $V = \dfrac{C^3}{6\pi^2}$

20. Explanations may vary.

21. b) As the radius increases, the slope increases, which means the ratio $\dfrac{V}{A}$ increases.

c) Larger animals can maintain body heat better; explanations may vary.

22. a) Volume is the same; explanations may vary.
b) Radius is greater; explanations may vary.
c) $\sqrt[3]{2}$: 1   **d)** $0.75\sqrt[3]{4}$: 1

23. a) 343 : 1   **b)** 3.7×10^{10} : 1

Linking Ideas: Mathematics and History
Archimedes of Syracuse, page 332

2. $V_{cyl} = \pi r^3$; $V_{hem} = \dfrac{2}{3}\pi r^3$; $V_{cone} = \dfrac{1}{3}\pi r^3$; The volume of the cone is one-third the volume of the cylinder. The volume of the hemisphere is two-thirds the volume of the cylinder, and two times the volume of the cone.

6.2 Exercises, page 335

1. a) $6a^2$ **b)** $-6a^2$ **c)** $-6a^3$ **d)** $-6a^4$
e) $6a^4$ **f)** $6a^5$ **g)** $-6a^6$ **h)** $12a^7$

2. a) $35x^6$ **b)** $12a^2$ **c)** $-54m^5$ **d)** $10n^3$
e) $56y^9$ **f)** $-28b^4$ **g)** $6t^8$ **h)** $72p^9$

3. a) $18xy$ **b)** $32p^2q$ **c)** $35m^3n$ **d)** $14a^2b$
e) $-45r^2s^3$ **f)** $-44c^2d$ **g)** $6s^2t^3$ **h)** $2b^3c^5$

4. a) $-8ab^3$ **b)** $-12m^5n^7$ **c)** $21a^3b^5$
d) $24p^3q^7$ **e)** $24x^6y^3$ **f)** $-6a^4b^4$

5. a) $4x^3$ **b)** $9m^4$ **c)** $-9a^6$
d) $4b$ **e)** $4d^2$ **f)** $-4n^2$

6. Answers may vary.
a) $(4ab)(9a^3b)$; $(6a^2b)(6a^2b)$; $(4a^2b)(9a^2b)$
b) $(-3x)(5yz)$; $(3x)(-5yz)$; $(xy)(-15z)$
c) $(7pqr)(p)$; $(-7pqr)(-p)$; $(pq)(7pr)$
d) $(-6)(8a)$; $(6)(-8a)$; $(3a)(-16)$
e) $(5cd)(5cd)$; $(5c^2d^2)(5)$; $(-5cd)(-5cd)$
f) $(2m^2)(2m^2)$; $(-2m^2)(-2m^2)$; $(4m^2)(m^2)$
Parts a, e, and f can be written as the product of two equal factors.

7. Answers may vary.
a) $\dfrac{72a^4b^2}{2}$; $\dfrac{72a^5b^2}{2a}$; $\dfrac{72a^4b^3}{2b}$ **b)** $\dfrac{-45x^2y^2z^2}{3xyz}$; $\dfrac{30x^2yz}{-2x}$; $\dfrac{-60xyz^2}{4z}$
c) $\dfrac{28p^2q^2r^2}{4qr}$; $\dfrac{-28p^2q^2r^2}{-4qr}$; $\dfrac{21p^2qr^2}{3r}$ **d)** $\dfrac{-48a^2}{a}$; $\dfrac{48ab^3}{-b^3}$; $\dfrac{-48a^2b}{ab}$
e) $\dfrac{100ac^2d^2}{4a}$; $\dfrac{-50c^2d^3}{-2d}$; $\dfrac{25c^2d^2f}{f}$ **f)** $\dfrac{4m^5}{m}$; $\dfrac{12m^4n}{3n}$; $\dfrac{8abm^4}{2ab}$

8. Explanations may vary.

9. a) $\dfrac{9}{2}r^4s^2$ **b)** $-3x^4y^3$ **c)** $-8m^6$ **d)** $2a^2b$ **e)** $6x^5y^6$

10. a) $54x^{11}$ **b)** $-12a^{10}b$ **c)** $45s^{10}t^5$
d) $320p^6q^3$ **e)** $-24m^5n^9$ **f)** $675x^{10}y^{14}$

11. a) $\dfrac{3}{2}x^4y$ **b)** $-6q^7$ **c)** $\dfrac{7}{2}a^2b^3$
d) $-\dfrac{2x}{5y^5}$ **e)** $-16m^8n^8$ **f)** $a^{12}b^{13}$

12. a) $\sqrt[3]{2}$ **b)** $\sqrt[3]{4}$

13. a) $\sqrt[3]{4}$ **b)** $\sqrt[3]{16}$

14. a) π : 4; no   **b)** π : 4; same ratio as in part a

15. a) 6 : π; no   **b)** 6 : π; same ratio as in part a

16. $\dfrac{\pi}{6}$

17. a) $\frac{1}{60}$　**b)** 15 m　**c)** 86.4 m^2　**d)** 38.9 m^3

18. a) 586.3 cm^3　**b)** 879.4 cm^3　**c)** $\frac{2}{3}$

19. a) Answers may vary. 311 cm^2; 398 cm^3
　b) i) 250　**ii)** 1944 m^2　**iii)** 6219 m^3

20. 2 : 1

Mathematical Modelling: Could a Giant Survive?, page 338

1.

Edge length (cm)	Volume (cm^3)	Surface area (cm^2)	Volume / Surface area
1	1	6	$\frac{1}{6}$
2	8	24	$\frac{1}{3}$
3	27	54	$\frac{1}{2}$
4	64	96	$\frac{2}{3}$
5	125	150	$\frac{5}{6}$
6	216	216	1
7	343	294	$\frac{7}{6}$
n	n^3	$6n^2$	$\frac{n}{6}$

2. a) Volume　**b)** It increases.　**c)** $n : 6$

3. a) Surface area would be 144 times as great; volume would be 1728 times as great.
　b) Giant's ratio would be 12 times as great.

4. a) 1024; 32 768　**b)** 32 times as great

5. , 6. , 7. Answers may vary.

6.3 Exercises, page 341

1. a) The answer is twice the larger number you started with.
　b) The answer is twice the larger number you started with.
　c) Yes, explanations may vary.

2. a) Predictions may vary.
　b) The answer is twice the smaller number you started with.

3. a) $10a + 4$　**b)** $10x - 10y$　**c)** $-6m^2 - 3$
　d) $3p + 7$　**e)** $11x^2 - 4x$　**f)** $-5k + 11l$

4. $12a + 10$; explanations may vary.

5. Answers may vary.
　a) $(2a + 5) + (6a + 5)$; $(4a + 5) + (4a + 5)$; $(3a + 12) + (5a - 2)$
　b) $(-3b + 7) + (-3b + 8)$; $(-4b + 6) + (-2b + 9)$; $(-8b + 12) + (2b + 3)$
　c) $(-2m - 7n) + (-2m - 4n)$; $(-6m - 3n) + (2m - 8n)$; $(-m - 5n) + (-3m - 6n)$
　d) $(8 - 7x) + (8 - 10x)$; $(7 - 6x) + (9 - 11x)$; $(6 - 5x) + (10 - 12x)$

6. Answers may vary.
　a) $(10a + 12) - (2a + 2)$; $(11a + 13) - (3a + 3)$; $(12a + 14) - (4a + 4)$
　b) $(-4b + 16) - (2b + 1)$; $(-3b + 17) - (3b + 2)$; $(-2b + 18) - (4b + 3)$
　c) $(-3m - 10n) - (m + n)$; $(-2m - 9n) - (2m + 2n)$; $(-m - 8n) - (3m + 3n)$
　d) $(17 - 16x) - (1 + x)$; $(18 - 15x) - (2 + 2x)$; $(19 - 14x) - (3 + 3x)$

7. $9x^2 + x - 2$; explanations may vary.

8. a) $11m^2 - 3m + 2$　**b)** $4a^3 + 4a^2 + 3a$
　c) $-3x^2 - 20x + 17y$　**d)** $9t^3 - 18t^2 - 15t - 1$
　e) $8a^3 - 18$　**f)** $-5x^2 - 8x + 5$

9. a) $P = 15.5n - 30\,000$
　b) i) $47\,500　**ii)** $125\,000　**iii)** $280\,000

10. Explanations may vary.
　a) 31; 19　**b)** 132; 72

11. a) $10x^2 - 6x^2y$; third　**b)** $p^2q^2 - 3p + 2q$; fourth
　c) $3m^2 - 3m - 2n^2 + 2n$; second
　d) 3; zero　**e)** $-3xy^2 + 2xy$; third

12. a) $10x^2 - 8x + 16$　**b)** $9z^3 - 2z^2 + 1$

13. a) $2x^2y - 7xy + 5xy^2$　**b)** $4m^2n^2 - 6mn + 7mn^3$

14. Answers and explanations may vary. Possible answer is $-3x^2 - x + 1, -6x^2 + 1$.

15. a) $-3x^2 - y^2 - 2$; second　**b)** $3m^2$; second
　c) $-5x^2y^2 + 7x^2y - xy - 6xy^2$; fourth
　d) $2a^2b^2 - 2a^2b - 2ab^2 - 2b^2$; fourth

16. Answers may vary.
　a) $x^2y^2 + x + y$　**b)** $x^3 + x^2 + y + 1$
　c) $x^2 + y^2 + xy + 1$　**d)** $x^4y + y^3$

17. Explanations may vary.

Investigate, page 344

1. a) $2(x + 3)$　**b)** $x + x + 1 + 1 + 1 + 1$
　c) Explanations may vary.

2. a) $x(x + 3)$　**b)** $x^2 + x + x + x$
　c) Explanations may vary.

3. Answers may vary.

4. a) $3x + 6$　**b)** $x^2 + 4x$　**c)** $2x^2 + 5x$

5. a) $4(x + 2)$　**b)** $x(x + 6)$　**c)** $3x(x + 4)$

6.4 Exercises, page 347

1. a) $20x^2 + 40$　**b)** $14a - 35$
　c) $16k - 24k^2$　**d)** $24b^2 - 36b + 108$
　e) $45m^2 - 63m + 27$　**f)** $24p^2 - 15p + 21$
　g) $60x^2 - 48x$　**h)** $6a - 21a^2$
　i) $12p^2 - 6pq$　**j)** $135n^3 - 90n^2$
　k) $21m^4n + 42m^3$　**l)** $-56x^2y - 40x^3$

2. a) $5(y - 2)$　**b)** $8(m + 3)$　**c)** $x(6 + 7x)$
　d) $5a(7 + 2a)$　**e)** $9d(5d^2 - 4)$　**f)** $7b(7b - 1)$
　g) $3x(x + 2)$　**h)** $4y(2y - 1)$　**i)** $5p^2(p - 3)$
　j) $8mn(3m + 2n)$　**k)** $6ab(2 + 3a)$　**l)** $-7xy(4y + 5x)$

3. Explanations may vary.

4. a) $3x + 19$　**b)** $-5a + 24$　**c)** $-2y + 10$
　d) $28m - 33$　**e)** $-13p^2 - 12p$　**f)** $-8x - 21y$

5. a) $5x - 6$　**b)** $-x - 3$
　c) $42b - 36c$　**d)** $16m - 37n + 14$

6. a) $3(x^2 + 4x - 2)$　**b)** $x(3x + 5x^2 + 1)$
　c) $a(a^2 + 9a - 3)$　**d)** $3x(x + 2x^2 - 4)$
　e) $8y(2y - 4 + 3y^2)$　**f)** $8xy(x - 4 + 2y)$

7. $200 - 4x^2$

8. a) $0 < x < 5$; descriptions may vary.
   b) No, descriptions may vary.

9. Values of x may vary.

10. About 3.5 cm

11. a) $x(20 - 2x)$, $20(10 - 2x)$, $x(20 - 2x)$; $10(20 - 2x)$, $x(10 - 2x)$, $x(10 - 2x)$
   b) $200 - 4x^2$; the polynomials are equal.

12. a) $3(2b^2 - b + 4)$ **b)** $y(5y^2 + 6y + 3)$
   c) $16x(1 + 2x + 3x^3)$ **d)** $12y^2(y^2 - y + 2)$
   e) $a(9a^2 + 7a + 18)$ **f)** $5z(2z^2 - 3z + 6)$

13. a) $5x(5y + 3x)$ **b)** $7mn(2m - 3n)$
   c) $3a^2b^2(3b - 4)$ **d)** $4xy(x - 4y)$
   e) $6pq(2p + 3q)$ **f)** $3m^2n(9m - 5n^2)$

14. a) $(a + b)(3x + 7)$ **b)** $(2x - y)(m - 5)$
   c) $(x + 4)(x^2 + y^2)$ **d)** $(a + 3b)(5x - 9y)$
   e) $(x - 3)(10y + 7)$ **f)** $(w + x)(7w - 10)$

15. Explanations may vary.

16. a) $(x - 7)(3x^2 + 2x + 5)$ **b)** $(2x + y)(4m^2 - 3m + 7)$
   c) $(x^2 + y)(5a^2 - 7a + 8)$ **d)** $(a - b)(2m - 3n - 7)$
   e) $(3a - 2b)(2x^2 + 5x - 9)$ **f)** $(b - a)(6a + 4b - 7)$

17. a) $9x^3 + 13x^2y$ **b)** $-2a^4 - 27a^3b$
   c) $4p^3 + 51p^2q$ **d)** $-34a^4 + 66a^3b - 80a^3$
   e) $2a^3b^2 - 2a^2b^3 - ab^3$ **f)** $3x^3y^3 + 3x^3y^2 - 3x^2y^2$
   g) $5m^2n^3 - 5m^3n^3 - 15m^3n^2$ **h)** $5x^2 - y^2 - 7xy + 2y$
   i) $2b^3 - 2b^2c - 2c^2 + 5bc$
   j) $7x^3 - 7xy^2 - 2xy - 2x^2y - 2y^3$

18. a) $2\pi r$ by h
   b) $A = 2\pi rh + 2\pi r^2 = 2\pi r(h + r)$
   c) $2\pi r$ by $(h + r)$; this rectangle has a greater area than the rectangle in part a.

19. a) 290π cm^2 **b)** $8\pi r^2$ units^2
   c) $(8\pi r^2 - 10\pi r)$ cm^2 **d)** $(3\pi r^2 + 14\pi r)$ cm^2

20. Explanations may vary.

21. a) $8a^2b - 7a^2 + 15a$; third   **b)** $21mn^2 - 33m^2n$; third
   c) $2xy^2 - 15x^2y$; third   **d)** $43s^3 + 21s^2 - 38s$; third
   e) $5x^3 - 14x^2y - 53xy^2$; third   **f)** $55xy - 15y^2 + 2x^2$; second

22. a) $\pi(x^2 - y^2)$ or $\pi(x - y)(x + y)$ **b)** $\pi x(x + 2y)$

23. $x^2\left(1 + \dfrac{3\pi}{4}\right)$

24. a) $\pi r^2\left(h + \dfrac{2r}{3}\right)$ **b)** $\pi r(3r + 2h)$

25. a) $(a - b)(2m + 3n - 7)$ **b)** $(3a - 2b)(2x^2 + 5x + 9)$
   c) $(a - b)(-6a + 4b - 7)$

26. a) $(x + 3)(x + y)$ **b)** $(x + 1)(x^2 + 1)$
   c) $(a + 2b)(5m + 1)$ **d)** $(x - 2y)(3x + 5)$
   e) $(5m - 3)(m + 2n)$ **f)** $(a - 3b)(2a - 3)$

Problem Solving: Round Robin Scheduling, page 352

1. a) 10   **b)** 15

2. a) $\dfrac{n(n - 1)}{2}$ **b)** $\dfrac{n^2 - n}{2}$

3. a)

n	1	2	3	4	5	6	7	8	9	10
$\frac{1}{2}(n^2 - n)$	0	1	3	6	10	15	21	28	36	45

4. a) 36; 66   **b)** $\dfrac{n(n - 1)}{2}$

5. a) 0, 2, 5, 9, 14

Number of sides	3	4	5	6	7
Number of diagonals	0	2	5	9	14

   b) **i)** $\dfrac{n(n - 3)}{2}$ **ii)** $\dfrac{n(n - 1)}{2} - n$

6. 8   **7.** 28

8. a) $6 \le x + y \le 15$ **b)** $-9 \le x - y \le 0$
   c) $5 \le xy \le 50$ **d)** $\dfrac{1}{10} \le \dfrac{x}{y} \le 1$

Investigate, page 354

2. x^2, x, 1
   a) $(x + 4)(x + 1)$ **b)** $x^2 + 5x + 4$

3. Yes   **4.** Answers may vary.

5. a) $x^2 + 8x + 15$ **b)** $2a^2 + 7a + 3$
   c) $4t^2 + 4t + 1$ **d)** $t^2 + tv + 3t + 2v + 2$
   e) $2m^2 + 2mn + 5m + 2n + 3$ **f)** $3x^2 + 4xy + 9x + y^2 + 3y$

6.5 Exercises, page 356

1. a) $x^2 + 5x + 6$ **b)** $2x^2 + 8x + 6$
   c) $x^2 + 2xy + 3x + y^2 + 3y$ **d)** $x^2 + xy + 3x + 2y + 2$
   e) $a^2 + 2ab + ac + b^2 + bc$ **f)** $6 + 5t + 3s + t^2 + st$

2. a) $x^2 + 7x + 12$ **b)** $2a^2 - 8$ **c)** $6 - 3b - 3b^2$
   d) $t^2 - 10t + 9$ **e)** $2x^2 + x - 15$ **f)** $24 + 2k - k^2$
   g) $4y^2 + 23y - 35$ **h)** $k^2 - 11k + 24$ **i)** $m^2 + m - 30$

3. a) $k^2 + 4k + 4$ **b)** $9m^2 + 18m + 9$ **c)** $a^2 - 12a + 36$
   d) $4b^2 + 28b + 49$ **e)** $y^2 - 2y + 1$ **f)** $z^2 + 12z + 36$
   g) $64 - 64x + 16x^2$ **h)** $a^2 - 6ab + 9b^2$

4. Explanations may vary.

5. a) Explanations may vary.
   b) Answers may vary.   **c)** 4, 4

6. a) $x^2 + 5x + 6$; $x^2 - x - 6$; $x^2 + x - 6$; $x^2 - 5x + 6$
   b) $x^2 + 5x + 4$; $x^2 - 3x - 4$; $x^2 + 3x - 4$; $x^2 - 5x + 4$
   c) $x^2 + 7x + 10$; $x^2 + 3x - 10$; $x^2 - 3x - 10$; $x^2 - 7x + 10$
   d) $x^2 + 9x + 18$; $x^2 + 3x - 18$; $x^2 - 3x - 18$; $x^2 - 9x + 18$

7. a) $x^2 + 2x + 1$; $x^2 + 3x + 2$; $x^2 + 4x + 3$
   b) $x^2 - x - 2$; $x^2 - 4$; $x^2 + x - 6$
   c) $x^2 + 3x + 2$; $x^2 + 4x + 4$; $x^2 + 5x + 6$

8. a) Descriptions and explanations may vary.
   b) $(x + 1)(x + 4) = x^2 + 5x + 4$,
      $(x + 1)(x + 5) = x^2 + 6x + 5$,
      $(x + 1)(x + 6) = x^2 + 7x + 6$;
      $(x + 4)(x - 2) = x^2 + 2x - 8$,
      $(x + 5)(x - 2) = x^2 + 3x - 10$,
      $(x + 6)(x - 2) = x^2 + 4x - 12$;
      $(x + 2)(x + 4) = x^2 + 6x + 8$,
      $(x + 2)(x + 5) = x^2 + 7x + 10$,
      $(x + 2)(x + 6) = x^2 + 8x + 12$
   c) $(x + 1)(x) = x^2 + x$, $(x + 1)(x - 1) = x^2 - 1$,
      $(x + 1)(x - 2) = x^2 - x - 2$; $x(x - 2) = x^2 - 2x$,

$$(x-1)(x-2) = x^2 - 3x + 2,$$
$$(x-2)(x-2) = x^2 - 4x + 4; \ (x+2)(x) = x^2 + 2x,$$
$$(x+2)(x-1) = x^2 + x - 2, \ (x+2)(x-2) = x^2 - 4$$

9. a) $t^2 - 1, \ t^2 - 4, \ t^2 - 9; \ (t+4)(t-4) = t^2 - 16,$
$(t+5)(t-5) = t^2 - 25, \ (t+6)(t-6) = t^2 - 36;$
$(t)(t) = t^2, \ (t-1)(t+1) = t^2 - 1, \ (t-2)(t+2) = t^2 - 4$
b) $t^2 - v^2, \ t^2 - 4v^2, \ t^2 - 9v^2; \ (t+4v)(t-4v) = t^2 - 16v^2,$
$(t+5v)(t-5v) = t^2 - 25v^2, \ (t+6v)(t-6v) = t^2 - 36v^2;$
$(t)(t) = t^2, \ (t-v)(t+v) = t^2 - v^2,$
$(t-2v)(t+2v) = t^2 - 4v^2$

10. a) $4x^2 + 2xy - 4y^2$ **b)** $10x^2 + 8xy$ **c)** $12x^2 + 38x + 20$

11. a) $20 - 2x$ **b)** $10 - 2x$
c) $200 - 60x + 4x^2$ **d)** $200x - 60x^2 + 4x^3$

12. $0 < x < 5$; boxes will have different volumes; explanations may vary.

13. Values of x may vary.

14. a) Yes; when x is about 1.2 cm and 3.1 cm
b) About 192 cm^3 when x is about 2 cm

15. a) $10 - 2x; \ 20 - 2x$ **b)** $10 - x; \ 5 - x$
c) Both $100 - 30x + 2x^2$ **d)** Both $100x - 30x^2 + 2x^3$

16. Descriptions and explanations may vary.
a) $x^2 + 2x + 1, \ x^3 + 2x^2 + 2x + 1, \ x^4 + 2x^3 + 2x^2 + 2x + 1;$
$(x+1)(x^4 + x^3 + x^2 + x + 1)$
$= x^5 + 2x^4 + 2x^3 + 2x^2 + 2x + 1,$
$(x+1)(x^5 + x^4 + x^3 + x^2 + x + 1)$
$= x^6 + 2x^5 + 2x^4 + 2x^3 + 2x^2 + 2x + 1,$
$(x+1)(x^6 + x^5 + x^4 + x^3 + x^2 + x + 1)$
$= = x^7 + 2x^6 + 2x^5 + 2x^4 + 2x^3 + 2x^2 + 2x + 1$
b) $x^2 - 1, \ x^3 + 1, \ x^4 - 1;$
$(x+1)(x^4 - x^3 + x^2 - x + 1) = x^5 + 1,$
$(x+1)(x^5 - x^4 + x^3 - x^2 + x - 1) = x^6 - 1,$
$(x+1)(x^6 - x^5 + x^4 - x^3 + x^2 - x + 1) = x^7 + 1$

17. a) $6x^2 + 5xy + y^2$ **b)** $9a^2 + 12a + 4$
c) $6x^2 + 5xy - 4y^2$ **d)** $x^2 + 7xy + 12y^2$
e) $35m^2 - 19mn + 2n^2$ **f)** $18x^2 - 48xy + 14y^2$

18. a) $6x^3 + 7x^2 - 15x + 4$ **b)** $4n^3 - 19n^2 + 33n - 36$
c) $6a^3 + a^2 - 58a + 45$ **d)** $15p^3 - 8p^2 - 6p + 4$
e) $6m^3 + 5m^2 - 12m + 21$
f) $8y^3 - 2y^2 - 6y - 12y^2z + 3yz + 9z$

19. a) $x^3 + 3x^2 + 3x + 1$ **b)** $x^3 + 3x^2y + 3xy^2 + y^3$
c) $8x^3 + 12x^2y + 6xy^2 + y^3$ **d)** $8x^3 + 36x^2 + 54x + 27$
e) $27x^3 - 54x^2y + 36xy^2 - 8y^3$

20. a) $-2x^2 + 14x - 20$ **b)** $x + 2$ **c)** $-4x + 12$

21. a) $6x^3 + 2x^2 - 128x - 160$ **b)** $3b^3 - b^2 - 172b + 224$
c) $18x^3 + 3x^2 - 88x - 80$ **d)** $50a^3 - 235a^2 + 228a - 63$
e) $125m^3 - 150m^2 + 60m - 8$ **f)** $8k^3 + 12k^2 - 18k - 27$

22. Descriptions and explanations may vary.
a) $x^2 + 2x + 1, \ x^3 + 3x^2 + 3x + 1, \ x^4 + 4x^3 + 6x^2 + 4x + 1;$
$(x+1)^5 = x^5 + 5x^4 + 10x^3 + 10x^2 + 5x + 1,$
$(x+1)^6 = x^6 + 6x^5 + 15x^4 + 20x^3 + 15x^2 + 6x + 1,$
$(x+1)^7 = x^7 + 7x^6 + 21x^5 + 35x^4 + 35x^3 + 21x^2 + 7x + 1$
b) $x^2 - 2x + 1, \ x^3 - 3x^2 + 3x - 1, \ x^4 - 4x^3 + 6x^2 - 4x + 1;$
$(x-1)^5 = x^5 - 5x^4 + 10x^3 - 10x^2 + 5x - 1,$
$(x-1)^6 = x^6 - 6x^5 + 15x^4 - 20x^3 + 15x^2 - 6x + 1,$
$(x-1)^7 = x^7 - 7x^6 + 21x^5 - 35x^4 + 35x^3 - 21x^2 + 7x - 1$

c) $x^2 - 1, \ x^4 - 2x^2 + 1, \ x^6 - 3x^4 + 3x^2 - 1,$
$(x+1)^4(x-1)^4 = x^8 - 4x^6 + 6x^4 - 4x^2 + 1;$
$(x+1)^5(x-1)^5 = x^{10} - 5x^8 + 10x^6 - 10x^4 + 5x^2 - 1,$
$(x+1)^6(x-1)^6$
$= x^{12} - 6x^{10} + 15x^8 - 20x^6 + 15x^4 - 6x^2 + 1$

23. a) $60 - w$ **b)** $60w - w^2$ **c)** $-2w + 59$
24. a) $100\left(1 + \dfrac{r}{100}\right)^2$
b) Explanations may vary. $100 + 2r + \dfrac{r^2}{100}$

Exploring with a Graphing Calculator, page 361

1. c) About 3.5 cm
2. c) 1.22 cm, 3.14 cm **d)** About 2.1 cm
3. b) Explanations may vary.

6.6 Exercises, page 364

1. a) $(x+1)(x+2), \ (x+1)(x+3), \ (x+1)(x+4);$
$x^2 + 6x + 5 = (x+1)(x+5),$
$x^2 + 7x + 6 = (x+1)(x+6), \ x^2 + 8x + 7 = (x+1)(x+7)$
b) $(x+1)(x+2), \ (x+2)(x+2), \ (x+3)(x+2);$
$x^2 + 6x + 8 = (x+4)(x+2),$
$x^2 + 7x + 10 = (x+5)(x+2),$
$x^2 + 8x + 12 = (x+6)(x+2)$
c) $(x+1)(x+2), \ (x+2)(x+3), \ (x+3)(x+4);$
$x^2 + 9x + 20 = (x+4)(x+5),$
$x^2 + 11x + 30 = (x+5)(x+6),$
$x^2 + 13x + 42 = (x+6)(x+7)$

2. Answers may vary.

3. a) $(x+6)(x+4)$ **b)** $(m+3)(m+2)$ **c)** $(a+8)(a+2)$
d) $(p+4)(p+4)$ **e)** $(y+7)(y+6)$ **f)** $(d+2)(d+2)$
g) $(x-4)(x-3)$ **h)** $(c-9)(c-8)$ **i)** $(a-6)(a-1)$
j) $(x-1)(x-8)$ **k)** $(s-10)(s-2)$ **l)** $(x-7)(x-7)$

4. Explanations may vary.

5. a) $(y+9)(y-1)$ **b)** $(b+20)(b-1)$ **c)** $(p+18)(p-3)$
d) $(x+14)(x-2)$ **e)** $(a+6)(a-4)$ **f)** $(k+6)(k-3)$
g) $(x+2)(x-4)$ **h)** $(n+3)(n-8)$ **i)** $(a+4)(a-5)$
j) $(d+5)(d-9)$ **k)** $(m+6)(m-15)$ **l)** $(y+6)(y-8)$

6. a) $(x+12)(x+2)$ **b)** $(m-5)(m-10)$ **c)** $(a-6)(a-5)$
d) $(x+3)(x+3)$ **e)** $(x+8)(x+9)$ **f)** $(a-6)^2$
g) $(x-2)(x+10)$ **h)** $(d-8)(d+2)$ **i)** $(b+9)(b-2)$
j) $(x^2+5)(x+2)(x-2)$, or $(x^2+5)(x^2-4)$
k) $(a^2+9)(a^2+5)$ **l)** $(m^2-12)(m^2+5)$

7. a) $(p+13)(p+2)$ **b)** $(x+41)(x-1)$ **c)** $(m-9)(m+8)$
d) $(t-1)(t-12)$ **e)** $(y+4)(y+9)$ **f)** $(m+12)(m-8)$
g) $(s+25)(s-4)$ **h)** $(c-4)(c-4)$ **i)** $(p-10)(p+9)$
j) $(k^2+6)(k^2+5)$
k) $(r-3)(r+3)(r-1)(r+1)$, or $(r^2-9)(r^2-1)$
l) $(c^2+4)(c^2-20)$

8. a) $(x+2)^2$ **b)** Not possible **c)** $(x+1)(x+4)$
d) Not possible **e)** $(x+5)(x-1)$ **f)** $(x-1)(x-4)$

9. Explanations may vary.

10. a) $-4, 0, 6, 14, 24$ **b)** $(x+4)(x-1); \ -4, 0, 6, 14, 24$
c) The results are the same.

11. a) $x^2 + 2x + 1 = (x + 1)(x + 1)$, $x^2 + x + 0 = (x + 1)(x)$, or
$(x + 1)(x - 0)$, $x^2 + 0x - 1 = (x + 1)(x - 1)$,
$x^2 - x - 2 = (x + 1)(x - 2)$
b) $x^2 + 2x + 0 = x(x + 2)$, or $(x + 0)(x + 2)$,
$x^2 + x - 2 = (x - 1)(x + 2)$, $x^2 + 0x - 4 = (x - 2)(x + 2)$,
$x^2 - x - 6 = (x - 3)(x + 2)$
c) $x^2 + x + 0 = (x + 0)(x + 1)$, or $x(x + 1)$,
$x^2 - x + 0 = (x - 1)(x + 0)$, or $(x - 1)(x)$,
$x^2 - 3x + 2 = (x - 2)(x - 1)$, $x^2 - 5x + 6 = (x - 3)(x - 2)$

12. a) $(x + 4)(x + 3)$, $(x + 6)(x + 2)$, $(x + 12)(x + 1)$,
$(x - 4)(x - 3)$, $(x - 6)(x - 2)$, $(x - 12)(x - 1)$
b) ±7, ±8, ±13

13. a) ±7, ±11   **b)** 0, ±8   **c)** ±6, ±9
d) ±1, ±4, ±11   **e)** ±9, ±11, ±19   **f)** 0, ±6, ±15

14. a) The product of any two integers whose sum is 1.
b) The product of any two integers whose sum is −1.
c) The product of any two integers whose sum is 2.
d) The product of any two integers whose sum is −2.
e) The product of any two integers whose sum is 3.
f) The product of any two integers whose sum is −3.

15. Explanations may vary.

16. a) $2(x + 1)(x + 3)$ **b)** $5(a - 1)(a + 4)$ **c)** $3(y + 2)(y - 6)$
d) $5(y - 4)(y - 4)$ **e)** $2(m + 8)(m - 7)$ **f)** $9(x + 5)(x + 1)$

17. a) $4(x + 3)(x + 4)$ **b)** $5(a + 2)(a - 6)$ **c)** $3(n - 5)(n - 4)$
d) $3(p - 4)(p + 9)$ **e)** $7(x + 2)(x - 14)$ **f)** $3(a - 5)(a - 1)$

18. a) $(x + 3y)(x + 2y)$ **b)** $(ab + 2)(ab + 3)$ **c)** $(1 + 3c)(1 + 2c)$

19. a) $4(y + 2)(y - 7)$ **b)** $3(m + 2)(m + 4)$
c) $4(x - 3)(x + 4)$ **d)** $10(x + 2)(x + 6)$
e) $5a(m - 1)(m - 7)$ **f)** $7d(c - 3d)(c - 2d)$

20. a) $(x^2 + 2)(x^2 + 5)$ **b)** $(a^2 + 2)(a^2 + 7)$ **c)** $(m^2 + 4)(m^2 + 9)$
d) $2(b^2 + 3)(b^2 + 5)$ **e)** $3(c^2 + 1)(c^2 + 7)$ **f)** $5(x^2 + 2)(x^2 + 3)$

21. a) $(x + y + 10)(x + y - 1)$ **b)** $(p - 2q - 8)(p - 2q - 3)$
c) $3(y + 1)(3y - 13)$ **d)** $(x + 1)(x + 3)(x^2 + 4x + 5)$
e) $(2m - n - 5p)(2m - n + 4p)$ **f)** $12(x + 3y + 2)(x - y + 2)$

22. Explanations may vary.   **23.** c is the square of $\frac{1}{2}b$.

24. a) $4a^2$ **b)** $2a^2 - 10ab + 12b^2$
c) $-7a^2 + 26ab - 15b^2$ **d)** $a^4 - 2a^2b^2 - a^2 + b^4 - 2 + b^2$

25. a) $(x + 4)(x + 6)$, $(x - 4)(x - 6)$, $(x - 2)(x + 12)$,
$(x + 2)(x - 12)$
b) Answers may vary. For example,
$(x^2 + 5x + 6) = (x + 2)(x + 3)$,
$(x^2 - 5x + 6) = (x - 2)(x - 3)$,
$(x^2 + 5x - 6) = (x + 6)(x - 1)$,
$(x^2 - 5x - 6) = (x - 6)(x + 1)$

26. a) $(x + 2)(x + 3)$, $(x + 5)(x + 1)$
b) Answers may vary. $x^2 - 9x + 8 = (x - 1)(x - 8)$,
$x^2 + 8x - 9 = (x + 9)(x - 1)$

Investigate, page 369

2. a) $(2x + 3)(x + 1)$ **b)** $(3x + 2)(x + 1)$
c) $(2x + 3)(2x + 3)$ **d)** $(4x + 3)(x + 2)$

4. $(2x + 4)(x + 1)$ and $(x + 2)(2x + 2)$; explanations may vary
- the trinomial has a common factor.

6.7 Exercises, page 371

1. a) $(2x + 1)(x + 2)$, $(2x + 1)(x + 3)$, $(2x + 1)(x + 4)$;
$2x^2 + 11x + 5 = (2x + 1)(x + 5)$,
$2x^2 + 13x + 6 = (2x + 1)(x + 6)$,
$2x^2 + 15x + 7 = (2x + 1)(x + 7)$
b) $(2x + 1)(x + 2)$, $(2x + 3)(x + 2)$, $(2x + 5)(x + 2)$;
$2x^2 + 11x + 14 = (2x + 7)(x + 2)$,
$2x^2 + 13x + 18 = (2x + 9)(x + 2)$,
$2x^2 + 15x + 22 = (2x + 11)(x + 2)$
c) $(2x + 1)(x + 2)$, $(3x + 1)(x + 2)$, $(4x + 1)(x + 2)$;
$5x^2 + 11x + 2 = (5x + 1)(x + 2)$,
$6x^2 + 13x + 2 = (6x + 1)(x + 2)$,
$7x^2 + 15x + 2 = (7x + 1)(x + 2)$

2. Answers may vary.

3. a) $(2x + 3)(x + 2)$ **b)** $(2a + 3)(a + 2)$
c) $(2d + 1)(d + 1)$ **d)** $(3s + 1)(s + 1)$
e) $(2y + 3)(3y + 1)$ **f)** $(2x^2 + 1)(4x^2 + 3)$

4. a) $(5x - 2)(x - 1)$ **b)** $(3n - 2)(n - 3)$
c) $(2c - 1)(7c - 3)$ **d)** $(2x - 5)(x - 3)$
e) $(3x - 1)(x - 7)$ **f)** $(2a^2 - 1)^2$

5. a) $(3t - 2)(t + 3)$ **b)** $(3k + 4)(2k - 1)$
c) $(2r - 1)(4r + 3)$ **d)** $(4m - 5)(m + 2)$
e) $(5y - 1)(y + 4)$ **f)** $(2d^2 - 3)(2d^2 + 5)$
g) $(5a + 3)(a - 2)$ **h)** $(3x + 2)(x - 5)$
i) $(2m - 7)(m + 3)$ **j)** $(4k + 3)(k - 3)$
k) $(2x - 3)(3x + 4)$ **l)** $(3a^2 + 1)(5a^2 - 2)$

6. Answers may vary.

7. a) $(3x + 1)(x + 4)$ **b)** $(m - 4)(2m - 3)$
c) $(2s - 5)^2$ **d)** $5(x + 2)(x + 1)$
e) $(3a + 4b)(2a + 3b)$
f) $(2x + y)(2x - y)(2x^2 - 3y^2)$, or $(4x^2 - y^2)(2x^2 - 3y^2)$
g) $(5x - 3)(2x + 1)$ **h)** $(6k - 5)(k + 1)$
i) $(5g + 1)(3g - 2)$ **j)** $3(2x - 1)(x - 1)$
k) $(2c + 5d)(4c - d)$ **l)** $(5x + 2y)(3x - 2y)$

8. a) $(2x + 5)^2$
b) Answers may vary. The terms of the factors are the
square roots of the first and third terms in the trinomial.
The roots are equal.
c) Answers may vary.

9. a) $(2x - 1)^2$ **b)** $(2h + 1)(h + 2)$
c) $(3q - 5)^2$ **d)** $(5u^2 - 2)(2u^2 - 5)$
e) $(10m + 3n)(m - 2n)$ **f)** $(4 - 3d)(2 + 5d)$

10. a) $(4h - 3)^2$ **b)** $(5r^2 - 1)(2r^2 + 3)$
c) $(w + 5)(2w + 3)$ **d)** Not possible
e) $(5x + y)(2x - 7y)$ **f)** $(3a - 4b)^2$

11. a) $(2x - 1)(x + 1)$, $(2x - 1)(x + 2)$, $(2x - 1)(x + 3)$;
$(2x - 1)(x + 4)$, $2x^2 + 9x - 5 = (2x - 1)(x + 5)$,
$2x^2 + 11x - 6 = (2x - 1)(x + 6)$,
$2x^2 + 13x - 7 = (2x - 1)(x + 7)$
b) $(2x - 1)(x + 2)$, $(3x - 2)(x + 2)$, $(4x - 3)(x + 2)$;
$(5x - 4)(x + 2)$, $6x^2 + 7x - 10 = (6x - 5)(x + 2)$,
$7x^2 + 8x - 12 = (7x - 6)(x + 2)$,
$8x^2 + 9x - 14 = (8x - 7)(x + 2)$
c) $(2x + 1)^2$, $(3x + 2)^2$, $(4x + 3)^2$; $(5x + 4)^2$,
$36x^2 + 60x + 25 = (6x + 5)^2$,
$49x^2 + 84x + 36 = (7x + 6)^2$,
$64x^2 + 112x + 49 = (8x + 7)^2$

12. Answers may vary.

13. For exercise 1.a): $2x^2 + 3x + 1 = (2x + 1)(x + 1)$,
$2x^2 + x + 0 = (2x + 1)(x + 0)$, or $(2x + 1)(x)$,
$2x^2 - x - 1 = (2x + 1)(x - 1)$,
$2x^2 - 3x - 2 = (2x + 1)(x - 2)$; for exercise 1b):
$2x^2 + 3x - 2 = (2x - 1)(x + 2)$,
$2x^2 + x - 6 = (2x - 3)(x + 2)$,
$2x^2 - x - 10 = (2x - 5)(x + 2)$,
$2x^2 - 3x - 14 = (2x - 7)(x + 2)$; for exercise 1c):
$x^2 + 3x + 2 = (x + 1)(x + 2)$, $0x^2 + x + 2 = (0x + 1)(x + 2)$,
$-x^2 - x + 2 = (-x + 1)(x + 2)$,
$-2x^2 - 3x + 2 = (-2x + 1)(x + 2)$; Answers for exercise 2
may vary. For exercise 11a): $2x^2 - x + 0 = (2x - 1)(x + 0)$,
or $(2x - 1)(x)$, $2x^2 - 3x + 1 = (2x - 1)(x - 1)$,
$2x^2 - 5x + 2 = (2x - 1)(x - 2)$,
$2x^2 - 7x + 3 = (2x - 1)(x - 3)$; for exercise 11b):
$x^2 + 2x - 0 = x(x + 2)$, or $(x - 0)(x + 2)$,
$0x^2 + x + 2 = (0x + 1)(x + 2)$, or $x + 2$,
$-x^2 + 0x + 4 = (-x + 2)(x + 2)$,
$-2x^2 - x + 6 = (-2x + 3)(x + 2)$; for exercise 11c):
$x^2 + 0x + 0 = (-x - 0)(-x - 0)$, $4x^2 + 4x + 1 = (-2x - 1)^2$,
$9x^2 + 12x + 4 = (-3x - 2)^2$, $16x^2 + 24x + 9 = (-4x - 3)^2$

14. a) $(5x + 6)(x + 1)$, $(5x + 3)(x + 2)$, $(5x + 2)(x + 3)$,
$(5x + 1)(x + 6)$, $(5x - 6)(x - 1)$, $(5x - 3)(x - 2)$,
$(5x - 2)(x - 3)$, $(5x - 1)(x - 6)$
b) ±11, ±13, ±17, ±31 **c)** Answers may vary.

15. a) ±7, ±8, ±13 **b)** ±20, ±25, ±29, ±52, ±101
c) ±3, ±15, ±25, ±53
d) ±22, ±23, ±26, ±29, ±34, ±43, ±62, ±121
e) ±6, ±10 **f)** ±1

16. a) $10(2x + 3)(x + 2)$ **b)** $5(3a - 1)(a - 4)$
c) $3(3a - 2)(2a + 3)$ **d)** $2(8 - 6r + 5r^2)$
e) $6(2x - 3y)(2x - 3y)$ **f)** $4(3a^2 + 2b^2)(a^2 - 5b^2)$

17. Yes

18. a) $x(3y^2 - 22y + 6)$ **b)** $n(3m - 4)(m - 3)$
c) $y(x - 5)(4x + 3)$ **d)** $xy(2x - 3)(x + 5)$
e) $mn(2m - 7)(m + 3)$ **f)** $xy(3x + y)(2x - 3y)$

19. a) $3x(2x + 5)(x + 3)$ **b)** $2a(3a - 2)(a + 5)$
c) $3y(6x^2 - xy - 15)$ **d)** $5m(2m + 3)(m - 4)$
e) $3a(3a - 7)(a - 2)$ **f)** $7a(6b^2 + 7b - 4)$

20. a) $(4x - 1)(8x - 3)$ **b)** $(8s + 1)(3s - 2)$
c) $(4a + 7)(a + 3)$ **d)** $(4x + 3y)(x - 6y)$
e) $(5a + 3b)(2a - 5b)$ **f)** $(7x - y)(3x + 4y)$
g) $(7x - 6)(3x + 5)$ **h)** $(9x - 2)(8x + 3)$
i) $(5x + 4)(3x - 8)$

21. b is twice the product of the square roots of a and c.

Investigate, page 374

2. The expressions are equal.

3. a) $(a + 5)(a - 5)$
b) The coefficients of the terms of the binomial are the
squares of the coefficients of the terms of the factors.
c) There is a relationship; explanations may vary.

6.8 Exercises, page 376

1. a) The results differ by 1.
b) Conclusions may vary. When two natural numbers differ
by 2, the square of their mean is 1 more than their
product.

2. The conclusion is still true.

3. a) $(x + 7)(x - 7)$ **b)** $(2b + 11)(2b - 11)$
c) $(3m + 8)(3m - 8)$ **d)** $(9f + 4)(9f - 4)$
e) $(5y + 12)(5y - 12)$ **f)** $(7x + 6)(7x - 6)$
g) $(4 + 9y)(4 - 9y)$ **h)** $(13 + 4t)(13 - 4t)$
i) $(10m + 7)(10m - 7)$ **j)** $(8b + 1)(8b - 1)$
k) $(11a + 20)(11a - 20)$ **l)** $(6b + 5)(6b - 5)$
m) $(5p + 9)(5p - 9)$ **n)** $(12m + 7)(12m - 7)$
o) $(6 + 11x)(6 - 11x)$ **p)** $(1 + 5q)(1 - 5q)$

4. a) $(2s + 3t)(2s - 3t)$ **b)** $(4x + 7y)(4x - 7y)$
c) $(9a + 8b)(9a - 8b)$ **d)** $(11c + 10d)(11c - 10d)$
e) $(p + 6q)(p - 6q)$ **f)** $9(4y + 3z)(4y - 3z)$
g) $(5m + 13n)(5m - 13n)$ **h)** $(2e + 15f)(2e - 15f)$
i) $(m^2 + 1)(m + 1)(m - 1)$ **j)** $(x^2 + 4)(x - 2)(x + 2)$
k) $(1 + 4y^2)(1 - 2y)(1 + 2y)$
l) $(4a^2 + 9b^2)(2a + 3b)(2a - 3b)$

5. a) $8(m + 3)(m - 3)$ **b)** $6(x + 5)(x - 5)$
c) $5(2x + y)(2x - y)$ **d)** $2(3b + 8)(3b - 8)$
e) $3(2a + 5)(2a - 5)$ **f)** $2(3p + 7)(3p - 7)$
g) $5(4s + 9)(4s - 9)$ **h)** $3(2p + 11)(2p - 11)$
i) $3x(2x + 3)(2x - 3)$ **j)** $2m(4m + 7)(4m - 7)$
k) $7b(3a + 2)(3a - 2)$ **l)** $3t^2(5s + 3)(5s - 3)$
m) $(x - y - z)(x - y + z)$ **n)** $(2a + b + 9)(2a + b - 9)$
o) $(6a - b)(12a + b)$
p) $(4x - 2y + 5z)(4x - 2y - 5z)$

6. Explanations may vary.

7. a) $-5(2x + 9)$ **b)** $4(m + 1)(4m - 3)$
c) $24a$ **d)** $96yz$
e) $-(11p - 5)(5p + 9)$ **f)** $-15(x + 1)(3x + 1)$
g) $(x - 3)(x + 3)(x - 2)(x + 2)$
h) $(a + 4)(a - 4)(a + 1)(a - 1)$
i) $(y + 3)(y - 3)(y^2 + 4)$

8. a) $(x + 9y)(x - 4y)$, $(x + 12y)(x - 3y)$, $(x + 18y)(x - 2y)$;
$x^2 - 36y^2 = (x + 6y)(x - 6y)$,
$x^2 + 35xy - 36y^2 = (x + 36y)(x - y)$,
$x^2 - 35xy - 36y^2 = (x - 36y)(x + y)$,
$x^2 - 5xy - 36y^2 = (x - 9y)(x + 4y)$,
$x^2 - 9xy - 36y^2 = (x - 12y)(x + 3y)$,
$x^2 - 16xy - 36y^2 = (x - 18y)(x + 2y)$
b) 0, ±5, ±9, ±16, ±35 **c)** Examples may vary.

9. Answers may vary. $a = -1$, $b = 0$, $c = 1$:
$-x^2 + 1 = -(x - 1)(x + 1)$

10. 6.53 m^2

11. a) $V = \frac{4}{3}h(s^2 - h^2) = \frac{4}{3}h(s + h)(s - h)$ **b)** 2 591 000 m^3

12. a) $x(x - y) + (x - y)y$
b) $(x - y)(x - y) + y(x - y) + y(x - y)$

13. $\frac{1}{4}(x - y)(x + y)$

14. a) $8(d + 2e)(d - 2e)$ **b)** $\left(5m + \frac{1}{2}n\right)\left(5m - \frac{1}{2}n\right)$
c) $2y^2(3x + 5y)(3x - 5y)$ **d)** Not possible
e) Not possible   **f)** $\left(p + \frac{1}{3}q\right)\left(p - \frac{1}{3}q\right)$
g) $5(x + 2)(x - 2)(x^2 + 4)$ **h)** $\left(\frac{x}{4} + \frac{y}{7}\right)\left(\frac{x}{4} - \frac{y}{7}\right)$

15. Explanations may vary.

16. a) 3, 7   **b)** 2   **c)** Proofs may vary.

17. a) $h = \frac{\sqrt{l^2(x^2 - 1)}}{2}$
b) 6.13 cm; 7.91 cm; 8.67 cm
c) Less than; explanations may vary.

6.9 Exercises, page 384

1. a) ±3   **b)** ±5   **c)** ±2   **d)** ±2
e) ±1   **f)** $\pm\sqrt{6}$ **g)** ±7   **h)** $\pm\sqrt{7}$

2. a) −3, −5   **b)** 3, 4   **c)** 5, −4
d) −8, 3   **e)** −6, −2   **f)** 9, −4
g) 6, 4   **h)** −7, −8   **i)** 7, −6

3. a) 4, 5   **b)** 7, 9   **c)** 8, −2
d) −7, −3   **e)** −2, 7   **f)** 3, −5
g) 4   **h)** 2, 6   **i)** 11, −4

4. 9, 3 or −3, −9   **5.** 1 s and 5 s   **6.** 0.9 s and 5.5 s

7. a) $V = 4\pi r^2$ **b)** Approximately 8.9 cm, 17.8 cm

8. Answers may vary.

9. a) $V = 4(x - 8)^2$ **b)** 23.8 cm by 23.8 cm
c) 15.8 cm by 15.8 cm by 4.0 cm

10. a) 420   **b)** 10

11. a) −180   **b)** 4 or 7
c) Explanations may vary. The terms in the series in exercise 11 change sign from positive to negative, so the sum 28 occurs twice.

12. a) $t = \sqrt{\frac{d}{4.9}}$
b) i) 1.4 s   **ii)** 2.0 s   **iii)** 2.5 s

13. a) $x = \frac{2\sqrt{A}}{\sqrt[4]{3}}$
b) i) 4.8 cm   **ii)** 6.8 cm   **iii)** 9.6 cm

6.10 Exercises, page 389

1. a) Quotient is the same, remainder is 5.
b) $3x^2 + 8x + 4$; the divisor is a factor of the dividend.
c) Quotient: $3x - 1$; remainder: 14

2. a) Quotient is the same, remainder is 6.
b) $-4x^3 + 6x^2 + 4x - 6$
c) Quotient: $-2x^2 - x$; remainder: −7

3. $4t^4 + 25t^2 - 33t + 10 = (2t - 1)(2t^3 + t^2 + 13t - 10)$;
$4x^3 + x^2 - 2x + 1 = (2x^2 - 3)(2x + 0.5) + 4x + 2.5$

4. a) $x^2 + 7x + 14 = (x + 3)(x + 4) + 2$
b) $x^2 - 3x + 5 = (x - 2)(x - 1) + 3$
c) $x^2 + x - 2 = (x + 3)(x - 2) + 4$
d) $n^2 - 11n + 6 = (n + 5)(n - 16) + 86$

5. Explanations may vary.

6. a) $\frac{x^3 - 5x^2 + 10x - 15}{x - 3} = x^2 - 2x + 4 - \frac{3}{x - 3}$
b) $\frac{x^3 - 5x^2 - x - 10}{x - 2} = x^2 - 3x - 7 - \frac{24}{x - 2}$
c) $\frac{3x^3 + 11x^2 - 6x - 10}{x + 4} = 3x^2 - x - 2 - \frac{2}{x + 4}$
d) $\frac{2x^3 + x^2 - 27x - 36}{x + 3} = 2x^2 - 5x - 12$

7. Division statements may vary.
a) $2x^2 + 5x - 1 = (x + 1)(2x + 3) - 4$;
$\frac{2x^2 + 5x - 1}{x + 1} = 2x + 3 - \frac{4}{x + 1}$
b) $3x^2 + 2x - 5 = (x - 2)(3x + 8) + 11$;
$\frac{3x^2 + 2x - 5}{x - 2} = 3x + 8 + \frac{11}{x - 2}$
c) $25u^2 + 1 = (5u + 3)(5u - 3) + 10$;
$\frac{25u^2 + 1}{5u + 3} = 5u - 3 + \frac{10}{5u + 3}$
d) $6x^2 - 3 = (2x + 4)(3x - 6) + 21$;
$\frac{6x^2 - 3}{2x + 4} = 3x - 6 + \frac{21}{2x + 4}$
e) $8x^2 - 6x + 11 = (2x - 3)(4x + 3) + 20$;
$\frac{8x^2 - 6x + 11}{2x - 3} = 4x + 3 + \frac{20}{2x - 3}$
f) $9m^2 - 5 = (3m + 2)(3m - 2) - 1$;
$\frac{9m^2 - 5}{3m + 2} = 3m - 2 - \frac{1}{3m + 2}$

8. a) $c^2 + 4c + 3$, R−7; not a factor
b) $x^2 - 6x - 11$, R15; not a factor
c) $-2n^2 + n + 1$; a factor
d) $x^2 - 2x - 8$, R−4; not a factor
e) $m^2 + 3m - 10$, R−54; not a factor
f) $2x^3 - x^2 + x - 2$, R1; not a factor

9. a) $x^2 - 9 = (x - 3)(x + 3)$ **b)** $x^2 - 9 = (x + 3)(x - 3)$
c) $x^2 + 9 = (x + 3)(x - 3) + 18$
d) $x^2 + 9 = (x - 3)(x + 3) + 18$

10. a) Explanations may vary.
i) $x + 4$ **ii)** $x + 3$ **iii)** x, R12   **iv)** $x + 7$, R12
v) $x + 5$, R2   **vi)** $x + 2$, R2   **vii)** $x + 6$, R6   **viii)** $x + 1$, R6
b) Patterns and explanations may vary.

Divisor	Quotient	Remainder
x	$x + 7$	12
$x + 1$	$x + 6$	6
$x + 2$	$x + 5$	2
$x + 3$	$x + 4$	0
$x + 4$	$x + 3$	0
$x + 5$	$x + 2$	2
$x + 6$	$x + 1$	6
$x + 7$	x	12

c) i) $x - 1$, R20   **ii)** $x + 8$, R20   **iii)** $x - 2$, R30   **iv)** $x + 9$, R30

11. a) Descriptions may vary.

Divisor	Quotient	Remainder
x	$x + 5$	6
$x + 1$	$x + 4$	2
$x + 2$	$x + 3$	0
$x + 3$	$x + 2$	0
$x + 4$	$x + 1$	2
$x + 5$	x	6
$x + 6$	$x - 1$	12
$x + 7$	$x - 2$	20

b)

Divisor	Quotient	Remainder
x	$x + 8$	12
$x + 1$	$x + 7$	5
$x + 2$	$x + 6$	0
$x + 3$	$x + 5$	-3
$x + 4$	$x + 4$	-4
$x + 5$	$x + 3$	-3
$x + 6$	$x + 2$	0
$x + 7$	$x + 1$	5
$x + 8$	x	12

c)

Divisor	Quotient	Remainder
x	$x + 6$	9
$x + 1$	$x + 5$	4
$x + 2$	$x + 4$	1
$x + 3$	$x + 3$	0
$x + 4$	$x + 2$	1
$x + 5$	$x + 1$	4
$x + 6$	x	9
$x + 7$	$x - 1$	16

12. a) $2x + 1$, R3; $2x + 2$, R3; $2x + 3$, R3

b) $x^2 + 4x + 1$, R3; $x^2 + 4x + 2$, R3; $x^2 + 4x + 3$, R3

13. a) Descriptions may vary. The constant term in the quotient increases by 1, while the remainders are the same.

b) For exercise 12a): $2x + 4$, R3, $2x + 5$, R3, $2x + 6$, R3; for exercise 12b): $x^2 + 4x + 4$, R3, $x^2 + 4x + 5$, R3, $x^2 + 4x + 6$, R3

c) For exercise 12a): $2x$, R3, $2x - 1$, R3, $2x - 2$, R3; for exercise 12b): $x^2 + 4x$, R3, $x^2 + 4x - 1$, R3, $x^2 + 4x - 2$, R3

14. Descriptions may vary.

a) $x + 1$; x, R1; $x - 1$, R4; $\frac{x^2 + 2x + 1}{x + 4} = x - 2 + \frac{9}{x + 4}$;
$\frac{x^2 + 2x + 1}{x + 5} = x - 3 + \frac{16}{x + 5}$; $\frac{x^2 + 2x + 1}{x + 6} = x - 4 + \frac{25}{x + 6}$

b) $x + 1$; $\frac{1}{2}x + \frac{3}{4}$, R$\frac{1}{4}$; $\frac{1}{3}x + \frac{5}{9}$, R$\frac{4}{9}$;
$\frac{x^2 + 2x + 1}{4x + 1} = \frac{1}{4}x + \frac{7}{16} + \frac{9}{16(4x + 1)}$;
$\frac{x^2 + 2x + 1}{5x + 1} = \frac{1}{5}x + \frac{9}{25} + \frac{16}{25(5x + 1)}$,
$\frac{x^2 + 2x + 1}{6x + 1} = \frac{1}{6}x + \frac{11}{36} + \frac{25}{36(6x + 1)}$

15. a) $\frac{1}{2}x + \frac{7}{4}$, R$\frac{3}{4}$; not a factor

b) $3x^2 - 2x + \frac{3}{2}$, R$-\frac{7}{2}$; not a factor

c) $-\frac{1}{2}x + 1$, R$-\frac{1}{2}x$; not a factor

d) $2x^3 - \frac{3}{4}x + \frac{31}{16}$, R$\frac{15}{16}$; not a factor

16. Explanations may vary.

17. a) $3x^3 + 4x^2 + 3x + 1 = (5x + 2)\left(\frac{3}{5}x^2 + \frac{14}{25}x + \frac{47}{125}\right) + \frac{31}{125}$

b) $6x^3 - 2x^2 + 7x - 11 = (3x^2 - 2)\left(2x - \frac{2}{3}\right) + 11x - \frac{37}{3}$

c) $7 + 8x + 10x^2 + 6x^3 = (4x^2 + 3)\left(\frac{3}{2}x + \frac{5}{2}\right) + \frac{7}{2}x - \frac{1}{2}$

d) $-11x^3 + 15x^2 + 12x^4 - 7x + 12$
$= (3x^2 - 2)\left(4x^2 - \frac{11}{3}x + \frac{23}{3}\right) - \frac{43}{3}x + \frac{82}{3}$

18. Descriptions may vary.

a) $x + 1$, $x^2 + x + 1$, $x^3 + x^2 + x + 1$;
$\frac{x^5 - 1}{x - 1} = x^4 + x^3 + x^2 + x + 1$,
$\frac{x^6 - 1}{x - 1} = x^5 + x^4 + x^3 + x^2 + x + 1$,
$\frac{x^7 - 1}{x - 1} = x^6 + x^5 + x^4 + x^3 + x^2 + x + 1$

b) x, R1; $x^2 + 1$; $x^3 + x$, R1;
$\frac{x^5 + x^4 + x^3 + x^2 + x + 1}{x + 1} = x^4 + x^2 + 1$,
$\frac{x^6 + x^5 + x^4 + x^3 + x^2 + x + 1}{x + 1} = x^5 + x^3 + x$, R1,
$\frac{x^7 + x^6 + x^5 + x^4 + x^3 + x^2 + x + 1}{x + 1} = x^6 + x^4 + x^2 + 1$

19. $\frac{x^2 - 1}{x + 1} = x - 1$
$\frac{x^3 - 1}{x + 1} = x^2 - x + 1$ R-2
$\frac{x^4 - 1}{x + 1} = x^3 - x^2 + x - 1$
$\frac{x^5 - 1}{x + 1} = x^4 - x^3 + x^2 - x + 1$ R-2

20. $4x + 3$, $x + 5$ **21.** -8

6 Review, page 393

1. 38.5 m^2; 22.4 m^3 **2.** 0.94 m; 11.1 m^2

3. Explanations may vary.

4. a) $-40xy^2$ **b)** $-6a^3b^3$ **c)** $24x^4y^4$

5. a) $-7m^3n$ **b)** $4x^4y^4$ **c)** $-a^3b^2$

6. a) $-9x^2 + 20xy$ **b)** $3m^2 - 8mn + 7n^2$
c) $-4x^2 + 11xy$ **d)** $3a^2 - 12ab$
e) $-5x^2y - 20xy^2$ **f)** $-12m^2n + 5mn^2$

7. a) $23x^2 - 69x$ **b)** $-14c + 12d - 57e$
c) $8m^2 - 17mn + 13m$ **d)** $8x^3 + 16x^2y - 26x^2$

8. a) $4x^3 - 24x^2 + 27x + 20$ **b)** $24y^3 + 41y^2 - 9y - 28$

9. a) $12x + 12y$; $7x^2 + 15xy + 7y^2$
b) $18x + 8y$; $16x^2 + 13xy + 2y^2$

10. Explanations may vary.

11. a) $4m^2(2m - 1)$ **b)** $4(2y^2 - 3y^4 + 6)$
c) $7a^2(4 - a)$ **d)** $3a^2b^2c(2b - 5c)$
e) $10x^2y(3 - 2y + xy)$ **f)** $4mn(2n - 3 - 4m)$

12. a) $(m + 4)(m + 4)$ **b)** $(a - 4)(a - 3)$
c) $(y - 4)(y + 2)$ **d)** $(n - 9)(n + 5)$
e) $(s - 3)(s + 3)(s^2 - 6)$ **f)** $(k^2 - 15)(k^2 + 6)$

13. a) $(a + 12b)(a + 2b)$ **b)** $(m + 6n)(m + 3n)$
c) $(s + 18t)(s + 2t)$ **d)** $(x - 5y)(x + 4y)$
e) $(c + 25d)(c - 4d)$ **f)** $(p - 24q)(p + 5q)$

14. a) $(x - 1)(4x - 3)$ **b)** $(3a + 1)(2a - 5)$
c) $(7n - 2)(3n + 2)$ **d)** $(6r^2 - 1)(r^2 - 5)$
e) $3(4t + 3)(t - 2)$ **f)** $2(7x + 4)(4x - 1)$

15. Explanations may vary.

16. a) $(b + 6)(b - 6)$ **b)** $(9k^2 + 1)(3k - 1)(3k + 1)$
c) $(6x + 7y)(6x - 7y)$ **d)** $(2a + 3b)(2a - 3b)$
e) $(5m + 9n)(5m - 9n)$ **f)** $(1 + 4s^2)(1 - 2s)(1 + 2s)$
g) $(14x + 5z)(14x - 5z)$ **h)** $(16p + 25q)(16p - 25q)$
i) $(17s + 18t)(17s - 18t)$

17. a) $8(a + 3)(a - 3)$ **b)** $6(5 + n)(5 - n)$
c) $7(x + y)(x - y)(x^2 + y^2)$ **d)** $3m(3m + 2)(3m - 2)$
e) $\left(\frac{a}{6} + \frac{b}{7}\right)\left(\frac{a}{6} - \frac{b}{7}\right)$ **f)** $5q^2(5p + 6)(5p - 6)$

18. a) $4(c + 3)(c - 8)$ **b)** $(x + y + z)(x - y - z)$
c) $(a + b - c)(a - b + c)$

19. a) $2m(4 + 7m)(4 - 7m)$ **b)** $5y^2(5x + 6)(5x - 6)$
c) $2y(8x + 5y)(8x - 5y)$

20. a) $(x - 5y)(x + 5y)(x - 2y)(x + 2y)$
b) $(m - 6n)(m + 6n)(m^2 - 2n^2)$

21. a) $-3, 7$　　**b)** $7, -8$　　**c)** $\frac{3}{2}$

22. a) $2a^3 - 5a^2 - 9a + 18 = (a + 2)(2a^2 - 9a + 9)$
b) $2x^3 - 13x + 5x^2 - 28 = (x + 3)(2x^2 - x - 10) + 2$
c) $6x^3 + 17x^2 - 26x + 8 = (x + 4)(6x^2 - 7x + 2)$
d) $32x^3 - 18x + 16x^2 + 9 = (2x + 1)(16x^2 - 9) + 18$

6 Cumulative Review, page 395

1. a) 4223　　**b)** 9130　　**c)** 26 562　　**d)** 43 215

2. a) i) 25　　**ii)** 110　　**iii)** 123
iv) 234　　**v)** 345　　**vi)** 726
b) Multiply the number of the term by the first term (or common difference).

3. a) $3, -6, 12, -24, 48$　　**b)** $12, 6\sqrt{2}, 6, 3\sqrt{2}, 3$
c) $5, \pm 10, 20, \pm 40, 80$　　**d)** Explanations may vary.

4.

Year	Population	5-year growth rate (%)	Increase in population	Population 5 years later
1995	2 747 000	11	302 170	3 049 170
2000	3 049 170	11	335 409	3 384 579
2005	3 384 579	11	372 304	3 756 883
2010	3 756 883	11	413 257	4 170 140

5. a) 3.7 m, 14.8 m　　**b)** 24 mm, 96 mm

6. a) 2　　**b)** 3　　**c)** 4　　**d)** 5
e) 6　　**f)** 2　　**g)** 3　　**h)** 1

7. a) 0.025 m³/min　　**b)** About 71 cm, 142 cm
c) 6.3 m²

8. a) 10　　**b)** 18　　**c)** $12\sqrt{2}$

9. $(6 + 6\sqrt{2})$ units; isosceles triangle; explanations may vary.

10. a) $(7, -4)$　　**b)** $(2, -2)$　　**c)** $(-1, 0.5)$

11. a) 6　　**b)** 10

12. a) 8　　**b)** -12.5　　**c)** 6　　**d)** 9　　**e)** 2.5

13. a) -4.5　　**b)** 2　　**c)** $-\frac{8}{3}$　　**d)** -4　　**e)** -1.6

14. a) $-2, 7$　　**b)** $\frac{3}{4}, -3$　　**c)** $-1, -2$
d) $-\frac{5}{2}, -3$　　**e)** $-2, 2.5$　　**f)** $-2, 3$

15. a) $y = -\frac{2}{3}x + 4$　　**b)** $y = \frac{6}{5}x - 3$

16. a) i) $2x - 7y + 48 = 0$　　**ii)** $16x + 9y - 10 = 0$
iii) $x + 5y + 10 = 0$
b) Explanations may vary.

17. Tables may vary.
i) a)

x	y
1	1
2	2
3	3
4	4

c) All input numbers can be used.

ii) a)

x	y
1	4
2	5
3	6
4	7

c) All input numbers can be used.

iii) a)

x	y
1	-2
2	-1
3	0
4	1

c) All input numbers can be used.

18. a) i) $y = x$　　**ii)** $y = x + 3$　　**iii)** $y = x - 3$
b) Explanations may vary.

19. Explanations may vary.
a) Travelling at constant speed in one direction
b) Person has turned around and is travelling in the opposite direction at constant speed.
c) Person is standing still.

20. a) $y = 2x + 1$　　**b)** $y = x^2 - 1$　　**c)** $y = \frac{1}{2}x + 3$

21. a) $2x^3 + 5x^2 + 4x + 3$　　**b)** $2x^4 + x^3 + 3x^2 - 4x - 4$
c) $7x^2 - x - 5$　　**d)** $3x^2y - 2xy - 2y^2$

22. a) $x^2 - 3x + 2$　　**b)** $2y^3 + y^2 + 9y - 7$
c) $x^2 + 3xy + 4y^2$　　**d)** $3a^2b^2 + 8a^3b - 4ab^3$

Chapter 7

7.1 Exercises, page 402

1. Explanations may vary.
a) Rational　**b)** Rational　**c)** Rational　**d)** Not rational
e) Rational　**f)** Rational　**g)** Not rational **h)** Not rational

2. a) Rational　**b)** Not rational **c)** Rational　**d)** Rational
e) Not rational **f)** Not rational **g)** Rational　**h)** Rational
i) Rational　**j)** Rational　**k)** Not rational **l)** Rational

3. Explanations may vary.

4. a) 2　　**b)** $-\frac{3}{4}$　　**c)** 0　　**d)** 37
e) 13　　**f)** $-\frac{3}{7}$　　**g)** $\frac{1}{32}$　　**h)** $-\frac{14}{3}$

5. a) $-\frac{1}{2}$　　**b)** 3　　**c)** $\frac{5}{3}$　　**d)** -3
e) 0　　**f)** 6　　**g)** $-\frac{1}{3}$　　**h)** $\frac{1}{3}$

6. a) -2　　**b)** 3　　**c)** $\frac{5}{2}$　　**d)** -2　　**e)** 0
f) 0　　**g)** $-\frac{3}{2}$　　**h)** 2　　**i)** $\frac{5}{4}$　　**j)** $\frac{1}{3}$

7. $\frac{x^3 + 1}{y - 3}, \frac{x^2 - 9}{x^2 - x - 6}$; explanations may vary.

8. a) $s = \frac{72}{t}$; the rational expression is $\frac{72}{t}$
b) i) 18 km/h　　**ii)** 16 km/h　　**iii)** 13.7 km/h

9. a) Answers may vary. You might expect the average speed to decrease the longer the distance raced.
b) 100-m butterfly: 1.71 m/s (women), 1.91 m/s (men);
200-m butterfly: 1.58 m/s (women), 1.72 m/s (men);

400-m freestyle: 1.64 m/s (women), 1.78 m/s (men); from the butterfly results, the average speed does decrease as the distance increases, however, the stroke swimmers use also affects the speed

c) Answers may vary. You might expect the speeds would be faster than those for the butterfly races of the same length. The data support this prediction; the winners' speeds at the 1996 Olympic Summer Games were 1.83 m/s (women's 100-m freestyle), 2.05 m/s (men's 100-m freestyle), 1.69 m/s (women's 200-m freestyle), 1.86 m/s (men's 200-m freestyle).

10. a) 0   **b)** Undefined   **c)** Explanations may vary.

11. a) 0   **b)** -1 **c)** 0

d) Expression is defined for all values of x.

e) $\frac{7}{2}$ **f)** 2   **g)** ± 3

h) -1 **i)** $0, -4$ **j)** 0

12. Explanations may vary.   **13.** Explanations may vary.

14. Explanations may vary.

a) $\frac{x^2 - 1}{4}$ **d)** $\frac{x^2 + 5x - 24}{x^2 + 9}$ **e)** $\frac{x^2 + 3x + 2}{x^2 + 5}$

15. Explanations may vary.

16. a) 0   **b)** Expression is defined for all values of x.

c) $-4, 1$ **d)** 2   **e)** $\frac{4}{3}$ **f)** $0, -5$ **g)** ± 3

h) $6, -2$ **i)** Expression is defined for all values of x.

j) ± 6 **k)** 0   **l)** 1

m) $4, -1$ **n)** $-2, -\frac{3}{2}$ **o)** $\frac{4}{3}, 2$

17. a) 1   **b)** 6

c) Tables may vary. $(2, 1), (6, 3), (0, 0), (8, 4), (4, 2)$

e) $y = \frac{x}{2}$

f) The equation represents the ordered pairs for which the denominator of the rational expression equals 0.

18. a) $x = 3y$ **b)** $y = 2x$

c) $x = 0, x = \frac{7}{2}y$ **d)** $x = 4y, y = 2x$

e) Expression is defined for all values of the variables.

f) $x = \pm 7y$ **g)** $x = 0$ and $y = 0$ together

h) $x = \pm \frac{3}{2}y$ **i)** $x = 2y, x = 5y$

19. Answers may vary.

a) $\frac{1}{x-5}$ **b)** $\frac{1}{x+3}$ **c)** $\frac{1}{x(x-4)}$ **d)** $\frac{1}{(x-2)(x+1)}$

20. Explanations may vary.

Investigate, page 405

1. a) 2, 5   **b)** 2, 7   **c)** 2

2. a) $\frac{5}{7}$ **b)** $\frac{4}{5}$ **c)** $\frac{2}{5}$ **d)** $-\frac{3}{4}$

3. Explanations may vary.

4. a) $2 \times 2 \times 2 \times 3 \times x \times y \times y$

b) $2 \times 2 \times 3 \times 5 \times x \times x \times x \times y$ **c)** $\frac{2y}{5x}$

5. a) $(x+2)(x-2)$ **b)** $(x+2)(x+4)$ **c)** $\frac{x-2}{x+4}$

7.2 Exercises, page 407

1. a) $8x$ **b)** $\frac{-9x}{y}$ **c)** $11a^2$ **d)** $-4m^2$

e) $\frac{x}{3}$ **f)** $\frac{4}{7}$ **g)** $\frac{6a^2}{b}$ **h)** $\frac{-12m}{n}$

2. a) $\frac{b}{2c}$ **b)** $\frac{3ac}{4b}$ **c)** $-\frac{6x}{y}$ **d)** $\frac{3ac^2}{4b}$

e) $\frac{x}{3z}$ **f)** $-\frac{5c^2}{9b^2}$ **g)** $\frac{9n}{m^3}$ **h)** $-\frac{5a^2c}{8b^5}$

3. a) 2   **b)** $\frac{1}{3}$ **c)** $\frac{2}{3}$ **d)** $-\frac{1}{2}$

e) $-\frac{3}{5}$ **f)** $\frac{1}{2}$ **g)** $\frac{1}{2}$ **h)** $\frac{7}{5}$

4. a) $\frac{2x+6}{5}$, all real values of x except $x = 0$

b) $\frac{x}{2}$, all real values of x except $x = 5$

c) $\frac{5x+7}{3}$, all real values of x except $x = 0$

d) $4x$, all real values of x except $x = 3$

e) $-\frac{3x}{7}$, all real values of x except $x = 2$

f) -3, all real values of x except $x = 3$

g) $x + 3$, all real values of x except $x = -4$

h) $-x - 3$, all real values of x except $x = 2$

i) $5 - x$, all real values of x except $x = 5$

5. Explanations may vary.

6. a) 4   **b)** $\frac{2x}{x-1}$, $x \neq -3, 1$ **c)** 4

d) The answers are the same.   **e)** Explanations may vary.

7. a) $-\frac{5}{3}$ **b)** -1 **c)** $\frac{-2}{3}$

d) $\frac{-3}{2}$ **e)** $\frac{-5y}{2}$ **f)** $\frac{-3}{2}$

8. a) $\frac{a+7}{a-4}$, $a \neq 4, 2$ **b)** $\frac{1}{x+6}$, $x \neq -6, 3$

c) $m - 5$, $m \neq 2$ **d)** $\frac{r-3}{r+3}$, $r \neq -3$

e) $\frac{1}{x-4}$; $x \neq -4, 4$ **f)** $\frac{x-3}{x^2+9}$; all values permissible

9. a) $\frac{m^2 - 7m + 10}{m-2}$ **b)** $\frac{a^2 + 5a - 14}{a^2 - 6a + 8}$

10. a) $\frac{x+3y}{2x}$ **b)** $\frac{m-4n}{3m}$ **c)** $\frac{3a+4b}{2a}$

d) $\frac{a-3b}{a}$ **e)** $\frac{3x+2y}{x}$ **f)** $\frac{m-n}{3m}$

g) $\frac{x+5}{x}$ **h)** $\frac{3m}{3m-1}$ **i)** $\frac{4t-8}{t+4}$

11. a) $\frac{1}{2x-3}$ **b)** $\frac{1}{x+3}$ **c)** $\frac{1}{x+3}$

d) $\frac{c-8d}{c+4d}$ **e)** $\frac{x-5y}{x+5y}$ **f)** $\frac{a+4b}{a-6b}$

g) $\frac{(x-1)(x+2)(x-2)}{x-4}$

h) $(x-6)(x+5)$ **i)** $(e+9)(e-8)$

12. a) $\frac{x}{x-1}$ **b)** $\frac{5x-10}{x+12}$ **c)** $\frac{x-4}{x+5}$

d) $\frac{a+4}{14-2a}$ **e)** $\frac{x-5}{2x}$ **f)** $\frac{x-17}{3-x}$

g) $\frac{2m-4}{3m-12}$ **h)** $\frac{a+6}{2a-8}$ **i)** $\frac{3b+15}{2b+12}$

13. Explanations may vary.

14. Explanations may vary.

a) $2x + 1$ **b)** 2   **c)** $4x$

d) $4x$ **e)** 1   **f)** -1

g) Cannot be simplified   **h)** Cannot be simplified

15. Answers may vary. $\frac{x^2 - 2x - 8}{2x^2 - 8x}$

16. Saleha; explanations may vary.

17. a) $\frac{2x+y}{3x-y}$ **b)** $3a^3 - 2a^2 + 5a$

c) $\frac{8 - 4n + 2n^2 - 2n^3}{n}$ **d)** $\frac{x-2y}{(x^2+4y^2)(x+2y)}$

e) $\frac{x+y}{x-4y}$ **f)** $\frac{2x+y}{(4x^2+y^2)(x+2y)}$

Investigate, page 411

1. a) $\frac{7}{5}$ **b)** $\frac{21}{10}$ **c)** $\frac{14}{5}$ **d)** $\frac{7}{2}$

2. a) , **b)** Explanations may vary.   **c)** $\frac{7x}{10}$

3. a) $\frac{6}{5}$ b) $\frac{8}{5}$ c) 2 d) $\frac{12}{5}$

4. a), b) Explanations may vary. c) $\frac{2x}{5}$

7.3 Exercises, page 413

1. a) $\frac{5a}{12}$ b) $\frac{m}{2}$ c) $\frac{-x^2}{6}$ d) $\frac{2c}{9}$
   e) $\frac{-4t^2}{5}$ f) $6r$ g) $16t^2$ h) $10b$

2. a) $\frac{2x}{3}$ b) $\frac{2}{3x}$ c) $\frac{3x}{2}$ d) $\frac{3x}{2}$
   e) $-2x$ f) $-\frac{1}{2x}$ g) $-\frac{2}{x}$ h) $2x$

3. a) $\frac{5a}{6b}$; $a, b \neq 0$ b) $\frac{x}{7}$; $x \neq 0$ c) $\frac{-y}{x}$; $x \neq 0$
   d) $\frac{3a}{2b}$; $b \neq 0$ e) $\frac{14}{5}$; $m \neq 0$ f) $\frac{9a^3}{10c^2}$; $a, c \neq 0$
   g) $\frac{4a}{3bc}$; $b, c \neq 0$ h) $\frac{t^4}{20}$; $s, t \neq 0$

4. a) $\frac{2t}{7s}$ b) $\frac{9x^2}{2}$ c) $\frac{3f^2}{7e^2}$ d) $-x$
   e) $\frac{2n}{3}$ f) $\frac{7x}{6y}$ g) $\frac{a}{10b}$ h) $\frac{21t^4}{2}$

5. a) $\frac{2x}{3}$; $x, y \neq 0$ b) $2b$; $a, b \neq 0$ c) $\frac{n}{-6}$; $m, n \neq 0$
   d) $\frac{2x}{5}$; $x, y \neq 0$ e) $\frac{20n^5}{3m^3}$; $m, n \neq 0$ f) 1; $c, d \neq 0$

6. Explanations may vary.

7. a) $\frac{a^5b^2}{3}$ b) $\frac{-5}{4xy}$ c) $6x^2y$
   d) $\frac{8n}{9m}$ e) $\frac{2}{5}$ f) $\frac{16m^2}{15}$

8. Answers may vary.
   a) $\frac{1-x}{x^2} \times \frac{x}{1-x}$ b) $\frac{x^2}{y^2} \times \frac{y}{x}$
   c) $\frac{3m(m+2)}{m+3} \times \frac{m+3}{-2n(m+2)}$ d) $\frac{a^2}{2b} \times \frac{b}{a}$
   e) $\frac{3b^2}{b+2} \times \frac{b+2}{b}$ f) $-\frac{5x^2}{xy} \times \frac{y}{x^2}$
   g) $\frac{1-b}{abc^2} \times \frac{c}{b-1}$ h) $\frac{4x^2}{3y} \times \frac{y}{x}$

9. Answers may vary.
   a) $\frac{4}{x^2} \div \frac{4}{x}$ b) $\frac{2x^2y}{3} \div \frac{2y^2x}{3}$
   c) $\frac{3m^2n}{5} \div \frac{-2mn^2}{5}$ d) $\frac{6a^2}{3b} \div \frac{4a}{b}$
   e) $\frac{12b^2}{a} \div \frac{4b}{a}$ f) $-\frac{5y}{2x^2} \div \frac{y^2}{2x}$
   g) $\frac{-ab}{c^2} \div \frac{(ab)^2}{c}$ h) $\frac{2xy}{9} \div \frac{y}{6}$

10. a) $\frac{7}{2}$; $a \neq 0, 3$ b) $\frac{1}{12}$; $x \neq 0, 2$
    c) $x - 3$; $x \neq -1$ d) $8m$; $m \neq -4, 0$
    e) $\frac{1}{12}$; $s \neq -5, 2$ f) $\frac{-2(a+1)}{a}$; $a \neq -1, 0, 5$

11. a) $\frac{2b}{5}$ b) $\frac{xy^3}{2}$ c) $\frac{3y}{x+3}$
    d) $\frac{3x+4}{12}$ e) $\frac{2m}{15n}$ f) $\frac{4(a+3b)}{3}$

12. a) $\frac{9}{4}$; $x \neq -3, 0$ b) $\frac{x+11}{x-2}$; $x \neq -2, 2, 11$
    c) $\frac{5}{12}$; $x \neq -1, 2$ d) 1; $a \neq b, y \neq 0, y \neq -2$
    e) $\frac{3x(x-2)}{4y(x+2)}$; $x \neq \pm 2, y \neq 0$ f) $\frac{x-2}{x-1}$; $x \neq \pm 1, -2$

13. Answers may vary.
    a) $\frac{x-1}{x} \times \frac{x}{x^2-1}$ b) $\frac{x^2-4}{x(x+2)} \times \frac{x}{x+2}$
    c) $\frac{a(a+1)}{b(a+b)} \times \frac{b}{a+1}$ d) $\frac{x(x+y)^2}{xy} \times \frac{y}{x}$

14. Answers may vary.
    a) $\frac{x+3}{(x+1)^2} \div \frac{x+3}{x+1}$ b) $\frac{x^2-4}{4} \div \frac{(x+2)^2}{4}$
    c) $\frac{2ab^2}{4(a+b)} \div \frac{b^2}{2}$ d) $\frac{3(x+y)}{y} \div \frac{6x}{2y(x+y)}$

15. a) -1 b) 1 c) $\frac{x+2}{x-1}$
    d) -1 e) $\frac{a-3}{a-1}$ f) $\frac{(x+3)^2}{(x-3)^2}$

16. a) $\frac{2xy(x+4y)}{3(x+5y)}$ b) $\frac{1}{2}$
    c) $\frac{x+2y}{x-y}$ d) $\frac{4m(m-2n)}{3(m+3n)}$

17. a) $\frac{(x-1)(x-6)}{2x^2}$ b) $\frac{x+y}{x-y}$
    c) 1 d) $3a + 7b$

18. a) $\frac{2a-b}{2b-a}$ b) $\frac{4b-3a}{8a-6b}$

Mathematical Modelling: Should Pop Cans be Redesigned?, page 416

1. a) $h = \frac{355}{\pi r^2}$ b) $A = 2\pi r^2 + \frac{710}{r}$

2. a), b)

Radius (cm)	Height (cm)	Surface area (cm²)
1.0	113.00	716.28
2.0	28.25	380.13
3.0	12.56	293.22
4.0	7.06	278.03
5.0	4.52	299.08
6.0	3.14	344.53
7.0	2.31	409.31
8.0	1.77	490.87

3. a) $r \doteq 4$ cm, $h \doteq 7$ cm b) About 278 cm²

4. b) 3.84 cm c) 7.67 cm

5. The height is twice the radius. The diameter and the height must be equal.

6. b) About 286 cm² c) About 8 cm² d) About 2.8% e) About $56 million

7. Answers may vary. 8. Answers may vary.

Investigate, page 418

1. a) $\frac{13}{12}$ b) $\frac{13}{18}$ c) $\frac{13}{24}$ d) $\frac{13}{30}$

2. a), b) Answers may vary. c) $\frac{13}{6a}$

3. a) $\frac{5}{9}$ b) $\frac{7}{16}$ c) $\frac{9}{25}$ d) $\frac{11}{36}$

4. a), b) Answers may vary. c) $\frac{2a-1}{a^2}$

7.4 Exercises, page 421

1. a) $\frac{-16}{24mn}$ b) $\frac{15m}{24mn}$ c) $\frac{-20n}{24mn}$ d) $\frac{-13mn}{24mn}$ e) $\frac{-6mn}{24mn}$
   f) $\frac{-9mn}{24mn}$ g) $\frac{14m^2n}{24mn}$ h) $\frac{-21m^2}{24mn}$ i) $\frac{-22n^2}{24mn}$ j) $\frac{-20m^2n}{24mn}$

2. a) i) $\frac{4x}{6}$ ii) $\frac{8x^2}{12x}$ iii) $\frac{2x^3}{3x^2}$
   b) i) $\frac{-9m-3}{-12}$ ii) $\frac{6mx+2x}{8x}$ iii) $\frac{12mx^2+4x^2}{16x^2}$
   c) i) $\frac{3a-21}{3a}$ ii) $\frac{a^2-7a}{a^2}$ iii) $\frac{5a^4-35a^3}{5a^4}$
   d) i) $\frac{15y-9}{6y}$ ii) $\frac{10y^3-6y^2}{4y^3}$ iii) $\frac{-5y^2+3y}{-2y^2}$
   e) i) $\frac{-8x-20}{-4x^2}$ ii) $\frac{40x^2+100x}{20x^3}$ iii) $\frac{2x+5}{x^2}$

3. Answers may vary. $\frac{6}{2x+8}$, $\frac{3x}{x^2+4x}$, $\frac{9x}{3x^2+12x}$

4. Answers may vary.

a) $\dfrac{3x}{6}, \dfrac{2x}{6}$

b) $\dfrac{10}{5x}, \dfrac{x^2}{5x}$

c) $\dfrac{3}{2a}, \dfrac{4}{2a}$

d) $\dfrac{15}{6a}, \dfrac{8}{6a}$

e) $\dfrac{5n}{n^2}, \dfrac{2}{n^2}$

f) $\dfrac{4}{24x}, \dfrac{15}{24x}$

g) $\dfrac{4x+4}{20x^2}, \dfrac{5x^2-5x}{20x^2}$

h) $\dfrac{2x+4}{6x^2}, \dfrac{3x^2-9x}{6x^2}$

5. a) $\dfrac{3}{x}, x \neq 0$

b) $\dfrac{2}{x}, x \neq 0$

c) $\dfrac{7-5x}{4x}, x \neq 0$

d) $\dfrac{2-9m}{3m^2}, m \neq 0$

e) $\dfrac{6b}{a}, a \neq 0$

f) $\dfrac{x}{y^2}, y \neq 0$

g) $\dfrac{2a}{b^2}, b \neq 0$

h) $\dfrac{3s^2}{t^3}, t \neq 0$

6. a) $\dfrac{-2a}{15}$

b) $\dfrac{-2}{15a}$

c) $\dfrac{10a^2-12}{15a}$

d) $\dfrac{10-12a^2}{15a}$

e) $\dfrac{-2}{15a}$

f) $\dfrac{10-12a}{15}$

g) $\dfrac{9a^3-7}{6a^2}$

h) $\dfrac{9a-7}{6a}$

7. a) $\dfrac{9}{2x}, x \neq 0$

b) $\dfrac{47}{5x}, x \neq 0$

c) $\dfrac{29}{30x}, x \neq 0$

d) $\dfrac{13}{4a}, a \neq 0$

e) $\dfrac{41}{24m}, m \neq 0$

f) $\dfrac{-11}{18k}, k \neq 0$

g) $\dfrac{43}{45t}, t \neq 0$

h) $\dfrac{13}{56b}, b \neq 0$

8. Explanations may vary.

9. Answers and explanations may vary.

a) $\dfrac{1}{a} + \dfrac{1}{a}$

b) $\dfrac{x-3}{2} + \dfrac{3}{2}$

c) $\dfrac{1-2x}{3x} + \dfrac{2}{3}$

d) $\dfrac{3-2y}{x} + \dfrac{2y}{x}$

e) $\dfrac{2-m}{3n} + \dfrac{m}{3n}$

f) $\dfrac{3a-d}{4b} + \dfrac{d}{4b}$

g) $\dfrac{1-c}{9ab} + \dfrac{c}{9ab}$

h) $\dfrac{7ab-a}{10c} + \dfrac{a}{10c}$

10. Answers and explanations may vary.

a) $\dfrac{3}{a} - \dfrac{1}{a}$

b) $\dfrac{3x}{2} - x$

c) $\dfrac{5}{3x} - \dfrac{4}{3x}$

d) $\dfrac{5}{x} - \dfrac{2}{x}$

e) $\dfrac{5}{3n} - n$

f) $\dfrac{3a}{2b} - \dfrac{3a}{4b}$

g) $\dfrac{1+c}{9ab} - \dfrac{c}{9ab}$

h) $\dfrac{9ab}{10c} - \dfrac{ab}{5c}$

11. a) $\dfrac{3a}{5}$

b) $\dfrac{5m}{24}$

c) $\dfrac{11x}{18}$

d) $\dfrac{-29c}{36}$

e) $\dfrac{7e}{12}$

f) $\dfrac{13m}{24}$

12. a) $\dfrac{13}{12a}, a \neq 0$

b) $\dfrac{-7}{12x}, x \neq 0$

c) $\dfrac{13}{24m}, m \neq 0$

d) $\dfrac{-5}{3x}, x \neq 0$

e) $\dfrac{7}{12y}, y \neq 0$

f) $\dfrac{-5}{24y}, y \neq 0$

13. a) $\dfrac{2x-2}{x}$

b) $\dfrac{m-9}{m}$

c) $\dfrac{11a+9}{3a}$

d) $\dfrac{4x+26}{5x^2}$

e) $\dfrac{-5m-7}{2m}$

f) $\dfrac{-8x-1}{4x}$

14. a) $\dfrac{k-43}{20}$

b) $\dfrac{5c-14}{6}$

c) $\dfrac{x+10}{12}$

d) $\dfrac{5m-9}{12}$

e) $\dfrac{-14a+25}{24}$

f) $\dfrac{16x-35}{18}$

15. Explanations may vary.

16. a) $\dfrac{8x-11}{6x}$

b) $\dfrac{-6n-5}{24n}$

c) $\dfrac{9a-29}{18a}$

17. a) $\dfrac{10a-5}{2a}$

b) $\dfrac{4n-3m}{6mn}$

c) $\dfrac{4y+3}{xy}$

d) $\dfrac{2b+a}{2b}$

18. a) $\dfrac{1+4x}{1-4x}$

b) $\dfrac{4x-1}{4x+1}$

c) $\dfrac{4x-1}{4x+1}$

d) $\dfrac{1+4x}{1-4x}$

e) $\dfrac{9+12x}{8-6x}$

f) $\dfrac{50x+4}{30x-15}$

19. a) $\dfrac{3y-2x+xy}{xy}$

b) $\dfrac{2bc-3ac+4ab}{abc}$

c) $\dfrac{6yz+9xz-10xy}{12xyz}$

20. a) $\dfrac{7x^2+10xy}{6xy}$

b) $\dfrac{5m^2-4n^2}{3mn}$

c) $\dfrac{9a^2-20b^2}{15ab}$

d) $\dfrac{15x^2-8a^2}{10ax}$

21. Answers may vary.

a) $4 - \dfrac{a}{3}$

b) $\dfrac{1}{5} - \dfrac{1}{2x}$

c) $\dfrac{x}{y} + 1$

d) $7x + 1 + \dfrac{1}{x}$

22. a) $\dfrac{a^4-3a^2+2}{a^2}$

b) $\dfrac{k^4-2k^2-15}{k^2}$

c) $\dfrac{4a^4-12a^2+9}{a^2}$

23. a) 2

b) 7

24. a) $\dfrac{2}{3}$

b) $\dfrac{5}{2}$

c) $\dfrac{5}{3}$

d) 5

25. a) $\dfrac{x^4+3x^2+2}{x^2}$

b) $\dfrac{x^4+3x^2+2}{x^2}$

c) Yes; explanations may vary.

26. True; reasons may vary.

Linking Ideas: Mathematics and Science
Exploring the Lens Formula, page 424

1. b) Answers may vary. This table is for $f = 40$.

p	q
80	80
60	120
100	$\frac{200}{3}$
$\frac{200}{3}$	100
160	$\frac{160}{3}$
$\frac{160}{3}$	160
120	60

2. a) The image distance approaches the focal length.

b) The image distance approaches infinity.

c) The image is on the same side of the lens as the object.

3. Explanations may vary.   **4.** 120 mm

5. Answers may vary. The formula is $q = \dfrac{fp}{p-f}$.

6. 60.55 mm   **7.** 8.86 cm

7.5 Exercises, page 428

1. a) $\dfrac{4m-1}{m+3}$

b) $\dfrac{-4s+11}{s-5}$

c) 3

d) $\dfrac{4x+18}{x+6}$

e) $\dfrac{3m-5}{2m+1}$

f) $\dfrac{-4a-11}{a^2+4}$

2. a) $\dfrac{3a+3}{a(a-3)}$

b) $\dfrac{30-4y}{y(y-5)}$

c) $\dfrac{4m-28}{m(m-4)}$

d) $\dfrac{2c^2-5c+5}{c(c-1)}$

e) $\dfrac{3x^2-6x-12}{x(x+2)}$

f) $\dfrac{8x+6}{x(x+2)}$

3. a) $\dfrac{3-8a}{2a}, a \neq 0$

b) $\dfrac{5-2y}{y+1}, y \neq -1$

c) $\dfrac{4n-29}{n-5}, n \neq 5$

d) $\dfrac{x^2+4x-2}{x+4}, x \neq -4$

e) $\dfrac{3+16s-2s^2}{s-8}, s \neq 8$

f) $\dfrac{-10w-4w^2}{w+3}, w \neq -3$

g) $\dfrac{2+3x-x^2}{x-1}, x \neq 1$

h) $\dfrac{6-x-x^2}{x-1}, x \neq 1$

i) $\dfrac{x^2-8x+17}{x-3}, x \neq 3$

j) $\dfrac{x^2+x-1}{x-2}, x \neq 2$

k) $\dfrac{34-4x-x^2}{x-4}, x \neq 4$

l) $\dfrac{5+2x-x^2}{x+2}, x \neq -2$

4. a) $\dfrac{5x+19}{(x+5)(x+2)}$

b) $\dfrac{2x+10}{(x-3)(x+1)}$

c) $\dfrac{x^2-3x-2}{(x+1)(x-1)}$

d) $\dfrac{3x+2}{x+4}$

e) $\dfrac{3x^2+17x}{(x-1)(x+3)}$

f) $\dfrac{5x^2+9x}{(x+5)(x-3)}$

5. Explanations may vary.

6. a) $\dfrac{-2}{(x+1)(x-1)}$

b) $\dfrac{x^2-5x+7}{(x-2)(x-3)}$

c) $\dfrac{x^2-25x-4}{(x+7)(x-3)}$

d) $\dfrac{2x^2-3x+10}{(x+2)(x-4)}$

e) $\dfrac{2x^2-6x-18}{(x-3)(x-5)}$

f) $\dfrac{6x}{(x-2)(x+2)}$

7. a) $\dfrac{2x-4-3x^2}{x(x-2)}$

b) $\dfrac{27}{10(x+3)}$

c) $\dfrac{19y}{6(y+9)}$

d) $\dfrac{3a+19}{3(a+1)(a-7)}$

e) $\dfrac{4a-2}{(a+1)(a-1)}$

f) $\dfrac{-x^2-x}{(x-2)(x-3)}$

8. Answers may vary.

a) $\dfrac{2-a}{a+1} + \dfrac{a}{a+1}$
b) $\dfrac{x}{x-6} + \dfrac{2}{x-6}$

c) $\dfrac{b}{a+b} + \dfrac{b}{a+b}$
d) $\dfrac{xy}{2x-y} + \dfrac{2xy}{2x-y}$

9. Explanations may vary.

10. Answers may vary.

a) $\dfrac{2+a}{a+1} - \dfrac{a}{a+1}$
b) $\dfrac{x+4}{x-6} - \dfrac{2}{x-6}$

c) $\dfrac{3b}{a+b} - \dfrac{b}{a+b}$
d) $\dfrac{4xy}{2x-y} - \dfrac{xy}{2x-y}$

11. a) $\dfrac{2x+1}{x(x+1)}$, $\dfrac{2x+3}{(x+1)(x+2)}$, $\dfrac{2x+5}{(x+2)(x+3)}$;

$\dfrac{1}{x+3} + \dfrac{1}{x+4} = \dfrac{2x+7}{(x+3)(x+4)}$,

$\dfrac{1}{x+4} + \dfrac{1}{x+5} = \dfrac{2x+9}{(x+4)(x+5)}$

b) $\dfrac{3x+4}{(x+1)(x+2)}$, $\dfrac{5x+12}{(x+2)(x+3)}$, $\dfrac{7x+24}{(x+3)(x+4)}$;

$\dfrac{4}{x+4} + \dfrac{5}{x+5} = \dfrac{9x+40}{(x+4)(x+5)}$,

$\dfrac{5}{x+5} + \dfrac{6}{x+6} = \dfrac{11x+60}{(x+5)(x+6)}$

c) $\dfrac{2}{x^2-1}$, $\dfrac{2}{x^2-4}$, $\dfrac{2}{x^2-9}$; $\dfrac{1}{x(x-4)} + \dfrac{1}{x(x+4)} = \dfrac{2}{x^2-16}$,

$\dfrac{1}{x(x-5)} + \dfrac{1}{x(x+5)} = \dfrac{2}{x^2-25}$

12. Patterns may vary.

13. a) $\dfrac{6}{x+2}$, $x \neq -2$
b) $\dfrac{41}{10(x-2)}$, $x \neq 2$

c) $\dfrac{-17x}{6(x+3)}$, $x \neq -3$
d) $\dfrac{x}{2(2x-3)}$, $x \neq \dfrac{3}{2}$

e) $\dfrac{8x+5}{3(x-4)}$, $x \neq 4$
f) $\dfrac{5x+6}{6(x+4)}$, $x \neq -4$

14. a) $\dfrac{2x^2-3x+3}{(x-6)(x-3)}$
b) $\dfrac{-3x^2-8x-7}{(x-5)(x+3)}$
c) $\dfrac{2x^2+4x-3}{(x-2)(x+2)}$

d) $\dfrac{4x^2-10x-9}{(x-3)(x+3)}$
e) $\dfrac{3x^2+4x}{(x-1)(x+2)}$
f) $\dfrac{-2x^2+13x}{(x-3)(x-4)}$

15. Steps could vary.

16. a) $\dfrac{4m-m^2}{(m-1)(m+1)}$
b) $\dfrac{3x^2+5x-14}{x(x+5)(x-1)}$
c) $\dfrac{49a^2+17a+10}{15(a-2)(a+3)}$

d) $\dfrac{-11k^2+28k}{4(k-3)(k-4)}$
e) $\dfrac{13x^2+2x-11}{2(x-2)(x+7)}$
f) $\dfrac{18m^2+107m+27}{6(2m-1)(3m+7)}$

17. a) $\dfrac{-x-10}{(x+8)(x+6)}$
b) $\dfrac{4m-7}{(m-4)(m+5)}$
c) $\dfrac{1}{x+2}$

d) $\dfrac{-15m-20}{(2m+3)(m+2)}$
e) $\dfrac{12x^2-13x+21}{(x+7)(3x-4)}$
f) $\dfrac{9-x}{(3x-2)(x-3)}$

18. Answers may vary.

a) $\dfrac{1}{x-4} + \dfrac{1}{x}$
b) $\dfrac{1}{x-y} + \dfrac{1}{x+y}$
c) $\dfrac{1}{x+4} + \dfrac{2}{x+3}$

19. a) $2x$
b) $-5y$

c) $\dfrac{5x^2+14xy-66y^2}{6(x+y)}$
d) $\dfrac{2a}{(a+3b)(a-3b)}$

20. a) $\dfrac{1}{x-1} + \dfrac{1}{x} = \dfrac{2x-1}{x(x-1)}$; $\dfrac{1}{x-2} + \dfrac{1}{x-1} = \dfrac{2x-3}{(x-2)(x-1)}$;

$\dfrac{1}{x-3} + \dfrac{1}{x-2} = \dfrac{2x-5}{(x-3)(x-2)}$

b) $\dfrac{0}{x} + \dfrac{1}{x+1} = \dfrac{1}{x+1}$; $\dfrac{-1}{x-1} + \dfrac{0}{x} = \dfrac{-1}{x-1}$;

$\dfrac{-2}{x-2} - \dfrac{1}{x-1} = \dfrac{-3x+4}{(x-2)(x-1)}$

c) $\dfrac{1}{x(x)} + \dfrac{1}{x(x)} = \dfrac{2}{x^2}$; $\dfrac{1}{x(x+1)} + \dfrac{1}{x(x-1)} = \dfrac{2}{x^2-1}$;

$\dfrac{1}{x(x+2)} + \dfrac{1}{x(x-2)} = \dfrac{2}{x^2-4}$

7.6 Exercises, page 434

1. a) $a \neq 0$, $a = \dfrac{1}{2}$
b) $a = 21$

c) $n \neq 0$, $n \neq -\dfrac{1}{3}$
d) $c = -\dfrac{12}{5}$

e) $x \neq 0$, $x = -3$
f) $m = 10$

g) $a \neq 0$, $a = -\dfrac{1}{22}$
h) $x \neq 0$, $x = \dfrac{1}{8}$

i) $x \neq 0$; $x = 9$
j) $x \neq 0$; $x = 30$

k) $x \neq 0$; $x = 20$
l) $x \neq 0$; $x = 8$

m) $x \neq 0$; $x = \pm 6$
n) $x \neq 0$; $x = \pm 12$

o) $x \neq 0$; $x = \pm 10$
p) $x \neq 0$; $x = \pm 15$

2. Each equation is undefined when its variable is 0.

a) 4
b) 6
c) 3
d) $-\dfrac{1}{4}$
e) $-\dfrac{1}{3}$
f) 6.8

g) 3, -2
h) 3, -5
i) 8, -1
j) 3, -2
k) 3.5, -1
l) $\pm\sqrt{5}$

3. Explanations may vary.

4. a) $\dfrac{5}{7}$
b) 12
c) 4
d) $-\dfrac{1}{50}$

5. 1600
6. a) 144°
b) 15
7. a) 30.7 cm
b) 8

8. a) 2, $-\dfrac{1}{2}$; -13
b) -2, 1; 13
c) $-\dfrac{5}{3}$, $-\dfrac{3}{2}$; $-\dfrac{1}{2}$

d) 2; no solution
e) $\dfrac{2}{5}$, $\dfrac{5}{4}$; $\dfrac{11}{2}$
f) $-\dfrac{7}{2}$; no solution

9. Explanations may vary.

10. a) 5
b) No solution
c) No solution

d) -8
e) -1
f) $\dfrac{1}{4}$
g) $-\dfrac{3}{5}$
h) -1

i) $\dfrac{1}{5}$
j) $\dfrac{1}{7}$
k) No solution
l) $\dfrac{11}{23}$

11. a) $\dfrac{8}{7}$
b) 3
c) $\dfrac{5}{2}$

d) No solution
e) $\dfrac{17}{3}$
f) $\dfrac{5}{19}$

g) 0, 8
h) -6, 1
i) -4, 1

j) $-\dfrac{3}{2}$, 3
k) $\dfrac{7}{5}$, 4
l) No solution

m) 0, 4
n) 0, $-\dfrac{2}{3}$
o) 0, $\dfrac{15}{4}$

12. a) -4, 3
b) -3, -2
c) 1, 3

d) -3, 2
e) 0
f) -5, -3

13. a) -11, 2
b) $-\dfrac{5}{3}$, 3
c) 0, 5

d) -2, $-\dfrac{5}{7}$
e) $\dfrac{3}{4}$, 1
f) 4

7.7 Exercises, page 439

1. a) $y = 3.5 - \dfrac{1070}{(17.5+x)^2}$

b) i) 0.825 cm
ii) 1.386 cm

2. a) $y = \dfrac{535}{(12.4-x)^2} - 3.48$

b) i) 1.02 cm
ii) 2.57 cm

3. Explanations may vary. $t = \dfrac{500}{165-x} + \dfrac{500}{165+x}$

4. a) $t = \dfrac{500}{200+x} + \dfrac{500}{200-x}$

b) i) 5.1 h
ii) 5.2 h

6. a) $y = \dfrac{x}{1+x}$
b) i) 9 cm
ii) 20 cm

7. a) $l = \dfrac{200+20w}{w-10}$

b) i) $l = 60$ cm, $A = 1200$ cm^2
ii) $l = 46.7$ cm, $A = 1166.7$ cm^2

8. b) About 1.3 cm, about 3.4 cm
c) Predictions may vary.

9. a) $y = \dfrac{210}{\pi(4.2-x)^2} - 3.8$

b) i) 1.1 cm
ii) 2.7 cm

10. a) $y = 13.1 - \dfrac{725}{\pi(4.2+x)^2}$

b) i) 4.6 cm
ii) 7.1 cm

11. a) 120 knots
b) $t = \dfrac{145}{120-x} + \dfrac{145}{120+x}$
c) Almost 3 h

12. b) 6.15 h, 7.17 h
c) Predictions may vary.

13. Train: 80 km/h; plane: 400 km/h

14. Bus: 80 km/h; train: 120 km/h

Problem Solving: Exploring Averages, page 443

1. a) 66%   **b)** Explanations may vary.   **c)** About 94%

2. a) 825 km/h   **b)** Explanations may vary.   **c)** 566 km/h

3. a) 48 min   **b)** 2 h 48 min   **c)** 8.6 km/h

4. a) $\frac{12}{x}$ h   **b)** $\left(2 + \frac{12}{x}\right)$ h   **c)** $\left(\frac{24}{2 + \frac{12}{x}}\right)$ km/h

5. Tables may vary.   **6.** Graphs should be the same.

Exploring with a Graphing Calculator, page 444

1., **2.** Explanations may vary.

3. a) 2.5 h, 3.8 h, 6.3 h
b) i) 26.5 km/h   **ii)** 55 km/h

4. Explanations may vary.

7 Review, page 445

1. a) $13ab$ **b)** $3m + 2$ **c)** $5y - 2$ **d)** $\frac{3s - 4}{2}$

e) $-3b + 6$ **f)** $\frac{2y}{x + 5y}$ **g)** $\frac{3m}{2}$ **h)** $\frac{7s}{5}$

2. a) -3 **b)** $\frac{a + 5}{3a}$ **c)** $\frac{n - 6}{n - 2}$

d) $\frac{a - 3}{a + 3}$ **e)** $\frac{b + 4}{b - 2}$ **f)** $\frac{m - 3}{m + 2}$

3. a) $6a^2$ **b)** $\frac{-n}{m}$ **c)** 1

d) $\frac{x^4}{y}$ **e)** $6m^2n$ **f)** $\frac{5x^2}{8y^2}$

4. a) $8a$ **b)** $\frac{-4(a + 1)}{a}$ **c)** $\frac{n}{m}$ **d)** $\frac{x^2y}{2}$

5. Explanations may vary.

6. a) $\frac{3b(2a - b)}{2(a^2 - 9)}$ **b)** $\frac{m - 1}{m - 2}$ **c)** -1 **d)** $\frac{y - 3}{y - 1}$

7. a) $\frac{3}{a}$ **b)** $\frac{-1}{6x}$ **c)** $\frac{23}{4m}$

d) $\frac{5x}{24}$ **e)** $\frac{5}{12a}$ **f)** $\frac{-17}{15n}$

8. a) $\frac{2a - 2}{a}$ **b)** $\frac{x - 9}{x}$ **c)** $\frac{y - 33}{20}$

d) $\frac{8a - 11}{6a}$ **e)** $\frac{x - 5}{24x}$ **f)** $\frac{1 + 12x}{12x}$

g) $\frac{11x + 5}{12x}$ **h)** $\frac{11m - 13}{8m}$ **i)** $\frac{-8a + 17}{30a^2}$

9. Explanations may vary.

10. a) $\frac{2b + 15a}{3ab}$ **b)** $\frac{9n - 10m}{12mn}$ **c)** $\frac{t + s}{t}$

d) $\frac{4c + b}{2c}$ **e)** $\frac{54p^2 + 49q^2}{42pq}$ **f)** $\frac{9x^2 - 10y^2}{15xy}$

g) $\frac{9c^2 + 35d^2}{21cd}$ **h)** $\frac{40x - 27y}{45y}$

11. a) 3   **b)** $\frac{2a^2 - 5a + 5}{a(a - 1)}$ **c)** $\frac{5 - 2x}{x + 1}$

d) $\frac{m^2 - 3m - 2}{(m - 1)(m + 1)}$ **e)** $\frac{y^2 - 5y + 7}{y - 2)(y - 3)}$ **f)** $\frac{a^2 - 25a - 4}{(a + 7)(a - 3)}$

12. a) $\frac{a}{2(2a - 3)}$ **b)** $\frac{8m + 5}{3(m - 4)}$ **c)** $\frac{2k^2 + 4k - 3}{(k - 2)(k + 2)}$

d) $\frac{-2b^2 + 13b}{(b - 3)(b - 4)}$ **e)** $\frac{49x^2 + 17x + 10}{15(x - 2)(x + 3)}$ **f)** $\frac{18x^2 + 107x + 27}{6(2x - 1)(3x + 7)}$

13. a) $\frac{x^2 - 5x - 6}{(x + 2)(x - 2)}$ **b)** $\frac{17a}{12(2a - 3)}$ **c)** $\frac{33a}{10(3a - 5)}$

d) $\frac{-4x + 3}{2(x - 4)}$ **e)** $\frac{13m^2 + 8m + 4}{6(m - 2)(m + 2)}$ **f)** $\frac{7t^2 + 3t}{5(t + 1)(t - 1)}$

---

14. a) $\frac{5}{7}$ **b)** 12   **c)** 4

d) $-\frac{1}{50}$ **e)** $\frac{19}{3}$ **f)** $\frac{5}{2}$

15. a) 1, 8   **b)** 1, -12.5 **c)** 1, 5

16. Explanations may vary.

17. 220 km/h   **18.** 4.2 km/h   **19.** 3.3 km/h

7 Cumulative Review, page 447

1. a) 17, 26, 35, 44   **b)** 25, 20.5, 16, 11.5, 7

2. a) i) $t_n = 3n + 3$ **ii)** $t_n = -5n + 10$ **iii)** $t_n = -n + 3$
iv) $t_n = n - 1$ **v)** $t_n = n$ **vi)** $t_n = n + 1$
b) Explanations may vary.

3. a) i) 18th   **ii)** 52nd   **iii)** 57th
iv) 112th   **v)** 421st   **vi)** 27th
b) Explanations may vary.

4. a) 1024   **b)** 1024   **c)** 1024
d) 1025   **e)** 1026   **f)** 999

5. a)

Day	Mon.	Tues.	Wed.	Thurs.	Fri.	Sat.	Sun.
Pay ($)	63.00	85.50	0	99.00	162.00	54.00	90.00

b) $553.50, about $11.30

6. a) 3   **b)** 15.6   **c)** 21.9
d) 9   **e)** 4   **f)** 1.32

7. a) 1   **b)** 3   **c)** 6   **d)** 10   **e)** 15

8. a) Answers may vary. The difference between consecutive square roots increases by 1 each time.
b) $\frac{n(n + 1)}{2}$ **c)** 55

9. a) $(2.5, -0.5)$ **b)** $(4, 1.5)$ **c)** $(-6, 6.5)$

10. a) 5.5   **b)** $(3, -1)$ **c)** 5

11. Midpoints: $(-1, 2)$, $(2, -2)$, $(6, 0)$, $(3, 4)$; slopes of opposite sides: $\frac{-4}{3}$, $\frac{1}{2}$

12. a) $2x + 5y = 8$ **b)** $5x + 3y = 1$

13. a) $3x + 5y - 1 = 0$ **b)** $y = 5$
c) $x + 3y - 12 = 0$ **d)** $x = 5$

14. Tables may vary.
i) a)

x	y
0	0
1	1
2	2
3	3
4	4

c) All input numbers are valid.

ii) a)

x	y
0	0
1	3
2	6
3	9
4	12

c) All input numbers are valid.

iii) a)

x	y
0	0
3	1
6	2
9	3
12	4

c) All input numbers are valid.

15. b) $25 **c)** $19 **d)** $y = 19x + 25$
 e) i) $405 **ii)** $323.30

16. a) x-intercept: 0.5; y-intercept: -1; slope: 2; range: all real numbers
 b) x-intercept: -1.5; y-intercept: 3; slope: 2; range: all real numbers
 c) No x-intercept; y-intercept: 3; slope: 0; range: 3
 d) x-intercept: -2; y-intercept: 1; slope: 0.5; range: all real numbers
 e) x-intercept: -3; y-intercept: 2; slope: $\frac{2}{3}$; range: all real numbers
 f) x-intercept: 2; y-intercept: 2; slope: -1; range: $y \leq 2$

17. a) $-3, -1, 1, -5$ **b)** $1, 2, 9, 6$
 c) Not possible, 0, 1, not possible
 d) Undefined, 2, 2.5, -2 **e)** $1, 1.5, \frac{7}{3}$, undefined
 f) $-0.5, \frac{2}{3}, \frac{15}{4}, 0$

18. a) 6π cm, 9π cm^2
 b) Multiplied by 2; multiplied by 4
 c) 16π cm^2, $\frac{32\pi}{3}$ cm^3
 d) Multiplied by 4; multiplied by 8
 e) Explanations may vary.

19. a) 5 **b)** 0 **c)** -9 **d)** 3
 e) 1 **f)** 1.75

20. a) 3 **b)** -1 **c)** 2.5 **d)** $-3, 1$
 e) $1, -1$ **f)** $1, -2$

21. a) $\frac{1}{x+2}$ **b)** $x + 3$ **c)** $2x + 1$
 d) 0.5 **e)** -1.75 **f)** $\frac{1}{7}$

22. a) $\frac{x}{x+1}$ **b)** $\frac{1}{(x+2)(x+1)}$ **c)** $\frac{y^2}{(x+2y)(x+y)}$

Chapter 8
8.1 Exercises, page 457

1. a) 0.268 **b)** 0.649 **c)** 1.732 **d)** 11.430

2. a) 11° **b)** 51° **c)** 37° **d)** 68°

3. a) i) $\frac{2}{3}$ **ii)** 1 **iii)** $\frac{5}{3}$ **iv)** $\frac{4}{15}$
 b) i) 34° **ii)** 45° **iii)** 59° **iv)** 15°

4. Explanations may vary.

5. Angles are given to the nearest degree.
 a) $\tan P = \frac{2}{2.1}$, $\tan Q = \frac{2.1}{2}$, $\angle P \doteq 44°$, $\angle Q \doteq 46°$
 b) $\tan P = \frac{0.8}{1.5}$, $\tan Q = \frac{1.5}{0.8}$, $\angle P \doteq 28°$, $\angle Q \doteq 62°$
 c) $\tan P = \frac{80}{60}$, $\tan Q = \frac{60}{80}$, $\angle P \doteq 53°$, $\angle Q \doteq 37°$

6. a) i) 1.5 m **ii)** 2.3 m **iii)** 3.4 m
 b) Angles are given to the nearest degree.
 i) 63° **ii)** 72° **iii)** 76°

7. Diagrams may vary. **8.** Explanations may vary.

9. 14.5 m **10.** 191 m

11. Assumptions may vary. 77 m, with eyes at ground level

12. 24 km **13.** 26.0 m

14. a) 1219 m **b)** 10°

15. a) 123 ft **b)** 1260 ft/min; about 6 s
 c) Exercises and answers may vary.

16. 71° **17.** 38.7°, 38.7°, 102.6° **18.** 38°

19. a) 33.7° **b)** 36.9° **c)** 68.2°

8.2 Exercises, page 464

1. a) i) $\frac{8}{13}$ **ii)** $\frac{1}{3}$ **iii)** $\frac{4}{11}$
 b) i) 38° **ii)** 19° **iii)** 21°

2. Explanations may vary.

3. a) i) $\frac{14}{17}$ **ii)** $\frac{7}{9}$ **iii)** $\frac{5}{11}$
 b) i) 35° **ii)** 39° **iii)** 63°

4. a) $\sin A = \frac{7.7}{8.7}$, $\cos A = \frac{4.1}{8.7}$, $\angle A = 62°$; $\sin C = \frac{4.1}{8.7}$, $\cos C = \frac{7.7}{8.7}$, $\angle C = 28°$
 b) $\sin A = \frac{16.2}{20.1}$, $\cos A = \frac{11.9}{20.1}$, $\angle A = 54°$; $\sin C = \frac{11.9}{20.1}$, $\cos C = \frac{16.2}{20.1}$, $\angle C = 36°$
 c) $\sin A = \frac{56.2}{65.1}$, $\cos A = \frac{32.8}{65.1}$, $\angle A = 60°$; $\sin C = \frac{32.8}{65.1}$, $\cos C = \frac{56.2}{65.1}$, $\angle C = 30°$

5. a) i) 17.4 cm **ii)** 12.0 cm **iii)** 4.0 cm
 b) i) 73° **ii)** 58° **iii)** 40°

6. Diagrams may vary. **7.** Explanations may vary.

8. 80° **9.** 2.9 m; 7.7 m **10.** 8.6 m

11. a) 69 m **b)** 80 m **c)** 57 m

12. 63°, 27° **13.** 5.5 cm **14.** 6.6° **15.** 35.4 m

16. a) 73° **b)** 47.6°

17. 80 m **18.** 10°

19. a) 39 m **b)** 25 m

20. a) 61.5 m **b)** 66.2 m

21. a) Descriptions may vary.
 b) i) 0.5 **ii)** $\frac{\sqrt{3}}{2}$ **iii)** $\frac{\sqrt{3}}{2}$ **iv)** 0.5

22. 11.5 cm **23.** 17.3 cm

24. 31.7 cm **25.** 6367 km; 40 000 km

Linking Ideas: Mathematics and Geography
The Spiral Tunnels, page 468

1. 4.5% **2.** 2.58° **3.** 1.26° **4.** 13.5 km

5. 6.9 km **6.** About 90 **7.** About 20 m **8.** About 3 min

8.3 Exercises, page 474

1. a) $\sin A = \frac{20}{29}$, $\cos A = \frac{21}{29}$, $\tan A = \frac{20}{21}$

b) $\sin A = \frac{8}{17}$, $\cos A = \frac{15}{17}$, $\tan A = \frac{8}{15}$

c) $\sin A = \frac{3.6}{3.9}$, $\cos A = \frac{1.5}{3.9}$, $\tan A = \frac{3.6}{1.5}$

2. a) JL = 21.0 m, ∠K = 51°, ∠J = 39°

b) UV = 26.1 m, ∠T = 42°, ∠V = 48°

c) GH = 20.6 km, ∠G = 51°, ∠H = 39°

3. a) ∠C = 43°, BC = 11.7 cm, RC = 8.6 cm

b) ∠L = 25°, TA = 14.9 m, TL = 35.3 m

c) ∠S = 59°, DS = 9.8 m, DW = 16.3 m

d) ∠N = 63°, PN = 15.8 cm, NE = 34.8 cm

e) ∠U = 35°, RG = 14.3 cm, UR = 20.5 cm

f) ∠Q = 68°, QM = 8.2 cm, MF = 20.4 cm

4. a) YZ = 25 m, ∠X = 47°, ∠Z = 43°

b) ∠Z = 63°, YZ = 8 m, XZ = 18 m

c) XY = 49 cm, ∠X = 15°, ∠Z = 75°

d) ∠X = 38°, XY = 57 cm, YZ = 44 cm

e) ∠Z = 26°, XY = 16 mm, XZ = 36 mm

f) XZ = 49 mm, ∠X = 24°, ∠Z = 66°

5. Explanations may vary.

6. Short wire: slope $= \frac{15}{22}$, angle = 34°; long wire: slope $= \frac{30}{22}$, angle = 54°

7. a) $\frac{4}{15}$, 15°   **b)** $\frac{4}{7}$, 30°   **c)** 0.9, 42°

8. 12°   **9.** 650 m   **10.** Explanations may vary.

11. a) 24.6 m   **b)** 4.3 m   **c)** 5.67

12. 4072 m   **13.** 3191 m   **14.** 41°

15. a) 6 cm, 5 cm; 4 cm, 7 cm; 2 cm, 4 cm   **b)** 41°, 60°, 76°

16. a) i) 126 nm   **ii)** 152 nm

b) 150 knots, 428 lb/h, 978 lb, $444

c) About 1.8 h, or 1 h 52 min

d) About 770 lb   **e)** $823

f) No, at sea level cruise speed, the 85 nm trip would take 0.57 h and require 243 lb of fuel. The helicopter has only 185 lb of fuel remaining.

17. 32°

Mathematics File: Indirect Measurement, page 478

Answers may vary.

1. a) 90°, 73°   **b)** 185 m, 177 m

2. a) 30°, 134°   **b)** 142°, 32°

Problem Solving: Calculating the Speed of Earth's Rotation, page 480

1. b) Answers may vary.

2. Explanations may vary.   **3.** Explanations may vary.

4. Answers may vary.   **5.** Letters may vary.

8.4 Exercises, page 484

1. a) 14 cm   **b)** 23 cm   **c)** 23 m

2. a) 53°   **b)** 88°   **c)** 21°

3. a) i) 45°   **ii)** 26.6°   **iii)** 18.4°

b) Answers may vary; their tangents are consecutive unit fractions.

c) Yes; explanations may vary.

4. 35.5 m, 52.0 m

5. a) 9.0 m, 9.3 m   **b)** 3.6 m

c) Explanations may vary.

6. a) $x = 1.7$ **b)** $x = 4.2$ **c)** $x = 3.5$, $y = 5.4$

7. 4.7 cm   **8.** 28.9 cm^2 **9.** 3.7 m   **10.** 2215 m

11. a) AC = 8 cm, CD = 16 cm, AD = $8\sqrt{3}$ cm; BC = $4\sqrt{3}$ cm, CE = 12 cm, BE = $8\sqrt{3}$ cm

b) 4 : 3   **c)** 1 : 1

12. a) 1148 m   **b)** 3431 m   **c)** 2259 m

13. 194 km   **14.** Answers may vary.

15. Answers may vary.   **16.** 176 m

17. 32 m, 24 m   **18.** 58.7°, 58.7°, 62.6°

19. 13.7 m

Linking Ideas: Mathematics and Technology
Sines and Cosines of Obtuse Angles, page 489

1. a) 50°   **b)** 2

2. a) 80°   **b)** 60°   **c)** 30°

3. $\sin A = \sin (180° - A)$

4. a) Explanations may vary.   **b)** 50°

5. a) 80°   **b)** 60°   **c)** 30°

6. $\cos A = -\cos (180° - A)$

Investigate, page 490

1. 0.87   **2.** 0.50

3. Descriptions may vary. The x-coordinate goes from 1 to 0 and back to 1. The y-coordinate goes from 0 to 1 and back to 0.

4.

∠A	cos A	sin A	P(x, y)
0°	1	0	(1, 0)
15°	0.97	0.26	(0.97, 0.26)
30°	0.87	0.5	(0.87, 0.5)
45°	0.71	0.71	(0.71, 0.71)
60°	0.5	0.87	(0.5, 0.87)
75°	0.26	0.97	(0.26, 0.97)
90°	0	1	(0, 1)

5. cos A goes from 1 to 0; sin A goes from 0 to 1; cos A = sin (90° − A); sin A = cos (90° − A)

6. , 7.

∠A	cos A	sin A	P(x, y)
105°	−0.26	0.97	(−0.26, 0.97)
120°	−0.5	0.87	(−0.5, 0.87)
135°	−0.71	0.71	(−0.71, 0.71)
150°	−0.87	0.5	(−0.87, 0.5)
165°	−0.97	0.26	(−0.97, 0.26)
180°	−1	0	(−1, 0)

8. a) −0.26, 0.97   **b)** −0.5, 0.87   **c)** −0.71, 0.71
    d) −0.87, 0.5   **e)** −0.97, 0.26

9. a) 1, 0   **b)** 0, 1   **c)** −1, 0

10. a) 0°, 180°; 15°, 165°; 30°, 150°; 45°, 135°; 60°, 120°; 75°, 105°; the angles are supplementary
   b) They are opposites.
   c) Supplementary angles have the same sines and opposite cosines.

8.5 Exercises, page 496

1. a) 0.866, −0.5   **b)** 0.259, −0.966   **c)** 0.966, 0.259
    d) 0.995, −0.105   **e)** 0.375, −0.927   **f)** 0.682, 0.731
    g) 0, 1   **h)** 0, −1

2. Sketches may vary.
   a) Positive   **b)** Positive   **c)** Positive
   d) Negative   **e)** Negative   **f)** Positive

3. Explanations may vary.

4. Sketches may vary.
   a) 0.940, −0.342   **b)** 0.996, −0.087   **c)** 0.669, −0.743
   d) 0.956, 0.292   **e)** 0.616, −0.788   **f)** 0.574, 0.819
   g) 0.139, −0.990   **h)** 0.809, 0.588

5. a, c, e, g   **6.** Explanations may vary.

7. a) 65°, 115°   **b)** 55°   **c)** 137°
   d) 25.4°, 154.6°   **e)** 35°, 145°   **f)** 112°
   g) 30°, 150°   **h)** 60°   **i)** 120°
   j) 42°, 138°   **k)** 14.5°, 165.5°   **l)** 146.4°

8. Explanations may vary.

10. Diagrams and explanations may vary.

8.6 Exercises, page 503

1. a) 2.6   **b)** 5.3

2. a) 1.5   **b)** 3.6

3. a) 10.8 cm   **b)** Explanations may vary.

4. 5.9 m   **5.** 3.0 km

6. a) 89°, 45.5°, 45.5°   **b)** 42°, 69°, 69°

7. a) $\angle Q = 46.4°$, $\angle R = 28.6°$, $r = 6.0$ cm
   b) $q = 21.0$ m, $\angle R = 46.4°$, $\angle P = 87.2°$

8.7 Exercises, page 509

1. a) 5.6   **b)** 7.4   **c)** 5

2. Explanations may vary.

3. a) 37°   **b)** 16°   **c)** 37°

4. a) 60°   **b)** 26°   **c)** 17°

5. Explanations may vary.   **6.** Drawings may vary. 12.5°

7. 45°, 21°, 14°, 10°   **8.** 54.3 m   **9.** 31.6 km; 22.8 km

10. 110 m   **11.** 36.2 m

12. a) 208.2 m, 133.1 m   **b)** 13 848 m²

13. a) 7.4 m   **b)** 12.4 m²

14. Answers may vary. **a)** 98 m, 141 m   **b)** 318 m, 274 m

15. a) 233°   **b)** 2.9 km   **16. a)** 5.1 m

17. a) 1.9   **b)** 2.7

18. 31.5   **19.** Explanations may vary.

20. a) $\angle C = 59°$, $\angle A = 81°$, $a = 9.2$ cm; $\angle C = 121°$, $\angle A = 19°$, $a = 3.0$ cm
   b), c) Answers may vary.

21. Answers may vary.

8.8 Exercises, page 517

1. a) 3.6   **b)** 5.0   **c)** 6.1

2. Explanations may vary.

3. a) 83.6°   **b)** 48.5°   **c)** 46.6°

4. a) $\angle A = 71°$, $\angle B = 42°$ **b)** Answers may vary.

5. 46 m   **6.** 824 m

7. a) 30°   **b)** 19°   **c)** 14°

8. Explanations may vary.

9. 11°   **10.** 29°, 17°, 12°, 9°

11. a) It is a right triangle. 90°, 37°, 53°   **b)** 59°, 41°, 83°

12. 47°   **13.** 89°, 56°   **14.** 1844 m   **15.** 3090 m

16. Answers may vary.
   a) i) 162 m   **ii)** 216 m   **iii)** 352 m
   b) 25°, 137°, 18°   **c)** 11 935 m²

17. 97 km   **18.** 34 m

19. 440 m   **20.** 107 m

21. 12 900 m²   **22.** 1586 m; 107 680 m²

23. a) 11   **b)** About 105 nm
   c) About 45 min   **d)** About 423 lb; about 1489 lb

24. 10.8 km   **25.** 26 km

26. a) 63.5 cm   **b)** 49.3 cm

27. a) 50.2 cm   **b)** 46.9 cm

28. 1.6 cm

29. a) 1.9 units   **b)** 0.25 units²

31. a) i) 108 m   **ii)** 70.5 m   **iii)** 62.4 m
   b) 64.2°
   c) i) 5.6 m   **ii)** 6.7 m   **iii)** 7.0 m

32. Answers may vary.

Mathematical Modelling: How Far Is the Sun? How Large Is the Sun?, page 522

1. $e^2 = v^2 + 53.1^2 - 106.2v(\cos 31.8°)$

2. $e^2 = v^2 + 210.2^2 - 420.4v(\cos 29.3°)$

3. 1.5×10^8 km   **4.** 1.1×10^8 km   **5.** 1.4×10^6 km

7. a) 9.4×10^8 km   **b)** 1.1×10^5 km/h

8. Answers may vary.

8 Review, page 524

1. a) 0.5095   **b)** 0.7813   **c)** 2.145   **d)** 6.314

2. a) 55°   **b)** 60.3°   **c)** 4°   **d)** 71°

3. a) 5.5   **b)** 15.1   **c)** 11.2

4. a) 31°   **b)** 50°   **c)** 56°

5. 26.6°, 63.4°   **6.** 71 m

7. a) i) $\frac{3.3}{3.9}, \frac{2.1}{3.9}$ **ii)** $\frac{15}{37}, \frac{34}{37}$ **iii)** $\frac{15.0}{18.6}, \frac{11.0}{18.6}$
   b) i) 58°   **ii)** 24°   **iii)** 54°

8. a) 14°, 76°   **b)** 53°, 37°   **c)** 66°, 24°

9. a) 22.9, 32.8   **b)** 21.1, 11.3   **c)** 32.6, 14.9

10. a) 4.3, 9.1   **b)** 9.9, 5.8   **c)** 7.2, 10.8

11. 37.5 m

12. a) 5.8 m   **b)** 5.5 m

13. 29 m   **14.** 36 m

15. a) 37°   **b)** 122 m

16. a) 20°, 160°   **b)** 83°   **c)** 32°, 148°
   d) 114°   **e)** 22°, 158°   **f)** 150°

17. a) 1.1 km   **b)** 284°

8 Cumulative Review, page 526

1. a)

Year	Opening balance ($)	Interest rate (%)	Interest earned ($)	Annual investment ($)	Closing balance ($)
1	1 000.00	8	80.00	1000.00	2 080.00
2	2 080.00	8	166.40	1000.00	3 246.40
3	3 246.40	8	259.71	1000.00	4 506.11
4	4 506.11	8	360.49	1000.00	5 866.60
5	5 866.60	8	469.33	1000.00	7 335.93
6	7 335.93	8	586.87	1000.00	8 922.80
7	8 922.80	8	713.82	1000.00	10 636.62
8	10 636.62	8	850.93	1000.00	12 487.55
9	12 487.55	8	999.00	1000.00	14 486.55
10	14 486.55	8	1158.92	1000.00	16 645.47

   b) $11 000   **c)** $5645.47   **d)** $23 748.18
   e)

Year	Opening balance ($)	Interest rate (%)	Interest earned ($)	Annual investment ($)	Closing balance ($)
1	1 000.00	10	100.00	1000.00	2 100.00
2	2 100.00	10	210.00	1000.00	3 210.00
3	3 210.00	10	321.00	1000.00	4 531.00
4	4 531.00	10	453.10	1000.00	5 984.10
5	5 984.10	10	598.41	1000.00	7 582.51
6	7 582.51	10	758.25	1000.00	9 340.76
7	9 340.76	10	934.07	1000.00	11 274.83
8	11 274.83	10	1127.48	1000.00	13 402.31
9	13 402.31	10	1340.23	1000.00	15 742.54
10	15 742.54	10	1574.25	1000.00	18 316.79

2. a) 3, 12, 48, 192   **b)** 3, 6, 12, 24, 48, 96, 192

3. Explanations may vary.

4. b) (0, 0)   **c)** $y = 2x; y = -\frac{1}{2}x$

5. a) $x - 2y + 5 = 0; 2x + y - 6 = 0$ **b)** No
   c) Yes   **d)** No   **e)** No

6. b) i) $2000, $1500, $800   **ii)** $2000, $1800, $1200

iii) $2000, $2200, $1920   **iv)** $2000, $2500, $2400
   v) $2000, $3000, $3200
   c) C   **d)** Explanations may vary.

7. a) $6(x - y)$ **b)** $x^2(7x + 27y)$ **c)** $3x(x - y)$

8. a) $(x + 2)(x + y)$ **b)** $(x + 4)(x^2 + 2)$

9. a) No solution   **b)** 0, 2
   c) 3   **d)** −0.5, 4

10. a) $q = 11.8$, $\angle P = 64.2°$, $\angle R = 40.8°$
   b) $\angle Q = 44.7°$, $\angle P = 83.3°$, $p = 35.3$
   c) $\angle R = 37.0°$, $q = 50.2$, $r = 31.3$
   d) $\angle R = 36.8°$, $\angle P = 29.6°$, $\angle R = 113.6°$

11. 16

12. a) $y = 10$, $\angle X = 28°$, $\angle Z = 42°$
   b) $\angle Y = 70°$, $\angle Z = 52°$, $z = 2.9$

Chapter 9

Investigate, page 530

Percent who said thank you	
Women	84.7%
Teens	81.4%
Younger girls	40%

Percent who said thank you	
Men	75%
Teens	44.4%
Younger boys	23.1%

1. Older people

2. a) Men, women, teens, younger boys, younger girls
   b) The surveyors counted people going in and out of a mall.

3. a) 278   **b)** Answers may vary.   **c)** Answers may vary.

4. The surveyors held the door for people going in and out of a mall, and counted how many said thank you and how many said nothing. The data were sorted by sex and age group.

5. Answers may vary.

9.1 Exercises, page 533

1. Answers may vary.   **2.** Answers may vary.

3. Answers may vary.

4. Answers may vary.
   a) Public-transit riders
   b) Drivers and teens close to the minimum driving age
   c) Canadians

5. Answers may vary.
   a) Students who drive to school
   b) Students on athletic teams
   c) Students who eat cafeteria food

6. Answers may vary.
   a) Hunters, police officers, crime victims
   b) Catholics, Planned Parenthood employees
   c) Homophobes, homosexuals

7. Answers may vary.   **8.** Answers may vary.

9. Answers may vary.   **10.** Answers may vary.

11. Answers may vary.   **12.** Answers may vary.

13. a) Hungary, Canada, Wales; Austria, Norway, Sweden

b) Belgium, Canada, Finland, Norway, Poland, Scotland, Spain, Sweden, Wales

c) Answers may vary.   **d)** Answers may vary.

14. Answers may vary. "Do you smoke?"

15. Answers may vary.

Mathematics File: Random Numbers, page 537

1. The first three digits are the exchange. There is a limited number of exchanges in any area.

2. Answers may vary.

3. Otherwise the sample would not be random.

4. Answers may vary.

9.2 Exercises, page 540

1. a) No; it is a self-selected sample.
   b) No; only people who are really interested in the topic will respond.

2. a) Systematic sampling   **b)** Answers may vary.

3. No; it is a self-selected sample.

4. Answers may vary.   **5.** Answers may vary.

6. a) Convenience sampling   **b)** Answers may vary.

7. a) Yes   **b)** Answers may vary.

8. Random sample

9. Answers may vary.   **10.** Answers may vary.

11. a) BC: 15; AB: 12; SK: 5; MN: 6; ON: 47; PQ: 34; NB: 4; NS: 4; PEI: 1; NF: 3
   b) Answers may vary.

12. No   **13.** Answers may vary.

14. Answers may vary.   **15.** Answers may vary.

16. Answers may vary. For example, use a random number generator.

17. Answers may vary.

9.3 Exercises, page 549

1. Between 3 and 9   **2.** Between 13 and 19

3. Between 12 and 24

4. a) Between 0 and 4   **b)** Between 1 and 7
   c) Between 5 and 15

5. Between 24 and 40   **6.** Between 21 and 35

7. Yes   **8.** No   **9.** No   **10.** Yes   **11.** Yes

12. a) There is a 90% probability that between 2 and 8 girls and between 1 and 7 boys in the sample often have bad tempers.
   b) Girls say they have worse tempers than boys. The percents and 90% probability estimates are given in the table below.

Country	Students who often have bad tempers			
	Percent of girls	Girls in sample	Percent of boys	Boys in sample
Finland	8	0 to 4	4	0 to 2
Scotland	28	2 to 9	22	2 to 8
Poland	26	2 to 9	12	0 to 5

   c) Exercises and answers may vary.

13. Not likely   **14.** Not likely   **15.** Very likely

16. a) There cannot be less than 0 marked items.
   b) There cannot be greater than 20 marked items in a sample of 20.

17. Answers may vary.

18. a) Between 4 and 12   **b)** Between 4 and 12

19. Answers may vary.

9.4 Exercises, page 557

1. a) There is a 95% probability that between 12% and 18% of Canadians feel that deficit/government spending is the most important problem facing Canada.
   b) There is a 95% probability that between 6% and 12% of Canadians feel that national unity is the most important problem facing Canada.
   c) There is a 95% probability that between 2% and 8% of Canadians feel that health care is the most important problem facing Canada.
   d) There is a 95% probability that between 0% and 6% of Canadians feel that taxes are the most important problem facing Canada.

2. There is a 90% probability that between 12% and 44% of students like the food in the cafeteria.

3. There is a 90% probability that between 22% and 60% of students have cats.

4. a) There is a probability of 90% that between 22% and 60% of students spend less than one hour on homework.
   b) There is a probability of 90% that between 62% and 84% of students spend less than one hour on homework.
   c) There is a probability of 90% that between 68% and 80% of students spend less than one hour on homework.

5. There is a 90% probability that between 68% and 88% of students spent more than one hour preparing for their last math test.

6. a) The margin of error is ±2.5%, 19 times out of 20.
   b) Answers may vary.   **c)** Answers may vary.

7. a) There is a 90% probability that between 56% and 80% of students ride the bus to school.
   b) There is a 90% probability that between 459 and 656 students ride the bus to school.

8. a) There is a 90% probability that between 36% and 52% of British Columbians believe in the Sasquatch.
   b) There is a 90% probability that between 1.4 million and 2.0 million British Columbians believe in the Sasquatch.

9. There is a 90% probability that between 2.2 million and 2.5 million Albertans believe that life exists elsewhere in the universe.

10. Yes

11. a) There is a 95% probability that between 87.2% and 92.8% of Canadians are concerned about our health-care system.
b) Answers may vary.

12. Yes   **13.** Answers may vary.   **14.** Answers may vary.

15. As the size of the sample increases, the size of the confidence interval decreases.

16. Answers may vary.

Linking Ideas: Mathematics and Science Estimating the Size of a Wildlife Population, page 561

1. a) There is a 90% probability that there are 177 to 575 trout in the lake.
b) There is a 90% probability that there are 209 to 460 trout in the lake.
c) Explanations may vary.

2. Explanations may vary.

3. There is a 90% probability that there are 781 to 2083 trout in the lake.

4. There is a 90% probability that the number of buffaloes in the park is between 3750 and 15 000.

5. There is a 90% probability that the number of Canada geese that fly through the region is between 3098 and 4750.

6. Answers may vary.

Investigate, page 562

1. a) Approaches 0.5   **b)** Approaches 0

2. Table of values may vary.   **4.** Answers may vary.

5. Approaches 0.5   **6.** Answers may vary.

9.5 Exercises, page 565

1. No. Explanations may vary.

2. Number of possible outcomes, number of favourable outcomes, how likely is the event

3. a) $\frac{1}{4}$ **b)** $\frac{1}{2}$ **c)** $\frac{1}{13}$
d) $\frac{1}{26}$ **e)** $\frac{1}{13}$ **f)** $\frac{3}{26}$

4. a) $\frac{2}{11}$ **b)** $\frac{1}{11}$ **c)** 0
d) $\frac{4}{11}$ **e)** $\frac{7}{11}$ **f)** $\frac{3}{11}$

5. There is a 90% probability that tails should appear from 15 to 25 times.

6. There is a 90% probability that a 6 should appear from 3 to 10 times.

7. Bag B   **8.** 0.4

9. a) $\frac{1}{3}$
b) There is a 90% probability that a 1 or a 6 should appear from 26 to 42 times.

10. a) $\frac{1}{4}$
b) There is a 90% probability that two heads should appear from 6 to 15 times.

11. a) 40%
b) There is a 90% probability that the number of people with O type blood is between 37 and 53; There is a 90% probability that the number of people with A type blood is between 32 and 48; There is a 90% probability that the number of people with B type blood is between 6 and 17; There is a 90% probability that the number of people with AB type blood is between 1 and 7.

12. a) There is a 90% probability that the number of times a black card appears is between 42 and 58.
b) There is a 90% probability that the number of times a heart appears is between 18 and 32.
c) There is a 90% probability that the number of times an ace appears is between 4 and 13.
d) There is a 90% probability that the number of times a face card appears is between 16 and 30.

13. a) Answers may vary.
b) The difference between the numbers of heads and tails increased.
c) The difference between the numbers of heads and tails varied between about 10 and 20.
d) Answers may vary.

14. There is a 90% probability that the thumbtack lands point up between 22% and 36% of the time.

Exploring with a Graphing Calculator: Will You Flip with Me?, page 569

2. Answers may vary. The scores are relatively low.

3. Yes   **4. a)** Yes   **b)** Yes

5. About 14   **6.** 260

9. Event 2 is more likely.

Problem Solving: Coincident Birthdays, page 570

1. Explanations may vary.

2. Neither

3. Number of rows in table may vary.

6. a) Answers may vary; 0.71 for a class of 30
b) i) 23   **ii)** 32   **iii)** 41
d) Answers may vary.

7. a) Answers may vary.   **b)** 0.87 for a class of 20

9.6 Exercises, page 575

1. a) 20   **b)** Explanations may vary.

2. a) 65.6   **b)** Explanations may vary.

3. $2.70   **4.** $0.545   **5.** $933.33

6. a) 50, 50   **b)** Yes

7. $1.50   **8.** −$0.82   **9.** −$0.056

10. a) −$20 **b)** 0.3¢

11. a) Fair **b)** Not fair

12. Local event

13. a) Blue tag: 6%; yellow tag: 3%; white tag: 1%
 b) 22¢ **c)** Answers may vary.

9.7 Exercises, page 579

1. a) They cover small claims. **b)** Answers may vary.
 c) His premium would decrease.

2. Answers may vary. **3.** Answers may vary.

4. a) 400 **b)** $50 000 **c)** $6.25 **d)** No

5. a) 3000 **b)** $45 000 000
 c) $90 **d)** Answers may vary.

6. Answers may vary. **7.** Answers may vary.

8. a) Yes **b)** Answers may vary.

9. a) 750 **b)** $3 150 000
 c) $12.60 **d)** Answers may vary.

Mathematical Modelling: Should We Harvest Today or Wait?, page 582

1. a)

Date	Value of juice (¢/L)	Expected gain that day (¢/L)	Expected loss that day (¢/L)	Net expected gain (¢/L)
Oct. 1	200	13.65	4.5	9.15
Oct. 2	215	13.65	5.85	7.8
Oct. 3	230	13.65	7.2	6.45
Oct. 4	245	13.65	8.55	5.1
Oct. 5	260	13.65	9.9	3.75
Oct. 6	275	13.65	11.25	2.4
Oct. 7	290	13.65	12.6	1.05
Oct. 8	305	13.65	13.95	−0.3
Oct. 9	320	13.65	15.3	−1.65
Oct. 10	335	13.65	16.65	−3
Oct. 11	350	13.65	18	−4.35

 b) i) 91% **ii)** 15¢/L **iii)** 13.65¢/L
 c) i) 9% **ii)** 50¢/L **iii)** 4.5¢/L **iv)** 9.15¢/L

2. b) On Oct. 8

3. a) Arithmetic; $t_n = 1.35n + 3.15$
 b) Gain: $t_n = 13.65$; loss: $t_n = 1.35n + 3.15$
 c) $13.65 − (1.35n + 3.15) = 0$
 d) $n = 7.8 \doteq 8$; yes

4. Answers may vary. **b)** The slope of the line represents the rate at which the net expected gain decreases. The y-intercept is the net expected gain on the first day. The x-intercept is the day the net expected gain reaches 0.
 c) $y = 10.5 − 1.35x$

5. a) Oct. 30; $y = 13.5 − 0.45x$ **b)** Oct. 10; $y = 20.5 − 2.25x$
 c) Oct. 2; $y = 1.5 − 1.35x$

9 Review, page 585

1. a) Explanations may vary. **b)** Answers may vary.

2. There is a 90% probability that between 57 and 416 of students have part-time jobs.

3. a) Explanations may vary. **b)** Answers may vary.

4. Answers may vary.

5. a) $\frac{1}{5}$ **b)** $\frac{3}{10}$ **c)** $\frac{1}{2}$ **d)** $\frac{7}{10}$ **e)** $\frac{4}{5}$
 f) 0 **6.** $20

9 Cumulative Review, page 586

1. a) $t_n = 2n$ **b)** $t_n = 4n − 2$ **c)** $t_n = 10n$
 d) $t_n = 9n − 6$ **e)** $t_n = 7 − n$ **f)** $t_n = 2n − 25$

2. a) $5, 7, \frac{49}{5}, \frac{343}{25}, \frac{2401}{125}$ **b)** 4, 6, 9, 13.5, 20.25
 c) $4, 4\sqrt{\frac{3}{2}}, 6, 6\sqrt{\frac{3}{2}}, 9$ **d)** 7, 21, 63, 189, 567
 e) 6, 24, 96, 384, 1536 **f)** 8, 24, 72, 216, 648

3. a) $1762.34 **b)** $3524.68
 c) $17 623.42, explanations may vary **d)** $2931.63

4. a) 2 **b)** 3 **c)** 4 **d)** 5
 e) Explanations may vary.

5. a and e, a and f, e and f **6.** b and c, a and d, d and e, d and f

7. b) Equations may vary. $y = −0.036x + 33.1$
 c) −39°C **d)** 3475 W/m^2

8. a) $h = 3l + 70$ **b)** 184.6 cm

9. a) $\frac{1}{2}, 3$ **b)** 2, 4 **c)** 0, 3 **d)** −4, 8 **e)** 3, −7

10. a) Answers may vary.
 b) A: 268.1 cm^3, B: 134.0 cm^3, C: 134.0 cm^3
 c) B: 3.17 cm, C: 3.17 cm
 d) A: 201.1 cm^2, B: 126.7 cm^2, C: 126.7 cm^2
 e) B and C: 253.4 cm^2; the surface area of B and C is greater than the surface area of A

11. a) $\frac{y}{24x^3}$ **b)** $5x^5y^5z^3$ **c)** $432x^7y^{11}$
 d) $\frac{c^2}{b^2}$ **e)** $\frac{9}{800a^4b^5}$ **f)** $\frac{243a^7b^4c^4}{2}$

12. a) $(x + 2)(x + 3)$ **b)** $(x + 3)(x − 2)$
 c) $3(x + 4)(x + 1)$ **d)** $5(x + 5)(x − 4)$
 e) $3(2x + y)(x − 2y)$ **f)** $4(3a + 2b)(2a + 3b)$
 g) $(x + 5)(x − 5)$ **h)** $4(x + 5)(x − 5)$
 i) $(7x + 8y)(7x − 8y)$

13. a) $2x + 3, x \neq 0$ **b)** $\frac{3x}{2}, x \neq −5$ **c)** $−2x, x \neq \frac{5}{2}$
 d) $2(x + 2), x \neq −3$ **e)** $\frac{x − 1}{4x}, x \neq 0, −2$ **f)** $x + 2, x \neq 2$

14. a) $\frac{19}{30}$ **b)** $\frac{x}{15}$ **c)** $\frac{5y}{6x}$
 d) $\frac{9x − 1}{6x}$ **e)** $\frac{2x^2 − 3y^2}{6xy}$ **f)** $\frac{1}{2}$

15. 18.75 cm, 18.75 cm, 26.52 cm

16. 37.6°, 104.8°

17. a) There is a 90% probability that between 8% and 40% of students believe in flying saucers.
 b) There is a 90% probability that between 40% and 78% of students believe in flying saucers.
 c) Answers may vary. The answers are different.
 d) Answers may vary. Use a larger sample size.

GLOSSARY

absolute value: the distance between any real number and 0 on a number line; for example, $|-3| = 3$, $|3| = 3$

acute angle: an angle measuring less than 90°

acute triangle: a triangle with three acute angles

additive inverse: a number and its opposite; the sum of additive inverses is 0; for example, $+3 + (-3) = 0$

algebraic expression: a mathematical expression containing a variable: for example, $6x - 4$ is an algebraic expression

alternate angles: angles that are between two lines and are on opposite sides of a transversal that cuts the two lines

Angles 1 and 3 are alternate angles.
Angles 2 and 4 are alternate angles.

altitude: the perpendicular distance from the base of a figure to the opposite side or vertex; also the height of an aircraft above the ground

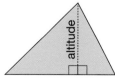

angle: the figure formed by two rays from the same endpoint

angle bisector: the line that divides an angle into two equal angles

approximation: a number close to the exact value of an expression; the symbol $\doteq$ means "is approximately equal to"

area: the number of square units needed to cover a region

arithmetic sequence: a sequence of numbers in which each term after the first term is calculated by adding the same number to the preceding term; for example, in the sequence 1, 4, 7, 10, …, each number is calculated by adding 3 to the previous number

arithmetic series: the indicated sum of the terms of an arithmetic sequence

array: an arrangement in rows and columns

average: a single number that represents a set of numbers; see *mean*, *median*, and *mode*

balance: the result when money is added to or subtracted from an original amount

bar graph: a graph that displays data by using horizontal or vertical bars whose lengths are proportional to the numbers they represent

bar notation: the use of a horizontal bar over decimal digits to indicate that they repeat; for example, $1.\overline{3}$ means 1.333 333 …

base: the side of a polygon or the face of a solid from which the height is measured; the factor repeated in a power

bias: an emphasis on characteristics that are not typical of the entire population

biased sample: a sample containing members of the population that are not representative

binomial: a polynomial with two terms; for example, $3x - 8$

binomial outcome: one of the outcomes of an event that has only two possible outcomes; for example, the toss of a thumbtack is an event, the outcomes "point up" and "point down" are binomial outcomes

bisector: a line that divides a line segment in two equal parts
The broken line is a bisector of AB.

box-and-whisker plot: a diagram in which a horizontal box represents the most likely 90% of outcomes of an experiment; the whiskers represent the least likely 10% of outcomes

boxplot: see *box-and-whisker plot*

broken-line graph: a graph that displays data by using points joined by line segments

centroid: the point where the three medians of a triangle intersect

circle: the set of points in a plane that are a given distance from a fixed point (the centre)

circumcentre: the point where the perpendicular bisectors of the sides of a triangle intersect

circumcircle: a circle drawn through each of the vertices of a triangle and with its centre at the circumcentre of the triangle

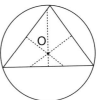

circumference: the distance around a circle, and sometimes the circle itself

cluster sample: a sample in which every member of a randomly chosen section of the population is selected

coefficient: the numerical factor of a term; for example, in the terms $3x$ and $3x^2$, the coefficient is 3

collinear points: points that lie on the same line

commission: a fee or payment given to a salesperson, usually a specified percent of the person's sales

common denominator: a number that is a multiple of each of the given denominators; for example, 12 is a common denominator for the fractions $\frac{1}{3}$, $\frac{5}{4}$, $\frac{7}{12}$

common difference: the number obtained by subtracting any term from the next term in an arithmetic sequence

common factor: a number that is a factor of each of the given numbers; for example, 3 is a common factor of 15, 9, and 21

common ratio: the ratio formed by dividing any term after the first one in a geometric sequence by the preceding term

commutative property: the property stating that two numbers can be added or multiplied in any order; for example, $6 + 8 = 8 + 6$ and $4 \times 7 = 7 \times 4$

complementary angles: two angles whose sum is 90°

∠ABC and ∠CBD are complementary angles.

complex fraction: a fraction that has other fractions in the numerator and/or denominator; for example, $\dfrac{2 + \frac{1}{3}}{3 - \frac{1}{4}}$

composite number: a number with three or more factors; for example, 8 is a composite number because its factors are 1, 2, 4, and 8

compound event: a combination of two or more events

compound interest: see *interest*; if the interest due is added to the principal and thereafter earns interest, the interest earned is compound interest

cone: a solid formed by a region and all line segments joining points on the boundary of the region to a point not in the region

confidence interval: the interval of numbers represented by a box in a boxplot; for example, 90% confidence interval

confidence level: a range of possible values; for example, 90% confidence level

congruent: figures that have the same size and shape, but not necessarily the same orientation

consecutive integers: integers that come one after the other without any integers missing; for example, 34, 35, 36 are consecutive integers, so are −2, −1, 0, and 1

constant: a particular number

constant of proportionality: the constant k in a direct variation of the form $y = kx$; the slope of the graph of this equation

constant slope property: the slopes of all segments of a line are equal

constant term: a number

convenience sample: a sample in which every convenient member of the population is selected

coordinate axes: the x- and y-axes on a grid that represents a plane

coordinate plane: a two-dimensional surface on which a coordinate system has been set up

coordinates: the numbers in an ordered pair that locate a point in the plane

corresponding angles: angles that are on the same side of a transversal that cuts two lines and on the same side of each line

Angles 1 and 3 are corresponding angles.
Angles 2 and 4 are corresponding angles.
Angles 5 and 7 are corresponding angles.
Angles 6 and 8 are corresponding angles.

corresponding angles in similar triangles: two angles, one in each triangle, that are equal

cosine: for an acute $\angle A$ in a right triangle, the ratio of the length of the side adjacent to $\angle A$, to the length of the hypotenuse

cube: a solid with six congruent, square faces

cube root: a number which, when raised to the power 3, results in a given number; for example, 3 is the cube root of 27, and -3 is the cube root of -27

cubic units: units that measure volume

cylinder: a solid with two parallel, congruent, circular bases

data: facts or information

database: facts or information supplied by computer software

denominator: the term below the line in a fraction

density: the mass of a unit volume of a substance

diagonal: a line segment that joins two vertices of a figure, but is not a side

diameter: the distance across a circle, measured through the centre; a line segment through the centre of the circle with its endpoints on the circle

difference of squares: a polynomial that can be expressed in the form $x^2 - y^2$; the product of two monomials that are the sum and difference of the same two quantities, $(x + y)(x - y)$

digit: any of the symbols used to write numerals; for example, in the base-ten system the digits are 0, 1, 2, 3, 4, 5, 6, 7, 8, and 9

Distributive Law: the property stating that a product can be written as a sum or difference of two products; for example, for all real numbers a, b, and c: $a(b + c) = ab + ac$ and $a(b - c) = ab - ac$

domain of a function: the set of x-values (or valid input numbers) represented by the graph or the equation of a function

entire radical: an expression of the form $\sqrt{x}$; for example, $\sqrt{20}$ is an entire radical

equation: a mathematical statement that two expressions are equal

equidistant: the same distance apart

equilateral triangle: a triangle with three equal sides

evaluate: to substitute a value for each variable in an expression and simplify the result

even number: an integer that has 2 as a factor; for example, 2, 4, -6

event: any set of outcomes of an experiment

expanding: multiplying a polynomial by a polynomial

expectation: the average amount you would expect to win if you perform an experiment many times; it requires a payoff for each event in an experiment (also called expected value)

expected value: see *expectation*

experiment: an operation, carried out under controlled conditions, that is used to test or establish a hypothesis

exponent: a number, shown in a smaller size and raised, that tells how many times the number before it is used as a factor; for example, 2 is the exponent in 6^2

expression: a meaningful combination of mathematical symbols, such as a polynomial

extrapolate: to estimate a value beyond known values

extremes: the highest and lowest values in a set of numbers

factor: to factor means to write as a product; to factor a given integer means to write it as a product of integers, the integers in the product are the factors of the given integer; to factor a polynomial with integer coefficients means to write it as a product of polynomials with integer coefficients

fifth root: a number which, when raised to the power 5, results in a given number; for example, 3 is the fifth root of 243 since $3^5 = 243$, and -3 is the fifth root of -243 since $(-3)^5 = -243$

focal length: for a lens, it is the distance from the centre of the lens to the point where incoming parallel light rays converge on the other side of the lens

formula: a rule that is expressed as an equation

fourth root: a number which, when raised to the power 4, results in a given number; for example, -2 and 2 are the fourth roots of 16

fraction: an indicated quotient of two quantities

frequency: the number of times a particular number occurs in a set of data

function: a rule that gives a single output number for every valid input number

general term of an arithmetic sequence: used to determine any term by substitution; for example, t_n is the nth term in the sequence, and $t_n = a + (n - 1)d$, where a is the first term, n the number of terms, and d the common difference

geometric sequence: a sequence of numbers in which each term after the first term is calculated by multiplying the preceding term by the same number; for example, in the sequence 1, 3, 9, 27, 81, …, each number is calculated by multiplying the preceding term by 3

geometric series: the indicated sum of the terms of a geometric sequence

greatest common factor: the greatest factor that 2 or more monomials have in common; $4x^2$ is the greatest common factor of $8x^3 + 16x^2y - 64x^4$

hectare: a unit of area that is equal to 10 000 m^2

hemisphere: half a sphere

hexagon: a six-sided polygon

horizontal intercept: the horizontal coordinate of the point where the graph of a line or a function intersects the horizontal axis

hypotenuse: the side that is opposite the right angle in a right triangle

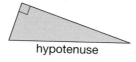
hypotenuse

incentre: the point at which the three angle bisectors of a triangle intersect

incircle: a circle drawn inside a triangle, with its centre at the incentre and with the radius the shortest distance from the incentre to one of the sides of the triangle

independent events: two or more events for which the occurrence or nonoccurrence of one does not affect the occurrence of the others

inequality: a statement that one quantity is greater than (or less than) another quantity

integers: the set of numbers... −3, −2, −1, 0, +1, +2, +3, ...

interest: money that is paid for the use of money, usually according to a predetermined percent

interpolate: to estimate a value between two known values

intersecting lines: lines that meet or cross; lines that have one point in common

interval: a regular distance or space between values

inverse: see *additive inverse* and *multiplicative inverse*

irrational number: a number that cannot be written in the form $\frac{m}{n}$, where m and n are integers ($n \neq 0$)

isosceles acute triangle: a triangle with two equal sides and all angles less than 90°

isosceles obtuse triangle: a triangle with two equal sides and one angle greater than 90°

isosceles right triangle: a triangle with two equal sides and a 90° angle

isosceles triangle: a triangle with at least two equal sides

kite: a quadrilateral with two pairs of equal adjacent sides

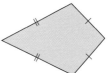

lattice point: on a coordinate grid, a point at the intersection of two grid lines

legs: the sides of a right triangle that form the right angle

lens equation: the formula that relates focal length (f), object distance (p), and image distance (q); that is, $\frac{1}{f} = \frac{1}{p} + \frac{1}{q}$

light-year: a unit for measuring astronomical distances; one light-year is the distance light travels in one year

like radicals: radicals that have the same radical part; for example, $\sqrt{5}$, $3\sqrt{5}$, and $-7\sqrt{5}$

like terms: terms that have the same variables; for example, $4x$ and $-3x$ are like terms

line of best fit: a line that passes as close as possible to a set of plotted points

line segment: the part of a line between two points on the line, including the two points

line symmetry: a figure that maps onto itself when it is reflected in a line is said to have line symmetry; for example, line l is the line of symmetry for figure ABCD

linear equation: an equation that represents a straight line

linear function: a function whose defining equation can be written in the form $y = mx + b$, where m and b are constants

mass: the amount of matter in an object

mean: the sum of a set of numbers divided by the number of numbers in the set

median: the middle number when data are arranged in numerical order

median of a triangle: a line from one vertex to the midpoint of the opposite side

midpoint: the point that divides a line segment into two equal parts

mixed radical: an expression of the form $a\sqrt{x}$; for example, $2\sqrt{5}$ is a mixed radical

mode: the number that occurs most often in a set of numbers

monomial: a polynomial with one term; for example, 14 and $5x^2$ are each a monomial

multiple: the product of a given number and a natural number; for example, some multiples of 8 are 8, 16, 24, …

multiplicative inverse: a number and its reciprocal; the product of multiplicative inverses is 1; for example, $3 \times \frac{1}{3} = 1$

natural numbers: the set of numbers 1, 2, 3, 4, 5, …

negative number: a number less than 0

negative reciprocals: two numbers that have a product of -1; for example, $\frac{3}{4}$ is the negative reciprocal of $-\frac{4}{3}$, and vice versa

numeracy: the ability to read, understand, and use numbers

numerator: the term above the line in a fraction

obtuse angle: an angle greater than 90° and less than 180°

obtuse triangle: a triangle with one angle greater than 90°

octagon: an eight-sided polygon

odd number: an integer that does not have 2 as a factor; for example, 1, 3, -7

operation: a mathematical process or action such as addition, subtraction, multiplication, or division

opposite angles: the equal angles that are formed by two intersecting lines

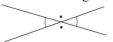

opposite number: a number whose sum with a given number is 0; for example, 3 and -3 are opposites

opposites: two numbers whose sum is zero; each number is the opposite of the other

order of operations: the rules that are followed when simplifying or evaluating an expression

ordered pair: a pair of numbers, written as (x, y), that represents a point on a coordinate grid

orthocentre: the point at which the altitudes of a triangle intersect

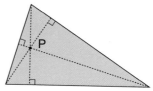

outcome: a possible result of an experiment or a possible answer to a survey question

parallel lines: lines in the same plane that do not intersect

parallelogram: a quadrilateral with both pairs of opposite sides parallel

pentagon: a five-sided polygon

per capita: for each person

percent: the number of parts per 100; the numerator of a fraction with denominator 100

perfect square: a number that is the square of a whole number; a polynomial that is the square of another polynomial

perimeter: the distance around a closed figure

period of a loan: the time it takes to pay back the loan

perpendicular: intersecting at right angles

perpendicular bisector: the line that is perpendicular to a line segment and divides it in two equal parts
The broken line is the perpendicular bisector of AB.

pi (π): the ratio of the circumference of a circle to its diameter; $\pi \doteq 3.1416$

plane geometry: the study of two-dimensional figures; that is, figures drawn or visualized on a plane

point of intersection: a point that lies on two or more figures

polygon: a closed figure that consists of line segments; for example, triangles and quadrilaterals are polygons

polynomial: a mathematical expression with one or more terms, in which the exponents are whole numbers and the coefficients are real numbers

population: the set of all things or people being considered

population density: the average number of people for each square unit of land

positive number: a number greater than 0

power: an expression of the form a^n, where a is called the base and n is called the exponent; it represents a product of equal factors; for example, $4 \times 4 \times 4$ can be expressed as 4^3

prime number: a whole number with exactly two factors, itself and 1; for example, 3, 5, 7, 11, 29, 31, and 43

principal: the amount of a loan or an investment

prism: a solid that has two congruent and parallel faces (the *bases*), and other faces that are parallelograms

probability: if the outcomes of an experiment are equally likely, then the probability of an event is the ratio of the number of outcomes favourable to the event to the total number of outcomes

proportion: a statement that two ratios are equal

pyramid: a solid that has one face that is a polygon (the *base*), and other faces that are triangles with a common vertex

Pythagorean Theorem: for any right triangle, the area of the square on the hypotenuse is equal to the sum of the areas of the squares on the other two sides

quadrant: one of the four regions into which coordinate axes divide a plane

quadratic equation: an equation in which the variable is squared; for example, $x^2 + 5x + 6 = 0$ is a quadratic equation

quadrilateral: a four-sided polygon

radical: the root of a number; for example, $\sqrt{400}$, $\sqrt[3]{8}$, and so on

radical sign: the symbol $\sqrt{}$ that denotes the positive square root of a number

radius (plural, **radii**): the distance from the centre of a circle to any point on the circumference, or a line segment joining the centre of a circle to any point on the circumference

random numbers: a list of digits in a given range, selected so that each digit has an equal chance of occurring

random sample: a sampling in which all members of the population have an equal chance of being selected

range: the difference between the highest and lowest values (the *extremes*) in a set of data

range of a function: the set of *y*-values (or output numbers) represented by the graph or the equation of a function

rate: a certain quantity or amount of one thing considered in relation to a unit of another thing

ratio: a comparison of two or more quantities with the same unit

rational expression: an algebraic expression that can be written as the quotient of two polynomials $\frac{3x^2 + 5}{2x + 7}$

rational number: a number that can be written in the form $\frac{m}{n}$, where m and n are integers $(n \neq 0)$

rationalize the denominator: write the denominator as a rational number, to replace the irrational number; for example, $\frac{6}{\sqrt{2}}$ is written $\frac{6\sqrt{2}}{2}$, or $3\sqrt{2}$

real numbers: the set of rational numbers and the set of irrational numbers; that is, all numbers that can be expressed as decimals

reciprocals: two numbers whose product is 1; for example, $\frac{3}{4}$ and $\frac{4}{3}$ are reciprocals, 2 and $\frac{1}{2}$ are reciprocals

rectangle: a quadrilateral that has four right angles

rectangular prism: a prism that has rectangular faces

rectangular pyramid: a pyramid with a rectangular base

reflex angle: an angle between 180° and 360°

regular decagon: a polygon that has ten equal sides and ten equal angles

regular hexagon: a polygon that has six equal sides and six equal angles

regular octagon: a polygon that has eight equal sides and eight equal angles

regular polygon: a polygon that has all sides equal and all angles equal

relation: a rule that produces one or more output numbers for every valid input number

relative frequency: the ratio of the number of times a particular outcome occurred to the number of times the experiment was conducted

rhombus: a parallelogram with four equal sides

right angle: a 90° angle

right circular cone: a cone in which a line segment from the centre of the circular base to the vertex is perpendicular to the base

right triangle: a triangle that has one right angle

rise: the vertical distance between 2 points

rounding: approximating a number to the next highest (or lowest) number; for example, rounding 3.46 to the tenth is 3.5, and rounding 4.34 is 4.3

run: the horizontal distance between 2 points

sample/sampling: a representative portion of a population

scale: the ratio of the distance between two points on a map, model, or diagram to the distance between the actual locations; the numbers on the axes of a graph

scale factor: the ratio of corresponding lengths on two similar figures

scalene triangle: a triangle with no two sides equal

scatterplot: a graph of data that is a series of points

scientific notation: a number expressed as the product of a number greater than −10 and less than −1 or greater than 1 and less than 10, and a power of 10; for example, 4700 is written as 4.7×10^3

self-selected sample: a sample in which only interested members of the population will participate

semicircle: half a circle

significant digits: the meaningful digits of a number representing a measurement

similar figures: figures with the same shape, but not necessarily the same size

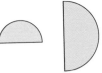

simple random sample: see *random sample*

sine: for an acute ∠A in a right triangle, the ratio of the length of the side opposite ∠A, to the length of the hypotenuse

slope: a measure of the steepness of a line; calculated as slope $= \frac{\text{rise}}{\text{run}}$

slope y-intercept form: the equation of a line in the form $y = mx + b$

sphere: the set of points in space that are a given distance from a fixed point (the centre)

spreadsheet: a computer-generated arrangement of data in rows and columns, where a change in one value results in appropriate calculated changes in the other values

square: a rectangle with four equal sides

square of a number: the product of a number multiplied by itself; for example, 25 is the square of 5

square root: a number which, when multiplied by itself, results in a given number; for example, 5 and −5 are the square roots of 25

standard form of the equation of a line: the equation of a line in the form $Ax + By + C = 0$

statistics: the branch of mathematics that deals with the collection, organization, and interpretation of data

straight angle: an angle measuring 180°

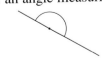

straightedge: a strip of wood, metal, or plastic with a straight edge, but no markings

stratified random sample: a sample in which all members of different segments of the population have an equal chance of being selected

sum of an arithmetic series: the formula for the sum of the first n terms is $S_n = \left(\frac{a + t_n}{2}\right) \times n$, where a is the first term, t_n the nth term, and n the number of terms

supplementary angles: two angles whose sum is 180°

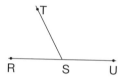

∠RST and ∠TSU are supplementary angles.

survey: an investigation of a topic to find out people's views

symmetrical: possessing symmetry; see *line symmetry*

systematic sample: a sample in which every nth member of the population is selected

30–60–90 property: in a right triangle with angles 30°, 60°, and 90°, the hypotenuse is twice as long as the shorter leg, and the larger leg is $\sqrt{3}$ times as long as the shorter leg

tangent: for an acute ∠A in a right triangle, the ratio of the length of the side opposite ∠A, to the length of the side adjacent to ∠A

term: of a fraction is the numerator or the denominator of the fraction; when an expression is written as the sum of several quantities, each quantity is called a term of the expression

tessellation: a tiling pattern

tetrahedron: a solid with four triangular faces

three-dimensional: having length, width, and depth or height

transversal: a line crossing two or more lines

trapezoid: a quadrilateral that has only one pair of parallel sides

tree diagram: a branching diagram used to show all possible outcomes of an experiment

triangle: a three-sided polygon

triangular number: a natural number that can be represented by arranging objects in a triangle; for example, 1, 3, 6, 10, 15, …

10

trinomial: a polynomial with three terms; for example, $3x^2 + 6x + 9$

truncating: approximating a number by cutting off the end digits; for example, 3.141 8 is truncated to 3.141, or to 3.14, or to 3.1, and so on

two-dimensional: having length and width, but no thickness, height, or depth

unit fraction: a fraction that has a numerator of 1

unit price: the price of one item, or the price for a particular mass or volume of an item

unit rate: the quantity associated with a single unit of another quantity; for example, 6 m in 1 s is a unit rate

unlike radicals: mixed radicals that have different radical parts; for example, $3\sqrt{5}$ and $7\sqrt{11}$

unlike terms: terms that have different variables, or the same variable but different exponents; for example, $3x$, $-4y$ and $3x^2$, $-3x$

variable: a letter or symbol representing a quantity that can vary

vertex (plural, **vertices**): the corner of a figure or a solid

vertical intercept: the vertical coordinate of the point where the graph of a line or a function intersects the vertical axis

vertical line test: if no two points on a graph can be joined by a vertical line, then the graph represents a function

volume: the amount of space occupied by an object

whole numbers: the set of numbers 0, 1, 2, 3,…

x-axis: the horizontal number line on a coordinate grid

x-intercept: the x-coordinate where the graph of a line or a function intersects the x-axis

y-axis: the vertical number line on a coordinate grid

y-intercept: the y-coordinate where the graph of a line or a function intersects the y-axis

zero exponent: any number, a, that has the exponent 0 $(a \neq 0)$, is equal to 1; for example, $(-6)^0 = 1$

Zero principle: the sum of opposites is zero

PHOTO CREDITS AND ACKNOWLEDGMENTS

The publisher wishes to thank the following sources for photographs, illustrations, articles, and other materials used in this book. Care has been taken to determine and locate ownership of copyright material used in this text. We will gladly receive information enabling us to rectify any errors or omissions in credits.

p. 2 (left) Agence France Presse/Corbis–Bettmann, (right) Ken Fisher/Tony Stone Images; **p. 3** (from top) Hunter Freeman/Tony Stone Images, Dan Bosler/Tony Stone Images, Chad Slattery/Tony Stone Images, David Young Wolff/Tony Stone Images, Bill Aron/Tony Stone Images, Dale Durfee/Tony Stone Images; **p. 16** (top) Corbis–Bettmann, (bottom) Musee de la Reine, Bayeux, France/Bridgeman Art Library, London/SuperStock; **p. 17** Ian Crysler; **p. 22** Neither Jeopardy Productions, Inc. nor Columbia TriStar Television wrote any of the answers and/or questions contained herein and neither Jeopardy Productions, Inc. nor Columbia TriStar Television makes any representation as to the content thereof. This material is not for duplication. "Jeopardy!" is a registered trademark of Jeopardy Productions, Inc.; **p. 66-67** Martine Mouchy/Tony Stone Images; **p. 67** (top) SuperStock/Four By Five, (bottom) A & L Sinibaldi/Tony Stone Images; **p. 68** Ken Graham/Tony Stone Images; **p. 75** SuperStock/Four By Five; **p. 94** Herbert Zettl/Tony Stone Images; **p. 113** Peter Poulides/Tony Stone Images; **p. 114** (top) Corbis–Bettmann, (bottom) The Cummer Museum of Arts and Gardens, Jacksonville/SuperStock; **p. 145** (top left) Canapress/Andrew Vaughan, (bottom right) Canapress/Doug Mills, (background top right) Canapress/Lynne Sladky; **p. 150** Ian Crysler; **p. 152** UPI/Corbis–Bettmann; **p. 162** L.J. Lozano; **p. 172** Ian Crysler; **p. 222** SuperStock/Four By Five; **p. 223** SuperStock; **p. 244-45** (top) Canapress/Jeff McIntosh, (bottom) SuperStock; **p. 245** (top) Canapress/Ruth Fremson, (bottom) Canapress/Alessandro Trovati; **p. 248** Ian Crysler; **p. 255** (left) Hulton Getty/Tony Stone Images, (centre) Ron Sangha/Tony Stone Images, (right) Lori Adamski Peek/Tony Stone Images; **p. 256** Figure 5.2 The Death Rate is no Longer Declining from *Predictions: Society's Telltale Signature Reveals the Past and Forecasts the Future* by Theodore Modis. Reprinted by permission of the author. **p. 267** Stewart Cohen/Tony Stone Images; **p. 277** SuperStock; **p. 293** SuperStock; **p. 298** SuperStock; **p. 324** Canapress Photo Service; **p. 325** David Starrett; **p. 328** Hand with reflecting sphere by M.C. Escher. © 1997 Cordon Art – Baarn – Holland. All rights reserved. **p. 332** Corbis–Bettmann; **p. 337** Lyn Budd, Coca-Cola Ltd.; **p. 346** Canapress/Michael Probst; **p. 352** Ian Crysler; **p. 378** Copyright © Corel Corporation; **p. 398-99** Jerry Kobalenko/Tony Stone Images; **p. 399** (inset) Reuters/Paul Hanna/Archive Photos; **p. 403** (top) Canapress/Armando Franca, (bottom) Canapress/Tom Strattman; **p. 424** Ian Crysler; **p. 425** Garry Hunter/Tony Stone Images; **p. 436** David Starrett; **p. 442** Ian Crysler; **p. 450** Derke/O'Hara/Tony Stone Images; **p. 458** SuperStock; **p. 469** Canadian Pacific Archives, MNC.1535, photo by N. Morant; **p. 480-81** "The Middle Kingdom – How and why things happen" from *The Globe and Mail* February 6, 1997 and February 13, 1997 reprinted with permission from *The Globe and Mail.* **p. 488** Paul Souders/Tony Stone Images; **p. 522** Copyright © Corel Corporation; **p. 528-29** David Starrett; **p. 529** Canapress/Kip Frasz; **p. 531** The Toronto Star/M. Stuparyk; **p. 539** Ian Crysler; **p. 543** David Starrett; **p. 555** "What is the most important problem facing Canada?" from *Maclean's,* December 30, 1996, reprinted by permission from *Maclean's.* **p. 558** "Canadians satisfied with life" graphic adapted by permission of Southam News. **p. 560** SuperStock; **p. 572** Ian Crysler; **p. 573** Ian Crysler

ILLUSTRATIONS

Bernadette Lau: 90
Dave McKay: 6, 170, 382, 438, 468, 476, 478-79, 483, 488, 511, 577
Jun Park: 196-97, 238, 336, 399, 451
Pronk&Associates: 29, 44, 92, 115, 122, 236, 237, 331, 458, 465, 566